Camille Flammarion's The Planet Mars

Astrophysics and Space Science Library

More information about this series at http://www.springer.com/series/5664

William Sheehan
Editor

Camille Flammarion's The Planet Mars

As Translated by Patrick Moore

Written by Camille Flammarion

Springer

Editor
William Sheehan
Willmar, MN, USA

ISSN 0067-0057 ISSN 2214-7985 (electronic)
ISBN 978-3-319-34477-5 ISBN 978-3-319-09641-4 (eBook)
DOI 10.1007/978-3-319-09641-4
Springer Cham Heidelberg New York Dordrecht London

Softcover reprint of the hardcover 1st edition 2015

Printed on acid-free paper

Springer is part of Springer Science+Business Media (www.springer.com)

Editor's Preface

In November 1893, a wealthy Bostonian, Percival Lowell, was in Japan, as he had been off and on for a decade, renting an 18-room Tokyo house with no clear idea as to what he was going to do when he sailed from Yokahama Bay back to the United States. Toying with the idea of an Easter jaunt to Seville with a friend, he had no sense of imminent destiny, no idea that by then he would be headlong into what his biographer David Strauss called the "daring enterprise" of founding an observatory, at first a temporary and makeshift one, for the purpose of studying Mars and trying to answer the question of its being an abode of intelligent life.

Lowell was back in Boston by Christmas, and there, a great thumping book awaited him that would determine his fate. It was a gift of his old aunt Mary Traill Spence Lowell Putnam, the sister of the late poet James Russell Lowell: *La Planète Mars et ses Conditions d'habitabilité*, Volume 1, by the French astronomer and prolific writer Camille Flammarion. Seldom has a more fateful collision of a book and a man occurred.

The author of that book, Nicolas Camille Flammarion, had been born into modest circumstances in 1842 at Montigny-le-Roi in the department of Haute Marne. His passionate interest in astronomy was awakened at the age of 5 when he witnessed the July 8, 1842 total eclipse of the Sun. By 11, he was using a pair of opera glasses and discovering with them, "Mountains in the moon, as on earth! And seas! And countries! Perchance also inhabitants!" (Flammarion 1890). When he was 14, his parents faced financial ruin and had to surrender their property. They moved to Paris, where his father found employment in a photographic studio. There were no longer funds to keep Flammarion in school, and he went to work as an engraver.

In his spare time he was "taken up with cosmographical questions, and working on a big book on the origin of the world." The manuscript, entitled *Cosmogonie universelle*, grew to 500 pages long. It came to the notice of a physician who called to the home to treat Flammarion for an illness. Recognizing a precocious talent, the physician recommended him for employment as a human computer at the Paris Observatory, then directed by Urbain Jean Joseph Le Verrier, the mathematical genius whose calculations had pointed the way to a new world, Neptune. Flammarion later recalled his reverential feelings when he entered the sublime building dating from the reign of Louis XIV:

> I entered it as though it were a temple, or a holy sanctuary, with all the innocent ardor of the neophyte. There was not the least shade of skepticism; and despite my poverty, I did not think in terms of present payment or future gain. I lived alone for science.

His expectations of Le Verrier were at first romantic:

> Since my childhood I had seen the name of the illustrious astronomer recorded on maps of the sky, for this planet had borne his name for several years, in the classic books, before it was called Neptune. This scientist, this genius, who had discovered a star with the tip of his pen… and had boldly rolled back the frontiers of the System of the World, this man was for me a kind of saint, an inhabitant of the heavens.

He was soon disillusioned. Le Verrier was long on mathematics but short on interpersonal skills. At their first meeting, the Director gave him an elementary mathematics exam. He passed. He was then summoned to the Director's large office. The great man dismissed him with a brief statement: "You start on Monday. Farewell, and to your work." The work involved standing at a desk from 8 till noon and then from 1 till 4, doing mind-numbing computations. Flammarion's assignment was the correction of the apparent positions of stars observed with a telescopic instrument known as a meridian-circle for effects of atmospheric refraction, which required him to take columns of numbers and add or subtract them from the recorded positions, all day, day after day after day. "After the first week," he would bitterly recall, "one manages to carry out this work almost unconsciously—and while thinking of something else…. From my first days of service, I perceived that of my five colleagues, none cared the least for astronomy."

A poet at heart, he could never reconcile himself to the drudgery of his routine. He dreamed of other worlds, and their possible inhabitants.

He had lost the Catholic faith of his youth, finding it irreconcilable with post-Copernican astronomy. He substituted for it another faith, based on the teachings of Jean Reynaud, whose *Terre et ciel* (1854) promoted the idea of the immortality of the soul. Each soul, Reynaud taught, passed from planet to planet, progressively improving at each stage. Flammarion was also familiar with the works of such pluralist authors as Fontenelle, de Bergerac, and Huygens, and "consecrated the year 1861" to the writing of a book advocating the plurality of worlds, "enflamed with a fiery ardor as one has at age nineteen, not doubting for an instant that I would demonstrate to myself that my conviction in extraterrestrial life was well founded" (*Memoires biliographiques et philosophiques d'un astronomer*, Paris, 1911, pp. 218–219).

The result was a 56-page booklet, *La Pluralité des mondes habités* (the plurality of the inhabited worlds), which laid out the program of what became the preoccupation of his life. On the title page, he grandly introduced himself as "ancien calculateur a l'observatoire impérial de Paris, professeur d'astronomie, member de plusiers sociétés savantes, etc." Le Verrier disapproved of the extracurricular activities of his young computer, and after a curt meeting, dismissed him from the observatory staff. But the booklet, as Flammarion recalled long afterward, "at once made my reputation. It was published by Mallet-Bachelier, the printer to the observatory, at two francs a copy, and the first edition was immediately sold out. I was then approached by the publisher Didier, who offered to undertake the second edition, for which purpose I rewrote many of the chapters, and made important editions to the text." The second edition grew to 468 pages, and by 1835 had gone through a 35th edition.

Flammarion, still in need of a regular income, managed to catch on as a computer at the Bureau des Longitudes where he worked for a few more years, but he had discovered his real talent—science journalism. While employed at the Bureau, he joined the staff of the scientific magazine "Cosmos," founded by the well-known popularizer of science the Abbé Moigno, and also began collaborating on a scientific review, "Le Magasin Pittoresque." He also began contributing to "Le Siècle," the most important daily in Paris at the time. He moved into a house in the Rue Gay-Lussac, near the Luxembourg, and carried out astronomical studies of sunspots and the surface features of the Moon in a small private observatory, and was appointed a professor at the École Turgot. He resigned from the Bureau in 1867 in order to devote full time to writing and lecturing. That year he joined the French Aerostatic Society, of which he was named president, and commenced a series of balloon ascensions, which included one undertaken as a honeymoon with his 8 days' bride, Sylvie Pétiaux, on August 28, 1874. "What could be more natural," he asked, "than for an astronomer and his wife to fly away thus like birds?" The voyage lasted 24 hours, and ended in a descent at Spa.

All the while Flammarion remained a whirling dervish of literary activity. He published one or two titles a year, including: *Les Mondes imaginaires et les mondes réels* (1865), *Le merveilles célestes; lectures du soir* (1865), *Études et Lectures sur l'astronomie* (1867, 1869, 1872, etc.), *Dieu dans le nature* (1867), *Les Ballons et le Voyages aériens* (1867).

During the war with Prussia, Flammarion was appointed captain of engineers, and with other astronomers was involved in observing the Prussian batteries at Saint Cloud and Meudon from a small observatory established by the government at Passy.

After the Commune, he moved into an apartment on the fifth floor of a house at 16, rue Cassini (it still exists), close by the Paris Observatory. During a visit by Robert H. Sherard (Oscar Wilde's friend and early biographer), who visited Flammarion and his wife there in 1894 (see: R.H. Sherard, "Flammarion the Astronomer: his home, his manner of life, his work," *McClure's Magazine*, 1894),

> It is certainly splendidly situated, commanding views, on all sides, over Paris, and flooded with light. Outside the windows may be seen the tops of the trees of the Avenue de l'Observatoire, which is just opposite the astronomer's study… The apartment is modestly furnished. The dining-room, which is the first room into which the visitor enters, is remarkable only for an immense window which takes up the whole side of the room, reaching from the floor to the ceiling, and arranged so as to serve as a conservatory, in which are a number of plants and flowers. The walls are decorated with astronomical maps…. Against the wall, at the side of the tilted stove, is a bookcase … filled with cases fashioned to represent volumes. Each case is used for the classification of the papers which the savant receives from all parts of the world, referring to various branches of astronomical study. Thus "Mars" and

"Jupiter" each have their case, the other planets being disposed of in one labeled "Planets Other than Mars and Jupiter." The leather dining-room chairs are decorated each with one of the signs of the zodiac."

In 1873, Flammarion set up a small observatory with a 108^{mm} aperture refractor, and with this small glass observed Mars in 1873. His chief interest at the time was double stars, however. As he explained, "At the time I approached the subject, that is to say, during 1873, to my surprise I discovered a total absence of documents concerning them. A number of questions, which so far have been unanswered, presented themselves to my mind. How many double or multiple stars are known? What is their proportion to the number of single stars? In the total number of groups which have been discovered, how many are simply optical phenomena due only to the effect of perspective, and how many are genuine or physical systems?... The only way to solve [these] questions was to examine, in a detailed manner, each one of the eleven thousand double stars which up till then had been discovered; to compare my observations with those previously... This is what I did." By this time he and Le Verrier had become reconciled. Once more the Director of the Paris Observatory in 1873 (he had been pulled down by his staff in 1870 but was reinstated in 1873 after his successor drowned), Le Verrier was very interested in Flammarion's researches into double stars, and placed at his disposal the 15-in (0^{m}.38). equatorial in the east tower of the Paris Observatory (of poor optical quality, it was usually stopped to 8 in (0^{m}.20)). By the end of 1877, Flammarion had the materials for his *Catalogue des étoiles doubles et multiples*, which he noted with pride a few years later, "has since become a classic in every observatory throughout the civilized world" (for a review, see *The Observatory*, 2 (1878), 376–380).

Flammarion's real breakthrough came in 1880, with the enormously successful *Astronomie populaire*. Published, at his insistence, by his brother, Ernest, the book became a runaway bestseller by the standards of the time, selling 131,000 copies during the author's lifetime, and in addition was translated into numerous languages. Flammarion received one franc royalty per copy. (No doubt, as he later confessed, his brother made very much more out of the work than he had himself.)

Flammarion would never again have to worry about money. Also, he had many admirers. One woman, whose lovely shoulders he had praised at a dinner party, willed that on her death the skin be removed and made into leather. In due course, she died, and her wish was realized as a leather-bound copy of the *Astronomie populaire*, which still exists. More importantly, a nobleman named Méret, of Bordeaux, an admirer of Flammarion's researches and popular writings but totally unknown to him, wrote in 1882 asking the astronomer accept the fee simple of a chateau, complete with stables and horses, at Juvisy-sur-Orge. The chateau occupied the site where a lodge known as "Cour de France" had existed since the time of Charles VI as a stopover for the kings of France on their annual journeys to Fontainebleau. Louis XIV had reportedly enjoyed taking coffee in the hollow of a celebrated yew tree here. The present building dated to 1730. There, in the salon, on March 30, 1814, the emperor Napoleon learned of the capitulation of Paris and of the fall of the Empire. Flammarion accepted the generous bequest, and established a private observatory, the maintenance of which, together with the salaries of his employees, was provided for by subscription among still other admirers.

Henceforth, Flammarion would divide his time between Paris and Juvisy. From November to May each year he could be found at 16, rue Cassini; from May to November at Juvisy. The money realized by the success of *Astronomie populaire* was used to hire contractors and builders to transform the ancient chateau into an observatory, boasting a cupola housing a fine 24-cm (9.5-in.) Bardou refractor (recently restored). That same busy year, 1882, Flammarion founded the journal "Monthly Review of Astronomy" (*l'Astronomie*). Though it would never pay its way, Flammarion was glad to subsidize it as the only serial publication devoted to astronomy in France at the time. In 1887 he founded the Société Astronomique de France. It was quartered in the Hôtel des Sociétés Savantes (residence of learned societies) in the heart of the Quartier Latin of Paris, with Flammarion's 108^{mm} refractor housed in a cupola on top of the building. During this time Flammarion was competing with himself, as the Société's Bulletin and *l'Astronomie* were published separately for 8 years; but in 1895 the two were merged. The combined journal, *l'Astronomie*, has run continually to the present day.

The planet Mars had always been one of Flammarion's main interests. As early as the first edition of *La Pluralité des mondes habités*, he had written:

> The world of Mars is so like the world of the Earth that, if one day it were ever possible to make a trip there, and forget how we came, it would be almost impossible to recognize the one that is our homeland. But for the Moon, which would rise to charitably put to rest our uncertainty, we would be at great risk of arriving among the inhabitants, and supposing ourselves in Europe or some other such place.

Not only did he observe the planet with his 108^{mm} refractor at the opposition of 1873, but in 1876 he drew up a map, *Carte géographique provisoire de la planète Mars*, on which he offered a system of nomenclature; it was first inserted in his book *Les Terres du Ciel* (Lands of the Sky) and presented to the Académie des sciences on 27 August 1877, just as Mars was swinging around to one of its exceptionally favorable perihelic oppositions. At that opposition, Asaph Hall at the U.S. Naval Observatory in Washington, D.C. discovered the satellites of Mars, while Giovanni Schiaparelli, at the Brera Observatory in Milan, produced a sensational new map which introduced not only a new nomenclature (displacing other schemes, including Flammarion's own; it is the basis of that still in use today) but showed the enigmatic features he called *canali* (canals). At Juvisy, Mars henceforth became the chief object of research. Flammarion intended that the observatory should be a nerve center of Martian observations all over the world, and thought that by gathering together and impartially analyzing all existing document, it might be possible to solve what he referred to as the "Martian enigma." His writings strongly endorsed the idea that conditions on the planet were favorable to life, and that it was almost certainly inhabited—so much so that when in 1891 a Madame Guzman, of Bordeaux, decided to bequeath 100,000 francs as a prize to be named after her deceased son Pierre, yet another keen admirer of Flammarion's writings, "for the person of whatever nation who will find the means within the next 10 years of communicating with a star (planet or otherwise) and of receiving a response, Mars was specifically excepted. As Flammarion explained (*La Planète Mars*, vol. 2, 1909, p. 500):

> Unfortunately, M. Guzman followed my writings on the planet Mars with such passion that he was quite convinced … that the question had already been asked and was half-solved, so that his venerable mother had the bizarre idea of excluding Mars from this magnificent contest.

During the winter of 1891–1892, in his study at 16, rue Cassini, Flammarion was engaged in a systematic survey of the already vast literature of the red planet on a scale which has perhaps never been attempted and rarely been equaled by any other single author. His routine at about this time (for he was a man of regular habits) was described to Sherard by Sylvie Flammarion:

> He is an extremely methodical man. He gets up regularly every morning at seven o'clock, and spends quite a long time over his toilet. Savants, as a rule, are not very tidy, but Flammarion is an exception to the rule. At a quarter to eight every morning he has his first breakfast, at which he always takes two eggs. From eight to twelve he works. At noon he has his *déjeuner*, over which he spends a long time. He is a very slow eater. From one to two he receives, and as he knows everybody in Paris, and as he is constantly being consulted on all sorts of questions by Parisian reporters, he is usually kept very busy during this hour. From two to three he dictates letters to me, and as he receives thousands of letters from all parts of the world, especially when anything new in the branch of astronomical science is occupying public attention, my time is fully occupied. At three o'clock he goes out and attends to his business as editor of the monthly magazine which he founded, and to his duties as member of various societies. He is back home again at half-past seven, when he has dinner, and spends the rest of the day in reading. He is a great reader, and tries to keep himself *au courant* with all that is said on the important topics of the day. At ten o'clock he goes to bed, for he is a great sleeper.

What is described here is, of course, his winter routine. During the warmer months, when he was at Juvisy, he would frequently spend entire nights in the observatory when the seeing was fine.

Advancing steadily in his grand project of reviewing the entire literature of Mars published up to this time, the methodical and meticulous man produced what would be nothing less than an encyclopaedic work on the subject: *La Planète Mars et ses conditions d'habitibilité*. The first volume, containing all the observations from 1636— shortly after the invention of the telescope—through 1890, appeared just as the planet came to its favorable perihelic opposition in August 1892. Of all Flammarion's books—and he published over seventy titles, not counting the many editions some of them went through, this was his Martian masterpiece—*La Planète Mars*, vol. 1 is perhaps the one work, of all his voluminous corpus, that still repays close attention today. (A second volume of the work, but only covering observations made between 1890 and 1900, was published in 1909; it is of interest only to specialists. A third, covering the period between 1900 and 1910, was in preparation at the time of his death in 1925; it has never been published.)

The stimulus of this great work was decisive and immediate. Percival Lowell was only the most illustrious of its readers. The vagabond Bostonian's receipt of that gift from Aunt Mary at Christmas 1893 directly propelled him to launch the expedition to Flagstaff, Arizona Territory, for the purpose of observing Mars at the favorable opposition of October 1894. (I have examined all of Lowell's annotations of the book; they testify to the close attention he gave to the work.) Arriving at the mesa on Mars Hill at the end of May, where he caught up with borrowed telescopes sent ahead by his assistants W. H. Pickering and A.E. Douglass, Lowell penned a poem testifying to the impulse to adventure and to what he expected to find on the planet. It is entirely Flammarionesque in theme:

One voyage there is I fain would take
While yet a man in mortal make;
Voyage beyond the compassed found
Of our own Earth's returning round....
[And] when staid night reclaims her sphere
And the beshadowed atmosphere
Its shutters to sight once more unbars,
Letting the universe appear
With all its wonderworld of stars,
My far-off goal draws strangely near,
Luring imagination on,
Beckoning body to be gone,—
To ruddy-earthed blue-oceaned Mars.

Readers of *La Planète Mars* will recognize, especially in its later chapters summarizing his views about the changes in the Martian features, on the polar caps and atmosphere, the canals and conditions for life, all the ideas in embryonic form that Lowell would develop into his startling book *Mars*, published in 1895, in which he asserted the existence of a Martian civilization of canal-builders. Lowell had to do little more than connect the dots. Immediately after seeing *Mars* through the Press, on December 10, 1895 he sailed on the Spree with Alvan G. Clark, who was to create the lens for the famous refractor that Lowell set up on Mars Hill, and on arriving in Paris, enjoyed dinner with the Flammarions at their home in Paris. He described the affair to his father, "There were fourteen of us, and all that could sat in chairs of the zodiac, under a ceiling of a place blue sky appropriately dotted with fleecy clouds, and indeed most prettily painted. Flammarion is nothing if not astronomical. His whole apartment, which is itself *au cinquieme* [on the fifth floor], blossoms with such decoration."

Flammarion was just 50 when *La Planète Mars* appeared. Though he lived another third of a century, the book marks the high-water mark of his career. He continued to observe with the Bardou refractor, though in the domain of practical astronomy he was soon surpassed by others. Lowell, equipped with the 24-in (61-cm). Clark refractor at Flagstaff, was at once acknowledged as the world's leading expert on the red planet; while even at Juvisy itself Flammarion was overshadowed by the extravagantly gifted amateur astronomer E. M. Antoniadi, who arrived via the Oriental Express from Turkey in the latter part of 1893 to become Flammarion's assistant, made a series of marvelous observations of the planets with the Bardou refractor, and then, achieving independence by marrying into wealth, left Flammarion's employment in 1902. As an "observateur volontaire" with the Grand Lunette of the Meudon Observatory, at the perihelic opposition of September 1909,

Antoniadi made definitive observations of the surface markings on Mars so as to effectively demolish the canal illusion that had so puzzled Flammarion and others, once and for all.

As with H.G. Wells, after turning 50, Flammarion's best work was behind him. The habit of literary activity continued, of course, but he was now largely revered for his past achievements. He published his *Mémoires* in 1911. The following year he was named an officer in the Légion d'honneur. When war broke out in 1914, Paris was threatened by the advancing Germany army, he took up refuge first at Arachchon, then at Cherbourg.

Sylvie died in February 1919, and 7 months later he married Gabrielle Renaudot, who even during Sylvie's life had lived at the observatory as Flammarion's assistant and soul mate. She was 42; he was 77. From early days he had been as interested in psychic phenomena as in astronomy, and increasingly in these last years—with the losses of the war in everyone's mind, and his own death approaching—he returned with a will to the problem of life after death. Three volumes of *Avant la mort* appeared in 1920, 1921, and 1922, respectively. He was promoted to commander of the Légion d'honneur in 1922, and elected president of the Society for Psychical Research of London the following year. His last book, *Maisons hantées*, appeared in 1923. He died at Juvisy on 3 June 1925. He is buried, with Sylvie and Gabrielle, in the park of Juvisy.

The work of the observatory continued, under Gabrielle and assistant astronomer Fernand Quénisset, a skillful astrophotographer and noted comet discoverer, who had been Flammarion's assistant during 1892–1893 and again from 1906. Quénisset died in 1951, and Gabrielle—still serving as general secretary of the Société Astronomique de France—lived on until 1962. By then the chateau, too costly for her to keep up, was crumbling around her; it is now, alas, in ruins, but at least it has a new cupola under which the refurbished Bardou refractor is once more seeing active duty, under the auspices of the Société Astronomique de France. Through the courtesy of one of its officers, François Oger, I was able to make some of the first observations with the restored instrument, in February 2012, of Mars near its aphelic opposition; it was a cold night, the winds were high—or perhaps the revenants of Flammarion, Sylvie, and Gabrielle were restless. In the telescope, Mars was a small colored disk, which, in a moment of suspended disbelief, seemed the "ruddy-orbed and blue-oceaned" world of yore. It was a thing of beauty, and still able to stimulate the imagination.

La Planète Mars is a rare book today, and has never been translated into English. In 1980, an astronomy popularizer even more prolific than Flammarion, Patrick Moore, produced a translation, but until now the translation has been far rarer than the original. There were only four copies. I had one of them.

I came by it as follows. In 1993, I visited Patrick at his home in Selsey, Sussex. For several years we had corresponded about publishing translations of classic works of lunar and planetary astronomy—I had, for instance, undertaken at his request a translation from the German of part of Beer and Mädler's *Der Mond*. We began to discuss the Flammarion volume; Patrick's original plans to publish it were foundering, and he hoped that I might succeed where others had failed. But it proved more difficult than either of us expected. Patrick's transcript had been produced on his old 1908 Woodstock typewriter—this was before the computer era—so to begin with, the whole work had to be retranscribed. Also there were many editorial matters to attend to; by the end, the work was almost as much mine as Patrick's. Several times we thought we had found a publisher, but in each case they backed out when they realized the sheer scale of the project. Fortunately, and earning the gratitude of all lovers of things Martian, Springer has come to the rescue. I now present it at last, a work that will find "fit audience even if few" among the true lovers of the red planet.

I dedicate this book to the memory of my friend Patrick Moore, who unfortunately did not live to see it published, though he awaited it eagerly (and at times impatiently) for some three decades.

It is fitting that Sir Patrick's very first published book had appeared in 1950—*The Planet Mars* (London: Faber and Faber), a translation of a monograph by the French astronomer Gérard de Vaucouleurs. His last is a translation from French of Flammarion's book also entitled *The Planet Mars*.

Willmar, Minnesota, USA William Sheehan
May 23, 2014

Author's Preface

Astronomy should not stop at measuring the *positions* of the celestial bodies: it should also study *their nature*.

In acceding to the request that has been made to me to publish a specialist work about the planet Mars, establishing and defining the present state of our positive knowledge concerning the physical constitution of our neighbour world—of which studies are indeed already far enough advanced to merit a summary and a general discussion—I have hesitated for a long time before deciding upon the method which will give the best scientific results.

Two methods of presentation come naturally to mind.

One is to deal with our various observations and studies of Mars in special chapters, such as: distance from the Earth, revolution around the Sun, year, day, seasons, climates, calendar, light, heat, mass, density, gravity, volume, geography, continents, seas, polar snows, atmosphere, water and clouds, observed movements and changes, satellites, etc.—and to treat each of these subjects separately. The other is to take the planet as a whole, and give all the results, and the deductions arising from them, in simple chronological order.

I have chosen the second method, mainly because it seemed to me to be the more interesting inasmuch as it gives us an immediate picture of facts and deductions which in themselves provide a history of the planet; and also because it provides a better account of the gradual development of our knowledge—in particular that of the subject which dominates all our studies of our neighbour world: that of its physical geography, seas, continents and polar ices. This is undoubtedly the most essential part of our telescopic observations. Therefore, it seems more logical to give in chronological order the studies made, up to the present time, of this world—which because of its closeness to the Earth and its favourable position for observation seems likely to be the first to give us answers about the great and deep questions which thinking Man has been asking over the centuries, faced with the silent enigma of the starry heavens.

A technical work must explain *what* we know; the historical account tells *how* this information was obtained. This is an advantage: the progress of Science speaks for itself, and renders literary embellishment unnecessary.

We must agree, moreover, that this is the right moment to produce a work of this kind. The astronomical study of Mars is very advanced. We have had a very great number of excellent observations, begun two and a half centuries ago and continued ever since. But these observations have been heaped up in their hundreds and their thousands, and have not been compared with each other or analyzed in a way which will add to our knowledge of the planet.

Mathematical astronomy clearly leads on to physical astronomy, without which it would lose the greater part of its interest. We are looking at a great problem, not merely at the movements of stones in space. The masses of the celestial bodies are not all-important; the significance of the Sun, or of the Earth, does not lie only in its weight. The true philosopher looks higher and sees further; he looks out to fundamentals. He admires the mechanical bases of the system of the universe, but he does not stop there. When he uses a telescope to contemplate a world lost in the depths of immensity, he is naturally interested in its distance, its movements and its mass; but he wants to know more, and he asks about the nature of this world—what is its physical constitution from the point of view of habitability? This is what really interests him—everything leads to this end.

Physical astronomy was founded in the time of Galileo.[1] Its progress has been bound up with that of optics, and indeed it has gradually followed upon the improvements made in the construction of refractors and reflectors, above all with respect to increased magnification and—of paramount importance—clarity of image. But the enthusiasm of the observers, their patience, their perseverance, the practical perfection of their methods, and even their adaptation to the difficulties of their researches have made by no means been the least of contributions to the progress of practical optics itself.

This work presented here is divided naturally into two parts. The first is an account and a discussion of all the observations made of Mars, from the earliest, dating from the first half of the seventeenth century, up to the present time. The second part gives a résumé of our conclusions with regard to the general study of the planet.

---

[1] But rarely appreciated, even by astronomers who write books. To cite one example: take the bibliography of astronomy, the *Traité d'Astronomie physique*, in five volumes out of eight, by J.B. Biot, Member of the Academy of Sciences, the French Academy, the Academy of Inscriptions, the Bureau of Longitudes, Professor of the Faculty of the Sciences of the Collège de France, etc. These five volumes of "Physical Astronomy" include no less than 2916 pages—of which less than 100 really deal with the physical condition of the celestial bodies! The physical constitution of Mars receives a grand total of one page (Vol. V, 1857, p. 401). This work would have been better called "Traité d'Astronomie *mathematique*." The same is true of most other authors. Delambre, speaking of the observations made of the rotation of Venus, the physical constitution of Mars and the spots on the Sun, dismissed such work as a waste of time! Etc.

Our first section is itself divided into three parts. The first begins with the very earliest observation, that of 1636, and extends up to the year 1830. It therefore includes almost two centuries. The drawings made during this period were rudimentary, and absolutely insufficient to give much idea of the physical constitution of the planet. The second period begins in 1830 and lasts until 1877. It sees the start of the study of Martian geography—or, to be more accurate, areography. During this period, studies made during the most favourable oppositions of Mars—the times of closest approach to the Earth—have led to more extensive and precise knowledge of the state of our neighbour world. The third period begins in 1877, with the first geodesic (areodesic) triangulation of the continental and marine surface of the planet, and continues today with the surprising discoveries and details with regard to this bizarre geography—and particularly to the changes in this remarkable country.

In the first period, information was obtained about the volume of Mars, the mass, density, surface gravity, axial inclination, lengths of the year and seasons, rotation period, and therefore the lengths of Martian days and nights and the existence of polar patches and their variations in summer and winter. It was deduced that the snows were analogous to those of our poles; astronomers began to think that the dark areas represented seas, and that the continents were yellow. The atmosphere was recognized rather than studied.

In the second period we find the first geographical maps of the planet, confirming that the polar patches are snows, which melt regularly under the influence of the rays of the Sun. It was thought that the only way to account for the dark areas was to consider them as seas, and it was found that their contours were subject to variations—the gulfs and estuaries of great rivers were traced. The atmosphere was chemically analyzed by means of the spectroscope, and the existence of water vapour there was proved; it was shown that the atmosphere could not be the cause of the red colour of the planet, because this colour is more pronounced near the centre of the disk, where the thickness of atmosphere as seen from the Earth is less than that near the edges of the disk where the colour is less pronounced. It was found that the temperature depends principally not upon the distance from the Sun, but upon the state of the atmosphere (e.g. the summit and the foot of Mont Blanc); and that certain vapours, notably water vapour, absorb the solar rays to a greater extent than some gases such as oxygen and nitrogen and it was recognized that the conditions for life on the surface of Mars are not essentially different from those on our own planet.

In the third period, areographical details were better distinguished and studied; the seas, the lakes, the gulfs, the straits and the rivers were drawn, watched and followed, so that surprising variations were seen unmistakably—the discovery was made of an enigmatical réseau of dark lines crossing all the continents in the manner of trigonometrical outline. It was suggested that the changes might be due to water flow; it was also recognized that the atmosphere is generally clearer than that of the Earth, and that clouds are rare, particularly in summer and near the equatorial regions. Analogies with the Earth grew in certain aspects, while in other respects dissimilarities were confirmed.

These three periods; therefore, form natural divisions in the first part of the present work. The second part gives the results and ends with a discussion.

We Earth dwellers, accustomed to judging phenomena according to the evidence of our eyes, and unable to imagine the unknown, have extreme difficulty in explaining phenomena which are strange to our own planet, and consideration of them can even plunge us into hopeless embarrassment. For example, on Mars we observe variations, which are certainly real, and are not minor, in the tone of the dark patches, which are regarded as seas. There is nothing analogous on the Earth, at least to a comparable degree. On Mars we also see a geometrical réseau of straight lines, crossing each other at angles, and which have received—not without analogy—the name of *canals*. On Earth we have nothing comparable to guide us in explaining these features. We are dealing with a new world, incomparably more different from ours than the America of Christopher Columbus, differed from Europe. How can we give an exact interpretation of these telescopic discoveries? All our efforts should be directed toward this interpretation, without preconceived ideas and with complete independence of spirit.

I propose to deal here with all the observations, and with all the accounts which have been written in every language.

It is very clear that the only way to attain anything like a complete knowledge of the state of the planet is to make a comparison between all these observations. The historical method used here speaks for itself. Scholars who want to gain a precise knowledge of the planet will have all the evidence and all the documents in front of them.

I cannot end this Preface without thanking MM.Gauthier-Villars, who have published this scientific work so excellently. They also know that scientific research is the aim of modern man, and that it is valuable to give the widest possible dissemination to the intellectual public, of the great and brilliant concepts of the present-day astronomy.

Before going further, let us give an exact statement of the position of Mars in the Solar System. Later we will study the orbit from the viewpoint of its precise elliptical form and its relations with the orbit of our own Earth round the Sun. For the moment it will suffice to give a table of the distances of the planets from the Sun.

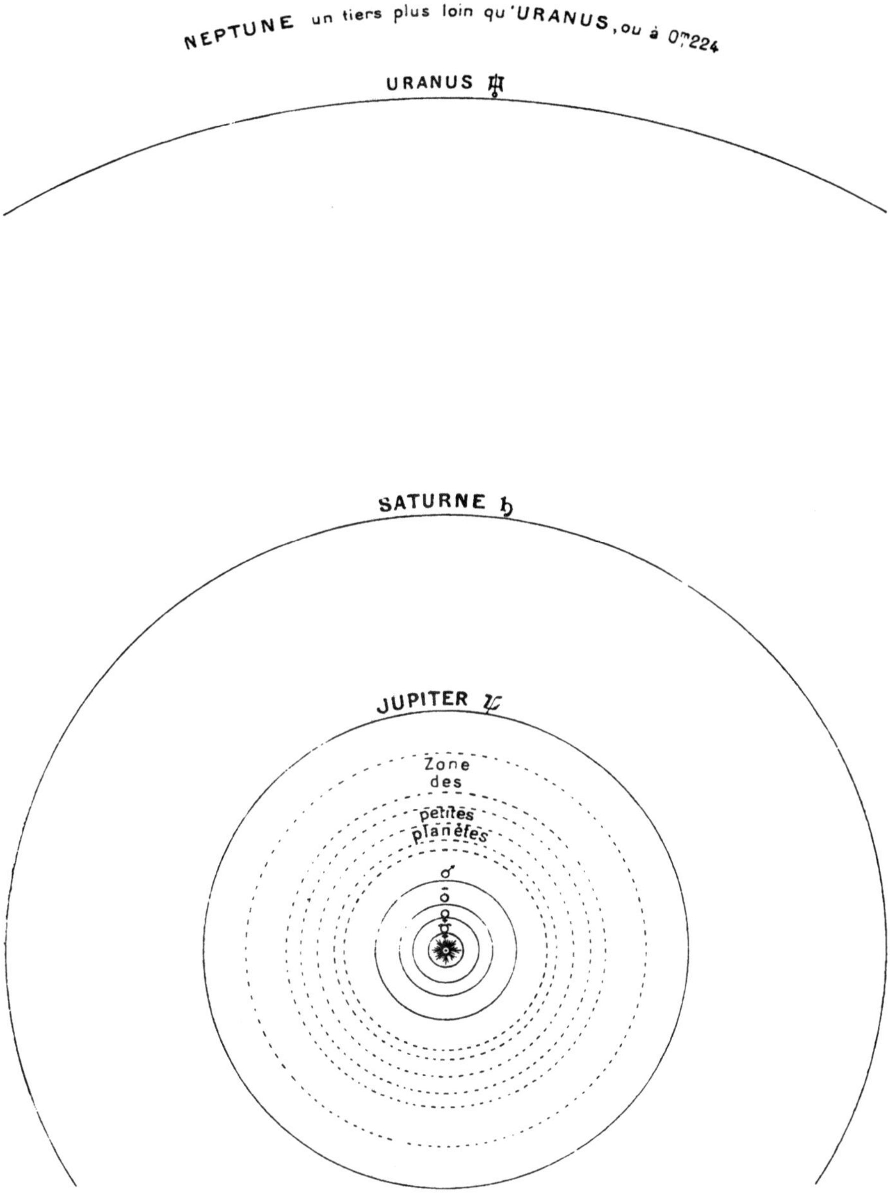

Plan of the Solar System, drawn at the precise scale of $0^{mm}.8$–20 million km

Planet	Distance from the Sun			
	Earth=1	Millions of km		
Mercury	0.387	57,678		
Venus	0.723	107,772		
Earth	1.000	149,000		
Mars	1.524	227,031		
			Millions of kilometers	
Minor planets	2.175 to 4.262	Maximum zones of asteroidal density:		324
		2.38		355
				408
		2.74		464
		3.12		510
		3.42		635
Jupiter	5.203	775,217		
Saturn	9.538	1,421,281		
Uranus	19.183	2,858,312		
Neptune	30.053	4,478,195		

The diagram has been constructed from the numerical data, on a scale of 1 mm–20 million km. This was the only way to show a plan of the Solar System in the format of this book and even then the orbit of Neptune would not fit on to the page. The diagram shows that Mars and the Earth are both comparatively close to the Sun. This is important, and it is interesting to take exact account of the position of our sister world compared with ours.

I have adopted a solar parallax of 8″.82, which is the most probable value. The distance corresponds to 149 million km.

We are now ready to begin telling the astronomical history of Mars, and to study our neighbour world without any preconceived ideas.

Juvisy Observatory, Juvisy-sur-Orge, France
August 4, 1892

Camille Flammarion

Translator's Preface

Camille Flammarion's great book is the most complete study of the history of observations of Mars ever written. It is now almost a century old, and for some strange reason it has never before been translated. I feel that it will be of use to historians of astronomy, which is why I have undertaken the translation.

Only four copies of this typescript exist. One is in the library of the British Astronomical Association, one in the Library of the Royal Astronomical Society, one at the Lowell Observatory in Arizona, and the last in my own library.

Selsey, Sussex, UK
23 August 1980

Patrick Moore

Contents

Part I
Presentation and Discussion of Observations

First Period, 1636–1830

The first period of what we may call the history of Mars began with the first telescopic views of the planet, obtained in the seventeenth century. The first drawing was made at Naples, by Fontana, in 1636. We are concerned here with physical astronomy rather than mathematical astronomy; otherwise this monograph would have had to begin with Kepler's work, *De motibus stellae Martis*, published in 1609.[1]

Until the invention of optical instruments the observation of the planets, as well as the stars, was limited to determinations of their apparent positions on the celestial sphere. With the naked eye we can indeed see these brilliant objects circling in the sky. Thinkers had deduced that the planets are celestial bodies without light of their own, that they are analogous to the Earth, and that they shine only by reflected sunlight. Copernicus, in his immortal astronomical reformation (1543), had announced that in the future men would probably invent instruments capable of proving that the planets show phases, and so must be non-luminous, as is the Earth. Similarly, we of today must hope that the day will come when by methods unknown to our science we will obtain evidence of the existence of inhabitants of other worlds, perhaps even putting us into communication with our brothers in space. People laughed disdainfully at Copernicus' temerity, just as skeptics of today laugh at ours; it is only too easy to follow calmly in the rut of the past. However, even in Copernicus' own century—in 1590, 47 years after the death of the canon of Thorn—a Middelbourg

[1] This work of Kepler's begins as follows: "Durissima est bodie conditio scribendi libros mathematicos, præcipue astronomicos." The same reflections are valid today for works on pure astronomy. How many readers will my present book have? Assuredly, very few. Most inhabitants of the Earth are not greatly concerned with the sky, and do not even know that their world is part of it; they are ignorant of where they are, and live in a remarkable ignorance of reality. This ignorance satisfies their native indifference.

C. Flammarion, *Camille Flammarion's The Planet Mars*, Astrophysics and Space Science Library 409, DOI 10.1007/978-3-319-09641-4_1

optician, Zacharie Jansen, invented the first telescope, if we accept the evidence of the earliest authority.[2]

The telescope was perfected 16 years later by Hans Lippershey, who was another optician in the same town. However, it was not at once turned toward the sky. In 1609, after receiving reports from Holland about the invention, Galileo constructed the first telescope to be pointed skyward, and immediately (in January 1610) discovered the satellites of Jupiter. Shortly afterwards he discovered the phases of Venus, thereby confirming Copernicus' prediction and providing direct proof of the new system. The first observations published by Galileo were those of the satellites of Jupiter, made on 7, 8, 10, 12 and 13 January 1610.

During the years 1610, 1611 and 1612, astronomical discoveries came in rapid succession: spots on the Sun, the geography and mountains of the Moon, the satellites of Jupiter, and the starry nature of the Milky Way. Galileo, Kepler, Fontana, Scheiner and Rheita earned reputations, and discovered in the mysteries of the heavens, things that had been previously hidden from the eyes of Earthmen.

The size of the lunar disk, the extent of the largest sunspots, the brightness of the satellites of Jupiter, the richness of the Milky Way: all these early discoveries were made with rudimentary telescopes of low magnifying power. Galileo's first refractor gave a magnification of only 3–4×. The immortal astronomer then managed, in succession, to achieve magnifications of 7, 10 and even 30 diameters; but he could go no further. His skill, patience, and perseverance enabled him to make marvelous discoveries with this modest instrument. This celebrated telescope of Galileo's has been carefully preserved, and is now in the Academy at Florence. The astronomer Donati put it into my hands one day, and also one of Galileo's rulers, which had been preserved at the Academy. It was not without emotion that I touched these venerable relics. It seemed to me that his first refractor, which had ushered in modern astronomy, had kept something of the glory of the past centuries. After sunset I recaptured the spirit of the Florentine astronomer on one of the beautiful Italian

[2]The invention of the first telescope has been lost in the unknown past. It is certain that in 1609 Galileo constructed a refractor, and then, on 7 January 1610, discovered the satellites of Jupiter. (I published a facsimile of his first drawings in *Les Terres du Ciel*, in the chapter dealing with the satellites of Jupiter.) It is equally certain that from 1606 to 1608 the name of Lippershey was known in Holland as making telescopes. But a work by Pierre Borel, physician to the King, member of the Academy of Sciences, author of *Discours prouvant la pluralité des Mondes* which I have quoted in *Les Mondes imaginaires*, claimed in 1655—that is to say, only half a century after the invention—that the "first inventor" was Zacharias Jansen, whose portrait he gave, while the second was Hans Lippershey (sic.), whose portrait was also given. The work is entitled *De vero telescopii inventore* (1655). Chapter XII of this treatise, entitled "De inventioris vero nomine," specifically discusses the point. The author wrote sometimes under the name of Zacharias Jansen, sometimes Zac. Joannides; the second sometimes Lipperhey, sometimes Lipperseim. All names at this time were Latinized, and on retranslating the Latin into French I had had to make new modifications. Thus, for example, Jean Muller took the name of his native town, Königsberg, and from its royal mountain called himself Regiomontanus. This name translated into French becomes Dumontroyal.

Whatever may be said about the first essays into optics, the year 1609 was the time of construction of the first astronomical telescope, by Galileo, and, practically speaking, the observation of 7 January 1610 was the first of all.

terraces, just as the stars were coming out. With feverish excitement I turned this marvelous tube toward the new worlds that he had discovered in the heavens. I recalled that he had shown these sights to those who were incredulous, and he still shows them to us today from his grave.

Because the disk of Mars is always very small, even when the planet is closest to the Earth, these primitive instruments, which magnify the disk only slightly and give poor definition, could show nothing whatever on the Martian surface.

Galileo observed Mars during his first year of observation, 1610. In his telescope the disk of the planet was barely measurable. On 30 December 1610, he wrote to P. Castelli: "I ought not to claim that I can see the phases of Mars; however, unless I am deceiving myself, I believe I have already seen that it is not perfectly round."[3] Kepler calculated the phases of Mars in his *Epitomes Astronomiæ*, Book V, Part V (1621), where he called the greatest phase of Mars, *perfectio phases dichotomæ*. However, he never observed it, and treated the problem only from a geometrical point of view.

Improvements in telescopes followed rapidly. There was tremendous enthusiasm at the time. People were eager to discover the inhabitants of the Moon, or at least to see their handiwork. There was feverish excitement. Lenses of great focal length were made, which were troublesome to use and full of imperfections; new combinations of lenses were invented to improve the clarity of the images. The skill and energy of opticians did not keep pace with the ambition of observers. Nevertheless, from 1636—that is to say, 37 years after Galileo's first telescope—a Neapolitan scholar, Fontana, emulated Galileo and Kepler; he made a better telescope, and under the clear sky of Naples was able to make good observations of the spots on the Moon, the Pleiades, the phases of Venus, and the planet of which I propose to write the history.

---

[3] When Mars is closest to the Earth, it shows a disk 30″ in diameter. To the naked eye it is a very brilliant point, a star of the first magnitude, glittering at night-time even though it shines only by light reflected from the Sun.

Galileo's telescope, magnifying 4×, showed Mars as large as a pea 7^{mm} in diameter seen from a 12^{m} distance.

A telescope magnifying 60 times shows it as a small pea seen from $0^{m}.80$ or a little larger than the Moon as seen with the naked-eye.

A power of 100, as a pea seen from $0^{m}.47$.

A power of 200, as a peach $0^{m}.06$ in diameter seen from $2^{m}.28$.

A power of 300, as the same peach seen from $1^{m}.42$.

A power of 500, as an orange $0^{m}.08$ in diameter seen from $1^{m}.12$.

A power of 1000, as the same orange seen from $0^{m}.60$.

A power of 1500, as the same orange seen from $0^{m}.36$.

An object will subtend an angle of 1° of arc when its distance from the eye is 57 times its own diameter.

An orange $0^{m}.08$, $4^{m}.56$ distant, subtends an angle of 1°, twice as large as the Moon seen with the naked eye, and equal to Mars seen with a magnification of 120×. A power of 1,200× corresponds to a distance of $4^{m}.56$ for the same object. A moment's reflection shows that powers of from 500× to 1200× are enough to make the disk of Mars conveniently large.

1636–1638.—Fontana

Here, then, are Fontana's observations. I will give all the observations in chronological order; I will discuss them, compare them, and from them draw out progressively the conclusions, which have led up to our present knowledge of the physical constitution of the planet.

The Neapolitan astronomer published his observations in a work entitled *Novae cælestium terrestriumque rerum observationes*, Naples 1655. I have this work in front of me, and I am happy to share its curiosities with my readers.

Here are the two oldest drawings of Mars, made by this lawyer, optician, and astronomer. The first is from 1636—he does not give the date exactly. The second is from 24 August 1638. Fontana wrote this legend (Figs. 1 and 2):

> 1636: Martis figura perfecte spherica distincte atque clare conspicebatur. Item in medio atrum habebat conum instar nigerrimae pilulae.
>
> Martis circulus discolor, sed in concava parte ignitus deprehendebatur. Sole excepto, reliquis aliis planetis, semper Mars candentior demonstratur.

I have translated this as follows:

> 1636. The form of Mars was observed to be perfectly spherical. In its centre was a dark cone in the form of a very black pill. The disk was of many colours, but appeared to be flaming in the concave part. Except for the Sun, Mars is much the hottest of all the stars.

Here is the second observation:

> Die 24 augusti, anno 1638.—Martis pilula, vel niger conus, intuebatur distincte ad circuli, ipsum anbientis, deliquim, proportionaliter deficere; quad fortarse Martis gyrationem circa proprium centrum significat.

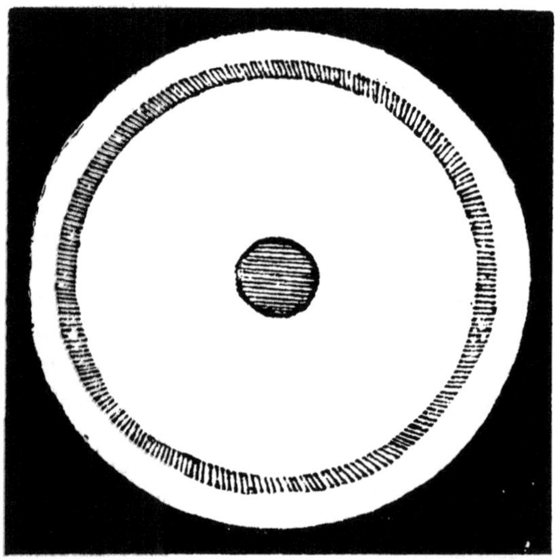

Fig. 1 First drawing of the disk of Mars, Fontana, 1636

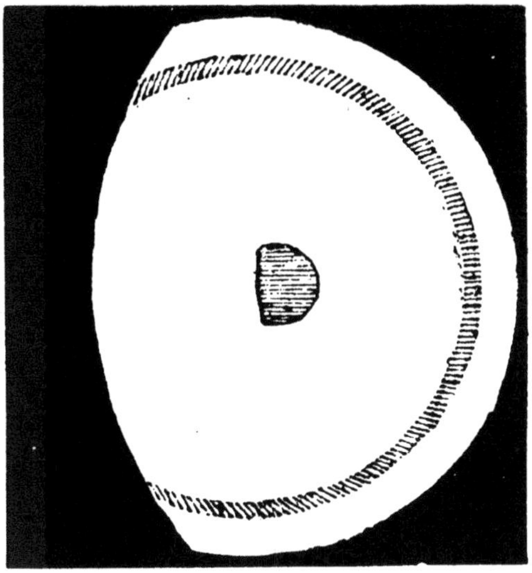

Fig. 2 Second drawing of Mars, done by Fontana in 1638

(Translation): "The pill, or black cone, showed up distinctly, with a phase proportional to that of the disk, which may perhaps indicate a rotational movement of Mars around its centre." (I admit that I cannot understand this phrase; does he mean that the patch was proportionally displaced?)

This "pill" or "small ball" seen at the centre of the disk of Mars was the first patch ever seen and drawn on this planet. These are the first two drawings ever made of Mars. I offer them to my readers in their naive aspect, if only as an historical curiosity.

The phase in the second figure is very exaggerated. Mars can never appear like that. The exact value of the phase has been given in this chapter. But at least Fontana did notice the phase of Mars. The spot is not, of course, real; it was an optical effect, due to a reflection, perhaps rather a sort of extinction of the light-rays, in the lenses of Fontana's telescope (Figs. 3 and 4).

Everything points to this interpretation: (1) the position of the round patch in the middle of the disk during the first observation: (2) the phase, corresponding with that of the planet, at the second observation. (3) the analogous effects in his drawings of Venus, here reproduced as a matter of curiosity, made on 11 November in 1645 and 22 January 1646, and in the description of which he even used the same term, "pill." These drawings of Venus are interesting to us now only because they show the phase.

But despite this, it is not without interest to publish here the very first drawings of Mars ever made—if only to preserve them.

Fontana began his book with a historical study of the invention of the telescope. He believed that the ancients had telescopes (though we know today that these were tubes without glass). He recalled what Porta had said about Ptolemy's mirror, which allegedly enabled Ptolemy to see ships at a distance of 700 miles. Fontana wrote that he had been unable to find out who had re-discovered optical instruments, and

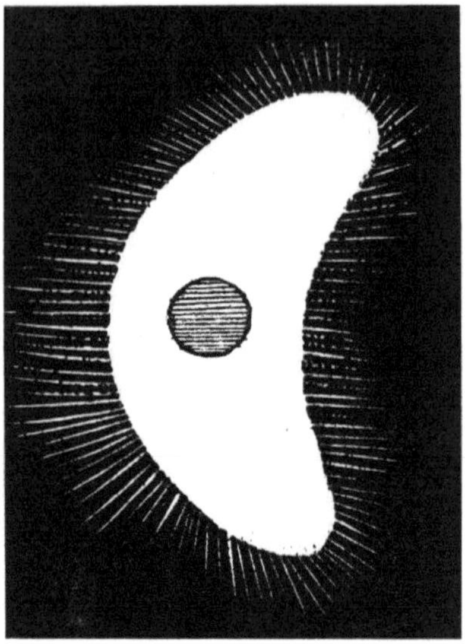

Fig. 3 Drawing of Venus by Fontana, in 1645

Fig. 4 Drawing of Venus by Fontana, in 1646

thought that it might have been Porta himself. Here is a passage from Fontana's *Magie naturelle*, published in 1589, Book XVII, Chapter X:

> Concave lenses clearly show distant objects, while convex lenses show nearby ones. This can be conveniently adapted to the eye." (Here he is evidently dealing with what we normally call eyeglasses for people who are either long-sighted or else myopic. Spectacles had been in use since the 12th century, and they had been known for a long time, even though

they were rare. Nero, who was myopic, had an emerald monocle with a concave lens.) But Porta then adds:

> Concavo, longe parva vides, sed perspicua: convexo propinqua majora, sed turbida: si utramque lentem recte componere noveris, et linginqua, et proxima majora, et clara videbis. Non parum multis amicis auxilii praestitimus, qui et loginqua obsoleta et proxima turbida conspicebant, ut omnia perfedtissime cernerent.

Here, unquestionably, we have the invention of the telescope—at least in theory.

Roger Bacon (died 1292) says that spectacles were in use in his time. Probably the two kinds of lenses were combined between the thirteenth century and the year 1570, because in a work published in 1570 (Euclid's *Elements*) an English author, Dee, recommends army commanders to use *perspective glasses*, and in a work by Digges, *Pantometria*, published in 1571, it is said that "by the combination of concave and convex mirrors and transparent lenses, objects can be made to seem much closer." These devices, whatever the details of their construction, were very rare. It was only in 1590 or even in 1606 that the Middelbourg opticians constructed the first really practical telescopes.

Fontana judged that: "The invention is also credited to Galileo, but in my view Galileo simply put Porta's theories into practice, or perfected a German invention."

Fontana said that he had himself made telescopes, and had done so since 1608. He had improved them considerably from year to year, particularly after 1611, when Kepler published his work on dioptrics. The first observation Fontana published was of the Moon, made on 31 October 1629. It is reproduced here as an historical curiosity, as it is, I believe, the first satisfactory drawing of the Moon that was ever made (Galileo's had been no more than a sketch). Readers will recognize the rays, which come from Tycho (C) and Copernicus (D). This drawing certainly drives home the rudimentary nature of those early refractors! (Fig. 5)

Fontana's book was adorned with an elegant frontispiece, which is given here as an astronomical and literary curiosity. Around the Fountain of Truth are grouped Geography, Mathematics, Cosmography, Poetry, Philosophy, Architecture and Astrology. To the right, Astronomy carries the Moon in his right hand and Ptolemy's book under his left arm.

The book is dated 1646. In the preceding year, that is in 1645, the Capuchin Schyrle de Rheita had published in Antwerp his bizarre work *Oculus Enoch et Eliæ*, which will be described later, and in which the same invention is related in the following terms (Fig. 6):

> In the year 1609, a Batavian optician named Joanne Lippersum, of Zealand, having by chance put a convex lens together with a concave lens, saw with admiration that this combination made objects appear larger and closer. Having therefore placed these two lenses in a tube at the most convenient distance, he could show passers-by the weathercock. The news of this invention spread, and curious people came in crowds to admire it: the Marquis of Spinola bought a telescope and presented it to the Archduke Albert. The magistrates, having sent for the optician, paid him well for a similar telescope, but with a strange proviso that he should not sell, or even make, another"—which explains (so Rheita tells us) how an invention of such power remained unknown for such a long time. Eventually it was disseminated and perfected, and Galileo—by his discoveries—made it widely known.

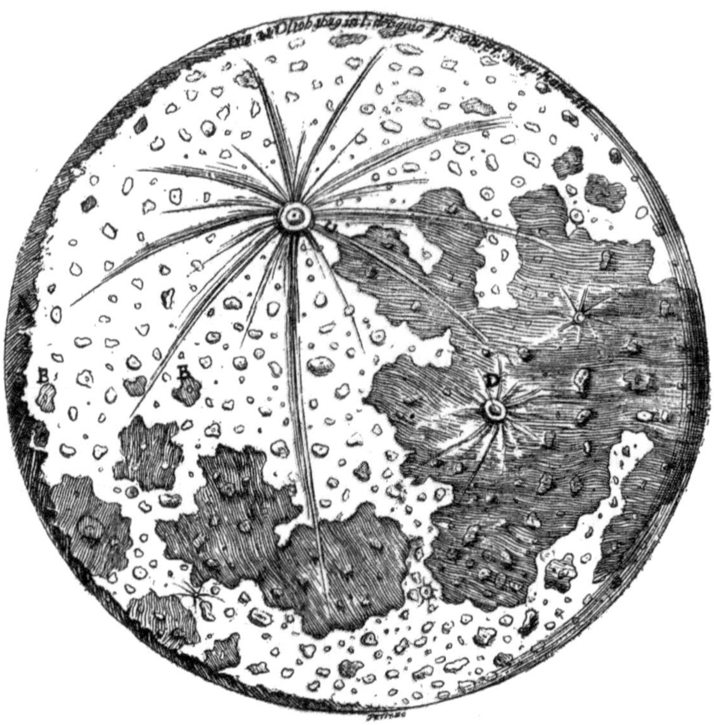

Fig. 5 The oldest telescopic drawing of the Moon

However, this kind of telescope was very inconvenient, because of its small field of view. Rheita saw the value of putting Kepler's ideas into practice, and used a combination of convex lenses. To him it did not seem that giving an upside-down image would matter much, and in any case the image could be made erect again easily enough by the addition of an extra lens. The field was increased enormously so that 40–50 stars could be seen at once, because the field was several hundred times larger than Galileo's. Encouraged by this success, he began to wonder if by using two telescopes, one for each eye, he could obtain a still greater improvement—and he succeeded in doing so. Le Gentil, who carried out new tests in the next century, found much the same; however, these binocular telescopes have remained rare, perhaps because few observers have eyes exactly equal.

Rheita explained his methods of cutting and polishing his glasses; he gave them a hyperbolic form, following an idea of Descartes. He also introduced the names *objective* and *ocular*, which are still in use.

Rheita's book was published in 1645. However, researches by M. Govi have shown that, in fact, the first binocular telescopes or field-glasses were presented to King Louis XIII by a Paris optician named Chorez, in 1620.

Let us now continue with our chronological account of the observations of Mars.

Fig. 6 Frontispiece of the work by Fontanta (Naples, 1646)

1640–1644.—Riccioli

In 1651 this prolific author published his great work *Almagestum novum*, which I have in front of me right now. On page 486 he reproduced Fontana's drawings, one-third reduced. He stated that his colleague Father Zucchi—like Riccioli himself, a Jesuit—had observed Mars on 23 May 1640, without being able to make out any patches, either black or red: "sine macula seu nigra seu rubra." Father Bartoli, his learned and eloquent Neapolitan colleague, observed Mars on 24 December 1644 and made two patches on the lower part of the disk. He believed that, God willing, future observers might be able to see them better: "Multa itaque observando supersunt, nobis aut vobis, a posteri!" He did not believe in the satellites of Mars observed by Rheita (described below); he was correct in believing them to be fixed stars.

1643.—Hirzgarter

In his work, *Detectio dioptrica planetarum verarum* (Francfort, 1643), written in German, this author writes at length about the planets. However, his observations are of very poor quality. He presents a drawing of Mars, which seems to be a caricature of Fontana's second drawing. Hirzgarter is mentioned here only to be dismissed.

1645.—Schyrle De Reita

This author, already mentioned above, was primarily interested in religion. He wrote with fervour about scientific studies and linked them with contemporary religion. In his bizarre book *Oculus Enoch et Eliæ, sive radius syderermysticus* (Anvers, 1645), dedicated to Jesus Christ, we find a bizarre chapter about Mars and a drawing which is more bizarre still—it is in the style of the rest of the book, and consequently devoid of any intrinsic value. However, this Capuchin was a relatively learned man, and he made some good telescopes, as mentioned above. Rheita's drawing, which is absolutely fantastical, is not reproduced here.

1645.—Hevelius

The hard-working and skillful observer Hevelius included a chapter about Mars, with particular reference to the phase, in his great work *Selenographica, sive lunae descriptio*, etc. (Gedani, 1647). He recalled an observation made on 26 March 1645, at 7 o'clock in the evening, together with another on the 28th of the same month. The phase shown in his drawing [Hevelius's *Pl.* G, *fig.* D; not reproduced by Flammarion.—WS]) is considerably exaggerated. It is almost like that of the Moon at quadrature, the eighth day of the lunation. Though his circle is 46^{mm} in diameter, the breadth of the gibbous is only 26^{mm}. In fact, Mars can never show a phase as pronounced as this. Hevelius made reference to Kepler's calculations about the phase of Mars, the observations by Fontana, and the treatise by Hirzgarter.

1656.—Huygens

In his *Systema Saturnium*,[4] the Dutch astronomer reported that during his observations of 1656,[5] he had once seen the globe of Mars crossed by a broad girdle, a sombre band hiding half the disk; and he gave the drawing reproduced here. Huygens was an eminent observer. However, this particular sketch is only of historical interest. This appearance of Mars could have been due to the effects of the polar caps. Huygens, like Galileo and Fontana, made telescopes for himself; with one of them he discovered Saturn's principal satellite, Titan, in 1655, and made out the ring in 1656 (Fig. 7).

(In passing, let us note that in this work by Huygens there is a charming medallion on the title-page, reproduced here for the benefit of readers who like bibliographical curiosities.) (Fig. 8)

Huygens made other, more important observations of the planet in 1659, 1672, 1683 and 1694. These pen-sketches are preserved in the library of Leyden University, where the Belgian astronomer M. Terby has examined them and compared them with modern drawings. For the first time he has been able to identify Huygens' principal features with those shown on modern maps. Though the 1656 sketch shows only a disk crossed by a broad darkened band, the later drawings were better, as will be described below.

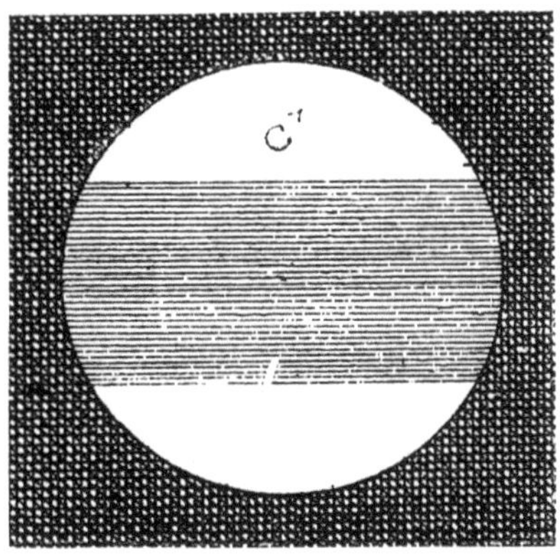

Fig. 7 Drawing of Mars by Huygens in 1656

[4] Christiani Hugenii a *Zutichem Opera Varia*, vol. 11, Hagæ Comitum, 1724, p. 540.

[5] This seems to have been not in 1656, but more probably in 1655 or 1657. In 1656 Mars was almost at its greatest distance from the Earth. However, we must accept the author's date. Probably it was in January or December.

Fig. 8 Medallion of the works of Huygens

1651–1657.—Riccioli

On page 372 of his work *Astronomia reformata*, etc. (Bononiæ, 1665), Father Riccioli stated that he and Father P. Grimaldi had observed patches on Mars on 4, 5, 6, 18 April and 29 May 1651, July 1653, July and August 1655, and September, October and November 1657. He recalled the observations by Fontana and de Bartoli, described above. He gave no drawings.

As will be realized, these first physical studies of Mars are only rudimentary. But with Huygens and Cassini, we come to a period that is more productive of results.

1659.—Huygens

In 1659, notably on 28 November and 1 December, Huygens observed Mars and made four sketches.

Reproduced here, after Terby,[6] is the drawing made on 28 November 1659 at 7 PM The patch which he shows has become the most characteristic feature of the geography of Mars, as will be seen later. Seeing that the patch was shifting its position, he wrote in his journal, on 1 December 1659: "Debet Martis conversio flori spatio circiter diurno, sive 24 horarum nostrarum quemadmodum item Telluris." (The rotation of Mars, like that of the Earth, seems to have a period of 24 hours.)

[6] Terby, *Aréographie* (Académie de Belgique, 1875), p. 8.

Fig. 9 Sketch of Mars by Huygens, on 28 November 1659

Some time afterwards, in 1666, Cassini independently discovered this rotational movement. Strangely, Huygens subsequently doubted it, as though he felt that he had placed too much importance on the changing appearance—a doubt he recorded in his journal on 9 April 1683: "Mars maculis aliter distinctus quam biduo ante, unde de conversaone 24 horarum quam Cassinus pordidit dubite."[7] Yet he did not remain doubtful for long, because in his *Cosmotheoros: a description of the celestial bodies and their habitability*, published posthumously in 1698, he stated that the rotations of Jupiter and Mars were established without question,[8] and that the inhabitants of these planets experienced days and nights which were very little different from ours (Fig. 9).[9]

Huygens made various other observations and drawings of Mars, notably in 1672, 1683 and 1694. These will be discussed later, in their chronological order.

1666.—Cassini

The brilliant Italian astronomer (he came from the vicinity of Nice, but by temperament was much more Italian than French) recorded his observations of Mars in two memoirs, entitled *Martis circa proprium axem revolubilis observationes Bononiæ habit*a (Bononiæ, 1666) and D*issertatio apologetica de maculis Jovis et Martis* (Bonoiæ, 1666) and in the *Journal des Savants* on 31 May 1666 and the *Philosophical Transactions* on 2 July of the same year.[10]

I have these four publications in front of me right now. The first is the most interesting from the viewpoint of the originality of the drawings. Though the figures published in the *Journal des Savants* and the *Philosophical Transactions* are modified

[7] *Ibid.*, p. 9.

[8] Huygens, *Cosmotheoros*, 1698, p. 16.

[9] *Ibid.*, p. 96.

[10] *Journal des Savants*, 2nd year, 1666, page 316. As we know, this publication still continues today. But strangely, it is scientifically much less interesting than it used to be 200 years ago. Whether this is due to a lesser number of astronomical authors is problematical.

copies of Cassini's original sketches, the drawings reproduced here are facsimiles of the originals.

Jean Dominique Cassini, as he was called in France by Louis XIV, became the first director of the Paris Observatory, then under construction. When in Bologna he had been the Papal astronomer, and was already well known for his measurement of the Bologna meridian and for a large number of other brilliant observations. Cassini's memoir is accurately summarized, as follows, in the *Journal des Savants* for 31 May 1666:

> These observations contain a new discovery regarding the planet Mars, not less curious than that reported last year upon Jupiter. It is described in this Journal for 22 February, and has already been commented on by scholars.
>
> M. Cassini, astronomer of Bologne [here, the editor mistakenly wrote Boulogne], having observed since the beginning of this year, 1666, with telescopes of focal length 25 palms or 16 ½ feet, constructed on the pattern of Signor Campani, has realized that Mars turns upon its axis, and has noted that there are several different patches on the two faces or hemispheres of the planet which appear in succession during this rotation.
>
> From the morning of 6 February, he began to see two dark patches on the hemisphere, and on the evening of 24 February he saw, on the second hemisphere, two more patches similar to the first, but larger. Since then, having continued these observations, he has seen these two hemispheres turning little by little from east to west, until finally coming back to the original aspect. Sr. Campani, who has also observed from Rome with telescopes of 50 palms, or 35 feet, has recorded the same phenomena on the planet. M. Cassini has engraved several figures, which represent these various positions.
>
> The *fig*. A (referring to Fig. 10) represents one of the hemispheres of Mars as M. Cassini observed it from Bologna on the evening of 13 March, with a 25 palm or 16 ½ ft. telescope.
>
> The *fig*. B represents the other face, as seen on the evening of 24 February.
>
> The *fig*. C represents the first hemisphere of the planet as Sr. Campani saw it from Rome on the evening of 13 March, with a 50 palm or 25 ft. telescope.
>
> The *fig*. D represents the second face as Sr. Campani observed it on the evening of 28 March.
>
> M. Cassini has made various comments about these figures. First, he says that on some occasions he has seen the two faces of Mars during the same night, one in the evening and the other in the morning. He noticed that the movement of these patches in the lower part of the apparent hemisphere of Mars was from east to west, as with all the other celestial bodies, with parallels of latitude inclined considerably to the equator and only slightly to the ecliptic.
>
> He found that on the next day the patches reached the same positions forty minutes later, so that after about 36 to 37 days they came back to the same places at the same times.
>
> In the near future he has promised to give special Tables of this movement and its inequalities with ephemerides, as he has already done in the case of Jupiter.
>
> From Rome, several other astronomers have also published observations made since 24 March with 25- and 45-palm telescopes made by Sr. Divini. From the way in which they represent these patches, they are very similar to those on the first hemisphere of Mars shown in the figure. The Roman observers judged that Mars turns on its axis in about thirteen hours.
>
> But M. Cassini considers that they are mistaken in their observations, because they stated that the patches seen on the planet on 30 March were small, distant from one another, and well away from the centre of the disk, and that the eastern patch was smaller than the western, as they are represented in the figure marked E, which seems to show the first hemisphere of Mars. However, from his observation made at Bologna at the same time, M. Cassini finds that the patches were very broad, close together, near the middle of the disk, and that the eastern patch was larger than the western, as seen on the figure marked F—which shows the second hemisphere of the planet. Moreover, he concludes that from only five or six observations it is premature to claim that the rotation period of Mars is 13 hours as do the Roman observers. Although he has been observing for a much longer

Fig. 10 Drawings of the planet Mars, made during February and March 1666. By Cassini and observers in Rome (*Journal des Savants* for 31 May 1666)

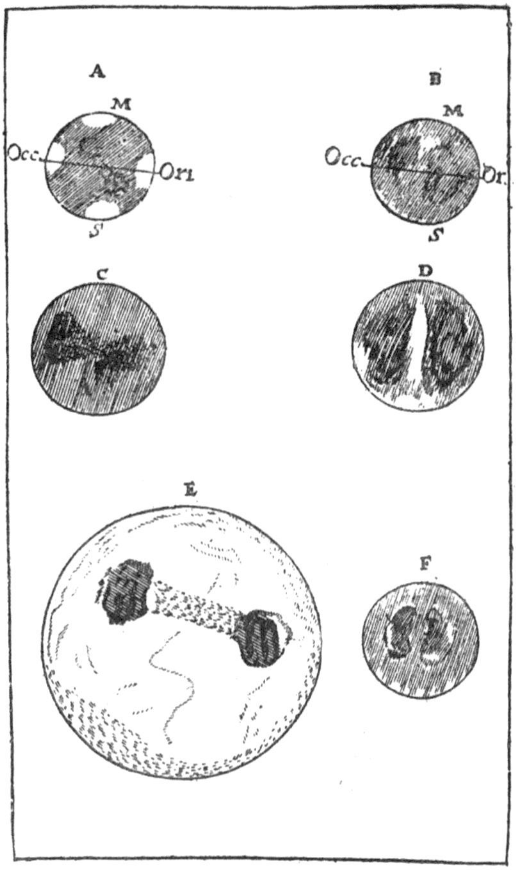

period, he will not dare to assert that Mars turns on its axis in $24^{h}\ 40^{m}$—all he will say for certain is that after 24^{h} 40, Mars again appears in the same guise.

But since these first observations, M. Cassini has published another work, in which for various reasons he concludes that Mars really does complete one turn in $24^{h}\ 40^{m}$ and that those who gave a 13-hour period saw not both hemispheres, but only the second one. They have not drawn the first hemisphere. He also states that when defining the rotation period of Mars we do not refer to the mean period, but only to that which is observed when Mars is in opposition to the Sun—which will be slightly shorter. He will give the reductions in the Tables, which he hopes to prepare.

This account is a complete résumé of Cassini's two Memoirs referred to above.[11] We provide our readers (Fig. 10) with a full-size facsimile of the page from *Journal des Savants* containing the six figures discussed in the text above.

---

[11]These observations were all made in 1666. Therefore it is surprising to read in Humboldt's *Cosmos*, generally so reliable, that "the first observation made by Cassini with regard to the rotation of a patch on Mars seems to have been made some time after the year 1670" (*Cosmos*, Vol. 111, p. 719). Humboldt quoted Delambre, *Histoire de l'Astronomie*, vol. 11, p. 694, as his source of information: but in fact Delambre is silent on the subjects, *des Savants* containing the six figures relevant to the text above.

Here, also (Fig. 11), are Cassini's original drawings, again reproduced in facsimile, from his memoir, *Martis circa proprium axem revolubilis observationes* (Bologne, 1666).

Also in this collection of Cassini's works are the writings referred to as the *Dissertation*, to be described below (Biblothéque de l' Observatoire de Paris, C.7, 15)—two editions, under two different titles, of a small work called *De planetarum facie, maculis et revolutione*; the other title is *Nuncii syderei interpres*. This is meant as a reply to Galileo's *Nuncius sidereus*. It has fifteen chapters. The first three differ between the two memoirs, but the other twelve are the same. In the first three chapters of the edition called *De planetarum maculis*, Cassini compares the planets with the Earth, showing that when seen from a great distance across space our globe resembles other planets; that the seas appear dark because they absorb the sunlight, while the continents appear bright[12]; that different varieties of soil ought to produce corresponding differences in aspect; also, that the figure of the Earth changes according to whether the Sun's rays strike the polar or the equatorial regions; and further, that the obliquity of the solar illumination, the clouds and their shadows, and the mountain chains and their shadows, are all able to cause variations in the aspect of our planet as seen from afar, and which must therefore affect the appearance of the Moon and planets as seen from Earth. Next, he passes on to the analogies between other planets and our own, and he considers astronomical observation from a philosophical point of view. He also shows that the irregularities of the surface of Venus had been suspected by Fontana on 22 January 1643, and had been observed by Cassini himself in Rome, with the Campani brothers, using their excellent telescope—no doubt in 1666.

On Mars, he notes that on 7 February 1666, at dawn, and also on the 17th and the 18th of the same month he had seen a white patch on the disk, near the terminator, extending down into the shaded part, and undoubtedly indicating a surface irregularity or roughness similar to what is observed on the Moon.

He then described the belts on Jupiter, observed from 1630 by Fontana, and the flattening of the Jovian disk. He compared the dark zones of Jupiter with mountain chains.

The rest of the work is devoted to the movements of the satellites of Jupiter. The book does not seem to have been completed, because both editions finish in the middle of a work of the last line of the last page (LXIII).

Same Year 1666.—Salvatore Serra

While Cassini was making these observations from Bologna, Salvatore Serra was working in Rome along the same lines, and in May 1666 he published his observations under the *title Martis revolubilis observations romanæ ab affictis erroribus*

[12] This was what Galileo had said in 1632, in his *Dialogo interno ai due massimi sistemi del monde*. Œuvres completes, 1842 edition, page 72.

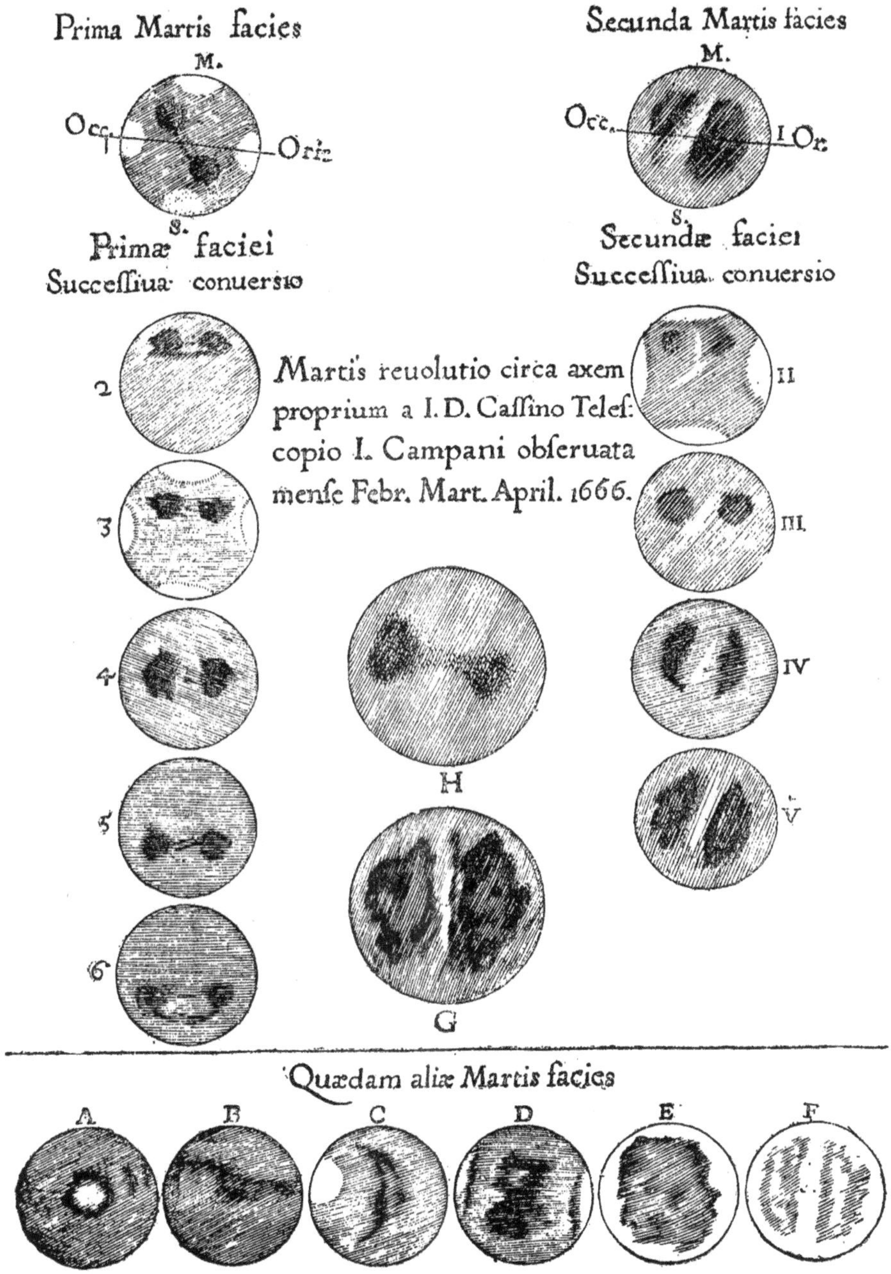

Fig. 11 Characteristic configurations of the two hemispheres of Mars, from the observations by Cassini in February, March and April 1666 (the *left row* and *fig*. H represent one hemisphere, the *right row* and *fig*. G the opposite hemisphere.). The change due to rotation is visible in the series on the left

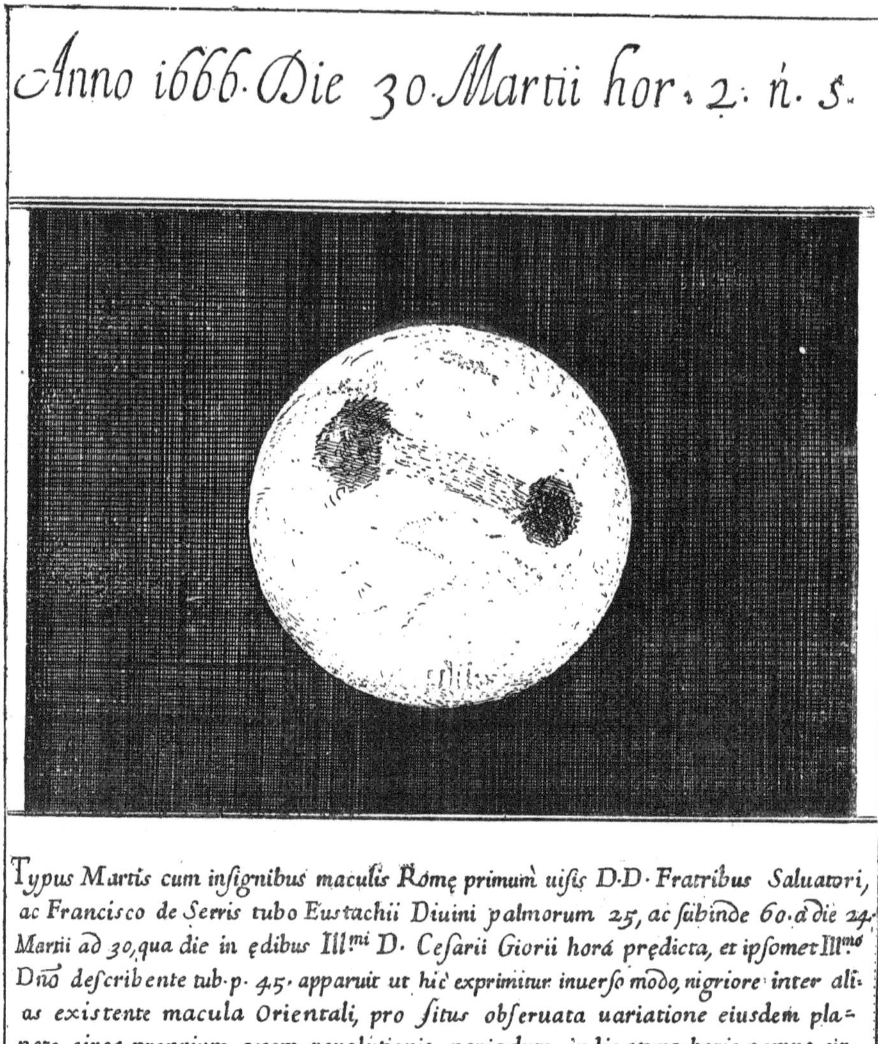

Fig. 12 Drawings of Mars, by Salvatore Serra, 30 March 1666

vindicatæ. Roma. Ex castro Sanctii Gregorii. Dated 1666: "Roman observations concerning the rotation of Mars, correcting some alleged errors." This is a reply to Cassini's declaration that the rotation period was $24^h\ 40^m$ and not 13 hours, as had been concluded by "some Roman observers."

The reply was accompanied by a drawing, reproduced here (Fig. 12).

The original of this drawing is in the Bibliothéque de l'Observatoire de Paris (C.7,3). As is stated in the accompanying legend, the drawing represents the telescopic view

Fig. 13 Sketch of Mars, by Cassini, 24 March 1666, at 7 hours

of the planet obtained from Rome with a 25-palm Divini telescope, by the Serra brothers on 30 March in 1666 at 2 a.m. The same aspect had been observed on the 24th and the 30th of March, which was consistent with a rotation period of 13 ^{h}.

The quarrel between Cassini and Serra was very lively, as can be seen from Cassiniís work entitled *Dissertationes astronomicae apologeticæ*, a collection including a 1665 memoir—about the shadow of the satellites of Jupiter, the discovery of which was disputed—and a 1666 memoir concerning the patches and rotations of Mars and Jupiter. In the latter memoir, he refuted the claims of Salvatore Serra, and even cast doubt upon the authenticity of his observations. He showed that although patches on Mars had been seen earlier by Fontana, Hevelius, Gassendi, Riccioli and Sirsalis, the discovery of the rotation period was due to Cassini himself; and from a long discussion of the patches observed in February and March 1666, he proved that the period could not be about 12–13 hours, as Serra had claimed. Instead, it had to be fixed at 24^{h} 40^{m}. In this memoir there is a small sketch of Mars, admittedly rather rudimentary, made on the evening of 24 March, showing that—contrary to Serra's assertions—the planet presented neither the first nor the second hemisphere as drawn in Cassini's observations published earlier. Instead, it showed another aspect—*aliam quemdam maculam semilunarem... qualem nos eodem die hora 1 noctis observavimus*. Cassini judged that what he had seen previously on 22 February, at six a.m., "corresponded to a retardation of forty minutes." (Fig. 13)

Cassini then referred to the observations of Mars made from Rome on the 3rd of March by Campani, and showed that they were consistent with his own observations at Bologne.

We see in Fig. 11—particularly the sketches to the left—the displacement of the patches caused by the planet's rotation. The Figs. 12, 13, and 14 are described in the text which accompanies them. The upper drawing of the latter plate is a reproduction made by Cassini of the drawing of the observation made by the brothers Salvatore and Francois de Serra. The first of the small drawings was made by Cassini on the same evening, 30th of March, and at the same time.

Serra's observations are shown next to Cassini's.[13]

[13] With Cassini's manuscripts, preserved in the Paris Observatory, there are several of Salvatore Serra's letters about this subject, written in Latin and Italian. In the first, dated 27 February 1666 (Rome), it is said that Serra observed the patches on Mars with a 25-palm telescope, and that the tube of the 50-palm telescope was inconvenient. In another letter, dated 24 March, we find the phrase to which Cassini replided "Maculas aliquot quarum una cæteris nigrier aliquantulum iam superarat diei medium"; in another, dated 27 March, we find analogous observations; in the last, dated 10 April, he discussed whether the rotation period is 12 or 24 hours. The same collection (C.7,3) contains a letter from Campani, dated 3 March: "Have observed the patches on Mars, and have recognized the movement of rotation."

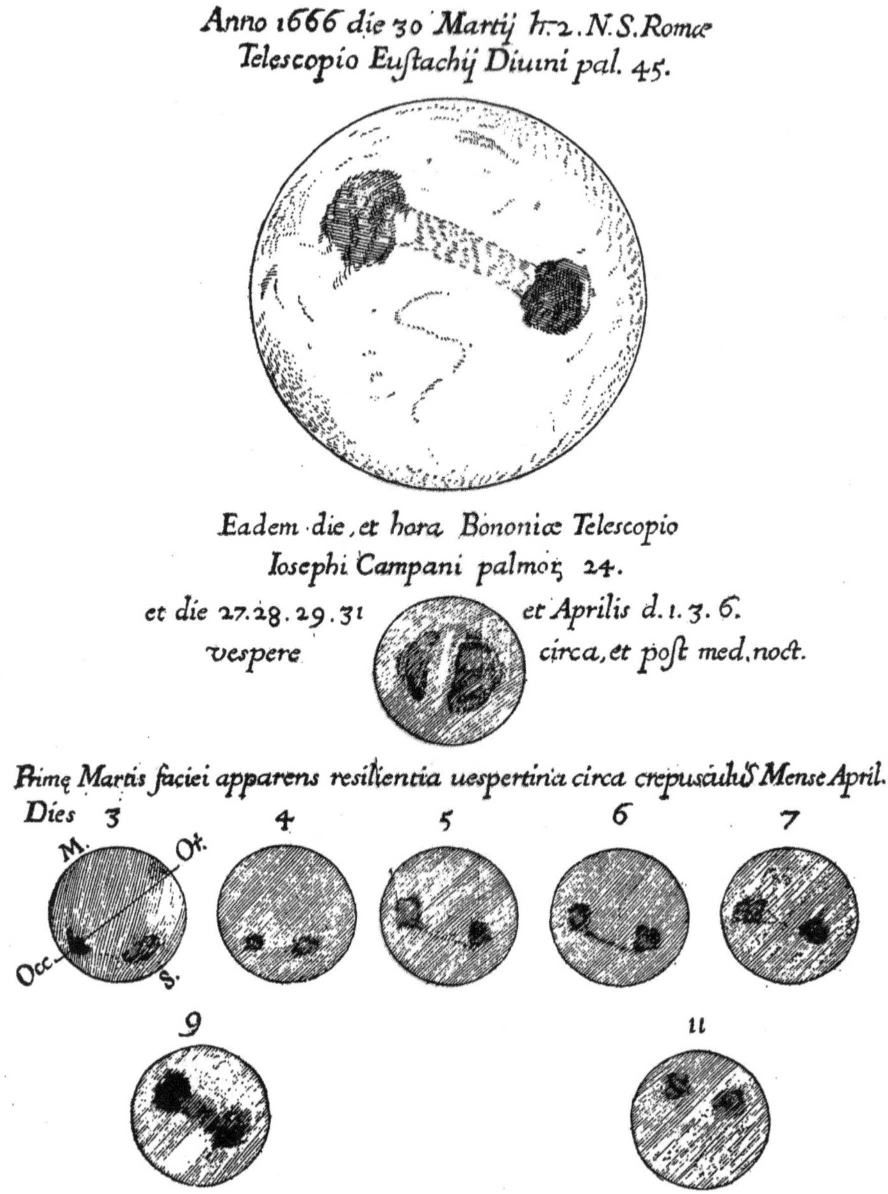

Fig. 14 Comparison made by Cassini in his drawings (all small), with the drawings by the observers in Rome

Cassini's observations confirm that he really did discover the rotation period of Mars ($24^h\ 40^m$; actually, as we now know, it is $24^h\ 37^m\ 22^s.6$), though it is true that he made the discovery without recognizing the precise forms of the patches.

Indeed, as a measure of the advance in our knowledge over the years, we have to admit that Cassini's drawings—published here in their original form—bear little resemblance to the actual geography of the planet. Should we therefore conclude that Mars has changed in appearance during the two centuries since Cassini's observations were made? The answer is "No," because in the same year—1666—the astronomer Hooke as we will see, made drawings much closer to the truth; and we have seen that in 1659 Huygens had made a sketch which is still useful today. When dealing with features near the limit of visibility, different people tend to distinguish and interpret things in different ways.

We have seen that Cassini's first observations of Mars were made on the 6th and the 24th of February in 1666. At the same time as he and other astronomers were observing from Italy, the English astronomer Hooke was observing from London, and he also discovered the planet's rotation by noting the movements of the surface patches.

Mars was then at an unfavourable opposition, almost at aphelion; general attention was directed to the planet because of the improvement in telescopic equipment. After discovering the patches on Jupiter, Hooke was led on to study those of Mars. The first announcement of his observations was a note published in the *Philosophical Transactions* for 2 April (p.198), announcing the existence of the patches on Mars, and the phenomenon of rotation. The following number, on 7 May, contained a memoir and the drawings. Here are these observations:

Same Year 1666.—Hooke

The English astronomer Hooke, contemporary and rival of Newton, published observations of Mars in the *Philosophical Transactions* for 1666, under the title *The Particulars of those Observations of the Planet Mars*, formerly intimated to have been made at London in the months of February and March, anni 1666, 5/6.[14] The memoir was translated in the *Journal des Savants* for the following 23rd of August, and this account is given here; it will provide a great deal of material for bibliographers. The observations were made with a 36-ft. telescope. The drawings reproduced here are taken not from the *Journal des Savants*, but from the originals themselves, published in the *Philosophical Transactions*. Here is the account (Figs. 15 and 16)[15]:

> Having a great desire to observe the Body of Mars, whilst Acronycal and Retrograde (having formerly with a Glass of about 12-foot long, observ'd some kind of Spots in the Face of it), though it be not at present in the Perihelium of its Orbe, but nearer its Aphelium, yet I found, that the Face of it, when neer its Opposition to the Sun (with a Charge, the 36 foot

[14] *Phil. Trans.* Giving some account of the present undertaking studies … of the world. Vol. 1, 1665–1666, p. 239.

[15] *Translator's note.* I am most grateful to Dr. M.A. Hoskin, of Churchill College, Cambridge, for providing me with a copy of Hooke's original memoir. PM.

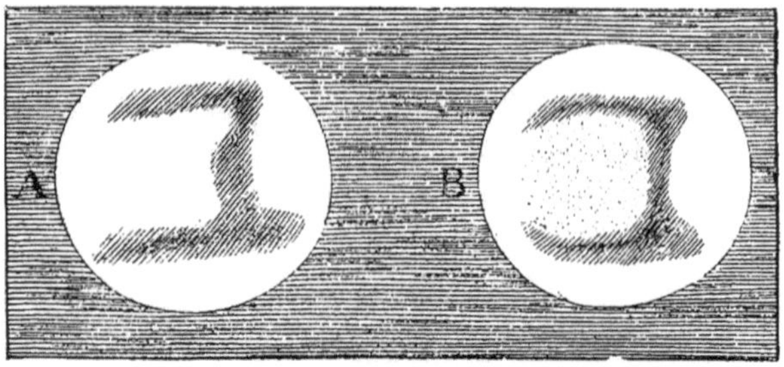

Fig. 15 Drawings of the planet Mars by Hooke at London, on the night of 12 and 13 March 1666, at 20 and 40 minutes after midnight

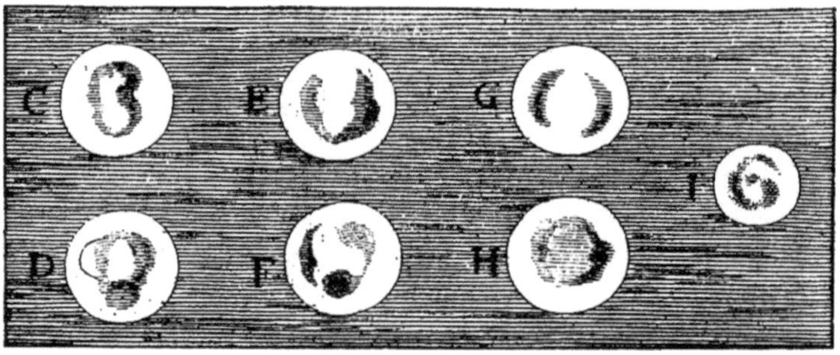

Fig. 16 Drawings of the planet Mars by Hooke at London, from 20 March to 7 April

glass, I made use off, would well bear), appear'd very near as big, as that of the Moon to the naked eye; which I found, by comparing it with the Full Moon, near adjoyning to it, on March 10.

But such had been the ill disposition of the Air for several nights, that from more than 20 Observations of it, which I had made since its being Retrograde, I could find nothing of satisfaction, though I often imagin'd, I saw Spots, yet the I*nflective* veins of the Air (if I may so call those parts, which being interspers'd up and down in it, have a greater or less Refractive power, than the Air next adjoyning, with which they are mixt) did make it so confus'd and glaring, that I could not conclude upon anything.

On the 3rd of March, though the Air were still bad enough yet I could see now and then the Body of Mars appearing in the form shown in A; which I presently described by a Scheme (see Fig. 15); and about 10 minutes after, as exactly representing what I saw through the Glass, as I could, I drew the Scheme B. This I was sufficiently satisfied (by very often observing it through the Tube, and changing my Eye into various positions, that so there might be no kind of Fallacy in it) could be nothing else, but some more Dusky and Spotted parts of the Face of this Planet.

March 10, finding the Air very bad, I made use off a very shallow Eye-glass, as finding nothing Distinct with the greater Charge; and saw the appearance of it as in C (see Fig. 16), which I imagin'd, might be a Representation of the former Spots by a lesser Charge. About 3 of the Clock of the same morning, the Air became very bad (though to appearance exceeding clear, and causing all the Stars to twincle, and the minute Stars to appear very thick) the Body seem'd like D; which I still suppos'd to be the Representation of the same Spots through a more confused and glaring Air.

But observing March 21, I was surprised to find the Air (though not so clear, as to the appearance of small Stars) so exceeding transparent, and the Face of Mars so very well defined, and round, and distinct that I could manifestly see it of the shape in E, about half an hour after Nine at night. The triangular spot on the right side (as it was inverted by the

Telescope, according to the appearances, as also all the preceeding Figures are drawn) appear'd very black and distinct, the other towards the left more dim; but both of them sufficiently plain and defin'd. About a quarter before 12 of the Clock the same night I observ'd it again with the same Glass, and found the appearance exactly, as in F; which I imagin'd to show me a Motion of the former triangular spot: But designing to observe it again about 3 of the Clock the same Morning, I was hindered by cloudy weather.

But March 22, about half an hour after 8 at night, finding the same Spots in the same posture, I concluded, that the preceeding Observation was only the appearance of the same Spots at another height and thickness of the Air: And thought myself confirm'd in this Opinion, by finding them in much the same posture, March 23, about half an hour after 9, though the Air was nothing so good as before.

And though I desired to make Observations, about 3 of the Clock those mornings; yet something or other interven'd, that hindered me, till March 28, about 3 of the Clock, the Air being light (in weight) though moist and a little hazy; when I plainly saw it, to have the form, represented in I; which is not reconcileable with the other Appearances, unless we allow a Turbinated motion of Mars upon its centre: Which, if such there be, from the Observations made March 21, 22 and 23, we may guess it to be once or twice in about 24 hours unless it may have some kind of Librating motion; which seems not so likely. Now, whether certainly so or not, I shall endeavour, as oft as I have the opportunity, further to observe.

Explication of the figures mentioned in the preceding discussion.

A. March 3—$00^h\ 20^m$ in the morning: the Air having many inflecting parts dispers'd up and down in it; by the Wheel Barometer, heavy.
B. Another Scheme—which I drew from my Observation, about 10 minutes after, the same morning. Both these were observ'd with a very deep Eye-glass.
C. March 10—$00^h\ 20^m$ in the morning: the Air heavy and inflective. Use was made of a shallow or ordinary Charge.
D. March 10—$3^h\ 00^m$ in the Morning: the Air very heavy and Inflective, which made if glare and radiate and be more confused, than about 3 hours before. A shallow Charge.
E. March 21—9 ½ hours past merid; the Air light (in weight) and clear, without inflecting parts; the Face appear'd most distinctly of this Forme. A shallow Charge.
F. March 22—11 ¾ hours past merid.: the Air continuing very light and clear, without inflecting vapours. A shallow Charge.
G. March 21—8 ½ hours past mer. The Air clear, with few inflecting veins in it, and indifferent light. A shallow Charge.
H. March 23—9 ½ hours past mer. The Air pretty light, but moist, and somewhat thick and hazy, but seem'd to have but few veins, or inflecting parts.
I. March 28—$3^h\ 00^m$ in the Morning: much the same kind of Air with that of March 23; light, moist, and a little hazy, with some very few veins.

This account of the English astronomer's observations was communicated to the Royal Society of London on 28 March 1666. Reproduced here, by photogravure, are authentic facsimiles—not retouched in any way, and of the original size—of Hooke's nine drawings. The dates of the observations given by Hooke should be increased by ten days, because of the reform of the calendar adopted in Italy in 1582 but not adopted in England until 1752. Thus 3 March corresponds to the 13th.

The same plate in the *Philosophical Transactions* from which these drawings have been reproduced also includes the drawings by Cassini and the Italian observers, but these are somewhat exaggerated—notably the large drawing at the top of Fig. 14, in which the two spots are shown as so massive that they take on the aspect of a dumb bell!

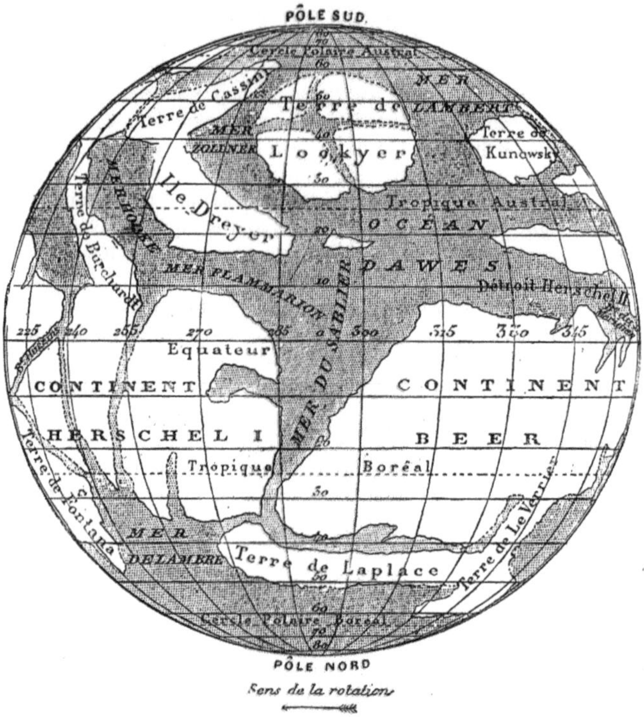

Fig. 17 The globe of Mars with the *mer du Sablier* (Hourglass Sea) at centre

At this stage, we must pause briefly to ask whether any preliminary conclusions can be drawn from these primitive sketches.

The sketches are very primitive indeed. We must agree that the telescopes of the period could not give good definition, because on all those drawings it is almost impossible to recognize any of the geographical configurations, which actually exist, on the globe of Mars. It is true that three drawings do provide a definite identification, even though distinctly vaguely: these are the drawings made by Huygens on 28 November 1659 (Fig. 9) and by Hooke on 13 March 1666, at $0^h\ 20^m$ and $0^h\ 40^m$ (Fig. 15) To introduce our real knowledge of Martian geography, I reproduce here (Fig. 17) an image of a globe of Mars that I constructed several years ago from all the available observations. One hemisphere—the most characteristic one—will suffice. The nomenclature followed in the globe is that of the general chart of the planet constructed by Mr. Green in 1877 and published by the Royal Astronomical Society of London, apart from the names of the Hourglass Sea (also called the Kaiser Sea) and the Meridian Bay (also called Dawes' Bay). I have kept the name of the *Hourglass Sea* for the characteristic triangular sea because: (1) it has been so called for a long time, (2) the name is very appropriate because of its shape, and (3) the patch has indeed served as an hour-glass for measuring the rotation period of Mars, because it is by means of its transits of the central meridian, and comparison

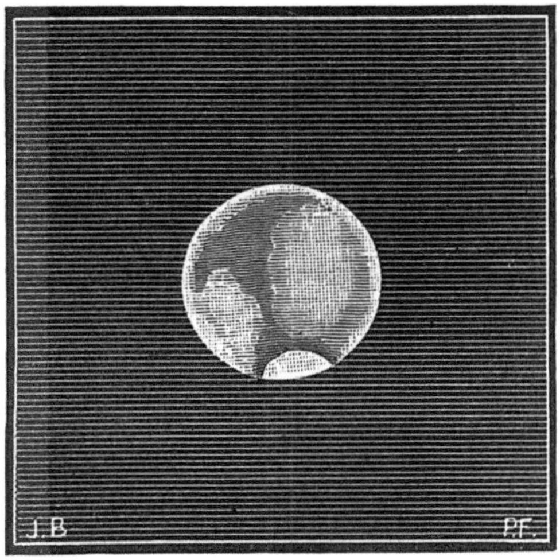

Fig. 18 Small drawing of Mars, obtained in 1884, with the aid of the small refractor of 75^{mm} aperture

of the old with the modern drawings, that we have been able to measure Martian time so exactly.

On this projection the globe of Mars is shown as perpendicular to its axis, with the south pole at the top and the north pole at the bottom. The globe does not often appear exactly upright, as we see it here, more often it is slightly inclined, sometimes the south pole being exposed and sometimes the north. However, this upright position will suffice in orienting ourselves. Other aspects can be discussed later.

The three drawings referred to above show the Hourglass Sea. If they are compared with Fig. 17, it is found that the first drawing shows that sea as very broadened and vague; the second shows it narrower, joined above to the Flammarion Sea and below to the Delambre Sea; the third gives a similar view. The representation is not precise or clear; it was drawn as seen from a great distance by imperfect instruments, but it is unmistakable, and calculations prove, from the $24^h\ 37^m\ 22^s.6$ rotation period, that the sea really was on the central meridian at the times when the observations were made.

To this chart may be added a low-power view of the planet (Fig. 18) made with a small telescope (75^{mm} aperture), to show the smallest features which can be seen with modest instruments—assuming good optics, an excellent eye and an experienced observer. With regard to the images given, modern telescopes of this size are comparable with the large but primitive instruments used by Hooke and Cassini. What progress has been made! An instrument of this type costs 200 francs today. Could not every primary school possess one? Citizens of the Earth have at last an idea of the scheme of the universe. But even in France, there is certainly not one person in a thousand who has ever seen these marvelous celestial bodies magnified ten times, let alone a hundred times!

This view of Mars was obtained on 18 February 1884. It is satisfactory. The sky was cloudy, cutting down the planet's light, but the clouds did not adversely affect the observation; in fact, the reverse.

On the disk (time of observation: $9^h\ 35^m$) may be seen a grey patch in the form of a champagne glass, considerably widened at the top, and reminding one of the outstretched wings of a seagull. This elongated patch is the Hourglass Sea, which was shown on Hooke's two drawings of 12 March 1666. Even by itself, this is enough to prove that the patches on the planet are permanent. At the lower edge of the disk, the north pole is marked by the polar snow, which forms a circular white patch.

As for the other drawings by Hooke—as with those of Cassini, Serra, etc.—we must admit that nothing can be recognized, let alone identified. Could the surface of Mars have been masked by clouds? Could the instrument have lacked the power of definition? Whatever the case, the patches have served to enable us to determine the rotation period. They genuinely exist; they are more or less precise, and cannot be false images, since the rotation period derived from them is accurate.

From these first observations—as we will see—Cassini, in 1666, was able to conclude that the rotation period of the planet was about $24^h\ 40^m$ (without taking into account the movement of the Earth). However, the opposition of 1666 was far from being one of the most favourable; it took place on 18 March, that is to say at a time when the planet was near aphelion, and hence a long way from the Earth. This opposition was analogous to that of 1886.

Let us continue our study.

1672.—Huygens

Huygens' observation of 1656 has already been described, and four made in 1659; several other drawings, including two made in 1662, are also preserved in the University of Leyden (Fig. 19).

One was made on 6 August and the other on 13 August. The first shows no dark patches, but only the white polar patch on the meridian; the second shows the pole and, in the lower part of the disk, the Hourglass Sea. The second drawing is reproduced here, after the facsimile, which has been published by Mr. Terby.

In 1672 Mars passed through a perihelic opposition, that is to say one of the most favourable of all. This happens about every 15 years. The oppositions of 1689, 1704

Fig. 19 Drawing of Mars by Huygens, on 13 August 1672, at $10^h\ 30^m$

Fig. 20 The globe of Mars, presenting its southern pole (perihelic oppositions). 1672–1689–1704–1719–1734–1751–1766–1783–1798–1813–1830–1845–1860–1877–1892

and 1719 were similar. At these times the planet appears tilted so that the upper pole is visible, with the Hourglass Sea very low down. Huygens' drawing is quite accurate, as can be seen from a comparison with the chart given here (Fig. 20).

Same Year 1672.—Flamsteed

The first Director of the Royal Observatory in England, founded in 1676, observed Mars, notably on 11 October 1672. He wanted to measure the position of Mars, and wrote: "Planet semper circa medium obscuritas aliqua apparuit, quam ut potui in figura adumbravi." This is all that he said about the planet. The sketch, which he gave, simply showed the interior of the disk, toward the central region; the *shading* of which he spoke was an irregular patch surrounded by a broad penumbra. This diagram does not seem interesting enough to reproduce here.[16]

At the same opposition, the planet was observed by LAURENTIUS,[17] but without any useful result in the progress of our knowledge of the physical state of Mars.

---

[16] It can be found in *Historia Coelestis*, 1725, Vol. 1, P. 17, Fig. 35.

[17] Joannis Francisci de LAURENTIIS, *Observationes Saturni et Martis Pisauriensis*. In. fol. Pisauri, 1672.

1683.—Huygens

To the observations by Huygens described above must be added those made in 1683, on 7 and 9 April and 7, 13, 17 and 23 May (Fig. 21).

Like the other drawings, these six observations are only vague sketches similar to those of 1659 and 1672; but they are enough for the identification of the characteristic Hourglass Sea with which we are already acquainted. Among the drawings showing its form well is that of 17 May, at $10^h\ 30^m$; like the rest, it has been made with a pen. Mars was then very distant from the Earth, since a perihelic opposition had occurred in 1672. On 4 February 1694 Huygens made another sketch of the same kind.

Such was the state of the study of Mars when Fontenelle published his *Entretiens sur la pluralité des Mondes*. One curious fact should be noted. Mars had passed very near the Earth in 1672, but had been observed only from the viewpoint of positional astronomy; the only physical observation had been that of the south polar patch, by Huygens.

1686.—Fontenelle

The ingenious author of the *Entretiens sur la pluralité des Mond*[18] was concerned with all the planets, as well as with the Sun and the fixed stars; and he summarized knowledge obtained up to that time, in the most elegant language. Although he wrote at great length about Venus—its rotation, its year, its climate and even its mountains—he seemed a little disdainful about Mars. He wrote:

Mars has no particular feature, so far as I know; its days are more than half an hour longer than our own, and its year is equal to two of ours, almost to within a month. It is five times smaller than the Earth, and sees the Sun as a little less large and much less vivid than we do. Quite different from Jupiter, with its beauty, its four moons or satellites!

Fig. 21 Outline of Mars by Huygens, on 17 May 1683, at $10^h\ 3^m$

[18] First edition; Paris, 1686.

This is all that Fontenelle said about Mars. He came back to it again a little later, but only to call attention to the lack of satellites, which he profoundly regretted from the point of view of logic. He wrote to his fictional Marquise: "We cannot disguise the fact that there are no moons, and there must be additional light-sources about which we know nothing. We have seen phosphorescent materials, either liquid or dry, which upon receiving light from the Sun, absorb it so that they can shine brilliantly when in shadow. Perhaps Mars has great, high rocks, naturally phosphorescent, which during the day can store up light, emitting it again during the night. Nobody can imagine a scene pleasanter than that of rocks illuminating the whole landscape after sunset, and providing a magnificent light without inconvenient heat. In America we know that there are birds, which are so luminous that in darkness we can read by their light. How do we know that Mars does not have great number of these birds, which when night comes, scatter on all sides and make a new day?"

This is charming. Even though Fontenelle made no advances in the technical studies that concern us here, his writing is at least interesting, and invites us to investigate further.

The two satellites of Mars were discovered 191 years later.

The seventeenth century ended several years after the appearance of Fontenelle's book—a book that marked, and indeed opened, a new era in the history of scientific literature. From our point of view, the 18th century opens with the researches of Maraldi (Cassini's nephew) at the Paris Observatory.

1704.—Maraldi[19]

In 1672 the planet passed through a very favourable opposition, and the parallax of Mars was successfully measured. In September–October 1704 there was another close approach. It was observed with particular care at the Paris Observatory for a new determination of the parallax, and Maraldi also took the opportunity to observe the patches and verify the movement of rotation. His conclusion was that the patches are variable. Here is his memoir together with the three drawings which accompany it:

> While Mars has been at its least distance from the Earth, I have observed the patches on the planet with a 34-ft. telescope by Campani, and I have been able to confirm the axial rotation which, as M. Cassini discovered, has a period of about $24^h\ 40^m$.
>
> The patches visible with large telescopes on the planets disk are not normally well defined, and they often change in form, not only from one opposition to another (that is to say, at the times most favourable for observation) but even from one month to the next. Notwithstanding these changes, the patches last for long enough for us to follow them for a time sufficient for a determination of the rotation period.

[19] Observations des taches de Mars pour vérifier sa révolution autour de son axe. *Histoire et Mémoires de l'Académie des Sciences*, 1706, P. 74.

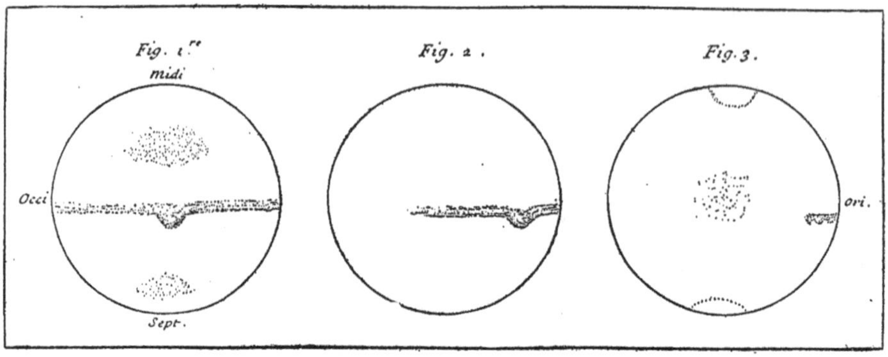

Fig. 22 Appearance of Mars on 14 October 1704, at $10^h\ 24^m$, 16 October 1704, at $9^h\ 5^m$, and the same date at $7^h\ 0^m$. Facsimiles of drawings by Maraldi

> Among these different patches, I have noted one in the form of a band near the middle of the disk, a little like one of the bands of Jupiter (Fig. 22) It does not go completely round the globe of Mars, but is interrupted, as is also the case with several of the bands of Jupiter, and occupies only a little more than one hemisphere of Mars; this is what I have found, after observing the planet at different hours on the same night and at the same hours on different nights. The band is not everywhere uniform, but at about 90° from the preceding limb of Mars it bends, and this bend forms a point well defined by comparison with the ordinary patches on the planet, which has enabled me to confirm the rotation.
>
> I saw the band when I made my first observation with the large telescope, in August, when Mars was approaching the Earth and had begun to look very big; however, I did not see the point to which I am referring until the following October. It came to the middle of the Martian disk at $10^h\ 24^m$ on 14 October; on the 15th, it did so at $11^h\ 9^m$.

On the 16th, at seven o'clock in the evening, near the two poles of rotation of Mars, I saw two bright patches, which have been observed many times during the past 15 years. Outside these two bright patches, I could see shading near the eastern edge; this was the extremity of the band, beginning to come on to the hemisphere of Mars, which was facing the Earth. On the same day, at $9^h\ 5^m$, the extremity of the band had already passed over the centre of Mars, and continued to the eastern edge, where I could see an indication of the patch attached to the band; this patch reached the central meridian at $11^h\ 38^m$. On the following days, further observations were made of the interrupted band, which was not so full on the disk as before; I observed that on 17 October the principal patch reached the central meridian at $11^h\ 18^m$. Comparing these observations with each other, it seems that the intervals between successive transits of the patch across the central meridian were not exactly equal; there were differences of a few minutes, which I attribute to the difficulty of making an exact timing of the moment when the patch reaches the central meridian. But comparing the observation of 14 October with that of the 17th—during which interval there have been three rotations—I find that the spot returns to the central meridian after $24^h\ 38^m$.

A better determination of the period could be made by comparing observations of the patch made over a longer interval, and this was done on 22 November, on which day, after having seen the preceding end of the band coming on to the disk at $7^h\ 0^m$, I observed that the patch reached the central meridian at $11^h\ 5^m$.

If we compare the latter observation with that made on 14 October at $10^h\ 24^m$, we find that the interval between these observations is 39 days 41 minutes, which when divided by 38, the number of rotations made during the interval, gives 1 day 39 minutes for each rotation—within one minute of the determination made by Cassini. These periods are derived from observations made over intervals so short—relatively speaking—that the movement of Mars itself is not very considerable. The mean rotation period will be a little longer than the apparent rotation period, but this can be virtually neglected—as can the difference in transit-time of the patch between the times when it appeared distinctly oval rather than circular; as was the case at the time of the observation made on 22 November.

I believe that it would be useless to take these differences into account, because the patches which we observe show changes, and I could not hope to time their transits with sufficient precision. The point attached to the band, which I observed for several days toward the middle of October, had greatly diminished by 22 November; the distance from the end of the band preceding it was the same, but otherwise it would not have been recognized as the same feature seen during previous observations. After 22 November I was unable to continue observations of the patch to see what further changes occurred, because the interval between the first two observations had been nearly a month, after which time Mars had receded too far from the Earth for the patches to be made out; but the observations made in the preceding September led me to believe that considerable changes had occurred—because taking as a starting-point the observation of the patch made on 14 October, and assuming that it subsequently returned to the central meridian after equal intervals of time, it follows that the patch would have come to the centre of the disk of Mars from 4 to 10 September at almost the same hours as it did in the middle of October. However, according to the observations which I had made on those dates, at various hours during that night, I saw no trace of the patch, though I easily distinguished the band and noted the point associated with it. At the beginning of September, instead of the point, I observed another patch at the centre of Mars, separated from the band toward the west, and evidently having an independent movement from north to south–so that it approached the band, and became the point which I observed toward the middle of October and had so greatly diminished by 22 November. These changes bear some resemblance to those observed by Cassini for the patches on Jupiter, and which are also sometimes observed with the spots on the Sun.

As we have seen, these observations made by Maraldi in 1704 confirmed the rotation period found by Cassini, and also the existence of patches of various kinds on the surface of the planet—some dark, others bright. These patches seemed to him to be *variable* in both size and position, as with the features on Jupiter.

In 1704, as in 1672, Mars came to perihelic opposition, and Maraldi's drawings should therefore bear comparison with my globe, showing the various aspects under such condition, and representing the whole surface of the planet. In Fig. 23A, for instance, the central meridian is at 270° and shows the Hourglass Sea, so that it almost corresponds to the hemisphere shown in Fig. 20 above.

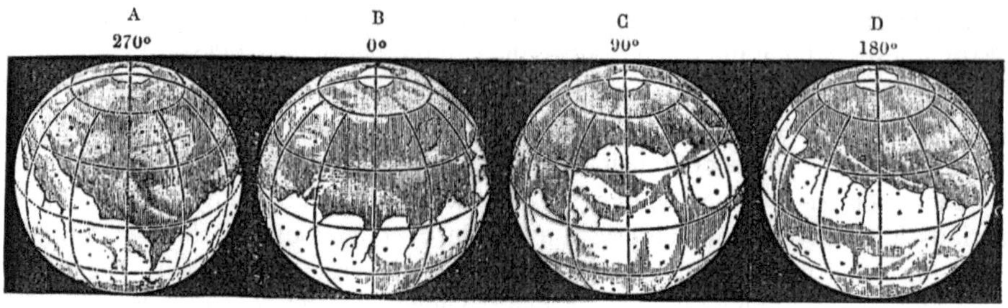

Fig. 23 Aspects of Mars during its perihelic oppositions

And yet the features on Maraldi's three drawings cannot be identified with certainty, or even probability. The band actually exists, more or less as shown in the drawings, and can serve for a determination of the rotation period, but it does not resemble the features seen by Cassini and Hooke, and Maraldi himself considered that there had been changes in the interim. Moreover, there may have been changes during the period when Maraldi himself was observing the planet. We can now list the few facts about Mars which had been established by 1704 following the date of the first observation by Huygens (1656), that is to say a period of 48 years:

- *The globe of Mars shows patches, as with that of the Moon;*
- *It has a rotation period of* 24^h 39^m*, analogous to that of the Earth;*
- *Unlike those of the Moon, the patches on Mars are variable;*
- *And the poles are marked by bright patches.*

At the following perihelic opposition, that of 1719, Maraldi made further observations at the Paris Observatory. His reports are given, as follows, together with the four drawings which accompanied them.

1719.—Maraldi[20]

During the autumn of 1719, Mars was again very favourably placed for observation.

When the planet reached opposition, on 27 August, it was only 2°30′ from perihelion, and because of its extraordinary brilliance a great many people believed it be an unexpected new star or a new comet. On 19 August, Maraldi observed the planet with a 34-ft. telescope, and noted two dark bands on the disk, making an obtuse angle with each other and presenting a very remarkable aspect. On 25 September he again observed the planet, and noted that this angular feature was in the same position on the disk. During the 37-day interval which had elapsed between these two observations, the planet had therefore completed 36 rotations on its axis, giving a period of 24^h 40^m, in perfect accord with the value given by Cassini.

[20] New observations of Mars. *Histoire et Mémoires de l'Académie of Sciences*, 1720, page 144.

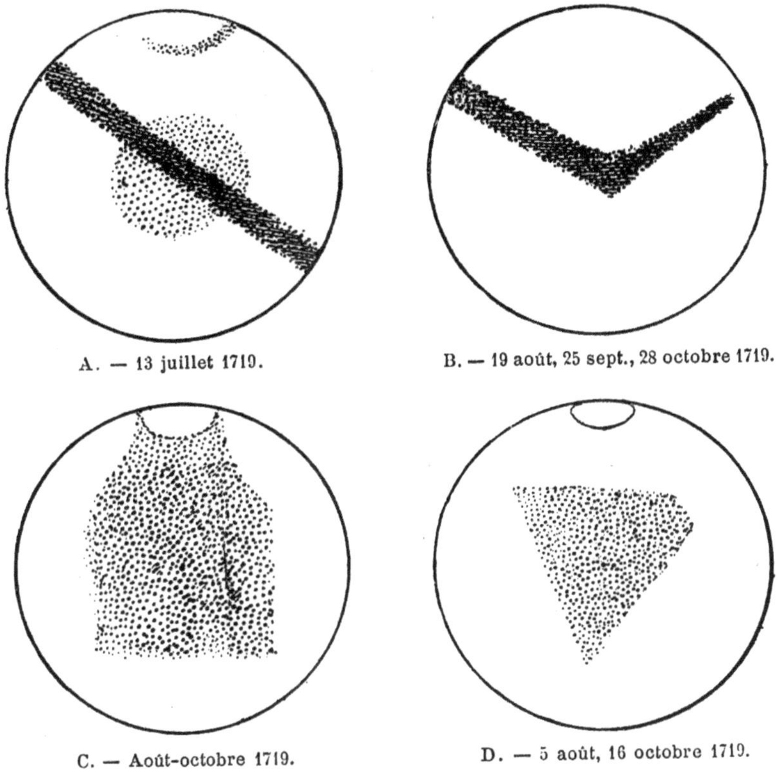

Fig. 24 Drawings of Mars by Maradi in 1719. A. 13 July 1719. B. 19 August, 25 Sept., 28 October 1719. C. August-October 1719. D. 5 August, 16 October 1719

As in 1704, Maraldi concluded that the patches were variable. Here is the rest of Maraldi's memoir, dated 29 May 1720:

At the end of 1719, the planet Mars was closer to the Earth than it had been for a long time.

As this was a favourable time to study the parallax of the planet, and to observe the patches—which can be distinguished only at close oppositions—I observed Mars whenever the sky permitted.

Using the 34-ft. telescope, I have noted different patches, which—due to the planets axial rotation—appear at different times on that part of the disk, which is turned toward the Earth. Among these patches there is a shaded band, not very broad, which takes up less than half a hemisphere of Mars. It is not perpendicular to the axis of rotation, as with most of the bands on Jupiter, but is strongly inclined, so that when it lies entirely on the Earth-turned hemisphere the extremity at the eastern edge lies between the north pole and the equinoctial, and the other extremity, at the western edge of the disk, falls near the south pole. Near the eastern end of the band it is joined by another, inclined to the first, and making an angle at the junction and producing a distinct point. The other end of the band runs in the direction of the south pole (Fig. 24B).

This angle, with its well-defined point, has enabled me to make a fresh confirmation of the rotation period of Mars.

On 13 July, at $3^h\ 40^m$ in the morning, I observed a large oblique band running in a straight line from one edge to the other (Fig. 24A) but I saw no angle, although the point ought to have been on the disk near the western edge; this makes me believe that it is not always visible, and that it was formed after this observation by some change in the components which produce the patches on the planet.

The startling oblique band is not the only patch, which I have noted on Mars. There is another, triangular in form and of considerable size, more than 130° around the planet's circumference from the site of the sharp bend in the former band. I observed it on 5 and 6 August, toward the middle of the disk—of which it occupied the greater part, with its point toward the south pole ((Fig. 24D).

It disappeared on the following days, since it lay on the hemisphere turned away from the Earth, but it returned on 16 and 17 October, after 72 revolutions—each of $24^h\ 40^m\ 10^s$, according to the observations of the other patch.

Besides these shaded patches lying in various positions on the surface of Mars, there is also a very bright, glittering patch near the south pole, which gives the impression of a polar zone ((Fig. 24C, D).

During my six months of observation, it was subject to different variations; at times it was very bright, but at other times feeble—and after having disappeared entirely, it reappeared with the same brilliance as before.

On every occasion when it was bright, the disk of Mars did not appear round, but the southern part of the limb bounding the cap appeared to protrude, making in this area an arc of a circle larger than the rest of the limb—so that in this aspect Mars as seen through a telescope looked a little like the Moon as seen with the naked eye when, in the crescent stage, a small part only of the disk lit by the rays of the Sun is turned toward us, the remainder being visible because of light reflected from the Earth. The part of the Moon's disk lit by direct sunlight then seems to make up a portion of a circle larger than that of the earth-lit section. Now, since this apparent difference is due only to the fact that the sunlit surface is so much brighter than the area lit only by the Earth, it may similarly be believed that the apparent protruding of the Martian polar zone is due merely to the fact that it is much more brilliant and more vivid than the rest of the disk.

In comparing all the observations of the bright patch, I have found that its changes of aspect are associated with the rotation of Mars on its axis—because starting from 17 May 1719, when the patch was very bright, we can add 37 days for each 36 complete rotations, bringing us to 23 June for the first return of the patch to the same position on the disk. Adding another 37 days brings us to 23 June; we have 30 July for the second return, 5 September for the third, 12 October for the fourth and 18 November for the fifth.

The patch has appeared very bright at the times of these different returns whenever the sky has been clear, and the patch has returned as predicted; if the sky has not been clear, the patch has been seen some days before and afterwards, because near the pole of Mars it takes up a large area on the globe and is therefore visible for several days consecutively. These appearances can therefore be explained by the rotation of Mars upon its axis, leading to the same bright patch being in the same part of the hemisphere directly facing us.

Now, if we take the same starting point—17 May, when the patch appeared very bright—and then add eighteen days, we will reach the time when the part of the disk directly opposite to the bright patch is facing us. This date falls on 4 June. On 1 June I saw, on this part of the disk, a definite brightening extending from one limb to the other; but it did not appear as brilliant as that on the opposite hemisphere, and I could see that the material making up the bright surface was spread all round the south pole of Mars—but that it was of unequal brilliancy.

To obtain the times of the next returns of the least bright part of the planet, we must add periods of thirty seven days to 4 June. We arrive at 11 July, 17 August, 23 September 30 October. On 12 July the region appeared rather as it had done at the beginning of June; but from 12 August, the time of the third return, it appeared less bright and less extended until 22 August—just as the third return had shown a diminution compared with 4 June and 12 July. However, at the end of August it appeared larger and more beautiful, for optical reasons, because Mars was then closer to us than it had been at preceding apparitions—though in fact the bright patch had in reality decreased.

At the fourth return, which fell on 23 September, not only had it diminished as on the preceding days, but it had even disappeared, being completely invisible between 16 and 26 September; however, 37 days later—that is to say, on 30 October—when the same part of the disk was exposed to view as had been the case on 23 September, it should have returned to the same place, as I had confirmed by the return of the dark patches. It ought therefore to have been invisible; but it appeared again, and I observed it on 28 October and 3, 5 and 9 November, i.e. two days before the day indicated from the rotation period and three days afterwards. Therefore, there is no doubt that it would also have been seen on 30 October, as well as on the preceding and following days, if the sky had been clear—because the patch was of great extent.

These observations therefore show that as the bright area spread round the pole there was a large portion, which for the 6 months' period of my observations, remained always very bright, while the brightness of the portion on the opposite hemisphere was subject to variations—being of considerable brilliance in June and July, and then declining in both brightness and size to disappear entirely during the months of August and September, at the time when Mars was at its closest to us.

These changing aspects in the portion of the patch lying near the south pole indicate that there have been physical changes in the material which is responsible for the brightness, or else that the inclination of the axis of rotation of Mars has been subject to some variation.

However, it is necessary to note that if the changing appearances and the disappearance of this part of the bright patch are caused by alterations in the axial inclination, the other dark patches lying near the centre of the disk ought at the same time to have moved closer to the limb; but nothing of the sort was found, so that it seems necessary to believe that the cause lies genuinely in *some physical change*.

It is true that these changes must be very great and sudden in order to explain the aspects which I have described, but there are other known examples—for instance, with the Sun and on Jupiter.

Although a large part of the bright patch is subject to the changes I have noted, it nevertheless remains the only patch which has persisted during the sixty years when Mars has been observed with large telescopes, though it does show changes in size and brilliancy. The other patches have altered in form or position, or have even disappeared entirely.

This is what happened to another bright patch lying near the north pole, and which, relative to that pole, was similar in appearance to the southern patch. I saw it during my observations made over several years; it showed differing degrees of brightness. It was again often seen during the 1704 opposition of Mars. Its appearances were rarer in 1717, and I only saw it once or twice. Nothing of it was seen in 1719; however, I noted that though the patch had vanished completely, the area to the side of the north pole was much brighter during 1719 than it had been in the preceding years.

The patches observed on Mars at various times have been subject to great changes, and have varied considerably in form, position and size. Here, it will suffice to report what has happened during the two latest oppositions, when Mars has been at its nearest to the Earth.

In 1704, I observed an extended band from east to west, occupying one hemisphere of Mars. It lay near the centre of the disk, and was fairly uniform except for a point turned toward the north pole in the middle of its length. During the several months when I observed it, it was subject to definite changes (refer again to Fig. 22 above). In other parts of the surface of Mars there were confused, ill-defined patches.

Among the patches seen near the opposition date in 1717 there was still a well-marked band, but it was more extended from east to west than that of 1704; it occupied more than one hemisphere, as was recognized by its appearance at different times in the same night. It was, above all, uniform—though the band of 1704 had a point in the middle. As well as these differences in form, there were considerable differences in position, because the band of 1717 lay between the apparent centre of Mars and the south pole, nearer to the pole than to the centre; the band of 1704 had been much closer to the centre.

From June until the beginning of September, I saw the band disappear over the eastern limb three times, passing on to the upper hemisphere and being hidden from us; each time it returned to the lower hemisphere at the same hour of the day, and in the same position, Mars having made over 70 rotations during the interval. In the other hemisphere of Mars there was a patch in the form of a crescent, with the points turned toward the two poles and the curve turned toward the west. These patches were subject to sensible changes during the several months over which I observed them in 1717; but in 1719 there was nothing more of the kind.

We see, then, that great changes occur on the surface of Mars, not only in the regions near the equator, where the movement ought to be greatest, but even around the poles, where the movement is much less noticeable.

Such, then, are Maraldi's observations. Readers will forgive the length of his report and his somewhat diffuse style, because of the honesty and interest of his observations. Maraldi's four drawings are reproduced here in facsimile (Fig. 24).

The first (A) represents the oblique band to which he refers at the beginning of his memoir, and which he observed particularly well on 13 July. The second (B) represents the angle, which enabled him to determine the rotation period, from 19 August to 28 October. The third (C) seems to have been made principally to show the south polar patch, and the quarter of the triangular patch lying at 130° from the curved band—observed notably on 5 and 6 August and 16 and 17 October. This patch recalls the observations made by Huygens on 28 November 1659 and 13 August 1672; it certainly represents the Hourglass Sea. The broad oblique band in (B) ought to be the Schiaparelli Sea, 130° from the Hourglass Sea. We must agree that instruments of this period gave poor definition.[21]

By 1704 four more facts had been established: Mars shows dark patches. (2) Mars rotates on its axis in about $24^h\ 39^m$. (3) Its patches are variable. (4) The poles are marked by bright patches. The observations of 1719 confirmed these four points, and added a fifth: the south polar patch is *eccentric* to the pole—sometimes it is turned toward us and sometimes it is hidden. Maraldi dared not seek an explanation for the nature of the polar caps, and did not suggest ice, show or even clouds. It was left to William Herschel to elaborate upon the fifth point by his accurate measurements, and to prove that the polar caps are due to polar snows analogous to our own, melting in summer and being re-formed in winter.

We see that gradual advances were being made in our knowledge of our neighbour world. However, one point still remained very mysterious: the variations in aspect, extent and even positions of the dark patches, which were quite genuine and which had been used to give a precise determination of the period rotation. These four new drawings resembled neither those of 1704, nor those of 1666. Could it be that the bands were cloudy and purely atmospheric in nature, as with Jupiter? Maraldi believed so. However, we have already noted that the geographical features such as the Hourglass Sea, drawn by Huygens (Figs. 9 and 19), Hooke (Fig. 15) and Maraldi himself (Fig. 24D) still exist today. Could the Martian seas produce dark fogs? Were the bands observed in 1704 and 1719 of a cloudy nature? But could high clouds, illuminated by the Sun, appear dark? From a balloon, passing above the clouds, we can see that they are always as white as snow. What, then, are these variable patches on Mars? Perhaps continued research will shed light upon the problem.

Another word about these somewhat vague drawings. I have taken care to reproduce them by photogravure, without any re-touching. This point is vitally important in our studies. It is essential to consult the original drawings, because all

[21] With regard to the old drawings of Mars and the optical quality of the instruments with which these observations were made, it is interesting to recall the following comment made by Cassini II, writing on 24 April 1720 about the telescopes at the Paris Observatory.

"The two stars which make up the double star of Gamma Virginis were occulted by the Moon on 21 April 1720. They are so close to each other that with an 11-ft. telescope they appear only in the form of a single elongated star; with a 16-ft. telescope the distance between them appears only about the same as the apparent diameter of each star taken separately."

Now, in 1720 the two components of Gamma Virginis, each of which is of the third magnitude, were separated by 6". Our present small telescopes, of aperture 57^{mm}, suffice to split the double.

The telescope used by Maraldi for his observations of Mars in 1719 was a 34-ft. Our present 108^{mm}. refractors are only $1^m.60$ long.

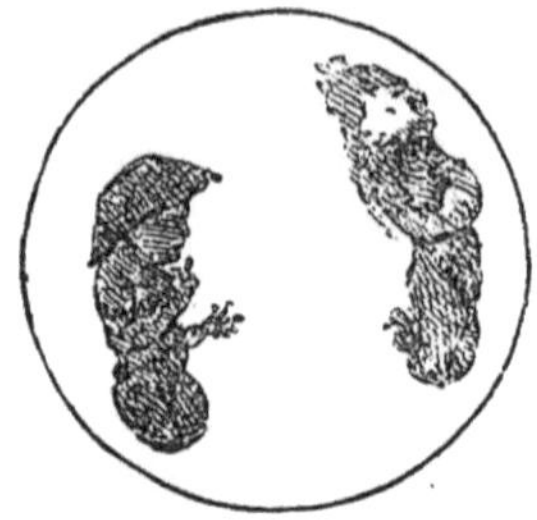

Fig. 25 What happens to the astronomical drawings

too often strange metamorphoses creep in, gradually and unobtrusively, from copy to copy. It is thus, for example, with the drawings by Cassini and Maraldi given in Fig. 25 reproduced from a work by Peirquin, *Œuvres physiques et géographiques*, very lavishly produced in Paris in 1744.

We read in this work, with reference to the drawing: "M. Cassini has discovered four shaded patches on the disk of this planet, similar to those of the Moon; three to one side representing *a baboon and the figure of a man*, while on the other side of the face there is the form of a drumstick which could be called 'le pilon d'Esculape'."

This fantasy shows that it is possible to distort interpretations, even when they do not go so far as this—we should take care never to see or draw anything, which does not exist! But let us continue our study. Instruments were not improved rapidly, because Æscalapius' drumstick which can be identified in Cassini's drawings seems to be shown again in the observations described below.

Same Year 1719: Bianchini[22]

Blanchinus (or Bianchini, in Italian), astronomer of Verona, friend of Popes Alexander VIII, Clement XI and Innocent XIII, and who made such curious observations of Venus, observed Mars during the apparition of 1719, and does not seem to have been better served by this planet than by the first! Readers can judge from the six drawings given here (Fig. 26). The first was made on 19 September, at $10^h 28^m$ (23-palm telescope, made by Campani). The second was made on the next day, 20 September, at $10^h 30^m$ in each case Mars was on the meridian. On 21 September, under the same conditions, Bianchini obtained the third drawing. The other was made on 21 September, 2 1/2 hours later; then came two drawings made on 24 September, at $7^h 0^m$ and $9^h 30^m$ respectively. These drawings do not prove much, and indeed seem to increase our perplexity. Cannot we see here bones against a blank disk?

Bianchini's observations and *geographical* drawings of Venus, which have been in part confirmed in our own century, were made with instruments which were undoubtedly more powerful and which gave better definition. In the same work we find a drawing of Venus made on 7 January 1728, with a 94-palm Campani refractor; conditions from 6 to 7 hours in the evening were excellent. Four observers testified

[22] Observations of Mars made in 1719, published in 1737. Francisci Bianchini veronensis astronomicæ ac geographicæ observationes, selectæ ex ejus autographies.—Vérone, 1737.

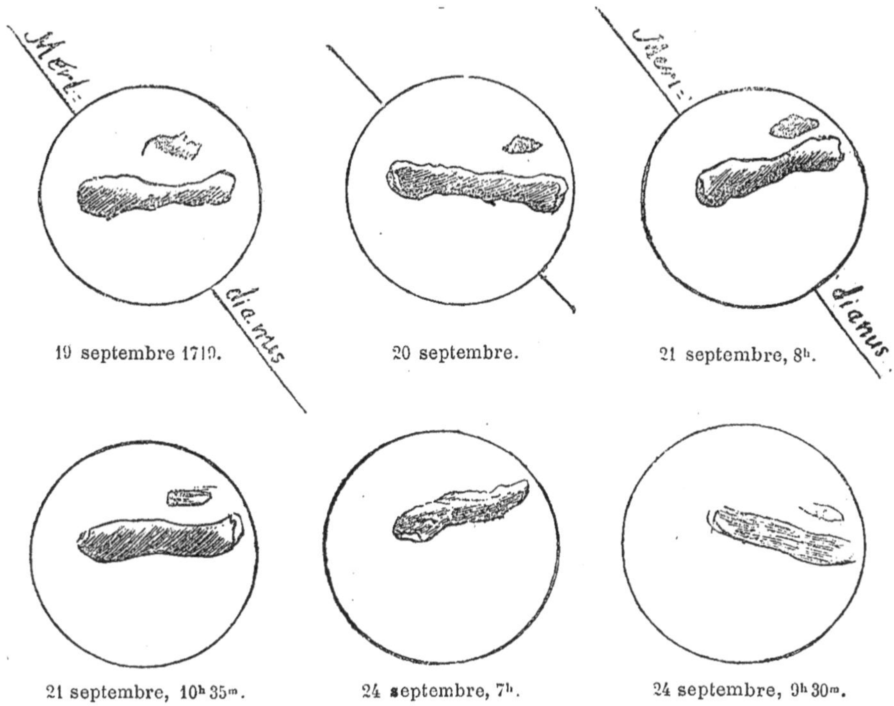

Fig. 26 Drawings of Mars by Bianchini in 1719

that they had unmistakably recognized the four seas represented; one among them even added *maxima voluptate*. The four dark patches seen on the disk of Venus at quadrature were named by Bianchini the seas of Vespucci, Galileo, Royal and the infant Henry. This observation is curious; it was not then certain that the geography of Mars would become known more rapidly then that of Venus.

This drawing is similar to that made by the same observer, which I have reproduced in the chapter on Venus in my book *Les Terres du Ciel*.

Let us note here, for the record, two publications about Mars—the first in 1731, by B. H. EHRENBERGER: *De Marte* (Coburgi), and the second in 1738 by G.M. BOSE: *De Marte Conglaciante* (Lipsiae).

1740.—Cassini II[23]

This astronomer, son of Dominique Cassini and his successor at the Paris Observatory, described in his book the observations made by his father and by Maraldi. He himself added nothing. He made no drawings for Mars, though he did so for Venus. It might seem then, that the geography of Venus was becoming known more rapidly than that of Mars.

[23] *Élements d'Astronomie, p. 457–461.*

1764–1766.—Messier

This great discoverer of comets observed Mars on 3 May 1764, at about 2 o'clock in the morning, from Paris; he had his observatory above the Hôtel de Cluny. On the disk he saw three bands, analogous to those of Jupiter, but very lightly shaded; the middle band was much broader than the others, and half of it was more strongly shaded. This drawing was published 41 years later in the *Connaissance des Temps* for 1807. It is not reproduced here, because it adds nothing.

Messier again observed Mars on 7 and 27 November of the same year, and made two drawings, also published in the *Connaissance des Temps* for 1807. Two feeble patches are shown. These observations are unimportant.

In the same year (1766), in August, the Cardinal de Luynes, at Nolon, and the Duke of Cheulnes, at Cheulnes, observed Mars and sent their drawings to Messier; they were published in the same report. The drawings are vague and indistinct, and they too add nothing.

1771.—Lalande

Here is what the great astronomer said about Mars in his book[24]:

> The globe of Mars never appears as a crescent, as do Venus and Mercury, because it is further from the Sun; but it can appear elliptical, and its shape then resembles that of the Moon three days before full.
>
> In 1636 Fontana observed a shaded patch on the disk of Mars. Father Bartoli, a Naples Jesuit, wrote on 24 December 1644 that with a good telescope by Sirfali he had seen Mars almost round, with two patches above the centre; however, there had been times when Zucchius had seen nothing of them, and he suspected that Mars must rotate on its axis. In 1666 Cassini observed the patches better than anyone else, and from them he found that Mars turns on it axis in $24^h\ 40^m$; he published a memoir on the subject, under the title *Martis circa proprium axem revolubilis observationes*, Bononiæ 1666. In this he showed that the axis of Mars is not very far from being perpendicular to its orbit, so far as he could judge from the patches used for the determination. He again observed the patches from Paris in 1670. Maraldi observed them in 1704 and 1706, and also found the rotation period to be $24^h\ 39^m$. These patches are very large, but they are not always well-defined, and change in form from one month to another; however, they are sufficiently definite for the rotation period of Mars to be established with confidence.

This is a summary—admittedly incomplete—of the previous observations. The existence of the dark patches, and the planet's rotation were all that the celebrated astronomer noted about Mars with regard to its physical constitution. He did not even refer to the white polar caps, which had been shown on drawings made earlier by Huygens and, in particular, by Maraldi; he did not realize that the polar ices affect the seasons, and he noted the inclination of the axis without being able to measure it. The surprising progress made earlier had come to an abrupt halt.

---

[24] *Astronomie*, Vol. III, p. 439.

1777, 1779, 1781 & 1783.—William Herschel[25]

This great astronomer was particularly concerned with Mars during the oppositions of 1777, 1779, 1781 and 1783, and published his observations in his memoirs. The first was called *Astronomical Observations on the Rotation of the Planets*, etc.; the second, *On the remarkable appearances at the polar regions of the planet Mars, the inclination of its axis*, etc. These observations were accompanied by drawings, reproduced here. In these two memoirs, William Herschel's special aim was to study the rotation period and the polar variations; he paid little attention to the Martian geography, and most of his drawings were simple sketches. It is perhaps possible to identify some of the principal seas; these drawings, like those preceding them, showed variations in the dark areas as well as in the bright ones.

The instruments used for these studies were much more powerful than those of his predecessors. The aspect, the whiteness and the variation of the polar caps led him to conclude that these patches represented masses of snow and ice which had accumulated near the poles, and he attributed their variations to the melting caused by the rays of the Sun, to which they were exposed during the planet's orbital revolution. The changes in the patches, carefully observed by him, brought immediate and punctual confirmation of his views. Thus during 1781, the south polar patch was seen to be very extensive, as Herschel had expected in view of the fact that the pole had been exposed to constant night for twelve months. In 1783 the cap had become considerably smaller, and continued to decrease throughout the series of observations, which extended from 20 May until the middle of September. During this interval, the south pole had already experienced the benefits of spring for eight months, and still continued to receive the rays of the Sun although, toward the end of the period, so obliquely that the rays can have had little power to melt the snows. On the other hand, during 1781 the north polar cap, which had been exposed to the solar warmth for twelve months and was due to enter its night, seemed to be small, and gradually grew. Herschel's explanation of the polar patches of Mars has been adopted since his time as the most natural, the simplest and the most logical answer, since it is identical with the explanation of our own polar patches. True, we must agree that the physical, climatological and meteorological conditions are not the same for other worlds as for our own. But the analogous explanation is clearly the first which Nature herself offers us. Since it completely satisfies the observed phenomena, there is no need to look around for alternatives.

During the same observing periods in 1777 and 1779, William Herschel concluded, from the movements of the patches, that the rotation period was $24^h\ 39^m\ 21^s.67$. We will see below that this period was corrected in 1840 by Mädler, using Herschel's own observations.

---

[25] *Philosphical Transactions* for 1781, Vol. LXXI, Part. 1, p. 115. –*Id.* for 1784, Vol. LXXIV, Part. II, p. 233.

At the same time, Herschel found that the inclination of the planet's equator to the ecliptic is 28°42′, and that its ascending node lies at $19^h\ 28^m$, in Sagittarius.

There follows a detailed analysis of William Herschel's two important memoirs.[26]

First Memoir[27]

Read 11 January 1781—sent to Bath Philosophical Society on 18 October 1780

As its title indicates, this work was aimed at determining the rotation periods of the planets, and using these results to find out whether the rotation period of the Earth always remains constant. Herschel begins by dealing with the movements of the Earth—in particular, its diurnal movement—and he suggests a method for determining the constancy of the diurnal movement of our planet by observing another. He was concerned principally with Jupiter and Mars.

The observations of the planet were begun on 8 April 1777, and were continued only for four days in that year: April 8, 17, 26 and 27. The great astronomer returned to the observations of Mars on 9 May 1779, and continued until 19 June of the same year.

Here are the principal results. [Note in what follows, *Figs*. 14, 15, 16, 17, 18, 19, 20, 21, and 22 are Herschel's, and refer to drawings in our Fig. 27 below.—WS]

8 April, $7^h\ 30^m$.—I observed two patches on Mars, separated by a brilliant band (see *Fig*. 14).

Same evening, $9^h\ 30^m$.—The spots have advanced across the disk, and are still visible (see *Fig*. 15). The rotation of Mars on its axis is now very evident (*Fig*. 16).

(Observations made with a 20-ft. Newtonian reflector, magnification 300×.)

17 April—Telescope a 10-ft. Newtonian reflector, magnification 211×. $7^h\ 50^m$. Mars appears as in *Fig*. 17. At *a* and *b* may be seen two brilliant patches, so luminous that they seem to project beyond the disk. At *c* and *d* there at two very dark patches joined by a black line, crossed in the direction *if* by a whitish division.

26 April.—Same instrument, same eyepiece. The patches are very feeble, and appear as in *Fig*. 18.

27 April.—Same instrument, magnification 324x. $8^h\ 40^m$. Very fine evening, telescope in good order; the patches appear as in *Fig*. 19.

As we have noted, William Herschel again observed Mars from 9 May to 19 June 1779. Here are the principal results:

9 May, $11^h\ 1^m$.—I find the positions of the patches to be as shown in *Fig*. 20; there is a very remarkable patch not far from the centre.

Same day, $11^h\ 30^m$.—The patches are further away from the centre.

11 May, $10^h\ 18^m$.—The patch seen on 9 May is visible on the disk, its darkest part being found to the south east of the centre (*Fig*. 21).

---

[26] *Philosophical Transactions*, 1781, page 134, and 1784, page 273.

[27] *Astronomical observations on the rotation of the planets round their axes, made with a view to determine whether the Earth's diurnal motion is perfectly equable.*

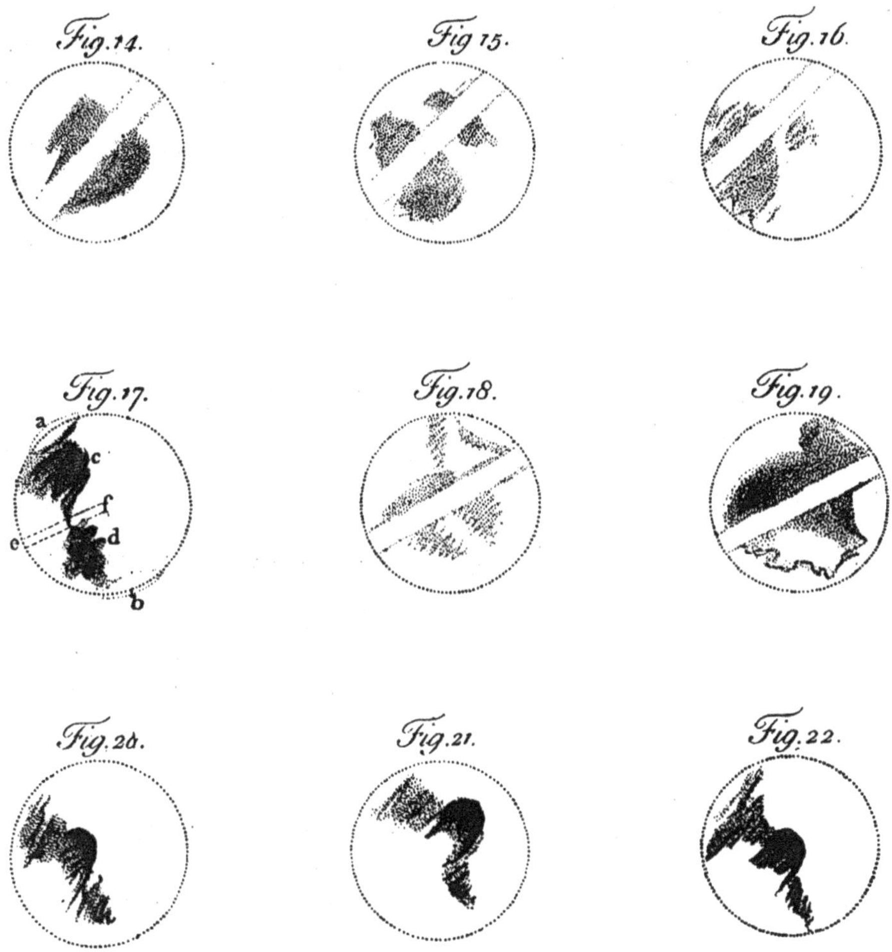

Fig. 27 Drawings of Mars by William Herschel, in 1777 and 1779 (facsimile)

Same day, $11^h\ 43^m$.—The darkest region has reached the centre (*Fig*. 22).

13 May, $11^h\ 26^m$. Mars looks the same as it did on the 11th, at $10^h\ 18^m$.

22 May, at $10^h\ 10^m$. I see on Mars the same configurations as on 8 April 1777, at $7^h\ 30^m$ (*Fig*. 14).

15 June, at $9^h\ 45^m$. The planet presents the same aspect as on 9 May at $11^h\ 1^m$. (*Fig*. 20), but the markings are further advanced across the disk.

17 June, from 9^h to 10^h.—Same aspect.

19 June, at $8^h\ 40^m$.—Same aspect as on 26 April 1777, at $9^h\ 5^m$, shown in *Fig*. 18.

Such were Herschel's observations; by combining his 1777 results with those of 1779 he was able to conclude that the sidereal rotation of Mars was:

$24^h\ 39^m\ 21^s.67$

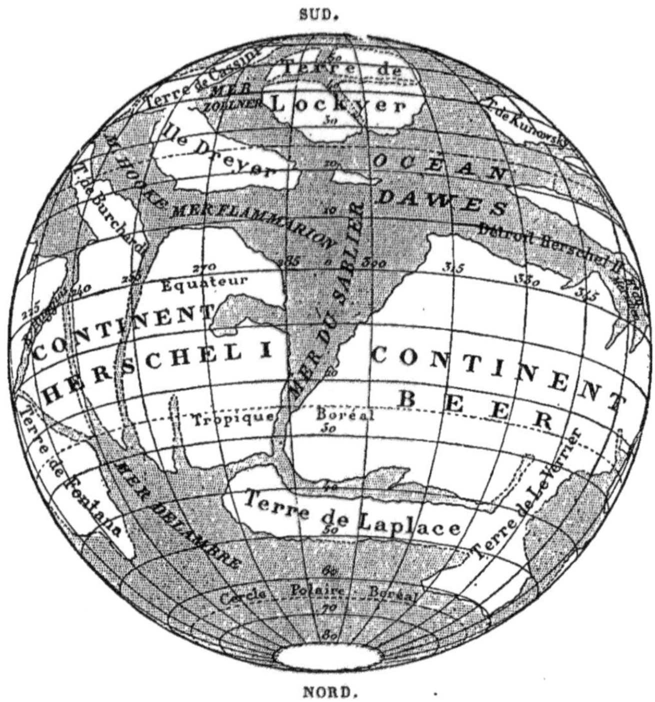

Fig. 28 The globe of Mars, presenting its northern pole (aphelic oppositions). 1777–1792–1807–1822–1837–1852–1869–1884–86

I have reproduced Herschel's drawings in facsimile as they appeared in *Pl. VI* of the *Philosophical Transactions*, from which the preceding descriptions have been taken. The facsimiles are not reduced in size.

The drawing Herschel designates as *Fig.* 17 (in our Fig. 27) certainly shows the Hourglass Sea, from *c* to the break at *ef.* I believe we can also recognize the sea on Herschel's *Fig.* 20, 21 and 22, and indeed one has the impression that every observer saw it and drew it in this form. The 1777 opposition of Mars was almost aphelic so that the drawings show not the upper south pole, as in 1672 (Fig. 20 above), but the lower north pole (Fig. 28.). The identification is not difficult for Herschel's *Fig.* 17, 20 21 and 22 in our Fig. 27, but it is not the same for the five others. The white band merits attention.

In all the drawings reproduced here, it is evident that the observer saw only a little, rather vague detail on the surface of Mars.

With regard to the comparison of the 1777 drawings with those of 1779, Herschel considered how to reduce the observations—which represented synodic rotations, seen from a moving Earth—to absolute or sidereal rotations:

> Suppose (he said) that the orbit of Mars, MABC (Fig. 29) is in the same plane as the orbit of the Earth EDFG, and that the axis of Mars is perpendicular to its orbit; let ME*me* be the respective positions of Mars and the Earth on 13 May and 17 June (1779), the line EM,

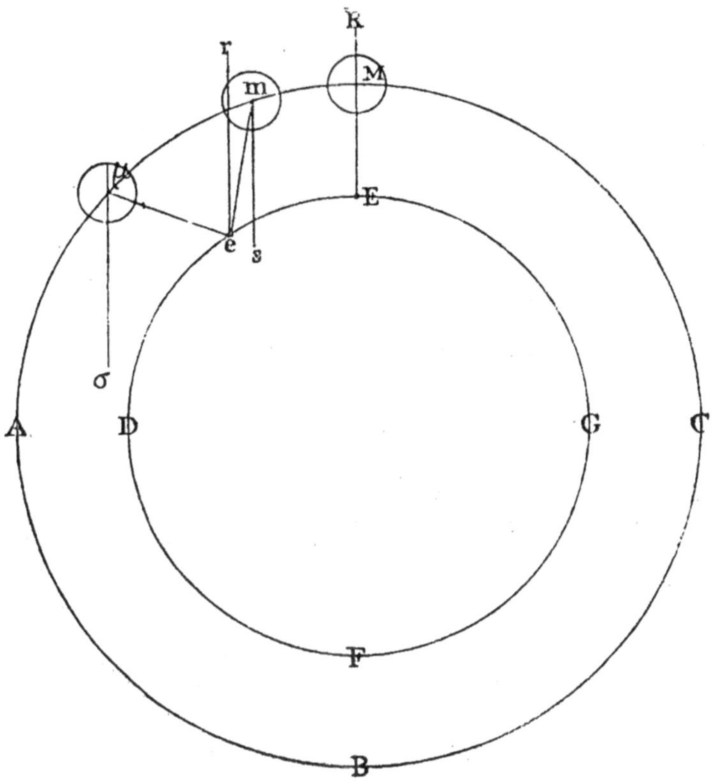

Fig. 29 Apparent variation in the duration of the rotational period of Mars, according to the position of the Earth

which joins the centres of Mars and the Earth, indicating the geocentric position of Mars on 13 May and the line *em* the geocentric position of Mars on 17 June. Now, take the lines *er* and *ms*, parallel to ER; then the line *er* indicates the geocentric position of Mars on 13 May, and the angle *sme* is equal to the angle *mer*. From the ephemerides, we know that the geocentric position of Mars at $11^h\ 26^m$ was, to seven figures, 20° 59′ 21″, and on the 17th of June, at $9^h\ 9^m$, 12° 27′ 22″. From this, we obtain the difference, or the angle rem = ems = 8° 31′ 59″.

Now, a patch on Mars lying in the direction ME will make one sidereal rotation when it comes back to the same direction, or on a parallel to the direction *ms*. From this, we conclude that the patch of 17 June, after having arrived at the line me at the end of the synodic revolution, had to cover an extra arc of 8° 31′ 59″ before arriving back in the direction of the line *ms*, where it completed its sidereal rotation.

The time taken to cross this arc at a sidereal rate of $24^h\ 39^m\ 20^s$ for 360°, or $4^s.109$ per minute of arc, would be $35^m\ 3^s.8$—a figure which when divided by 34 (the number of rotations) gives 1m $1^s.8$, which added to $24^h\ 38^m\ 20^s.3$, gives us $24^h\ 39^m\ 22^s.1$ for the sidereal rotation of Mars.

Note that the movement of Mars in the example given above is retrograde; this is why the measurement of the angle *ems* has been added to the synodical rotation to derive the sidereal rotation. But if the movement had been direct, or if the planet had been further along the ecliptic—if, for example, it had been at μ, then the line μ parallel to E*m*, would have indicated the direction to which the patch would have had to have returned before completing a sidereal rotation; and consequently the angle $\mu e = \mu e r$, or the difference of the geocentric positions, would have to have been subtracted from the synodical rotation in order to obtain the sidereal rotation.

Herschel's Second Memoir

Read 11 March 1784.[28]

In the second memoir Herschel, as he indicated in the title, was concerned principally with studying the poles and the axial inclination of Mars. He first referred to the drawing of 17 April 1777 (see Fig. 30; Herschel's *fig.* 17), and commented that during the 1779 observations neither polar cap had caught his attention. His new observations extended from 13 March [the night he discovered Uranus!—WS] to 7 September 1781, and from 20 May to 17 November 1783. Here are the principal results. [As before, the figure numbers below are Herschel's; see our Fig. 30.—WS]

17 April 1777, at $7^h\,50^m$.—*Fig.* 1 was drawn to show the polar patches observed. Herschel was not concerned with the other patches, as is evident from a comparison of Herschel's *Fig.* 17 of the preceding memoir with *Fig.* 1 of the second memoir, which represents, in fact, the very same observation.

13 March 1781, at $17^h\,40^m$.—*Fig.* 2 shows the south polar cap, of considerable extent, as observed on with the 20-ft. telescope. The following drawing (*Fig.* 3) refers to the observation of 25 June, at $11^h\,36^m$, made with the 7-ft. telescope. Herschel wrote: "Two brilliant patches at *a* and *b*, *a* being larger than *b*."

28 June, at $11^h\,5^m$.—The difference between the two polar patches was even greater, as is seen from *Fig.* 4.

30 June, at $10^h\,48^m$.—Only the upper patch could be seen (*Fig.* 5) but at $11^h\,25^m$ both were visible.

3 July, at $10^h\,54^m$.—The upper polar cap was very evident (*Fig.* 6); at $11^h\,24^m$, the lower patch was still not visible (*Fig.* 7); at $12^h\,36^m$ it could be seen (*Fig.* 8). On 4 July, Herschel commented that the two polar caps were not diametrically opposite to each other.

15 July, at $9^h\,54^m$.—The upper patch was visible (*Fig.* 9).

22 July, at $11^h\,4^m$ (*Fig.* 10). Both polar caps were well seen; the upper was the larger. Herschel wrote: "Very probably the south pole is turned toward us, while the north pole is hidden from us. If the patches are polar, then the upper or southern one seem to us, in effect, to be larger than the lower; and if the north patch extends a little to either side of the north pole, the apparent variations are due to the rotation of the planet on its axis."

8 August, at $10^h\,4^m$, only the upper patch was seen (*Fig.* 11).

---

[28] *On the remarkable appearances at the polar regions of the planet Mars, the inclination of its axis, the position of its poles, and its spheroidical figure; with a few hints relating to its real diameter and atmosphere*, by WILLIAM HERSCHEL, Esq. F.R.S.

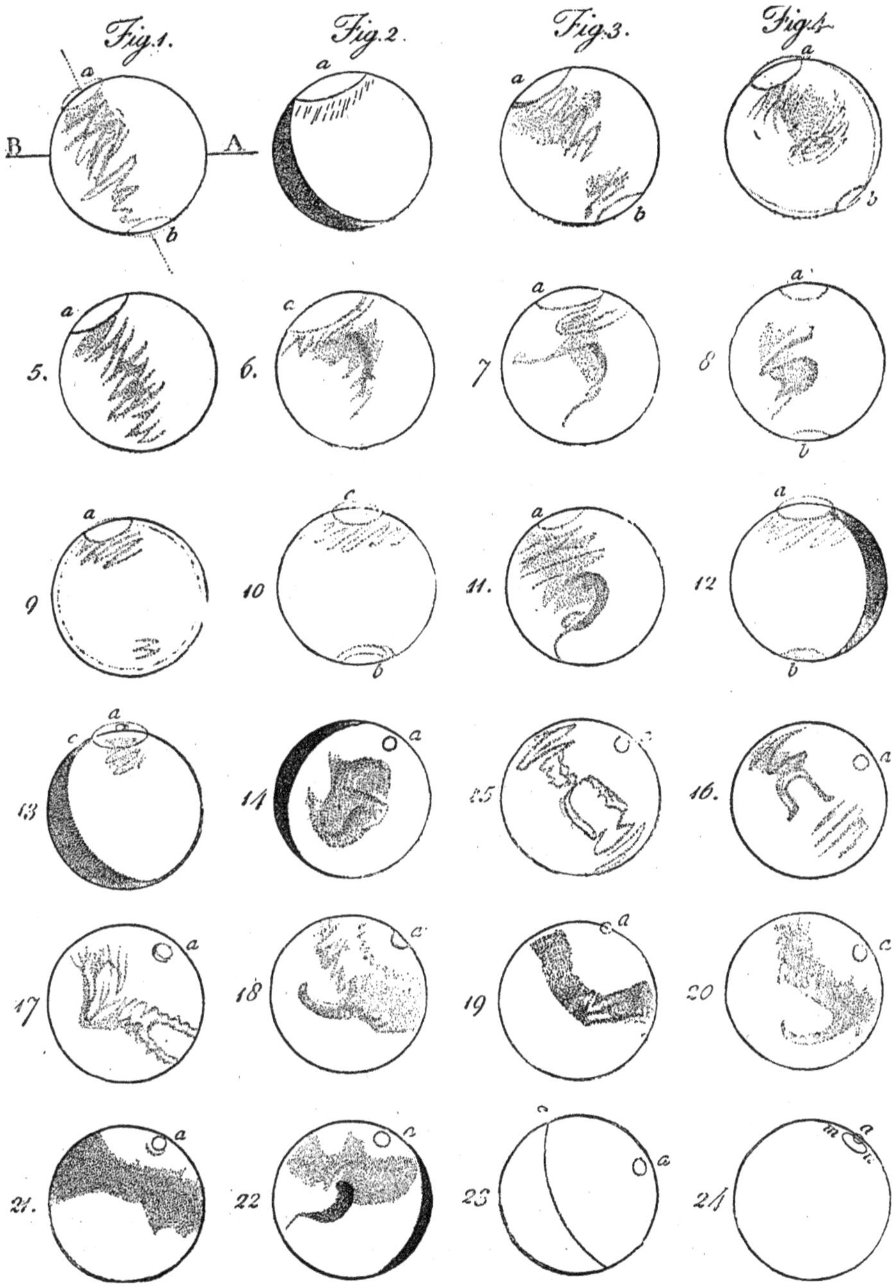

Fig. 30 Observations of the planet Mars by William Herschel in 1781 and 1783

23 August, at 8^h 44^m, the upper patch was very evident, and was easily seen, while a little of the lower patch was visible.

Such were the observations made by this eminent astronomer during the opposition of 1781; we see that they were made only to show the polar patches. Next, the 1783 observations can be summarized as completely as possible:

20 May.—The planet Mars presents a singular appearance; at *a* (*Fig.* 13) there is a brilliant polar patch, and its brilliance is such that it seems to protrude beyond the disk and be separated from it at the point c.

26 August.—The brilliant patch on Mars marks the south pole, because it remains fixed in the same place, while the dark equatorial patches show evidence of constant rotation. This south polar patch is sensibly circular.

22 September.—A magnificent view of the planet, which is near the meridian. A light mist cut down the brightness, and made the features show up very clearly. The polar cap was measured; its smaller diameter, in the direction of the equator, is 1°41″.

23 September.—At 9^h 55^m, polar patch easily visible (*Fig.* 14).

25 September.—At 12^h 30^m (*Fig.* 15), polar patch perfectly round, away from the edge of the disk. At 12^h 55^m it was realized that the track of the equatorial patches was curvilinear and convex toward the north, proving that the white patch does indeed indicate the south pole; and after a long period of close observation, I was able to recognize the edge of the disk on the far side of the polar patch; the distance between the patch and the limb was about a quarter the diameter of the patch.

26 September, 12^h 10^m, the polar patch is in line with the centre of the disk and the extremity of the crochet "hook" (*Fig.* 16).

30 September.—The planet appeared as shown in *Fig.* 17.

1 October.—I was led to think that the white patch had a slight movement due to rotation, and that in consequence it could not be centred exactly on the Martian pole. The real pole should be in the interior of the patch, but about one-third of its diameter out toward the edge. I hope to find this out within a few days.

Here, William Herschel suspended the description of his observations by declaring that neither of the two polar patches exactly marked the geographical pole, as was proved by their rotation. He added that they were not very elongated. He then continued in his diary:

9 October, at 10^h 35^m: the planet Mars appeared as shown in *Fig.* 18. The south polar patch was carried round by the planet's rotation and arrived at a position turned toward us.

10 October, at 6^h 55^m (*Fig.* 19).

Same day, at 9^h 55^m (*Fig.* 20).

17 October, at 7^h 47^m (*Fig.* 21).

23 October, at 7^h 11^m (Fig. 22); the polar patch should be at the very end of its parallel of latitude.

Herschel then passes on to examine the rotational movement of the polar caps, and their eccentricity. He reached the conclusion that the latitude of the north polar cap, studied during the opposition of 1781, was N. 76° or 77°, "because," he wrote, "I find that for the inhabitants of Mars the declination of the Sun on 25 June 1781, at 12^h 15^m our time, was about 9°56′ south, and thus the polar cap must have been sufficiently far from the pole to be several degrees on the sunlit side of the globe, and hence invisible to us. He added: The south pole of Mars was some way away from the centre of the brilliant south cap in the year 1781; this cap was large enough to cover all the polar regions as far as latitude 70° or even approaching latitude 60°."

In 1781 the south polar cap extended along the arc of a great circle of 45° 50′ or perhaps 60° of the globe of Mars; it did not extend to a great distance from the pole, but its centre was not coincident with the pole.

From this study it follows that the north pole of Mars is at 17°47′ in the constellation of Pisces, and that the inclination of the axis to the ecliptic is 59°42′. Herschel continues:

"Having thus determined what the inhabitants of Mars would call the obliquity of their ecliptic, and hence the positions of the equinoctial and solstitial points, we can take the seasons of Mars into account, and therefore the remarkable variations of the polar patches." To quote Herschel himself:

> The analogy between Mars and the Earth is certainly more evident than for any other planets in the Solar System. Their diurnal movement is almost the same; the obliquity of the ecliptic, causing the seasons, is analogous; of all the superior planets, the distance of Mars from the Sun most nearly resembles that of the Earth, and as a result, the length of the Martian year is not enormously different from ours—though this is not the case for Jupiter, Saturn and the Georgium Sidus [Uranus]. Since, therefore, we know that our planet has polar regions which are covered by mountains of ice and snow, which melt when exposed alternately to the rays of the Sun, it is permissible to suppose that the same causes probably produce the same effects on Mars; hence the polar caps, which are so brilliant, are due to the vivid reflection of light in these regions by ice and snow, and the shrinking of the patches can, as with the Earth, be attributed to the action of the solar rays.

Herschel then passes on to an examination of the spheroid of Mars, and the polar flattening. He first considers this flattening to be a certainty from the theoretical point of view, because of gravitational considerations, and he comments that it can seldom be measured, since only near opposition do we see the whole of the illuminated hemisphere of Mars—that is to say, only for three or four weeks in every two years.

His observations on this subject extended from 25 September to 9 October 1783. Later, the phase of Mars became more noticeable. From his measurements, Herschel found that the flattening of the planet was evident, and at first sight even as marked as that of Jupiter—which is indeed singular; personally I have never had this impression. "On 29 September," Herschel wrote, "the planet was only 37 hours from opposition, and on 28 September I watched it only 2 1/2 days from opposition. The flattening was recognized, not only by myself but also by three other observers, Messrs. Wilson, Blagden and Aubert; the micrometrical measurements gave 29″ 35‴ or 1355‴ for the equatorial diameter, and 21″ 29‴ or 1289‴ for the polar diameter, according to which the ratio of the equatorial to the polar diameter is 1355:1289‴.

In making the reductions of the polar diameter, taking the axial inclination into account, Herschel concluded that the two diameters were in the ratio of 1,355:1,272, or approximately 16:15. From these measurements, the flattening is found to be 1/16.

Let us now summarize the results of Herschel's memoir:

- The axis of Mars is inclined to the ecliptic by 59°42′.
- The node of the axis is at 17°47′, in Pisces.
- The obliquity of the ecliptic on the globe of Mars is 28°42′.
- The equinoctial point on the Martian ecliptic is at 19°28′, in Sagittarius.
- The figure of Mars is that of a flattened spheroid, whose ratio of equatorial to polar diameter is 1,355:1,272, or about 16:15.
- The equatorial diameter of Mars, reduced to the mean distance from the Earth to the Sun, is 9″8‴.
- The planet has a considerable but moderate atmosphere, so that in some respects its inhabitants probably enjoy conditions analogous to our own.

As we have seen, William Herschel was also concerned with the atmosphere of Mars. He thought that it must be quite considerable, because variations were often observed in certain relatively bright regions—variations which seemed to him to be due to clouds and vapours floating in the atmosphere. He even considered that a dark band drawn at a very high latitude (see Herschel's *Fig.* 18 in our Fig. 30) was also due to clouds. He recalled an observation by Cassini, who had seen a star in Virgo disappear at a distance of six minutes from the disk of Mars, though he did not believe that the disappearance could be due to anything but the dazzling brightness of the planet. On 26 and 27 October 1783, he observed two stars of magnitudes 12 and 13, at 3′9″ and 2′56″ from the limb of the planet, and found that their brightness did not seem to be reduced by the effect of the nearness of the light of Mars.

He therefore concluded that the atmosphere of Mars is not so extensive as it would have to be according to Cassini's interpretation.

This new series of drawings by Herschel confirmed what he had found in the preceding series: the appearance of Mars varies considerably. Some of the changes can be attributed to faulty observation, particularly with these drawings which were made only to show the polar caps, and paid no heed to the other markings on the planet; however when, as in 1777 and 1779, there are drawings which show definite details, we are forced to conclude that the planet did appear more or less as drawn. Yet the features are not the same as those shown by Cassini, Hooke, Huygens, Maraldi or Bianchini.

Each observer, in his own fashion, saw light, vague and ill defined features, as would be expected on a distant world surrounded by a vapour-laden atmosphere. This is why the interpretation of the features is often difficult, though the drawings have a sound basis.

The observations made by the distinguished discoverer of Uranus represent a great advance on our knowledge of Mars. Before Herschel, we know only that: (1) Mars shows dark patches; (2) its rotation period is about $20^h\ 40^m$; (3) the dark patches are variable; (4) the poles are marked by white patches; (5) the caps are not centred on the geographical poles, but are eccentrically placed. We now know that in addition (6) the caps are analogous to the terrestrial polar ices, melting in spring and reforming in winter; (7) that in October 1781 the centre of the north polar snow area was at latitude 76° or 77°; (8) that the atmosphere seems to be analogous to that of the Earth; (9) that the obliquity of the ecliptic on Mars is 28°42′. The polar flattening found by Herschel has already been discussed. Further comments are given below.

This is great progress indeed. Year, day, seasons and climate have been determined; the seasons are analogous in type to those of the Earth, though they are more than twice as long; as with our world, the pole of cold is not coincident with the geographical pole.

Assuredly these are interesting facts; they have been established over a period of more than a hundred years.

Herschel's work did not advance our knowledge of the geography of Mars. This was not his main object.

1783.—Messier

From Paris, Messier re observed Mars on 15 and 16 September of this year, and noted that the south polar cap was circular, well-defined, and equal in diameter to the first satellite of Jupiter when seen in transit on the Jovian disk. The great comet discoverer made similar observations on 3 August and 19 and 23 September 1798. The drawing published in the *Connaissance des Temps* for 1807 shows absolutely nothing except the polar cap, in the form of a small circle at the south pole.

1785.—Bailly[29]

The distinguished historian—later guillotined, as was Lavoisier, because of the idiocy of political parties—gave a summary of what was known in France about Mars at this time. He did not know about Herschel's work. The French results were confined to those of Maraldi in 1719.

"On the globe" (he wrote), "we see a patch near the south pole in the form of a polar zone; it is susceptible to changes in brilliancy, and when it is very bright Mars does not appear round. It seems that this is for the same reason that the bright part of the Moon appears to project beyond the limit of the dark side, making up a larger circle. It is due to the effect of irradiation of the sunlit part against the dark part. It is believed that we can trace a connection between the maximum brilliancy of the polar patch and the diurnal rotation of Mars, so that maxima occur after every 36 rotations. This bright patch is the only one to remain permanently visible, though with changes in brightness and clarity. The other changes in form and position, and may even disappear entirely. What is strange is that we can see the north pole of the planet just as clearly as the south pole, but the brightness there persists, while the

[29] *Histoire de l'Astronomie moderne*, vol. 2, p. 603.

other disappears. Thus, the two brilliant areas lie at the two poles, as though there is some link with the magnetic fluid or with the aurora borealis."

It is strange that Bailly, the philosopher-author whose ideas were not confined within narrow limits, did not rhapsodize about the polar snows. Moreover, as we can see, the work carried out by Herschel had made Bailly's résumé of our knowledge of Mars 60 years out of date.

We now come to one of the most eminent and devoted observers of Mars: Schröter. His work embraced eighteen years, from 1785 to 1803.

1785–1803: Schröter[30]

The observations made by this industrious astronomer are the most important and the most significant in the first period of Martian research. They made up a great book, with 447 pages and 230 drawings, which however was not published until 1881—by the diligence of M.Van de Sande Bakhuyzen, Director of the Leyden Observatory.[31]

The observations were begun in 1785 and were continued until 1803; they therefore carried on directly from William Herschel's researches, which ended in 1783.

Schröter's work, *Areographische Fragmente*, remained in manuscript form in the hands of the astronomer's family in Lilienthal. Our first knowledge of it came from Dr. Terby of Louvain, who in 1873 was able to study it in detail, and who appreciated its great value. I give here, as an excellent résumé of Schröter's work on Mars, an extract from the report presented by the Louvain astronomer to the Belgian Academy of Sciences.

Nature of the Dark Patches on Mars: According to Schröter

The Lilienthal astronomer recalled a comment made by William Herschel in a memoir about Venus, published in 1793, as follows:

> I assume that the bright bands of Jupiter, contained between darker bands, are the zones in which the planet's atmosphere is cloud laden. The dark bands correspond to regions in which the atmosphere is completely transparent, and allows the Sun's rays to reach through to the solid part of the planet, where, in my view, the reflective power is less than with the clouds.

[30] Areographic Observations made at his observatory in Lilienthal.

[31] *Areographische Beitrage zur genauern Kenntniss und Berurtheiling des Planeten Mars, in mathematisch Hinsicht*, von Dr. J.H. Schrœter; mit 16 Kupfertafeln. Nach dem manuscripte auf der Leidener Sternwarte, herausgegeben von H.-G. VAN DE SANDE BAKHUYZEN, Director der Sternwarte. 1 vol. in-8° with 230 drawings. Leyde, 1881.

The explanation given by Schröter for the dark patches on Mars was exactly the opposite; to him, the patches were clouds, less reflective than the solid body of the planet. Herschel did not agree, and regarded the idea as unacceptable. In support of his theory, Schröter cited the following observation, made during an ascent of the Brocken mountain:

> A thick mist prevailed before sunrise; when the Sun began to lift above the horizon, the vapours descended little by little into the valleys, below the level of the observer. Above the mist-level, the sky became perfectly clear. Now, compared with the splendour of the mountain-tops, the greyish appearance if the cloud reflecting the light of the Sun is very much as the dark patches on Mars are to the brightly illuminated surface.

Schröter wrote at length about the points of resemblance between the Earth and Mars. He said:

> We find a very great analogy between these two celestial bodies; their atmospheres are so similar that one is disposed to feel that the globes themselves are very alike. But we must be cautious in our conclusions, because we lack definite proof. I have never observed with certainty completely fixed dark patches, which like our seas and lakes, would have a lower reflecting power.

Schröter then went on to explain why, in his view, the features on the actual surface of the planet were not seen distinctly. However, the large patches ending in a point to the north side attracted the close attention of this distinguished astronomer; he devoted a special paragraph to them. He wrote:

In studying these observations closely, one is convinced that these masses of dark clouds, in the form of a pyramid, are produced over different parts of the planet's surface. What natural force can cause this form, and why does the base always support the principal band? Why is their point always directed toward the north? It is impossible to answer these questions. But on the surface of the Earth we also have phenomena, which are associated with the poles, and with the natural magnetic force. Perhaps we could understand these phenomena better if we could observe our Earth from a comparable distance.

The Rotation of Mars and Movements of Clouds in the Atmosphere: According to Schröter

The observations made in 1787 and 1792 gave different values, and Schröter gave a mean rotation period of

$24^h\ 39^m\ 50^s$

which, he said, came between the periods given by Herschel ($24^h\ 39^m\ 21^s$), Cassini ($24^h\ 40^m$) and Maraldi ($24^h\ 39^m$). Believing that he could not derive an absolutely precise period, because of the changes observed in the patches, he supported Cassini's period, and adopted it in all his calculations.

Attributing the dark patches to clouds floating in the Martian atmosphere, Schröter explained the apparent irregularities—which he had found in the rotation period—as being due to real cloud movements. If a patch yielded a period much too short, he concluded that it had a direct individual motion—that is to say, in the same sense as the rotation—and conversely. Schröter was thus led on to discuss winds in the atmosphere of Mars, and to speculate about their speeds and directions. He carefully calculated the displacement of any patch, which seemed to him to disagree with the known rotation, and he drew up a wind table in which he gave the speeds and directions of 45 atmospheric movements that he had studied during his long and laborious researches.

Schröter relied upon studying those atmospheric phenomena of Mars which took place on a grand scale and he believed in three causes: the absence of sufficiently precise reference points in the patches observed, the confused aspect of patches which resembled each other to a greater or lesser degree, and his complete faith in the apparent changes on the planet's surface.

If the patches were subject to such movements, then, on Schröter's theory, how could Cassini have measured the rotation period so exactly? This is the question which Schröter himself asked at the end of his book. He wrote:

> It is natural that the patches affected only by constant winds will agree with this determination; bands running northward or southward will not shift to the east or west. It is the same with the isolated patches characteristic of other regions of the planet; and it was under such conditions that Cassini and Maraldi found so close a value for the rotation.

Thus Schröter's observations led him to believe that the dark patches on Mars were clouds. This is certainly very strange. And do not forget that he was an excellent observer.

Schröter's Observations of the Polar Caps

On the night of 18–19 July 1798 Olbers, who was at the Lilienthal Observatory, observed Mars with Schröter's 13-in ($0^{m}.33$) reflector, and saw the southern polar patch. This was also the first time that Schröter showed it on his drawings. The two astronomers agreed that the edge of the planet was more brilliant than the centre; the latter was reddish and patchy, but the south polar region was very white, very bright and very broken up. Schröter never failed to observe the brilliant patch during his observations for the rest of the year.

According to him, the southern solstice of Mars was due on 27 September. Be that as it may, the observations covered a large part of the southern hemisphere spring. The south pole was tilted toward the Earth, and the brilliant southern region was shown on all the drawings made after 18 July, while the northern patch remained invisible for a long time.

At the time of its discovery by Olbers, the southern patch was remarkable because of its size; over the following days it showed variations in brilliancy and extent, but after 2 September it began to decrease; this was very obvious, and subsequently the patch became very small. After 8 October it was reduced to a small luminous disk, clearly separated from the limb of the planet; from that time its fixed position—despite the planet's rotation—allowed the position of the pole to be determined. On 25 October it again seemed to be approaching the limb; on the 26th, Schröter found it as small as one of the lesser satellites of Jupiter. On the following days he saw it approach nearer and nearer to the edge of the disk, almost blending with it on 15 November.

On 20 November, the skilful astronomer again observed the small cap in the same position, but its appearance was altered by a bright extension reaching westward from the tiny disk. This feature was soon to be displaced by the planet's rotation, and moved perceptibly. Schröter had established that the pole lay within the small cap.

After this observation, the cap was only very rarely shown in the polar zone; but Schröter constantly observed a decidedly brilliant patch, showing variations in appearance and extent. Finally, on 20 December, the north polar cap appeared in its turn; and from this time, from 20 December to 1 January 1799—that is to say, until the end of the series of observation—Schröter was able to see the snows at both poles.

These phenomena fully confirmed Herschel's observations described above; the south polar patch was reduced in size during summer in that hemisphere. Clearly separated from the edge of the disk, it appeared as a luminous point. It resumed its development when the Sun passed over the Martian equator once more. In fact, the planet's climate and seasons were well demonstrated during this period.

Résumé of the Observations

Schröter had therefore observed almost all the features of the polar caps which we note today: the variability in brightness and extent, the inequality of the extensions in different directions, and above all the fact that the caps do not have a regular, circular outline; the brilliant polar patch is surrounded by areas which are less vivid, and the bright zone is bordered by a dark collar, due apparently to irradiation.

Moreover, Schröter attached some importance to a difference in aspect, which he noted between the two polar patches; the southern appeared to him as whitish or yellowish, the northern as slightly bluish.

Recalling the observations of Cassini, Maraldi and W. Herschel, Schröter first commented that the constancy of these polar features must be linked with the particular climate of that region of the planet; according to him, the changes in the patches showed the influence of atmospheric phenomena. However, Schröter could not admit that the appearances and disappearances of the caps depended upon the seasons. Comparing the observations of Maraldi, W. Herschel and himself, he claimed that a given season on Mars did not always correspond with identical observations of the polar patches; in other words, the presence of a snowy cap did not depend upon the season. He found, for example, that the southern patch was sometimes observed during the southern summer, sometimes during the northern summer. However, there would be no need to look for regularity in the cap phenomena if the caps themselves were to be explained by the action of the Sun. At the times of the Martian solstices, there were two main influences upon the visibility of a polar cap; the cap was generally largest when most strongly tilted toward the Earth, while the major development of the cap in the opposite hemisphere took place with that pole tilted away from the Earth. The first situation (1) occurs during summer in the relevant hemisphere; the second (2) in winter. To take one example; the southern polar patch can be seen during southern summer because of condition (1), and during summer in the opposite hemisphere because of condition (2).

"The polar zones," Schröter wrote, "undoubtedly owe their brilliancy to a dazzling atmospheric precipitation. We can imagine a polar sky covered with a dazzling white layer, similar to our snow; we can also picture the surface liquids changed by cold into a solid reflecting surface, and this explanation establishes an even greater analogy between Mars and our Earth."

Spheroidal Form of Mars: Apparent and Random Deformations

By January 1788, Schröter was already turning his attention to the shape of the disk of Mars. His observational diary expressly mentions that he had not been able to establish any difference between the polar diameter and the equatorial diameter until 19 March 1792, when he noted a flattening which was less than that of Jupiter. On 20 March 1792 he measured the planet's diameter, and found a flattening of 1/15. However, the position of the maximum diameter did not agree with the displacement of the patches, and Schröter did not attach much importance to this result.

The most important observations were made in 1798, the year when Mars was at its closest to the Earth. Schröter–an excellent observer, carrying out all kinds of investigations and working under the most favourable conditions possible—found that the image of Mars was an almost perfectly circular disk and the ratio of the greatest and least diameters, was no more than 81 to 80. Therefore, if the planet were flattened at the poles, this flattening must be less than 1/81.

William Herschel had given 1/16 from his observations, and Schröter wrote a long dissertation upon this point. He paid tribute to the skill of the Slough astronomer, and regarded Herschel's result as correct for the time when the measurements had been made. But what caused the discrepancy? Schröter believed that the cause was to be found in the atmosphere of Mars; and he established a relationship between the flattening as measured at various epochs and the local deformations of the disk, which have been described above.

Schröter also discussed analogous observations of isolated deformations in the outlines of Jupiter and Venus. On 21 September 1798 he observed the same phenomenon on Mars for the first time. The outline of the planet seemed to be flattened from the south polar cap to a distance of about 70° to the west. Another appearance of this kind was seen on 12 November 1800, at 7^h 29^m in the evening. Light mists covered the sky, and cut down the planet's brilliance to some extent, but this only increased the sharpness of the image. In the region between the south and the west, the disk was bounded by a straight line instead of a continuation of the circumference. This fact was the target of Schröter's close attention. He put the planet in various different positions in the field of his 13-in ($0^m.33$) telescope, using a magnification 136x; but the illusion was still persisting at $7^h\ 35^m$. Several minutes later the effect became less evident, but the mist covering the sky thickened and prevented any further observation.

Schröter believed the cause to be due to the bending of the Sun's rays as they passed through various regions of the Martian atmosphere.

Schröter often mentioned the advantage of observing under slightly misty conditions, which cut down the light, but made the seeing very calm and clear. It was under these conditions that he liked to make his most delicate measurements. It was the same with observations when Mars was close to the Moon, or was seen against a light sky. All observers can indeed appreciate the advantages of such conditions.

Direction of the Axis, Obliquity of the Ecliptic, Positions of the Equinoctial and Solsticial Points, Apparent Diameter of Mars

The Lilienthal astronomer did not neglect any of these points. For measuring the position of the pole, he made special use of the perfectly fixed, very small south polar cap seen from 8 October to 16 November 1798, and this also enabled him to measure the direction of the axis. The results of his numerous measurements, made with the help of Harding, were summarized in calculations made by Olbers. They were as follows:

Celestial latitude of the south pole of Mars = 60°33′12″.
Celestial longitude of the south pole of Mars = 172°54′44″.
Obliquity of the ecliptic of Mars = 27°56′51″.
Longitude of the vernal equinox for the northern hemisphere = 264°53′35″.

On the morning of 1 September 1798, under very favourable conditions and with Mars at its closest to the Earth, Schröter measured the diameter of the planet, and after repeated observations found it to be

26″.17.

He regarded this result as very reliable, and from other measurements taken at the same epoch he found a mean value of

26″.04,

differing by only 0″.13 from the value obtained on 1 September under the most favourable conditions possible.

Subsequently, Schröter gave a value of

9″.84

for the apparent diameter of Mars seen from the mean distance between the Earth and the Sun. For this, W.Herschel had found a value of

9″.8.

As we have noted, all Schröter's observations have been given in Dr. Terby's memoir on the subject.[32]

We next come to a series of 65 drawings, chosen from the most notable of Schröter's 230. In spite of my wish to be as complete as possible, it would obviously be beyond the scope of this book to reproduce all 230 drawings. However, these observations are so important in the history of the planet that I propose to give as many as possible. They are reproduced directly from the 1881 book referred to above.

As the drawings are reproduced in facsimile by photogravure, so as to preserve their authenticity, the numerous figures in the originals have also been shown. These give the dates of the drawings, and a brief description of each of them.[33]

For it is here, despite Schröter himself—what a bizarre situation!—that we begin the real study of the geography of Mars. The three drawings reproduced above (Figs. 17, 20 and 28)], representing only one side of the planet and seen under three different inclinations, are not sufficient for the purpose of recognition. We need a chart of the entire planet, and I have produced a General Chart of Mars, based on modern observations. As has been noted, the nomenclature given here (Fig. 31) has been generally adopted ever since the publication of Green's chart by the Royal Astronomical Society of London.

But now let us turn to Schröter's own drawings, and comment on whatever is most interesting in them.

[Note regarding the figures: because Flammarion has inserted facsimiles, the numbers referred to in the text below refer to those given by Terby, and not to our figure numbers. Thus Figs. 1, 2, 3 below are those in our Figs. 31, Figs. 4, 5, 6, those in our Fig. 32, Figs. 9, 10 and 18 in our Fig. 22 so on.—WS]

Fig. 1, 12 November 1785, at $7^h\ 44^m$.

Fig. 2, 8 November 1785, at $6^h\ 49^m$.

Fig. 3, 21 November 1785, at $7^h\ 0^m$.

These three views of Mars were taken near the time of opposition (26 November). The Hourglass Sea can be recognized on these drawings. It is well shown in Figs. 1 and 2 but on Fig. 3, we see the Flammarion Sea and the Hooke Sea, and the point goes down to the Gruithuisen Bay—very much broadened, as happen on rare occasions.

Fig. 4, 10 December 1787, at $7^h\ 0^m$. The only observation of 1787, made 28 days before opposition. Features rather peculiar (Fig. 33).

[32] For the same elements, W. Herschel found:

Latitude of the south pole = 59°42′

Longitude of the south pole = 167°47′

Obliquity of the ecliptic = 28°42′

Longitude of the vernal equinox for the northern hemisphere = 259°28′.

[33] Memoir by Van de Sande Bakhuyzen, *Untersuchungen uber die Rotationszeit des Planeten Mars*, has been used for the identification of the patches.

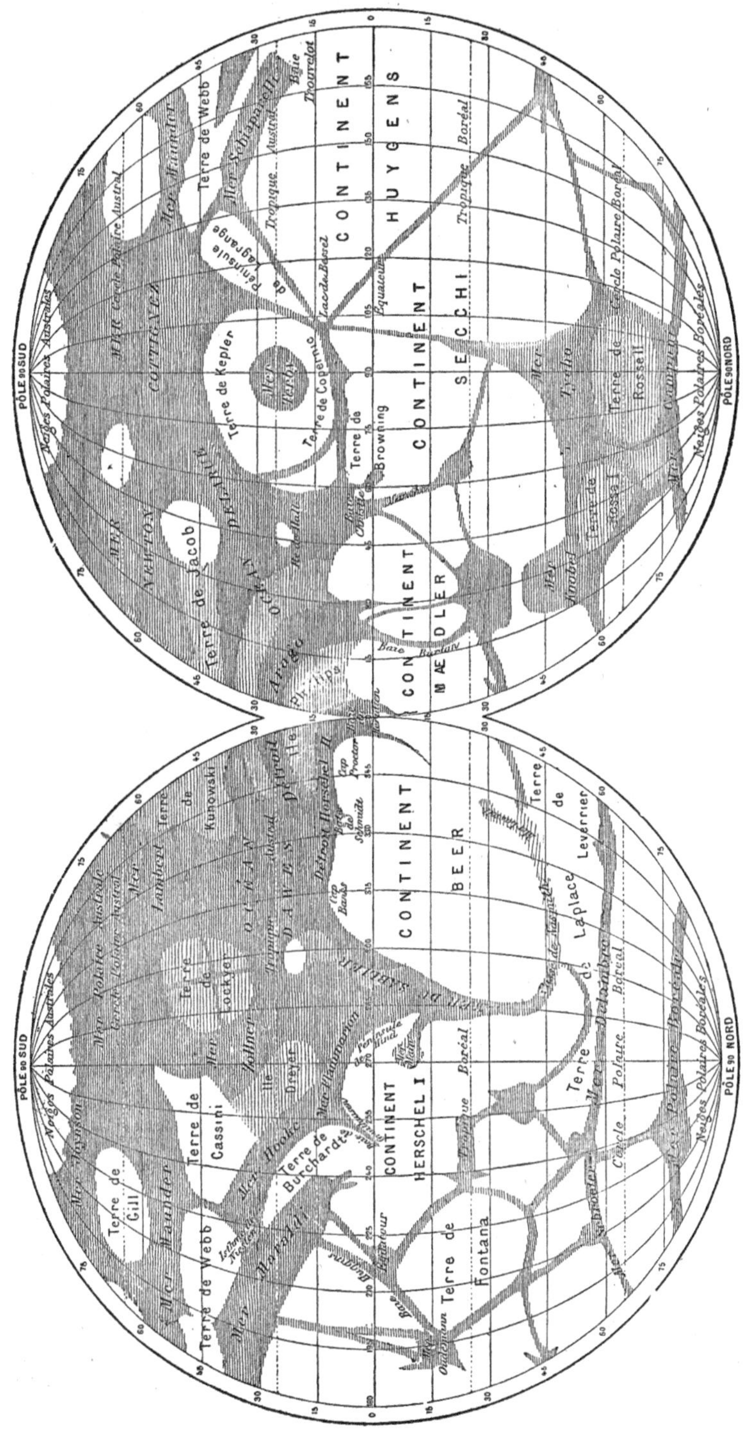

Fig. 31 General chart of the planet Mars

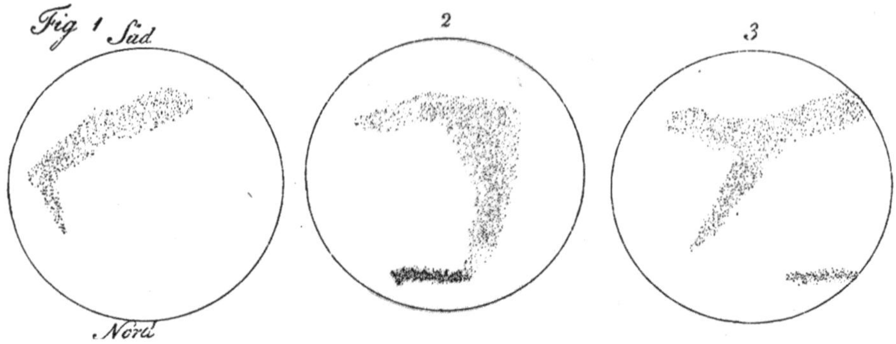

Fig. 32 Drawings by Schröter, November 1785

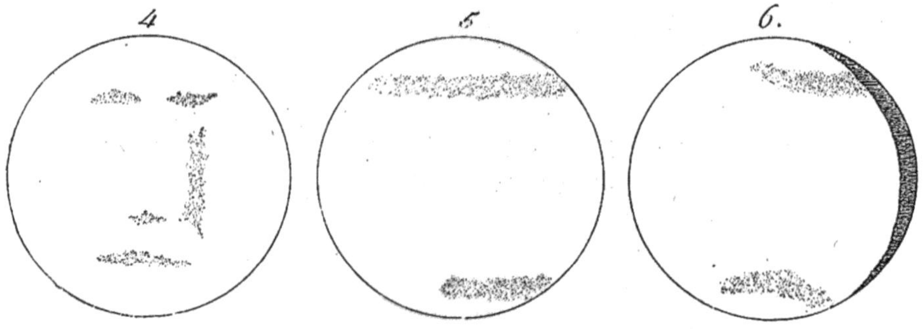

Fig. 33 Drawings by Schröter, December 1787-January 1788

Fig. 5, 15 January 1788, at 5^h 30^m. Eight days after opposition, which had fallen on 7 January.

Fig. 6, 28 January 1788, at 6^h 53^m.

Fig. 9, 19 March 1792, at 11^h 3^m. Two patches, *a* and *b*, were seen on the central meridian; they had arrived there since the beginning of observations at 7^h 48^m. They are difficult to identify. The longitude of the central meridian was 51° (Fig. 34).

Fig. 10, 20 March, at 6^h 50^m. The two spots *a* and *b* are not the same as those on the preceding drawing; the longitude of the central meridian is 348°, so *b* is perhaps the Meridian Bay. These spots shifted toward the centre, because of the planet's rotation. Opposition had taken place on 16 March. Schröter observed the planet to see if he could detect any flattening, but could not find any—in contrast to the results obtained by Herschel.

Fig. 18, 2 April, at 7^h 32^m.

Fig. 19, same evening, at 10^h 2^m. These new observations confirmed Schröter's opinion that the patches on Mars were variable, and of an atmospheric nature, as with those on Jupiter. In fact, the patches shown here are not easy to identify with known Martian features (Fig. 35).

Fig. 25, 24 March 1794, at 8^h 44^m. 30 days before opposition.

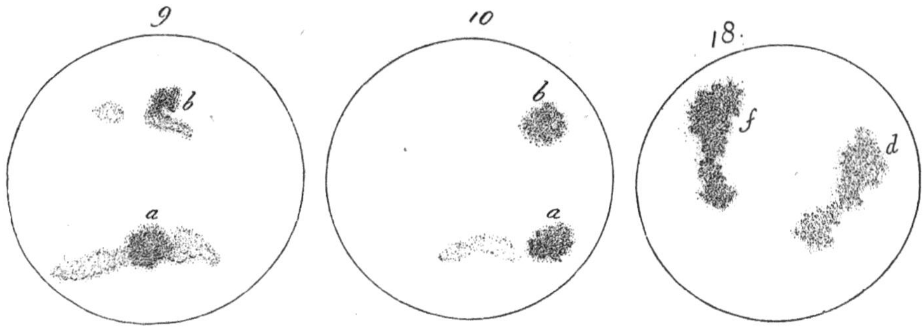

Fig. 34 Drawings by Schröter, March-April 1792

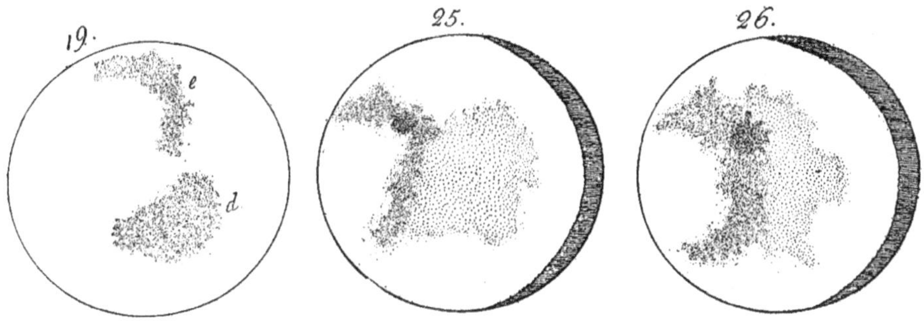

Fig. 35 Drawings by Schröter, April 1792-March 1794

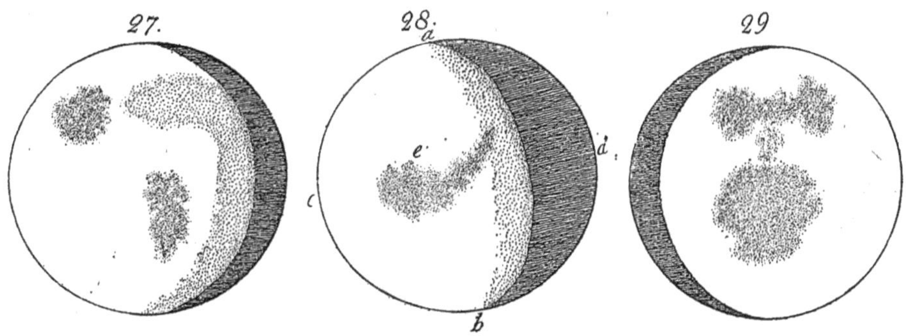

Fig. 36 Drawings by Schröter, June 1794, August 1796, Jul 1798

Fig. 26, 25 March 1794, at $8^h\ 25^m$ (Fig. 36).

Fig. 27, 1 June 1794, at 10^h.

Fig. 28, 17 August 1796. During this evening, two months after opposition; which had occurred on 15 June; the phase was very marked.

Fig. 29, 1798. *Excellent perihelic opposition*. 15 July, at 11^h in the evening.

Fig. 30, 18 July, at midnight. This observation, like the preceding one, was made in the company of Olbers. The south polar patch was very marked.

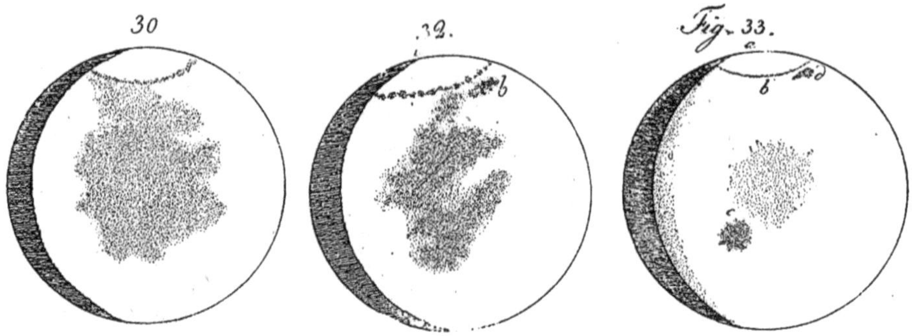

Fig. 37 Drawings by Schröter, July 1798

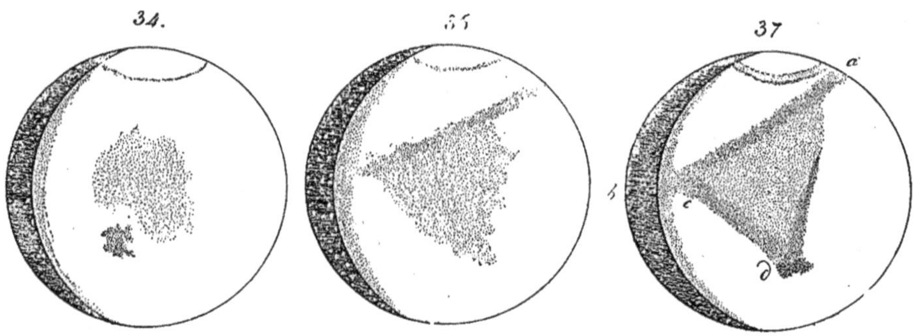

Fig. 38 Drawings by Schröter, July 1798

Fig. 32. 19 July, at $11^h\ 40^m$. Also with Olbers. South polar patch very marked at *a*; small sombre patch at *b*. Central meridian, 65°.

Fig. 33, 23 July, at $11^h\ 22^m$. The south polar cap was noted as being very brilliant at *a*; a sombre patch at *c*, and a small patch at *d*, recalling that shown in the previous drawing. Central meridian, 23° (Fig. 37).

None of the patches in these drawings can be identified with real features of the geography of Mars.

Fig. 34, 24 July, at $11^h\ 20^m$ (Fig. 38).

Fig. 36, 28 July, at $10^h\ 7^m$. Central meridian, 326°.

Fig. 37, 31 July. In the morning, there was an occultation of Mars by the Moon. Schröter observed the planet together with Harding. Central meridian, 332° (Fig. 39).

Fig. 38, 2 August, at $10^h\ 41^m$. South polar patch very brilliant. Much detail. The triangular patch is the Hourglass Sea.

Fig. 47, 26 August, at $10^h\ 3^m$. Large patch, with extensions at *a*, *c*, *d*, *b*, *e*.

Fig. 48, 27 August, $10^h\ 16^m$. Details just as well marked. Central meridian, 51° (Fig. 40).

Fig. 49, 30 August, at $10^h\ 24^m$. The dark patch is extended at *e* and *f* toward the extremities of the polar patch. Central meridian, 27°. Comparison with my chart shows no certain identifications.

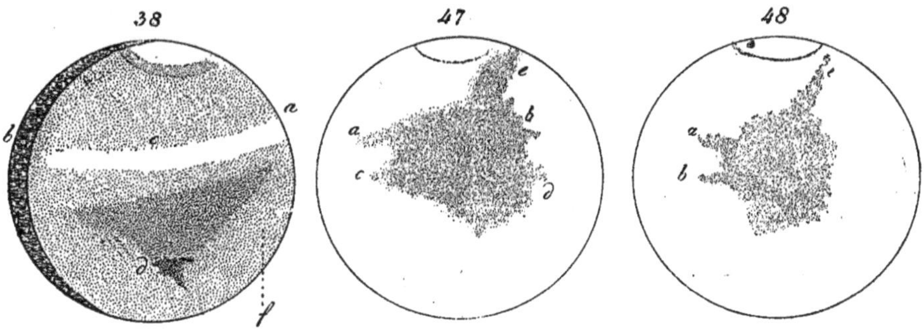

Fig. 39 Drawings by Schröter, August 1798

Fig. 40 Drawings by Schröter, 30 August, 2 September 1798

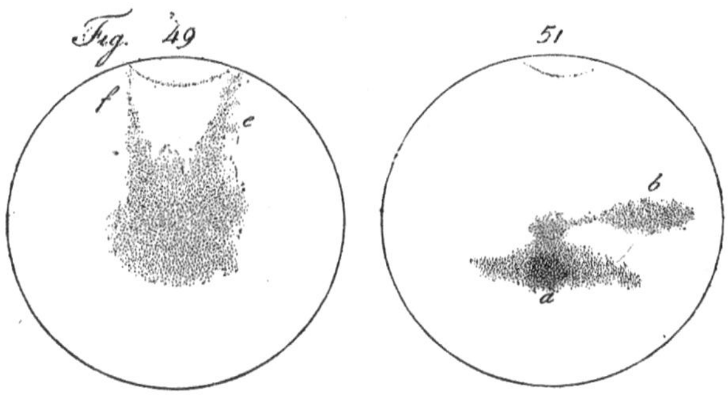

Fig. 41 Drawings by Schröter, 3 and 4 September 1798

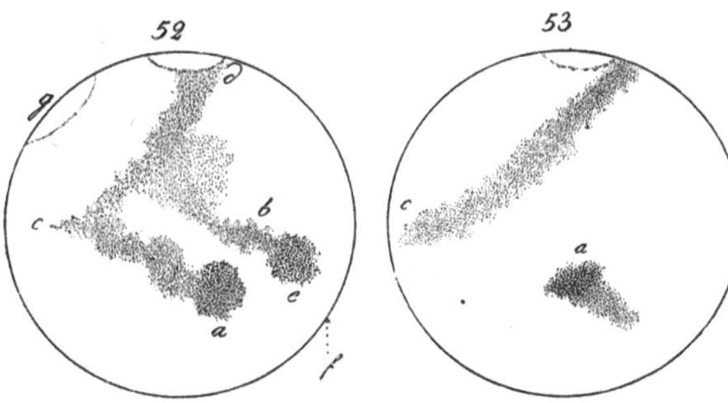

Fig. 51, 2 September, at $10^h\,47^m$. The day after opposition, which had occurred on 1 September 1798. Central meridian, 6° (Fig. 41).

Fig. 52, 3 September, at $10^h\,5^m$. In these two drawings on 2 and 3 September, *a* marks the Meridian Bay and *b* is the Arago Strait.

Fig. 53, 4 September, at $10^h\,46^m$. Again *a* is in the Meridian Bay (Fig. 42).

Fig. 54, 9 September, at $7^h\,55^m$.

Fig. 55, same day, two hours later; $9^h\,5^m$.

Fig. 56, same day, at $11^h\,8^m$. These three drawings are extremely valuable, because of their continuity; the Hourglass Sea is identifiable with certainty. It was darkest at its point, which is unusual.

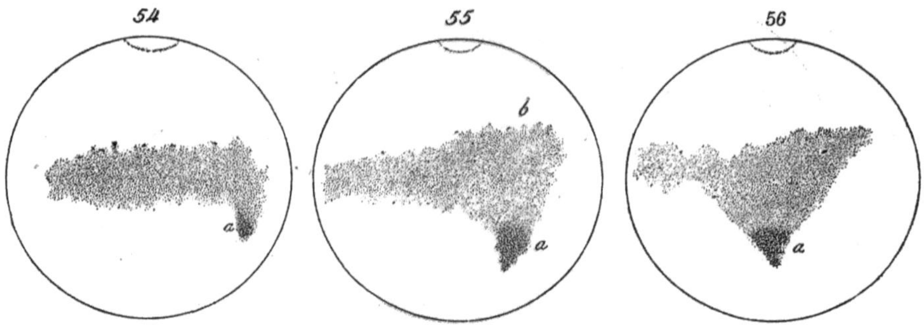

Fig. 42 Drawings by Schröter, 9 September 1798, the Hourglass Sea

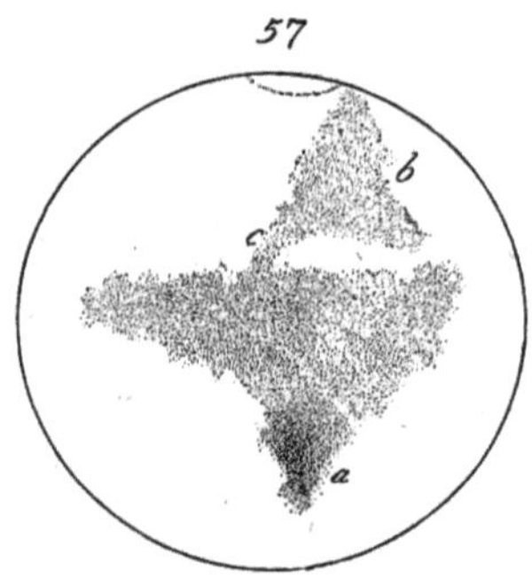

Fig. 43 Drawing by Schröter, 10 September 1798

Fig. 57, 10 September, at 10^{h} 15^{m}. Almost the same view as at 9^{h} 55^{m} on September 9, but an extra upper patch was noted (*bc*) (Fig. 43).

Fig. 65,19 September, at 7^{h} 31^{m}.

Fig.66, 20 September, at 7^{h} 27^{m}.

Fig. 67, same day, at 9^{h} 48^{m}. These figures also show almost the same face of the planet (the Maraldi Sea). 19 September at 7^{h} 31^{m} corresponds to 20 September at about 8^{h} 11^{m}, and the difference between the second and the third observation is only 2^{h} 21^{m}. The patch is the same one; it has moved toward the centre of the disk in the third observation, and the difference in shape is perceptible. The point, which goes down near longitude 193°, is the strait, which can be seen near the Oudemans Sea (Fig. 44).

Fig. 83, 8 October, at 6^{h} 40^{m}.

Fig. 84, 9 October, at 7^{h} 35^{m}.

Fig. 85, 10 October, at 7^{h} 55^{m}. Here are three drawings representing almost the same face of the planet (central meridian = 341°). The first two are similar; the third is different. The features are impossible to identify on the chart. One can understand Schröter's conclusions (Fig. 45).

Fig. 102, 15 November, at 6^{h} 50^{m} (Fig. 46).

Fig. 103, 16 November at 6^{h} 13^{m}. The same face again; the second drawing us 1^{h} 15^{m} earlier than the first. The difference is perceptible. The planet is a long way from the Earth, and shows a marked phase (Fig. 47).

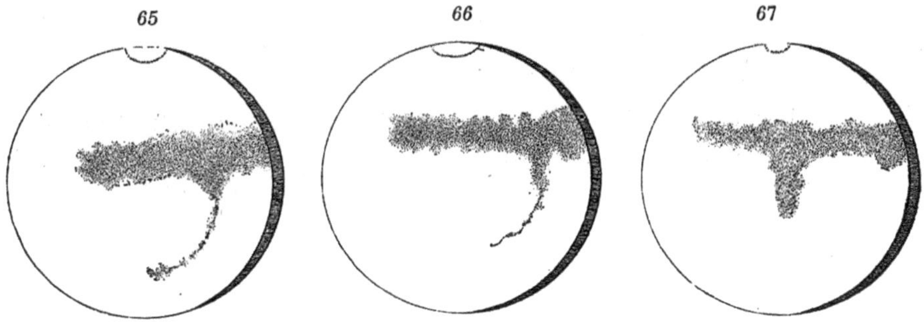

Fig. 44 Drawings by Schröter, 19 and 20 September 1798

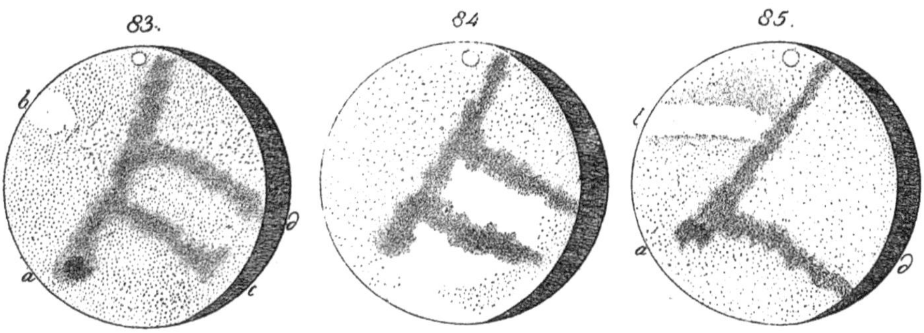

Fig. 45 Drawings by Schröter, 8, 9 and 10 October 1798

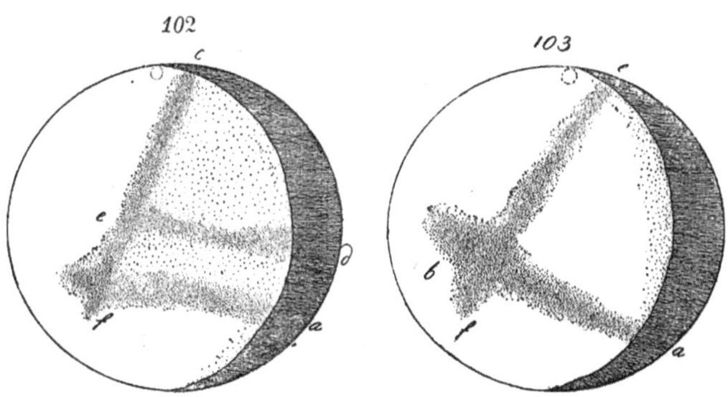

Fig. 46 Drawings by Schröter, 15 and 16 November 1798

Fig. 104, 20 November, at 6^h 16^m.

Fig. 105, same day, at 8^h 2^m. A very great overall change in less than two hours. (Schiaparelli has expressed the opinion that the lowest part of the patch is the extremity of the Hourglass Sea, to which he has given the name of Nilosyrtis; but this canal-like feature is slanted toward the right, while that in *fig*. 105 is slanted

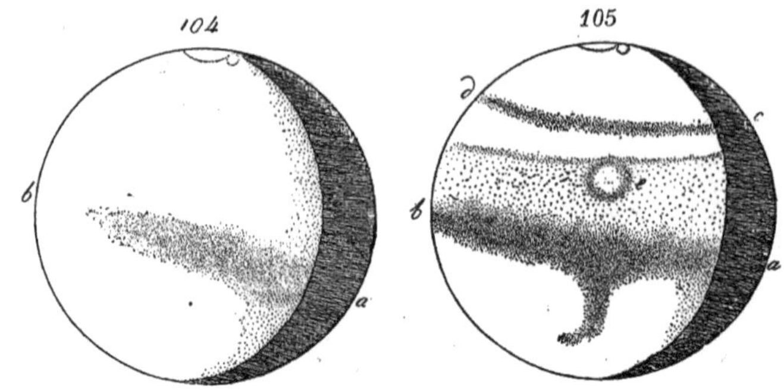

Fig. 47 Drawings by Schröter, 20 November 1798

Fig. 48 Drawing by Schröter, 10 December 1798

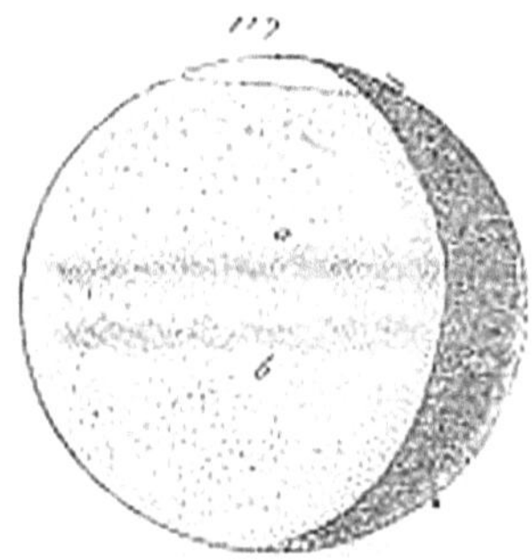

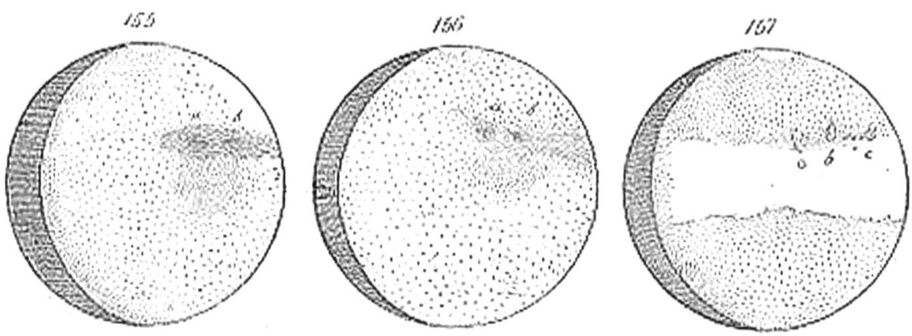

Fig. 49 Drawings by Schröter, October 1800

toward the left.) The white patch *e* is Lockyer Land; an island perhaps covered with snow.

Fig. 119, 10 December, at $4^h\ 43^m$. Two parallel equatorial bands *a* and *b*, as with Jupiter. Without the phase, the drawing could easily be mistaken for Jupiter instead of Mars. In the *Areographische Beitrag* there are 16 similar figures, showing parallel bands, made about this time (Fig. 48).

Fig. 155, 8 October 1800, at $10^h\ 20^m$.

Fig. 156, 9 October 1800, at $10^h\ 40^m$ (Fig. 49).

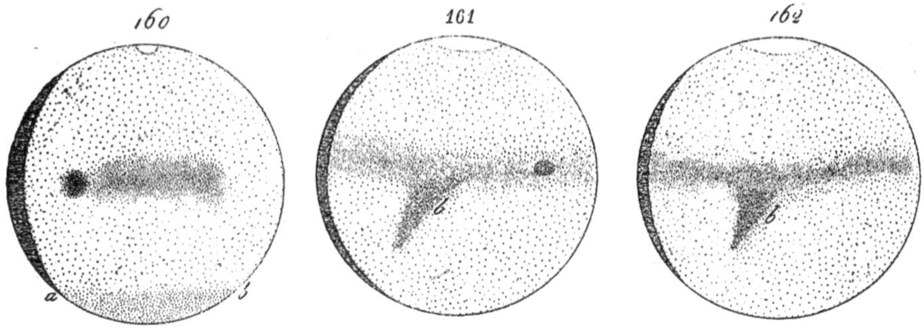

Fig. 50 Drawings by Schröter, Houglass Sea to Meridian Bay, October 1800

Fig. 51 Drawing by Schröter, 1 November 1800

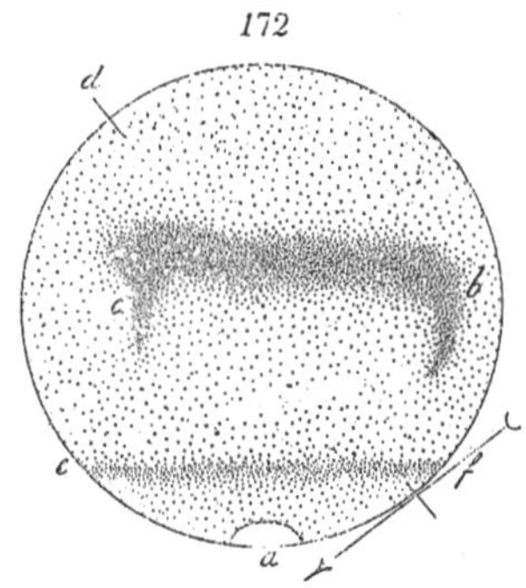

Fig. 157, 11 October, at $10^h\ 32^m$. The drawings are interesting because of the patches *a*, *b* and *c*, which confirmed Schröter's view, that constant changes were taking place on the planet.

Fig. 160, 20 October, at $10^h\ 22^m$.

Fig. 161, 24 October, at $8^h\ 17^m$.

Fig. 162, 25 October, at $9^h\ 32^m$. Observation made together with Olbers, at his observatory at Bremen. The black point is the Meridian Bay, seen in the form of a very black disk by Beer and Mädler in 1830. Patch *b* is the Hourglass Sea (Fig. 50).

Fig. 172, 1 November, at $8^h\ 10^m$ (Fig. 51).

Fig. 174, 2 November, at $7^h\ 42^m$.

Fig. 175, same day, at $11^h\ 20^m$. As Terby has already noted, these three drawings are particularly interesting. The second shows a triangular patch which covers the central meridian, and the third, made $3^h\ 38^m$ later, shows a patch which is of the same shape but is much more extensive, occupying almost the same position; a little further over on the disk, and which lies about one-sixth of the circumference further to the right; about 60° in longitude. *Fig*. 172, made on 1 November, confirms this interpretation, and also shows the two dark patches *b* and *c*. The regions marked *f*, *g*, and *d* on *figs*. 174 and 175 are very bright. Schröter regarded them as evidence of new variations. The large triangular patch in *fig*. 175 is the Hourglass Sea. That in *fig*. 174 is a point near 228°, that is to say the right-hand extremity of the Maraldi

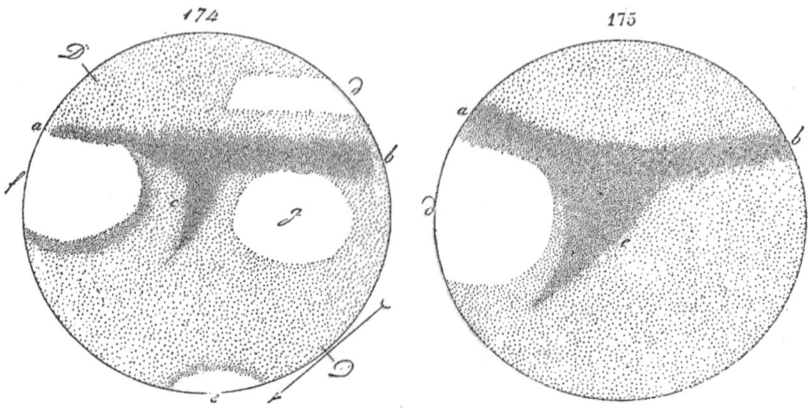

Fig. 52 Drawings by Schröter, Hourglass Sea (*fig.* 175) and another pointed sea (*fig.* 174), about 228° longitude, 1 and 2 November 1800

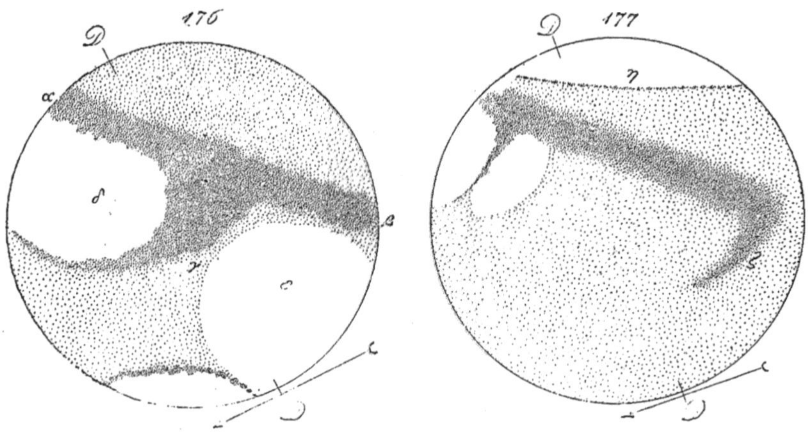

Fig. 53 Drawings by Schröter, November 1800

Sea; *fig.*172 shows these dark patches. The drawings, which follow, confirm this interpretation (Fig. 52).

Fig. 176, 4 November, at 8^{h} 20^{m}.

Fig. 177, same day, at 10^{h} 41^{m}. The first of these two drawings shows the same face of the planet as that in *fig*. 174, and the second shows the same as that in *fig*. 172, with the two patches significantly similar. Observing conditions were excellent. The 1800 opposition took place on 9 November (Fig. 53).

Fig. 195, 8 December, at 5^{h} 19^{m}.

Fig. 196, same day, at 6^{h} 45^{m}.

Fig. 197, same day, at 9^{h} 43^{m}. These three drawings lead to the same conclusion. *Fig*. 197 shows the Hourglass Sea (Fig. 54).

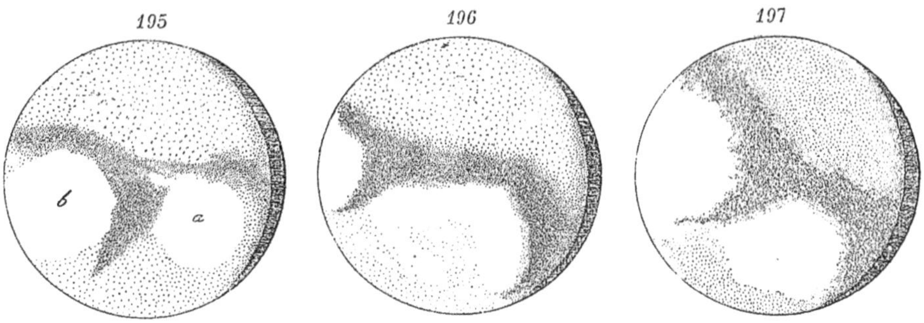

Fig. 54 Drawings by Schröter, 8 December 1800

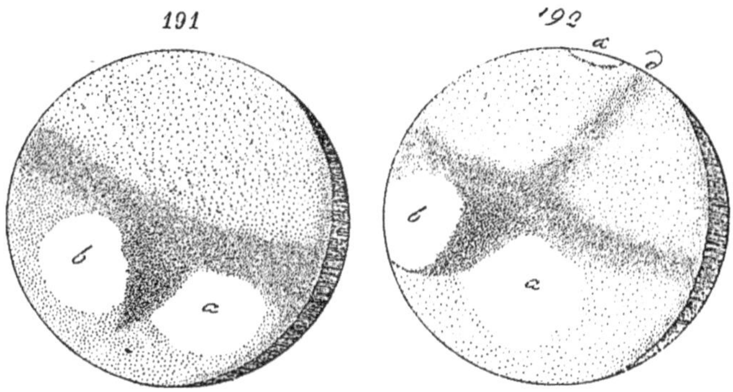

Fig. 55 Drawings by Schröter, 3 December 1800

Fig. 191, 3 December, at 6^{h} 27^{m}.

Fig. 192, same day, at 7^{h} 16^{m}. I have given these two drawings after the preceding ones because of the concordance of *figs*. 195, 196 and 197 with *figs*. 172–177. These are also interesting for another reason (Fig. 55).

Thirty drawings made between 24 October 1800 and 8 January 1801 are almost identical with *fig*. 191, and seven with *fig*. 192—that is to say, showing the grey streak which runs to the right side of the pole. It is easy to understand why Schröter became more and more convinced of the reality of the changes.

Fig. 182, 12 November, at 7^{h} 29^{m}. Central meridian, 224°, 18 December 1802 at 8^{h}. A curious example of the deformation of the disk referred to above (Fig. 56).

Fig. 217, 11 October 1802, at 11^{h} 5^{m}. Central meridian, 275°.

Fig. 218, 14 October, at 10^{h} 45^{m}. For October 10 there is a drawing absolutely identical with that of the 11th that is to say, showing the grey streak to the right of the patch *a*. On the 14th, this grey streak was no longer visible. On the 16th, only the round patch a, was seen. There is nothing like this to be found in modern observations, because this feature is not the Meridian Bay; the longitude of the central meridian was 242° (Fig. 57).

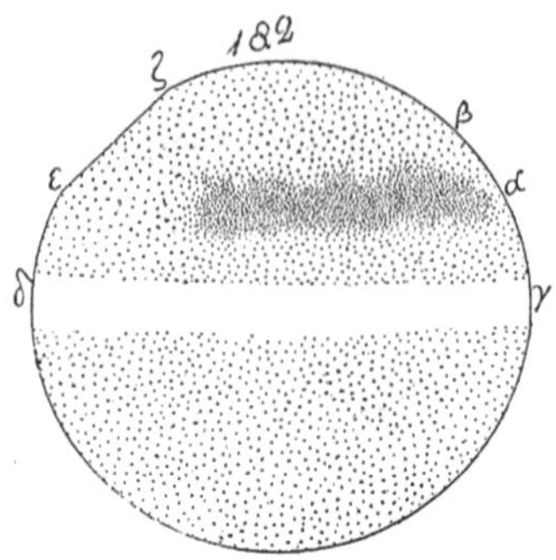

Fig. 56 Drawing by Schröter. Apparent deformation of the disc of Mars, 12 November 1800

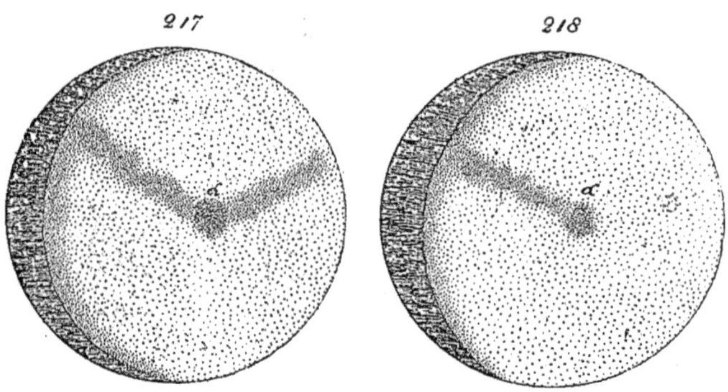

Fig. 57 Drawings by Schröter, October 1802

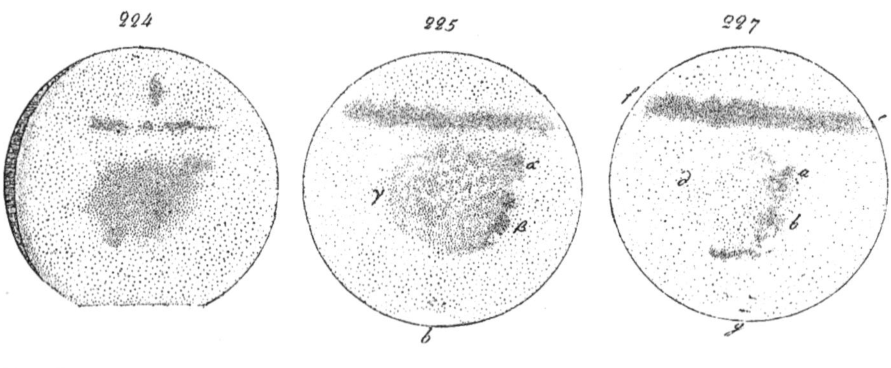

Fig. 58 Drawings by Schröter, December 1802

Fig. 224, 18 December, at $8^h\ 0^m$. Central meridian, 333°. The planet appeared as though cut-off at the bottom.

Fig. 225, 23 December, at $5^h\ 58^m$. Central meridian, 260°.

Fig. 227, 24 December, at $8^h\ 12^m$. Further new appearances. Drawings were made at the time of opposition, which occurred on 25 December. Central meridian, 284° (Fig. 58).

All these observations confirmed Schröter's conviction of *constant changes* on the surface of Mars; this eminent observer always believed that the patches on this strange planet were atmospheric in nature.

Such were Schröter's observations. They were more important than any of those we have examined in the earlier pages of the present book.

Everything that had been learned from the work of earlier observers was confirmed; diurnal rotation, axial inclination, seasons, polar ices, and atmosphere. Knowledge of Mars was gradually being accumulated. The polar flattening remained dubious.

Knowledge of Martian topography, however, had still not made great progress. As we have seen, Schröter, from his long series of observations, was even convinced that the sombre patches on the planet were not seas, but were formed by clouds. This was also the conclusion to which Maraldi had been led.

Despite the skill of these observers, and despite their excellence of judgment, this latter conclusion cannot be accepted. In fact, most of the patches observed and drawn by the Lilienthal astronomer are permanently fixed. Our famous Hourglass Sea, the most characteristic of all, is—as we have seen—shown on a great number of drawings; among others on Schröter's *figs*. 1, 2, 54, 56, 161, 162, 175 and 197. The Meridian Bay is shown on *figs*. 52, 53 and 161. These are the first definite observations of this very important feature, chosen by Beer and Mädler in 1830 as the zero for Martian longitudes. The Maraldi Sea is recognizable on *fig*. 67, and others—as in the form of a band resembling those on Jupiter. Moreover, modern observations prove the permanence of the principal patches. Thus Schröter was mistaken in his conclusions—as Maraldi also had been.

However, all these observations show that there are real and considerable changes on Mars. There is no longer any doubt about this. We must agree that the sombre patches are formed by fixed regions which are undoubtedly seas, since it is know that water—as with other liquids—absorbs a part of the incident light, while continental surfaces are better reflectors. Besides, if there is water on Mars, it has always been probable that we can observe polar snows and clouds, and today this has been made certain by spectral analysis.

Therefore, I maintain that we must admit that the sombre patches of Mars are formed in part by fixed seas, and in part also by an unstable element. This unstable or variable element may be of the same nature as the seas; it may also be water in another state.

This conclusion is absolutely demonstrated by the observations by Maraldi and Schröter, as described above. The sketches made by Huygens, Cassini and Hooke point to the same conclusion.

Sometimes perhaps, when the changes are slight, we can assume that the seas overflow onto the beaches, flooding vast plains and changing their coastlines.

Yet the differences between the drawings of Schröter, Herschel, Maraldi, Cassini, Bianchini, etc., are so great that it is impossible to use them in drawing up a rigorous representation of the planetís geography. All observers who draw Mars know that it is extremely difficult to make a faithful drawing of what can be seen, because the forms are nearly always indefinite, diffuse, vague, without sharp outlines, and sometimes quite uncertain. The aspects are vague, feeble, dubious and hard to draw; the instruments used are different, to the observer's eyes and methods of observation are perhaps even more different. Nevertheless, it is manifestly impossible to attribute all the discrepancies to errors of observation, and we must remember that the overall representation of the patches sufficed to give a value for the planet's rotation period and position of the axis. It is quite certain that the observations have a basis of reality.

Do the Martian seas produce by evaporation, dark mists—sombre when seen from above—when they are in full sunlight? Are these mists or clouds distributed in accordance with the shapes of the surface features? It seems difficult to avoid this double interpretation.

On Earth we do not see black clouds from above, on the same side as the Sun's rays. (Schröter's observation cited above was made when the sunlight was coming obliquely.) The upper surfaces of these clouds are as white as snow. But there must also be mists whose molecular constitution makes them poor reflectors of incident light. Our exclusively earthbound observations mean that we find it difficult to appreciate this point. Other worlds should differ from our own in various ways. Thus on Jupiter and Saturn we see sombre bands and dark patches, of which a certain number are unquestionably atmospheric in nature.

These variations are henceforth unquestionable.

Yet we cannot yet resolve all the problems posed by the surface of Mars. Let us sincerely note all the facts, and see what they produce. Now let us continue our study.

1794.—Von Hahn

In the *Astronomisches Jahrbuch* for 1797 there is a drawing by Von Hahn, which adds nothing to the preceding work, and which is noted here only for the sake of completeness.

1796, 1798, 1800, 1802, 1805, 1807, 1809 and 1813.—Flaugergues[34]

Honoré Flaugergues had his observatory at Viviers (Ardèche), from which he made a great number of interesting observations. In particular, he observed Mars from 1796 to 1809, and again in 1813. The first observations appeared in the *Journal de Physique*, Vol. LXXIL, 1809, p.126; the second in the *Astronomical*

[34] *Les taches de la planète Mars*. To the observations made by Flaugergues in 1796 we may add one, made on 18 April, when the planet was in conjunction with the 6th-magnitude star *b* Sagittarii; Mars ought to have occulted the star, but the star did not in fact vanish completely, though it

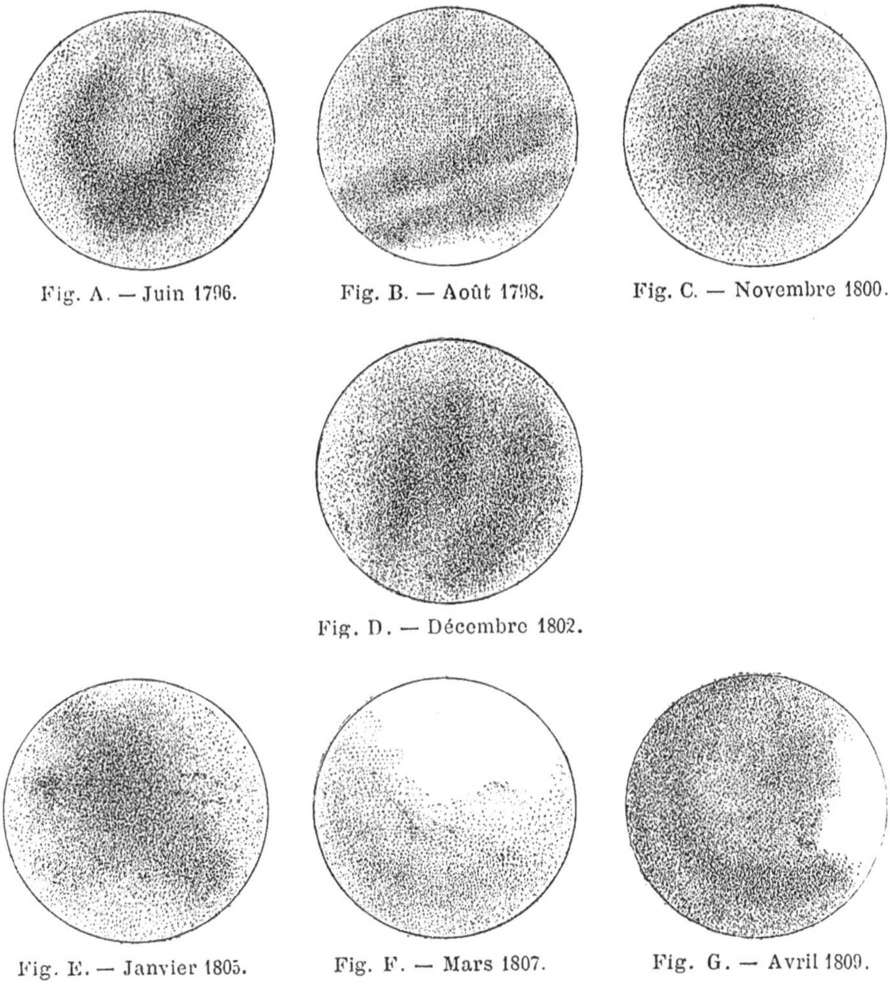

Fig. 59 Sketches of Mars, by Flaugergues

correspondence of Baron de Zach, Vol. I, 1811 p.180. Here is an extract from the first memoir, together with the seven drawings (Fig. 59) which accompanied it. Flaugergues stated that the patches were variable, and set out to decide whether they were due to the planet's surface or to its atmosphere.

I observed Mars several days before and after opposition, and with the greatest care. I drew the patches, which were visible; I gave a description; the figure was applicable to the time when Mars crossed the meridian on the day of opposition;

---

was within half a Martian diameter of the planet. Flaugergues very logically attributed its fading to the brilliance of Mars.

about midnight mean time. In all the drawings the axis of Mars is along the vertical diameter; the north pole at the top.

Opposition of 1796

(16-ft. refractor; magnification 105×)

On the south part of the disk I constantly saw a dark, horseshoe-shaped patch whose branches were turned toward the north (*Fig*. A in our Fig. 59).

Opposition of 1798

(Achromatic 44-ft. refractor; magnification 90x)

On the south part of the planet's disk, I constantly saw two fairly broad parallel bands—dark red—lying east to west and separated by a narrower bright band. Also in the south part I saw a white patch, oval and stationary, lying near the edge of the disk, and about 6° to the right of the vertical in the refractor, which gives a reversed image (*Fig*. B in our Fig. 59).

Opposition of 1800

(Same refractor)

I constantly saw a large round patch, darker red than the rest of the disk, whose centre was a little north of the centre of the planet. This patch was bounded in the south by a feature in the form of a hook, whose curvature was similar to that of the large patch (*Fig*. C in our Fig. 59).

Opposition of 1802

(Same refractor)

On the disk I consistently saw a large round patch of darker red, almost concentric with the planet's limb, and cut transversely at an angle of 45°with the vertical from east to west, following one of these diameters by a brighter band which extended up to two-thirds of the patch (*Fig*. D in our Fig. 59).

Opposition of 1805

(Same refractor)

On the disk I constantly saw a large patch, darker red than the rest of the disk, of an irregular and indefinite form, more extended and darker in the north part of Mars (*Fig*. E in our Fig. 59).

Opposition of 1807

(Same refractor)

On the south part of the planet's disk I constantly saw a feature in the form of a band—in tint slightly darker than the rest of the disk—long, narrow, ill-defined and lying east-west; this band was easy to see in (*Fig*. F in our Fig. 59). Above all, I noted that the north part of the disk was perfectly white and extremely brilliant, particularly around the area corresponding to the north pole.

Opposition of 1809

(Same refractor)

The western edge of the planet appeared white and brilliant, the eastern edge dark red; I could see two patches; one long and in the form of a band, lying east to west in the southern part of the disk, and the other smaller, irregularly rounded, lying in the northern part near the west edge; these two patches were darker red than the rest of the disk (*Fig*. G in our Fig. 59).

These patches seemed to me to be in general confused and badly-defined, to an extent that it was difficult to distinguish exactly their outlines and their full extent. I can say only, that normally the south part of the disk which was the region of Mars, contained the principal patches.

With regard to the oval patch, very remarkable for its brilliance and for its whiteness, which I observed in 1798, and which corresponds on the disk to the south pole of Mars. I may add that it was also observed by Messier, the Duc la Chapelle and Vidal.

These white oval patches—always corresponding to the poles of Mars—present exactly the same aspect as would be seen of the snow and ice caps of Earth if observed from Mars. Herschel has not hesitated to attribute the Martian caps also, to snow and ice, and one cannot criticize this suggestion, which appears to be well founded.

For the red and darker patches on Mars, whose appearance has always been different in the various observations I have made, it may perhaps be permissible to think that the changes are purely optical, caused because the rotational movement of Mars on its axis means that the visible hemisphere of the planet will not be the same from one observation to the other—so that the appearance also will be different.

To appreciate thus, and to evaluate the effect of the change produced by the rotation, I have taken as a reference the meridian of Mars whose plane passed through the centre of the Earth at the time of the first observation, on 14 June 1796, at midnight mean time. This meridian, which I take as the prime meridian of Mars should be regarded as fixed to the globe of the planet, turning with the globe just as the prime meridian of the Earth is regarded as fixed at the Île de Fer.

The author here makes a calculation of the central meridian of Mars, which he calls the "méridian gédiabenique," which he derives from γη {Greek}, *Earth*, and διαβαινω {Greek}, *I pass*. He compares the positions of the planet for the seven drawings given above, and finds that the first, fourth and sixth observations show much the same aspect, while the second and the seventh are not far from it. Then he adds:

The appearance of the patches on Mars ought to be about the same in the first, fourth and sixth observations, and the similarity should be recognizable in the second and the seventh, assuming that the forms of the patches on the planet are constant, and that their appearance varies only because of the rotational movement of the planet about its axis. Therefore, when the form, the number and the distribution of the patches appear so very different in each observation, we must conclude that the changes which we observe in the Martian patches are real, and that the patches show physical changes, increasing and shrinking, disappearing and reappearing once more, just as we see with the patches on the Sun. But at the same time, we must note that the variations which we observe are so great that to produce comparable effects on the Earth's globe, seen from the distance of Mars, we would have to reckon with the submersion of a continent such as America, or the drying up of a sea such as the Atlantic Ocean. These changes are too considerable to be accepted on the solid globe of Mars as a cause of the variations, which we have observed in the patches. It is not in accord with the constant state of equilibrium of the planets, which judging from the Earth has prevailed for a long time. It is much more probable that these patches and their great changes, lie in the atmosphere of Mars, whose existence has been indicated by many observations.

It seems that the fluid of which it is composed has much in common with our air, at least it resembles our atmosphere in one remarkable property—it absorbs the blue and violet rays, and sensibly transmits only the yellow and the red. This is indicated by the red colour of Mars. On this assumption, which appears to be proved, the large red patches which we have observed are great masses of clouds floating in the atmosphere of Mars, or perhaps immense fogs similar to those which for several months covered a large part of our globe in 1783—whose extent, form, number and situation—would obviously and considerably vary because of the effects of weather, winds, and other unknown causes and which for the same reasons would dissipate and be subsequently re-born, as we see on the Earth.

Such were the observations made by Flaugergues from 1796 to 1809. They add little to the preceding ones. The polar variations were confirmed, as were those of the sombre patches, which were still believed of an atmospheric nature; it can only be said—as we said also at the end of our discussion of Schröter's observations—that the hypothesis is untenable. The disk is redder near the centre than near the limb; therefore it cannot be the thickness of the atmospher, which causes the colour, because the light reflected from the centre of the planet passes through less of the Martian atmosphere than that coming from near the edge of the disk.

In 1813 Flaugergues made new observations. Here is an extract from his second memoir:

I observed Mars several times around the date of the last opposition, as I had done during several previous years, with the aim of drawing the patches on the planet and noting the considerable and singular variations which they show. This year I again noted a white oval patch lying at the south pole of Mars, and so brilliant that it appeared to project from the disk. This patch was particularly brilliant on the night of 31 July—the day of opposition—it diminished in size much more rapidly than it could have done if the shrinking had been purely optical, and due to the planet's increasing distance. On 22 August the patch was barely visible, and several days later I could no longer see it. In 1798 I saw a similar white patch at the south pole of Mars, but this was much less brilliant than the patch seen during the present year.

For the south part of Mars, spring began on 12 March, and the declination of the Sun, seen from the planet, was $21°0'$ on 31 July; consequently the patch or the polar cap which I observed was continuously lit up and warmed by the Sun for several consecutive days, because over this part of Mars the Sun does not set. Therefore, if the cap is made up of ice or snow, similar to the ice and snow of our own globe—as everyone believes—there is no doubt that it will melt very rapidly.

On Mars can be seen large irregular patches, which are variable, and present the same appearance as would be given by our clouds and mists to an observer on Mars. The two planets each have their poles surrounded by white caps, which shrink when the Sun approaches the pole which they cover, and which in consequence, seem to be of the same nature as our own polar caps—that is to say, snow or ice.

If this conjecture is correct, the melting of the polar ices on Mars is much more prompt and much more complete than with our own terrestrial ices, most of which persist up to the heat of summer; it therefore seems that the heat on Mars is greater than on the Earth, though, because of the planet's greater distance from the Sun, it ought to be less in the ratio of 43 to 100. This is an extra reason to add to those which have made the most skilful physicists believe that the rays of the Sun do not in themselves cause heat, but are only the indirect cause of heat.

As we have seen, Flaugergues was the first to note that the polar snows of Mars vary much more markedly than those of the Earth, and that the mean temperature of the planet could be higher than that of the Earth. This is perfectly correct, and we will see later that modern measurements confirm this very interesting fact about the Martian climatology.

1802–1807.—Fritsch[35]

In the books cited (p. 188 and 218), Pastor Fritsch published a summary of his observations of the planet. The observations were accompanied by five drawings, which are reproduced here (Fig. 60) (taken on 21 November, 24 November, 26 November, 19 December 1802 and 10 January 1803, and one on 17 March 1807, at 9^h). The latter drawing shows the polar patch protruding from the disk, undoubtedly because of irradiation, and two equatorial bands recalling those of Jupiter. Fritsch wrote about the atmosphere of Mars and the rotation of the planet, but gave no details.

His sketches, like those of Flaugergues, have south at the bottom.

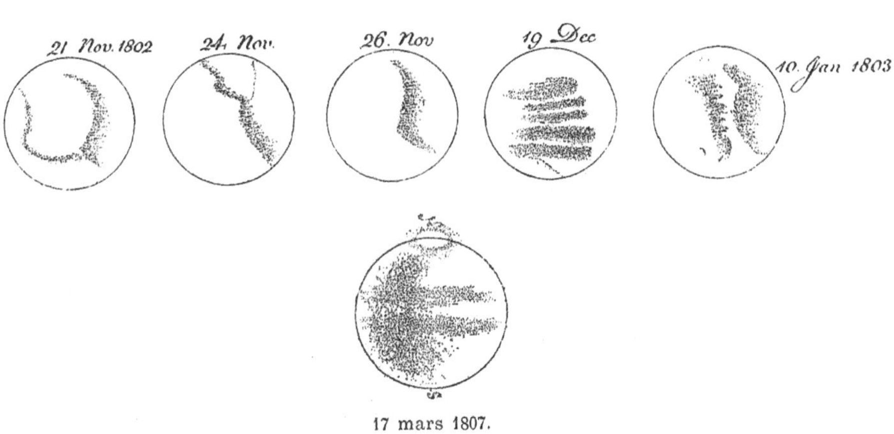

Fig. 60 Drawings of Mars by Fritsch, in 1802, 1803, and 1807

[35] Observations of Mars made at Quedlinburg in 1802, published in the *Astronomisches Jahrbuch* for 1806 and 1810.

1805.—Huth[36]

These observations are not of any great interest. They are accompanied by a drawing made on 22 February 1805, showing, at the north pole, a strong white oval patch, which by irradiation, seemed to protrude from the disk. An indication of the south cap can also be seen at the opposite pole. Apart from these two polar patches, the disk is blank. Huth's observations led him to give a value of $24^h\ 43^m$ for the rotation period. He discussed the analogy between Mars and the Earth with respect to atmosphere and meteorology.

1813, 1814, 1822, 1839 & 1847.—Gruithuisen[37]

The author first gives an account of his observations of Mars made in 1813, notably of the snowy patch (Schneeflecken) of the south pole.

He gave three drawings, reproduced here (Fig. 61), made on 1 July and 31 August 1813 and 14 January 1814. On the equatorial zone there may be noted some dark streaks, which can be identified with those observed by Maraldi.

Gruithuisen also gave notes on the rotation, the position of the axis and the patches; but these are not accompanied by any drawings.

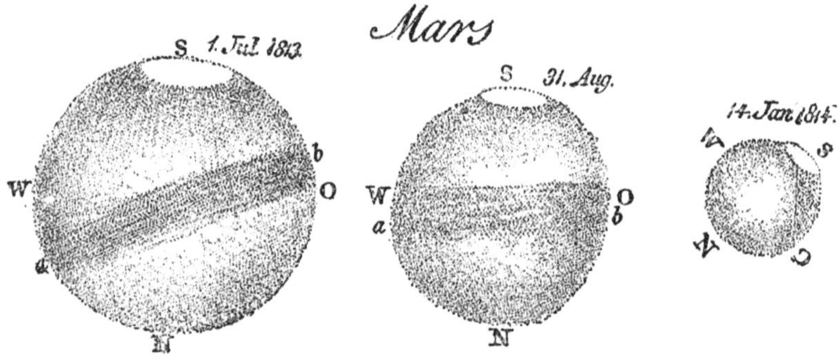

Fig. 61 Drawings of Mars made by Gruithuisen, in 1813 and 1814

[36] Observations made at Mannheim and Frankfort in 1805. *Astronomisches Jahrbuch* for 1808, p. 238 (Berlin: 1805).

[37] *Einige physisch astronomische Beobrachtungen des Saturns, Mars, des Mondes, des Venus*, etc. (*Astronomisches Jahrbuch* (of Berlin) for 1817, p. 185;--*Id.*, Munich, edited by Gruithuisen, 1839, p. 72; 1840, p. 98; 1842, p. 155; 1847, p. 149; 1848, p. 124.).

1811, 1813, 1815, 1817, 1845 & 1847.—Arago

Arago made a certain number of observations of Mars, which he collected in a Memoir read to the Académie des Sciences on 31 January 1853, the same year as his death. This Memoir is published in his complete works, Vol. XI, p.245–304.

The illustrious director of the Paris Observatory begins by stressing the value of physical astronomy, which is barely eclipsed by that of mathematical astronomy. After a quick historical sketch, he devotes a chapter to the flattening of Mars, and gives in detail his observations made from 1811. The results varied considerably over the years, from 1/29 to 1/100. After a discussion, Arago concluded that the flattening of Mars must be greater than 1/30.

We recall that Herschel had given 1/16, Schröter less than 1/81.

Applying to the flattening of Mars the theory that had given such good agreement with observation in the case of Jupiter, we arrive at a value of 1/230. This is in strong disagreement with observation. Arago commented that in order to give a satisfactory explanation, it was necessary to suppose the mass of Mars to be eight times less than the accepted value—which is inadmissible. He consulted Laplace, and Laplace replied: "Local anomalies, analogous to those whose effects can be seen in different parts of the Earth, particularly in the equatorial regions, must have had a greater influence on the figure of this small planet than on Jupiter or our globe."

The discrepancies in the values obtained for the flattening are worthy of attention. From several points of view, Mars is exceptional. Its first satellite moves round it more quickly than the planet itself spins; its revolution period is $7^h\ 39^m$, while that of the globe of Mars is $24^h\ 37^m$. The surface shows enigmatical variations. Mars is a world strikingly different from our own. At the end of this book, I will undoubtedly be able to present some definite conclusions.

For the diameter of Mars at unit distance (distance from the Earth to the Sun), Arago gave a value of $9''.57$.

His observations of the patches commenced in 1813. The telescope he used was a 4-in. (108^{mm}) Lerebours refractor, given to the Observatory by Napoleon; it was called "the Emperor's telescope."[38] It was perhaps the best instrument at the Observatory. Progress has been rapid since then; today, most student telescopes are of this standard. The telescope was used at magnifications of 150× and 200×.

When Arago began to study the disk of Mars, he first noted a white patch marking the upper or southern pole, and then a dark patch in the form of a hook (Fig. 62A). The gap *b* appeared smaller than 1/3 the disk of the planet: 16 July 1813.

[38] In 1804 Napoleon, in camp at Boulougne, called Delambre and asked for the best telescope of the Bureau of Longitudes, so that he could examine the English coast. "Sir," was the reply, "we can give you the Dollond refractor; Your Majesty would perform an act welcomed by the astronomers if you would give us in exchange an excellent 4-in. refractor which is to be made by M. Lerebours." "Then the Lerebours is the better?" asked the Emperor. "Yes, Sir." "Well, I will take it for myself." After returning from the Boulogne camp, Napoleon gave the telescope to the Observatory.

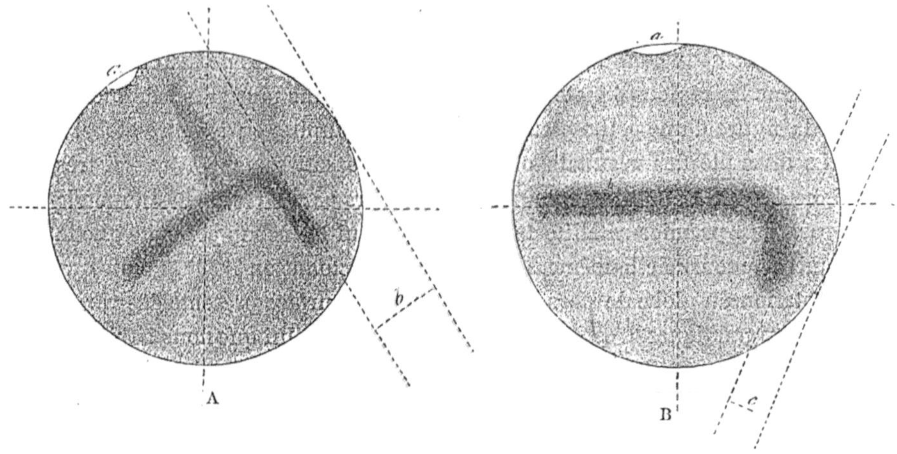

Fig. 62 Observations of Mars made by Arago, in July 1813

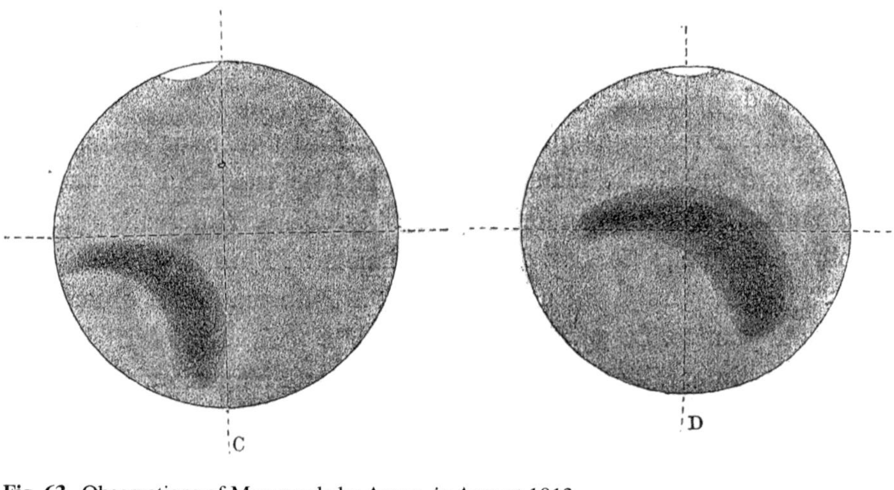

Fig. 63 Observations of Mars made by Arago, in August 1813

On 22 July, at about the same time (midnight to 1 a.m.), he again observed the planet. Figure 62B was drawn at $1^h\ 15^m$. He wrote: "I believe that the gap *c* is 1/7 of the disk; an hour ago I could not see the vertical portion of the black band."

But is not this vertical portion the Hourglass Sea? On 27 July, at $10^h\ 45^m$, Arago did not see the hooked band drawn on the previous nights, which seemed so suitable for determining the rotation period of the planet.

On 18, 19, 20, 23 and 24 August, Arago saw a dark marking in the form of a crescent.

The white polar patch was always very luminous. These five drawings are all very similar. Two of them (those of 20 and 23 August) are reproduced in Fig. 63.

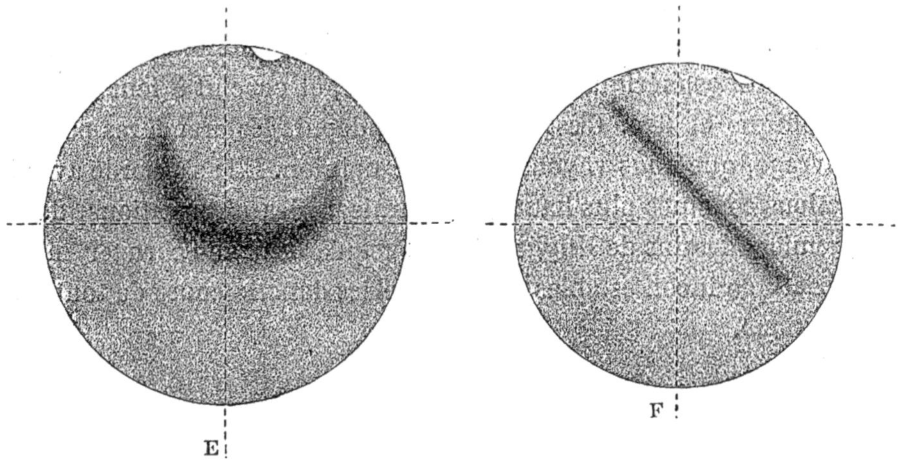

Fig. 64 Observations of Mars made by Arago, in October 1815

On 11 October the brilliant white polar patch was still very distinct, though Flaugergues indicated the contrary. On 19 October and 5 November, to 30 December, the patch was very small, and almost imperceptible.

In 1815, from 2 October to 6 November, the polar patch was very small. On 20 October, the planet appeared as drawn in Fig. 64E, and on 26 October as in Fig. 63D. In the first drawing, the crescent is facing in a direction opposite to that shown in the drawings made in 1813. In the second there can be seen a straight band which does not touch either limb of the planet.

In 1817, 1845 and 1847 Arago again made some observations but without adding much to our knowledge.

According to his observations, the polar patch had a diameter of 3″.66 on 7 July, 3″.60 on 12 July (planet diameter then 22″.86). 2″.25 on 22 July, (planet 24″.16) and 2″.63 on 2 August (planet 24″.96). As we have seen, the cap can often have a diameter more than 1/10 that of the planet. The other markings cannot be identified, and were apparently variable—but do not forget that Arago's telescope was only 4 in., or 108^{mm}, aperture.

1821–1822.—Kunowsky[39]

Kunowsky observed the planet from the autumn of 1821 to the spring of 1822, using a 4 1/3-in. Fraunhofer refractor.

On page 225 of the *Astronomisches Jahrbuch* he wrote about the patches and the rotation of the planet; he described the snowy zones (Schneezonen) and the sombre

[39] *Einige physiche Beobachtungen des Mondes, des Saturns und Mars*, etc. Astronomisches Jahrbuch für 1825. Berlin, 1822.

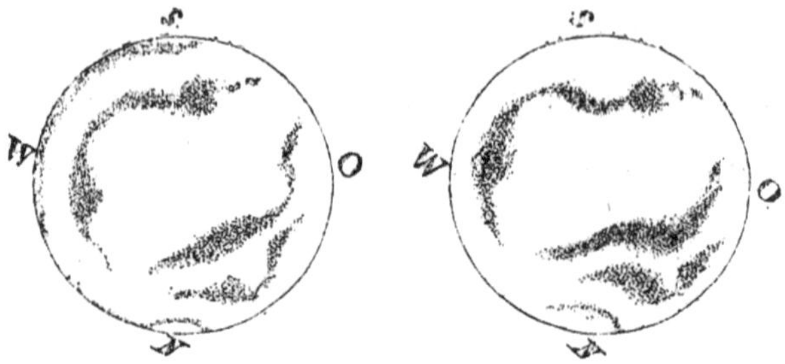

Fig. 65 Two drawings of Mars, made by Kunowsky, in 1821 and 1822

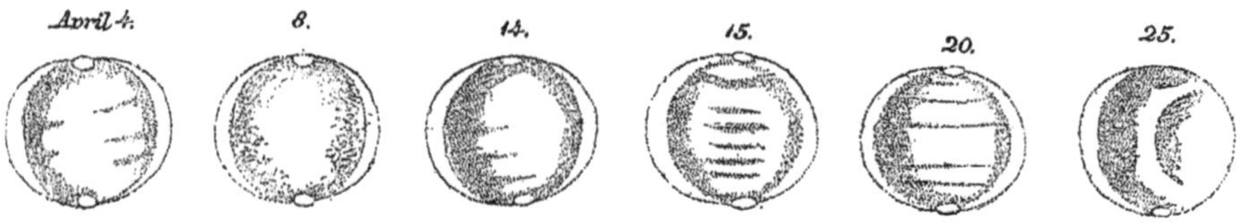

Fig. 66 Sketches of the form of Mars, made by Harding, in 1824

patches, and published two drawings (Fig. 65), of which the first was made in November 1821 and the second on 15 March 1822. He concluded that the patches were fixed. Indeed, his two drawings, made at an interval of four months, are very similar to each other. Kunowsky commented that the grey line extending along the west border of the disk was the beginning of the phase. He was the first to recognize, not far from the south pole, the round patch that Beer and Mädler chose in 1830 as the zero point for Martian longitudes (see below for their own comments upon this selection). Kunowsky challenged Schröter's views about the atmospheric and variable nature of the patches, and concluded that they were fixed surface features. This is the first case we have found of such an opinion since the early pages of the present book. It is in accord with the deductions drawn above from a comparison of the observations, from the first drawings by Huygens and Hooke, through to those of Schröter—despite the undoubted variations which have come to light during the whole series of observations.

1824.—Harding[40]

Harding—discoverer of Juno—reviewed the observations of Herschel and Schröter, and discussed the measurements of flattening. To him, the planet seemed to vary in form in the equatorial to the polar sense; this was undoubtedly an atmospheric effect. He published six drawings, made on 4, 8, 14, 15, 20 and 25 April 1824, reproduced here in Fig. 66, which are very curious.

[40] *Beobachtungen und Bermerkungen über den Mars vom Jahr 1824, vom Prof. HARDING in Göttingen (Astr. Jahrbuch für 1828*. Berlin, 1825.) – In 1824 PICTET, at Geneva Observatory, also observed Mars, but only to measure its position and its parallax. This was also the case with LALANDE in 1798. In this book I do not propose to discuss positional observations.

Here again is something new. Can it be valid? Certainly it would explain the discrepancies in the measurements of flattening. But such variations are hardly admissible. They would be much greater than the limits of possible error in the observations—particularly those made with the planet at its closest to the Earth—as in 1824.

Harding's observations add nothing to those preceding them. They conclude the first period in the history of Mars, during which much was learned about the planet, but which saw only the start of geographical studies of it.

From what has been said above, we know that Mars has years, seasons, days and nights, just as we have on our own world. That meteorological precipitation analogous to our snow occurs each winter over the poles, and that the centre of these icy regions does not coincide with the geographical pole, but is some distance from it. That there is an atmosphere containing clouds and snow and that the polar ices can melt more completely than with Earth; since the melting is made easier by the constitution of the clouds, by the nature of the atmosphere, or perhaps even because the temperature is higher than on our own planet. We know, moreover, that there are dark patches on the Martian globe; many of these patches are fixed and permanent, and should indicate seas and they are sometimes affected by extensive variations, visible from Earth. All in all, the appearance of Mars is essentially different from that of the Earth.

But the differences in the drawings are such that they must be attributed chiefly to the difficulties of making precise observations of so small a disk, to the lack of sharpness of the features, and—in brief—to the uncertainty of the observations. Nevertheless, some of the patches observed by Huygens, Cassini, Hooke, Maraldi, Herschel, Schröter etc. have given precise results with regard to the movement of rotation and the position of the axis; these observations have a sound basis. To complete the documentation of the first period, I append, after M. van de Sande Bakhuyzen, the longitude of the central meridian as derived from Herschel's best drawings (referring to our Figs. 27 and 30) The Hourglass Sea can be recognized with certainty on Herschel's *fig.* 17 in 1777; 20, 21 and 22 in 1779; 6, 7, 8, 11 and 17 in 1781. Herschel II. Strait and Arago Strait are recognizable on this last drawing.

Date	*Fig.*	Deg.	Date	*Fig.*	Deg.	Date	*Fig.*	Deg.	Date	*Fig.*	Deg.
1777	14	37	1779	20	310	1781	8	317	1783	18	230
"	15	66	"	21	282	"	11	319	"	19	167
"	16	74	"	22	303	1783	14	1	"	20	211
"	17	324				"	15	21	"	21	118
"	18	262	1781	6	292	"	16	8	"	22	55
"	19	247	"	7	300	"	17	308			

We must assume that the surface of the planet is subject to only minor changes; otherwise there would be no stable features. Whereas the observations prove that the general topography is in fact stable. Some of the sombre patches on Mars may therefore be atmospheric in nature.

We now enter a period of new discoveries.

Conclusions from the First Period

From the discussion of the numerous preceding documents, varied and often contradictory, we can already begin to form an opinion about the nature of the Martian world, defining the first elements of the knowledge of which our present work is the end product.

We can consider the following facts as established:

1. The revolution of Mars has been approximately known since early times. Since the days of Copernicus, we have known that it moves round the Sun. Today we know that the period is 687 days, or in terrestrial terms one year 322 days. The Martian year is nearly twice as long as ours.
2. The distance of Mars from the Sun, compared with that of the Earth, is in the ratio of 1.5237 to 1. The light, heat and radiation which Mars receives from the central star is thus less than the amount which we receive in the ratio of the squares of these two numbers, that is to say 2.32 to 1; more than twice less intense. But it is useful to remember that the temperature is regulated by the constitution of the atmosphere. The temperature on the surface of Mars should be equal to, or perhaps above, that of our world.
3. The diameter of Mars at unit distance—that is to say, the distance of the Earth from the Sun—is 9″.35, corresponding to 0.528 or a little more than half of that of our globe. This diameter gives a volume of 0.147.
4. The mass of Mars, as a function of that of the Sun, is given in the *Annuaire du Bureau des longitudes* for 1830 as 1/2,546,320. This is the mass obtained by Delambre by the perturbations of the Earth, and adopted by Laplace in his *Mécanique céleste* (1802). Today the mass is known with greater precision, from the movements of the Martian satellites, and we know it to be 1/3,093,500. Relative to the Earth, this works out at 0.105, or about 1/10.
5. The density, obtained by dividing the mass by the volume, is 0.711.
6. The surface gravity, computed from the mass and the radius of the planet, is 0.376.
7. By 1830 the rotation period was already known with great precision: it is $24^{h}\ 39^{m}$.
8. On Mars there are patches of different degrees of darkness. In drawing them, the observers made them more well marked than they actually are, so that their drawings cannot be taken literally. However, the variations observed are so great that we must conclude that the patches are certainly variable. We have already examined 191 views of Mars, drawn by very different observers; these views should contribute the first basis of our knowledge of the Martian world.
9. Also on Mars there are white patches, markedly at the poles. These patches vary with the seasons, being largest in winter and smallest in summer. They, like our polar ices, are affected by the Sun; we must regard them as being ice or snow.
10. Their polar caps are not situated exactly at opposite ends of a diameter, and do not exactly mark the geographical poles. The poles are generally covered; but at minimum size the cap is reduced to a white, sensibly circular patch, which is a definite distance away from the pole. In 1781 Herschel found the centre of the

polar cap to be 13°–14° from the north pole, though the difference became very small after the spring—when the southern ice was very extended, and its centre was in the neighbourhood of the pole—while in 1783 the south polar cap, also very small during summer, was 8°.8′ away from the south pole.[41] One degree of the meridian on Mars is equal to 60 km. We know that on Earth the pole of cold does not coincide with the geographical pole.

11. The inclination of the axis of Mars is not very different from that of the Earth, so that the seasons are analogous to ours, though almost twice as long.
12. There is on Mars a second order of seasons, due to the great eccentricity of the orbit. Mars is much nearer the Sun at perihelion, in the ratio of 1.3826 to 1.6658, or 10 to 12.
13. The planet is surrounded by an atmosphere in which snow can form, and in which float white clouds and also, probably, dark clouds.

Such is the natural conclusion, logically founded, which we have drawn after critically examining all the observations made during the first period of 193 years. Does Mars have a fixed surface topography, as does the globe on which we live? This seems impossible to decide by comparing the observations given above. Perhaps the progress of optics and of astronomy will allow us, in the period, which we now enter, to resolve this important question. We can, in fact, enter what we may call the *geographical* phase in the study of Mars.

[41] Here are William Herschel's observations on this important point. In 1781, the north polar snow was a long way from the pole, and was at latitude 76° to 77° (*Phil. Trans.* 1784, p. 245). In 1783 the latitude of the *south* polar cap was 81° 52′ (*Phil. Trans.*, p. 255). At this time (October) the cap was very small and round. In 1781 the centre of the south polar cap was not many degrees from the pole, and extended down to 70° or 65° (*Phil. Trans.*, p. 246), and was very extended after twelve months because of the inclination of the planet.

1781: South polar cap very large (after its winter)
North polar cap very small (after its summer)
1783: South polar cap very small (after its summer)
North polar cap invisible because of its inclination

Distances from the pole:
South cap: 1781, in the neighbourhood of the pole
1783, at 8°8′
North cap: 1781, at 13° or 14°
1783, invisible.

Second Period, 1830–1877

The second period of our study begins with the great areographic work of Beer and Mädler, the first continuous observations that allowed the authors to draw up a geographical chart of Mars. With these observations, we also begin to gain some physical knowledge of the planet. The difficulties and uncertainties do not disappear; but science provides a solid basis for what follows.

Christopher Columbus was happy when he was halted by the American continent during his voyage of circumnavigation beyond Asia. Mars does not have its Christopher Columbus. He achieved fame by the simple fact of touching America; a phalanx of astronomers has been busy for more than a century studying their celestial continent. But Beer and Mädler deserve to be remembered as the true pioneers in this new conquest—though preceded by the eminent observers already discussed, among whom William Herschel and Schröter hold pride of place. Beer and Mädler published their observations of Mars in the *Astronomische Nachrichten* in 1831, 1834, 1835, 1838 and 1839, and combined their studies in a work entitled *Fragments sur les corps célestes du systéme solaire*, French edition (Paris, 1840) and *Beitrage zur physichen Kenntniss der himmlischen Körper im Sonnensysteme*, German edition (Weimar, 1841). These two editions are identical. From the French edition I have taken all the important data and drawings.

The instrument they used for their studies of Mars, as well as their great map of the Moon, Beer and Mädler's refractor was a 3-3/4-in (95^{mm}) refractor, analogous to that mentioned above when describing Arago's observations. This is a relatively modest instrument, but it was made by Fraunhofer and was therefore excellent; moreover the observers were particularly skilful, meticulous, and patient. We often say that the man is as important as his telescope.

C. Flammarion, *Camille Flammarion's The Planet Mars*, Astrophysics and Space Science Library 409, DOI 10.1007/978-3-319-09641-4_2

1830–1841.—Beer and Mädler[1]

In 1830 Mars made one of its closest approaches to the Earth. The opposition was perihelic. This was the main reason why the observers began their series of observations, which we will now examine. Here is a résumé of their great work:

> Our chief aim (wrote the authors) has been to make an exact determination of the period of rotation, about which opinions differ appreciably. From his observations of 1778 to 1780, Herschel deduced a period of $24^h\ 39^m\ 21^s$. Huth of Mannheim gave $24^h\ 43^m$, and the observations made by Kunowsky in the winter of 1821–1822 gave $24^h\ 36^m\ 40^s$, though none of them achieved an exact determination of the period. In Herschel's observations, the number of complete rotations was doubtful, and in particular he did not take aberration and phase into account, while the data of the two others depended upon the results of one opposition only. From one opposition it is important to deduce the rotation period accurately enough to allow the number of complete rotations between one opposition and the next to be worked out with certainty. The mean error of the first result should not exceed 30 to 40 seconds, an accuracy which we hoped to exceed during the close approach of Mars to the Earth.
>
> At the same time, prolonged observations ought to show whether or not the patches on the surface of Mars are variable in form, size, and colour; if they have individual movement, and if they should be regarded as condensations or shadings similar to our cloud or as features fixed to the surface. The preceding observations have already provided the important data listed above. As early as 1716 Maraldi, in Paris, made out the white patch at the north border of Mars; this has been confirmed by almost all subsequent observations. A patch is also seen at the south border of the planet, and sometimes it even happens that these two patches are visible at the same time.
>
> Even before Herschel, it had been suggested that these patches are due to snows, similar to those at the poles of our planet. Several people have considered that the patches form on slight elevations which project from the mean border of the planet, while others have very reasonably regarded the great brightness of the patches as being due to this cause. Most observers also regard the other patches as variable; Kunowsky, at Berlin, maintains however that they are permanent. Several observers have called attention to particular brilliancy at the east and west limbs of the planet, and this suggests the idea of narrow menisci surrounding the globe, particularly in these areas. The contradictions shown by observations made with different instruments, or—if you like—the physical changes which take place over periods of time, are consequently very considerable.
>
> From 10 September to 20 October 1830, we made observations on 17 nights of varying degrees of good seeing; during this period all the areas of Mars visible at this opposition were studied several times. We obtained 35 disk drawings. We did not use the micrometer, because the weakness of the patches made accurate measurements impossible; and an appreciation after the different parts of the disk had been studied made us certain that the great white patch at the south pole, which has been excellently seen ever since the start of our observations, could be used to show the meridian dividing the disk in half. A definite time elapsed before the main patches, at first vague and indeterminate, presented perfectly distinct forms. The drawings were made immediately, at the telescope. The co-ordinates of the most distinct points were determined and represented on the drawing; the rest of the detail was filled in later.

The most characteristic patch which struck the observers was a small round one, which appeared to be hanging from an undulating ribbon, and which is shown on drawings 1, 2, 3, 14, 15, and 16 of 1830 and 4 of 1832 (see Fig. 1). This patch is the

---

[1] *Fragments sur les corps célestes du systeme solaire* (Paris, 1840), *Beitrage*, etc. (Weimar, 1841) and *Astr. Nach.*, 1831 to 1842.

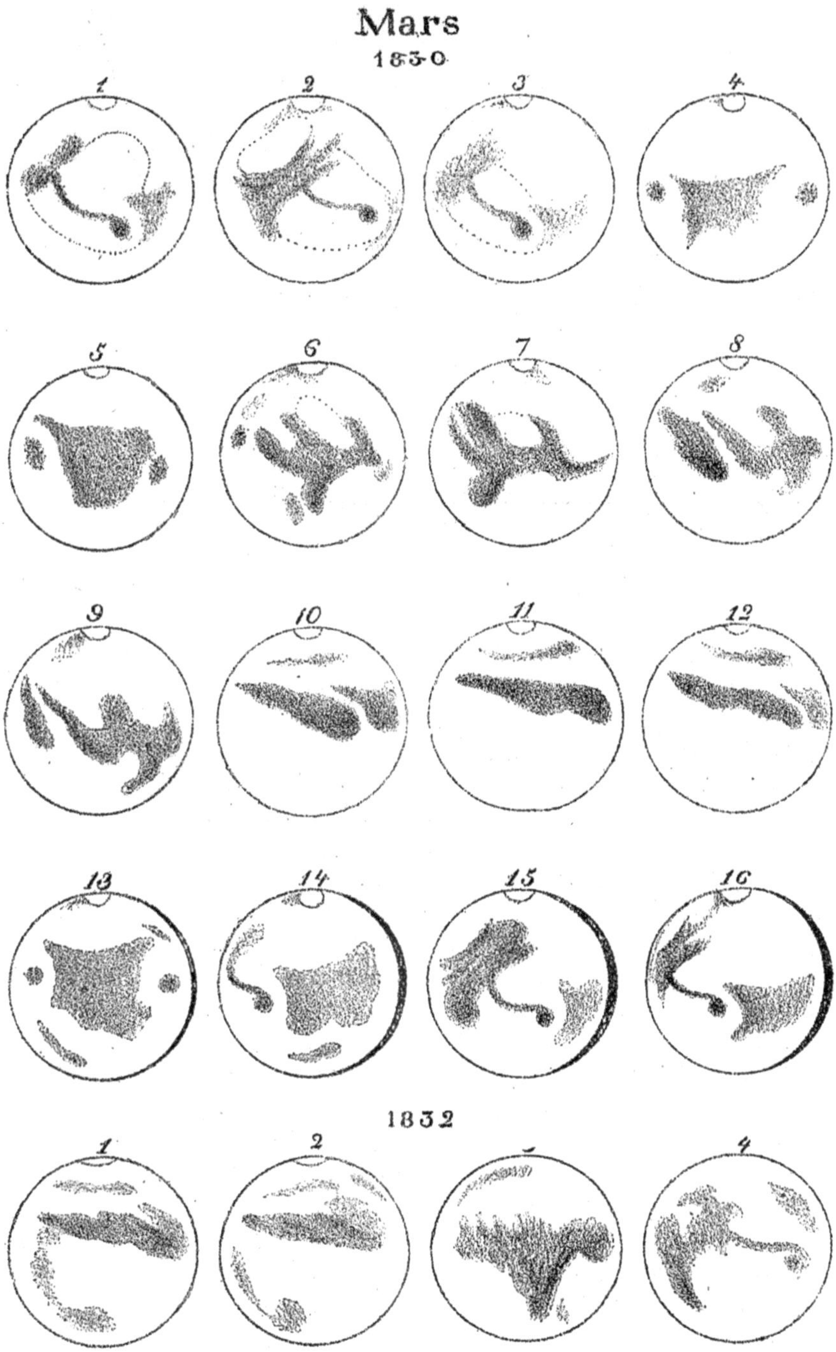

Fig. 1 Drawings of Mars by Beer and Mädler in 1830 and 1832

Meridian Bay of my own chart, near Herschel II Strait. But let us go back to the observers themselves.

[Note: the *fig*. numbers below refer to the small drawings in our Fig. 1.—WS]

A small patch of a very pronounced black was so strong—from the very first observations—and so well marked, and so near to the assumed equator, that we believe we should choose it as the reference point for the determination of the rotation period.[2] At $9^h\ 30^m$ (*fig*. 1) it appeared to be 7° of arc from the central meridian of Mars. On the 14th, we saw it from 10^h (*fig*. 2) to $15^h\ 15^m$ (*fig*. 4), advancing from the eastern hemisphere to the region of the western edge; five drawings of it were made. On the 15th, at $8^h\ 50^m$ (*fig*. 5). It was no longer visible; neither was it seen at $13^h\ 15^m$. On the 16th, at 9^h, it was not visible; but at midnight it had returned, and was very distinct. We were then able to deduce the rotation period: it was clear that from the 19th and the following evenings, until the middle of October, the patch would not be observable during suitable hours at night. On the 19th (*figs*. 6 and 7), with a perfectly distinct image, it showed up as two reddish areas (bounded by these points in the general drawing), similar to the beautiful red colour of twilight on Earth. After an hour it had already become weaker; later it was again very clear, but never regained its former distinct red colour. Moreover, at $10^h\ 6^m$ it showed up as a small patch *g*, not really dark, by the side of the point *f*, but slightly later it ceased to be visible. Probably it had been seen only because of the great steadiness of the atmosphere; and when it reappeared it was always joined to *f*, since the gap separating them was always very difficult to distinguish.

In the observations from 26 September to 5 October (*figs*. 10 to 12), several reasonably dark patches were seen extended over the disk in the form of a zone; they were sharply bounded, particularly to the north, where they made a very pronounced contrast with the areas which were free of patches and showed up as purely bright. A salient from these patches at the point m was broad and distinct, particularly on the north side; to the south side; on the other hand, it was so narrow that we could scarcely see it. The patch *PM* was very black, particularly at the western extremity *p*, which was rounded. Between this patch and the white cap at the south pole we consistently saw a band *q*, quite broad, but of a pale tint. From 5 to 12 October clouds prevented us from making any observations. It was only on the 13th that we again saw a small dark patch near the west border (*fig*. 13); and in our observation of the 14th, at $7^h\ 37^m$ (*fig*. 14) we were confident that it was in fact the patch *a* of our first observation. During the following evenings it was particularly important to time its transit across the central meridian as accurately as possible; this we were able to do on the 19th and 20th, under remarkably steady atmospheric conditions (*figs*. 15 and 16).

These observations truly represent the first methodical attempt at studying Martian geography. The chart which Beer and Mädler constructed from these invaluable observations is given in Fig. 2. It reproduces the two hemispheres as they drew them, and presents an overall picture of the planet as obtained from their previous observations between 1830 and 1839. (This is the same diagram as is shown in the memoir which Beer and Mädler wrote.) It is, in fact, the first geographical chart ever drawn of the Martian world. It remained the only one for 30 years, and thus became the classical reference for all later observers.

The northern hemisphere chart seems to contain an error. The extremity of the patch *ehf* (the patch which is none other than the Hourglass Sea), which emerges from the southern hemisphere between longitude 62° and 73°, is traced in the

[2]This very distinctive spot was first seen by Schröter on 3 September 1798 (see his drawings in Fig. 52 above). It was also observed on the following day, 4 September (Fig. 53) and on 24 October. It was also shown on two drawings by Kunowsky in 1821–1822 (Fig. 65)—but what differences of aspect!

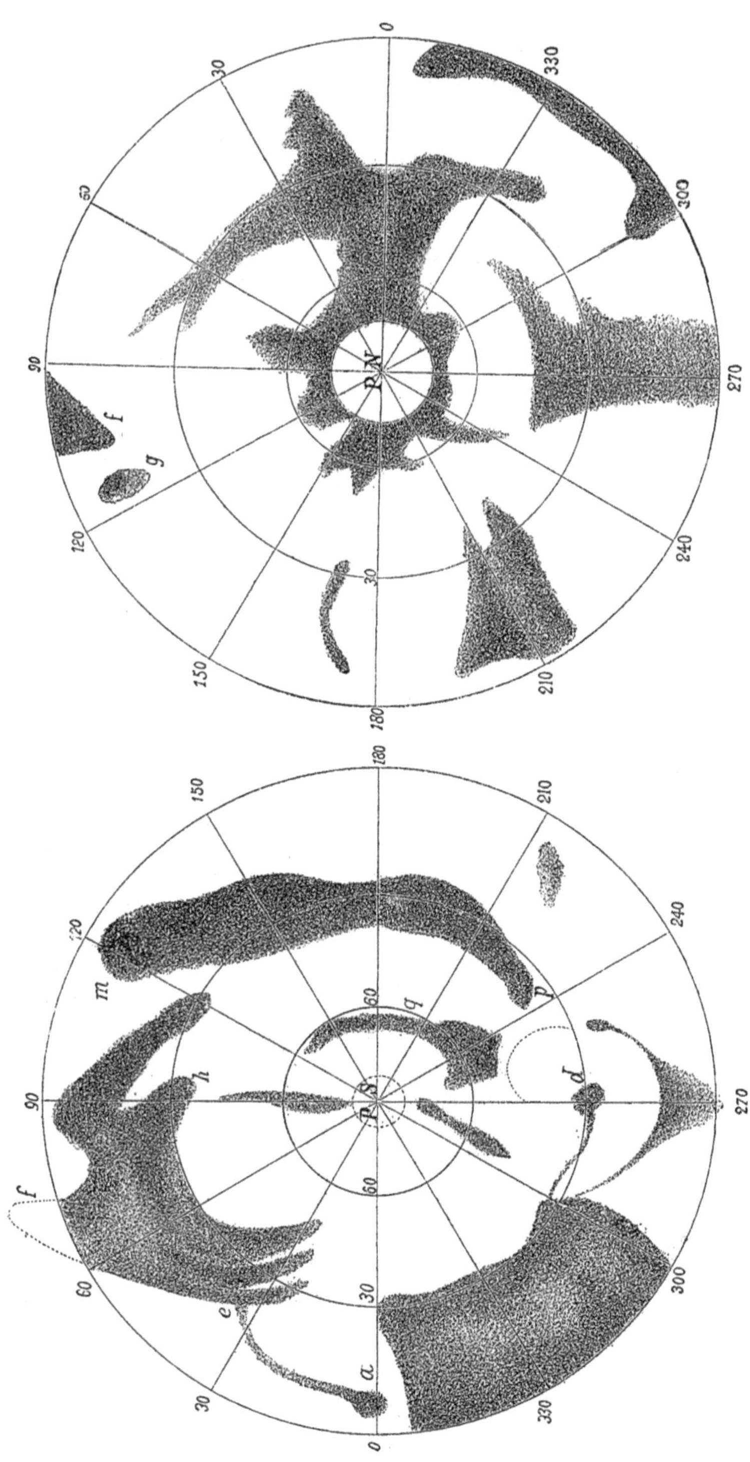

Fig. 2 General chart of the planet Mars, constructed by Beer and Mädler in 1840

northern hemisphere between 92° and 110°. Some mistake must have been made. It is necessary to trace the continuation between 62° and 73° to line up with the feature shown coming from the southern hemisphere.[3]

The Berlin astronomers chose the small round spot *a* as the zero meridian. I have followed this choice in drawing my chart, which is why I have proposed to name it the Meridian Bay.

Beer and Mädler reckoned longitude from right to left, assuming the south pole to be at the top. I reckon it in the contrary sense, i.e. from left to right, in the sense of the movement of rotation, meridian 0° passing in transit before meridian 10°.

These authors subsequently calculated the rotation period as derived from their observations. From the observations of 1830 they found $24^h\ 37^m\ 9^s.9$; from 1830 to 1832, $24^h\ 37^m\ 23^s.7$; 1830 to 1835, $24^h\ 37^m\ 20^s.4$. The second value seemed to them to be the best, and accordingly they adopted it. To resume their own account:

> It is impossible to make an accurate comparison with the observations made nine years earlier by Kunowsky, in which the same patch was very distinct. And yet Kunowsky's work evidently confirms the consistency of the patches which we have seen, at least for *a* and also for the strongly curved arc which extends in a serpentine fashion from *a* to *e*. The patches further south were not seen when unfavourably placed, and some were not seen at all; also, Kunowsky showed others to the north which were no longer visible in 1830, while the usual patch was seen between November 1821 and March 1822 just as we observed it. The same was true between 10 September and 20 October 1830. Therefore, there is no chance of it being due to clouds.
>
> For the rest—particularly if one observes the planet for the first time, and does not repeat the observations often—it is easy to imagine variations in the patches and attribute them to real physical changes in the atmospheric state of the Earth and perhaps also in that of Mars. This is why error of detection and drawing—small in themselves but considerable in relation to the object being studied—are inevitable. A patch approaching the limb may disappear before the moment when it actually reaches the limb (doubtless due to the effects of the planets atmosphere, as with Jupiter). Also, remember that during any opposition it is rare to have two absolutely identical views of the same hemisphere. Moreover, the opposition distance varies; the planet does not often approach the Earth as closely as it did in 1830, and with greater distances it is necessary to use higher magnifications and more powerful optics than most of the earlier observers had at their disposal.
>
> The white south polar patch was clearly shown in each observation, even when the atmospheric conditions were unfavourable, but its size has been very variable. Before 31 August, when we made a superficial observation, it extended to 1/8 to 1/10 of the diameter of Mars. On 10 September, an estimate (made in the east–west direction) gave 1/10; on 15 September 1/16; on 2 October 1/18; on 5 October 1/20; and on 20 October 1/15. Assuming a value of 1/9 for 31 August, we have, for the days indicated, in the season corresponding to the months of June and July in our northern hemisphere, the following boundaries for the polar cap, assuming that the south pole lies in the centre of the cap:

31 August	83°37	latitude corresponding to Earth	16 June
10 September	84°15	latitude corresponding to Earth	23 June
15 September	86°25	latitude corresponding to Earth	26 June
2 October	86°50	latitude corresponding to Earth	7 July
5 October	87°7	latitude corresponding to Earth	9 July
20 October	85°59	latitude corresponding to Earth	19 July

[3] All astronomical treatises written since 1840—even Aragos excellent *Astronomie populaire*—have reproduced this chart without noticing the error.

This shows that the cap shrank until the Martian season corresponding to our mid-July, and then it began to increase again, which lends strong support to the hypothesis that the Martian pole is truly snow-covered. Moreover, almost all observers agree that the size of the patch is variable, and at times of greatest cold is considerably larger than we saw it in 1830.

The northern hemisphere of the planet, or rather as much of it as could be seen in 1830, showed no sign of any white patch, although it was midwinter in that hemisphere. This can be explained by the strong inclination of the axis of Mars, which, at the same time, receives indirect confirmation.

The observers here give a table of their drawings, and the aerographic longitudes of the patches.

At the opposition of 1832, atmospheric conditions were so consistently unfavourable, and the greater distance of the planet had so mischievous an influence, that the observations obtained were few in number and imperfect in quality. Of 16 attempts to draw details on the disk, only four merited comparison with those of 1830 (they are shown at the bottom of Fig. 1). The patch *a*, so remarkable and so characteristic two years earlier, could be recognized at once, even when it was well away from the centre (16 December).

However, these observations, even though not numerous, seem to be sufficient to show that the visible patches have not changed their positions since 1830. This was quite evident from the three principal patches, in particular the region *PM* and the weak band *q*. The latter was so close to the southern border that it was seen only with the greatest difficulty, and the features, which were even closer to the poles, included in the drawings of 1830, could not be seen at all—for reasons which are easy to understand. The south pole was not on the limb on 20 November, according to Herschel's elements, but 10° from the apparent border; and thus most of the bright part, so glittering in 1830 [the text says 1840, but this is surely a misprint for 1830.—PM] was only very feeble; it was seen with certainty only twice (20 November, 9 h, and 23 November, 8^{h} 14^{m}). During all other evenings it was uncertain or invisible. On the northern hemisphere, between approximate longitude 180° to 230° and latitude 0° to 35°N., we twice saw a weak band, broad and concave toward *PM*, but only its northern extremity was distinct. Red glimmers were often seen between this band and *PM*. In general the light of the northern hemisphere, in the part containing no patches, did not seem so clear and uniform as it had done in the two preceding years. No trace could be seen of the whiteness in the region of the north pole (this pole was still hidden from view).

The oppositions of 1834–1835 and 1837 were just as unfavourable as the two preceding ones from the point of view of atmospheric conditions; moreover, the opposition distance of Mars was approaching its maximum (Fig. 3).

The results of our observations would have been insignificant had we not had the use of the large telescope established in 1835 at the Royal Observatory in Berlin.

This instrument, in size exactly the same as that at Dorpat, allowed the use of a magnification at least double that of our telescope, and it collected six times as much light; a very convenient mechanism gave it a movement which, without any help from the observer, followed the planet in the sky. From 12 January until 22 March, over 15 partly clear nights, we obtained 32 drawings which related particularly to the northern hemisphere, though much less detail was shown than had been recorded in the southern hemisphere in 1830. In all the observations, without exception, the white patch at the north pole was visible with a degree of clarity which we never recalled having seen at the south pole; it was also considerably larger than the cap in 1830, and, particularly during the months of January and February appeared so much more distinct than the other parts of the globe that at first glance it was easy to suppose that in this place the planet was covered by another planet.

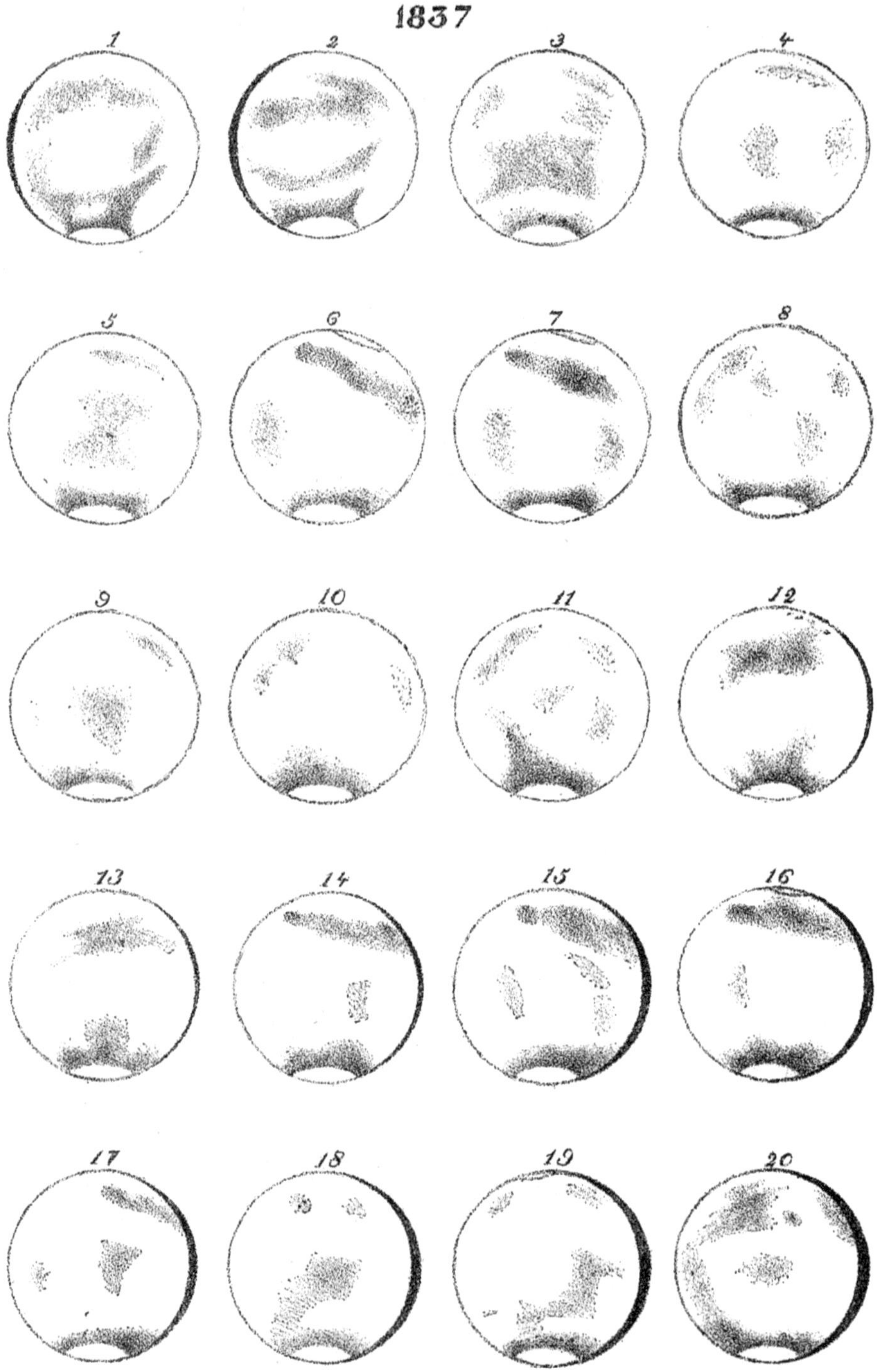

Fig. 3 Drawings of Mars, made by Beer and Mädler in 1837

The real size of the south polar patch, in February and March 1837, was several times greater than it had been in September and October 1830.

In the first observation, on 12 January, the north polar patch was so well marked that its size could be estimated with certainty; it covered 0.27 of the planets diameter along the border, and its breadth was 0.13. The first figure made us conclude that the semi-diameter was 15°.7 of the globe of Mars, or a limiting latitude of 74°.3; the second—adopting the elements of rotation given by Herschel, and assuming the pole to lie in the middle of the circular patch—gave a north latitude of 78°.7, since the north pole was 18°.13 inside. The first of these figures is less than half as reliable as the second. In any case, it is evident that at the opposition of 1837 the north polar patch was considerably larger than the south polar patch had been in 1830, and much smaller than the south polar patch in 1837. In the following observations its size does not seem to tend to lessen; this can be stated with certainty, and it is only the sharpness of its boundary in which there is any change; it becomes feeble after opposition.

With the micrometer, we set out to measure the position angle of the white patch, so as to obtain the data necessary for a direct examination of the position of the axis of Mars. The bad weather partly prevented this from being done. The few measurements, which we were able to make, showed only that in every case the eccentricity of the polar patch was very slight. The distance from the pole was estimated as 4° in 1837 for the northern patch and 8° for the southern, but with great uncertainty.

However, we cannot fail to mention that on the few occasions when we were able to distinguish a trace of the south polar patch, we found that is was not directly opposite to the north polar patch. On 7 February, at 16^h 14^m, it was displaced by about 12° from the point opposite to the north patch, and a 18^h 16^m it was some 8° to the east; on 7 March, at 10^h 34^m, it was displaced by 5° toward the east; finally, on 18 March at 7^h 56^m, it was 3° to 5° to the west.

Of all the patches in the southern hemisphere observed with considerable precision in 1830, only one, marked *PM*, could be recognized with certainty. We first saw it on 7 February, at 16^h 4^m (*fig*. 6) with certainty; subsequently, on 28 February, at 6^h 49^m (*fig*. 7), and on three observations made during the night of 7 March (*figs*. 14, 15, and 16); finally, it was slightly less well marked on 10 March, from 7^h 7^m to 9^h 22^m, and on 11 March, at 8^h 22^m (*fig*. 17). The aerographic latitude of the western extremity of *p* was determined, from eleven observations, as +43° 29; in 1830 we had found it to be between 39° and 42° from three observations, and this close agreement indicates that the two patches are identical. An attempt to link the longitude observed this year with that of 1830 gave a rotation period of 24^h 37^m $29^s.0$; This result, although sufficient to confirm the identity of the patches, cannot be used to correct the rotation period previously calculated, because of the strongly eccentric position of the patch. However, it is definite that there is no error in the number of complete rotations.

A second patch, marked *efh* on our chart, was recognized on 12 January and 22 February, and also on 12 March; but no point on it was sufficiently well marked for us to draw it with real accuracy.

This comment is curious, and worthy of close attention, because this patch *efh* is the Hourglass Sea, which is generally so clear and so distinctive. For instance, during the last opposition (1890) it was striking each time it lay on the hemisphere facing us.

As for the opposition of 1839, all the observations were made with the large telescope of the Royal Observatory. Mars requires a high magnification and, in consequence, great atmospheric calmness. This last condition was rarely met with in the winter of 1838–1839, which is why the observations were not numerous; 62 percent of the southern hemisphere was hidden from view, so that a great part of it was unobservable, and no patches in this hemisphere could be distinguished with precision.

The ten drawings published by Beer and Mädler for this year, 1839, are so pale and indefinite that to reproduce them here would be absolutely useless.

Here are the general conclusions, which they draw from their observations of the poles and of the seasons.

> The colour of the polar patches, whenever they could be seen distinctly, was always a pure, brilliant white, in no way similar to the colour of the other parts of the planet. In 1837 there was a time, during the observing period, when Mars was completely concealed by cloud—apart from the polar patch—which remained distinctly in view. This great difference also means that the extent and figure of the polar patch can be evaluated with much more certainty than for any other part of the planet, and it may even be possible to make successful micrometrical measurements of it with powerful instruments.
>
> We must also note the diminution and the growth of these patches, which however keep to the same form; this indicates that normally the polar patches are central over the true poles, or at least only a few degrees away. We have already indicated the variations of the southern cap; the seasons of Mars are repeated regularly, and we have expressed the Martian seasons in terms of the seasons of the Earth. The north polar cap showed the following variations:

1837	Jan. 12	Limit	74°18′	season corresponding to	4 May
"	Mar. 7	"	76°	"	4 June
1839	Feb. 26	"	78°33′	"	7 June
"	Apr. 1	"	80°48′	"	4 July
"	Apr. 16	"	82°20′	"	12 July
"	May 1	"	81°	"	20 July

> According to this, the minimum for the two patches falls about 1/18 of a year after the summer solstice, which corresponds to our 12 July (and 12 January) on the Earth. But while the southern cap shrank to 6° diameter, that at the north pole never decreased below 12° to 14°, so that the surface area of the north cap was then five times as great as that of its southern counterpart.
>
> Reciprocally, in 1837, during its winter (the days of observation corresponding to our 4 and 10 December) the south polar patch was so large that we could distinguish it even when the pole was 18° away from the limb. At that time the cap extended down to latitude 55° and had a diameter of 70°.
>
> We have never seen the same thing at the north border, when the southern hemisphere is having its summer. The variations of the southern cap are considerably greater than those of the north cap.
>
> Because of the position of the axis of Mars, the south pole is exposed more directly to the Sun during the period when the amount of light and heat which it receives is 0.52 of that which we receive on Earth; with the north pole, the corresponding figure is only 0.37. But over a full year this difference is completely compensated for by the greater length of the winter; but even for the other seasons we find partial compensation. The length of the summer half-year in the northern hemisphere is greater than that in the south in the ratio of 19 to 15; however at the points of greatest heat and greatest cold a very considerable difference still remains, because of the inclination of the axis. Accordingly, the south pole has hotter summers and colder winters than the north pole, and the difference is more marked than in the case of the Earth. With our world, the difference is scarcely appreciable—but the eccentricity of Mars orbit is five times greater than that of the Earth.
>
> The differences, which we have noted, are in perfect accord with the idea that the white patches represent a precipitate analogous to our snow; it is indeed impossible to reject this explanation, which is confirmed in so convincing a manner. Seen from the distance of another planet, our Earth would show similar phenomena; with us, however, the two hemispheres are less unequal.

The other patches on the planet appear to be surface features. At the distance of Mars, we could under no circumstances make out the shadows produced by mountains, however great their heights (the spherical form of the disk is always preserved). These shadings must therefore be due to differences in reflecting power, which again recall the differences between the reflectivity of different places on the Earth. It was during the opposition of 1830 that we found it possible to delineate the patches in the southern hemisphere—those between the equator and latitude 45°—with the greatest precision; however, the darkness and relative clarity of the patches does not remain constant, as was shown, at least, in 1837 and 1839. Thus although the patches themselves are not analogous to our clouds, they sometimes present certain optical analogies to our own cloudy condensations—because they are better defined, sharper, and more intense during Martian summer, vaguer, paler and more ill-defined during the winter.

We have sometimes seen reddish colouration in certain particular regions on the disk. To the naked eye Mars appears as the reddest star in the sky. With the telescope, this ruddiness does not show up to the same degree; rather the general colour is more or less yellowish-red. The colour in these regions recalls that of a beautiful sunset on our Earth.

If all this has already led us on with certainty toward admitting that Mars has a very appreciable atmosphere, similar to that of the Earth, then it also explains the comment we have made that the patches seem to soften or completely vanish when approaching the limb. As we have often seen, the limb brightness appears to be due to the presence of the Martian atmosphere.

There is no objection to this idea from the contention that the atmosphere of Mars would produce refraction when the planet occults a star or other celestial body. At the times when Mars is closest to us, an arc of longitude of 20° on its globe would appear as an angle of 0″.30. At such a distance the refraction would be completely undetectable, even if at the surface it were considerably greater than on the Earth.

The observations we have made indicate great variations in size, form, and intensity of the sombre patch adjoining the north polar zone, and there is probably a plausible explanation for this. If the polar patches are genuinely made up of snow, their shrinking with the onset of summer would be by continual melting and evaporation, if, as we may assume, the thickness of the polar snow is very considerable; but then those parts of the surface close to the evaporating snow will become extremely humid. Now, a wet, marshy soil is certainly liable to have low reflecting power, and will consequently appear to us as darker than the rest of the planet. The greatest darkness should occur at the time when the melting is most rapid, that is to say, in high latitudes between the equinox and the summer solstice. This explains why the dark patch which surrounds the north pole, which had not been seen in the summer, was so extended and prominent in 1837, whereas in 1839 it was very pale and at first very small.

It is not going too far to claim that Mars bears a very strong resemblance to the Earth, even with regard to physical conditions, and it appears as an image of the Earth in the firmament seen from a great distance. The most important differences between Mars and the Earth are the smaller volume of Mars and the strong eccentricity of its orbit. However, the length of the Martian day is practically the same as ours.

The orbital eccentricity causes inequalities in the lengths of the seasons, as follows; if we adopt Herschel's position for the axis and our own period of rotation:

A Martian year includes		669⅔ rotations
consequently,		668⅔ Martian solar days
The north hemisphere spring contains		191⅓ days on Mars
Summer		181 days on Mars
	total	372⅓
Autumn		149⅓ days on Mars
Winter		147 days on Mars
	total	293⅓

so that in the northern hemisphere spring and summer combined are 76 days longer than in the southern hemisphere. The two parts of the year separated by the equinoxes are in the ratio of 19 to 15.

Beer and Mädler ended their memoir with an examination of the length of the rotation of the planet, comparing their results with those obtained by Sir William Herschel. We have already seen that they derived $24^h\ 37^m\ 23^s.7$ for their most reliable estimate. They wrote:

The rotation period we have found (they remark) differs by two minutes from Herschel's period, and it is Herschel's value which has been accepted up to the present time; his period was also based on observations carried out over two oppositions, and therefore so great a difference is astonishing. However, the difference vanishes almost entirely if we admit that in one of the two years there is an error of one complete rotation—if we add one rotation to Herschel's estimate or subtract one rotation from ours. Since a period of $23^h\ 39^m\ 22^s$ is irreconcilable with our observations as compared with each other, and we do not consider that we can have made such errors, it will perhaps be of interest to reconsider Herschel's observations, and see what results can be drawn from them when they are reduced with greater accuracy.

In 1777, from 8 to 26 April, Herschel observed different patches, which however did not present any certain identifications—which is why he decided to re-observe during the following opposition. This fell on 12 May 1779, when the apparent diameter of Mars attained $13''.5$; this shrank, and by 19 June was reduced to $11''$. On 11 May, at $11^h\ 43^m$, he saw at the centre a patch that he had previously seen on 9 May at $11^h\ 43^m$, though on that occasion it had been slightly above the centre. The same patch was seen on 19 June, though Mars was then low. Here are his observations:

"June 19, $11^h\ 30^m$. The figure of 11 May has not come to the position; it was then at $11^h\ 43^m$, but cannot be far from it. I fear that as Mars approaches the horizon, I shall not be able to follow it until the marking comes to the centre.

"$11^h\ 47^m$. The state of the air near the horizon is very unfavourable. With much difficulty, I can just see that the marking is not quite so far advanced as it was on 11 May at $11^h\ 44^m$, but can certainly not be more than two or three minutes from it."

In three minutes a marking on Mars moves toward the centre by a distance equal to 1/153 of the diameter of Mars; this amounts to only 1/14 of a second of arc, and Mars was only 9° above the horizon. However, let us accept Herschel's estimate that the marking reached the centre at $11^h\ 49^m\ 30^s$. The calculations then are as follows:

June	19	11^h	49^m	30^s	
May	11	11^h	43^m	0^s	
Interval	39^d	0^h	6^m	30^s	
Correction I			$+3^m$	36^s	(because of the change in geocentric long.)
Correction II			-16^m	14^s	(because of the phase of Mars)
Correction III			-0^m	49^s	(because of aberration)
	39^d	0^h	27^m	3^s	
38 rotations of			24^h	38^m	$36^s.4$

Herschel observed another patch on 11 May at $10^h\ 17^m\ 41^s$, and 13 May at $11^h\ 25^m\ 51^s$, after which it reappeared on 17 June at $9^h\ 12^m\ 20^s$. He then said:

"June 17, $9^h\ 12^m$ (clock 20^s slow). The dark spot is rather more advanced than it was on 11 May at $10^h\ 18^m$."

Herschel also took into account a correction of three minutes, so that the true timing was given as $9^h\ 9^m\ 20^s$. This gives the following results:

June	17	9^h	9^m	20^s
May	11	10^h	17^m	48^s
	36^d	22^h	51^m	32^s
Correction I				
Correction II				
Correction III			$+37^m$	28^s
			-15^m	0^s
			-0^m	44^s
	36^d	23^h	13^m	16^s
	36^d	24^h	38^m	42.9^s
June	17	9^h	9^m	20^s
May	13	11^h	25^m	51^s
	34^d	21^h	43^m	29^s
Correction I				
Correction II				
Correction III			$+34^m$	31^s
			-15^m	0^s
			-0^m	43^s
	34^d	22^h	2^m	17^s
	34^d	24^h	38^m	53.4^s

The mean of these three *extremely uncertain* determinations is therefore:
$24^h\ 38^m\ 44^s.2$.

But Herschel, not having allowed for the corrections needed because of the change in longitude, and paying no attention to the others, gave in 1779, as his final result:
$24^h\ 39^m\ 22^s.1$.

However, we must also consider something else. We calculate the amount of the phase at the mean of the angle made by the Earth and the Sun with the centre of Mars, but experience with Venus and Mercury had shown that the dark part is always slightly greater in extent than theory predicts. Moreover, with an instrument giving irradiation effects as strong as those in Herschel's telescope, the whole bright border will seem to be shifted more to the dark part than to the opposite border. As on 11 and 13 May the full disk was visible, whereas on 17 and 19, June it showed a phase to the east from 28° 161′ to 29° 221′, it is necessary to increase Correction II in order to obtain a true value—that is to say, diminish the rotation period.

Taking his period of $24^h\ 39^m\ 22^s.1$ as his basis, Herschel observed on the following days when the same markings appeared:

1777 April 8	1779 June 6	if we cover 768 rotations
1777 April 17	1779 June 15	if we cover 768 rotations
1777 April 26	1779 June 19	if we cover 763 rotations

He derived a period:

	24^h	39^m	$23^s.03$
	24^h	39^m	$18^s.94$
	24^h	39^m	$23^s.04$
Mean	24^h	39^m	$21^s.67$

Hereby the two minutes' difference between Herschel's result and our own is reduced to 2½ seconds.

It is clear that to give an exact result from these observations this explanation is admissible, while to reduce by two the number of rotations for our observations of 1830 and 1832, and to assume a mean error of 1^h 15^m in the intervals observed in 1830, is unthinkable.

Far be it from us to doubt the accuracy and observational skill of Herschel; but the extremely favourable conditions under which we were able to observe in 1830, and strict precision with which we have made the calculation, seem to decide in favour of our result—which, as we have seen, is in fact in agreement with Herschel's observations.

All the recent, very precise determinations confirm the period given by Beer and Mädler. With no possible doubt, the rotation period of Mars is 24^h 37^m $22^s.6$, while the period given by Beer and Mädler was 24^h 37^m $23^s.7$. Their value was thus correct to $1^s.1$.

The observations were continued by Mädler from Dorpat Observatory during the opposition of 1841, and a résumé was published in the *Astronomische Nachrichten*, 1842, together with a plate of 40 drawings.

It is difficult to identify markings which had been shown on previous drawings. From this plate I have selected a series of nine sketches, which are among the best and which were made near opposition. These are here reproduced in facsimile. These are the ones that Mädler designates as *figs*. 6, 7, 8, 14, 15, 16, 22, 23 and 24 in his plate, and are reproduced here in our Fig. 4. The dates are as follows (opposition 1 April; distance from Earth = 0.591: diameter = 15″1):

Fig. 6 or 1: 1 April 0^h 8^m Paris mean time.
Fig. 7 or 2 to the left: 5 Apr. 9^h 13^m Paris mean time.
Fig. 8 or 3: Same day 10^h 13^m Paris mean time.
Fig. 14 or 1 of 2nd row: 26 April at 9^h 12^m.
Fig. 15: Same day at 9^h 52^m.
Fig. 16: 29 April at 8^h 50^m.
Fig. 22 or 1st of 3rd row: 8 May at 8^h 41^m.
Fig. 23: 9 May at 8^h 11^m.
Fig. 24 or last: 11 May, at 7^h 54^m

These observations completed the previous series, though without adding much new information.

Such were the researches of Mädler, in many of them conducted together with his friend Wilhelm Beer—brother of Meyerbeer, the composer—who was no less an enthusiastic student of the sky. These researches were the most fruitful ever undertaken up to that time, and they really inaugurated our knowledge of Martian geography, or *areography*.

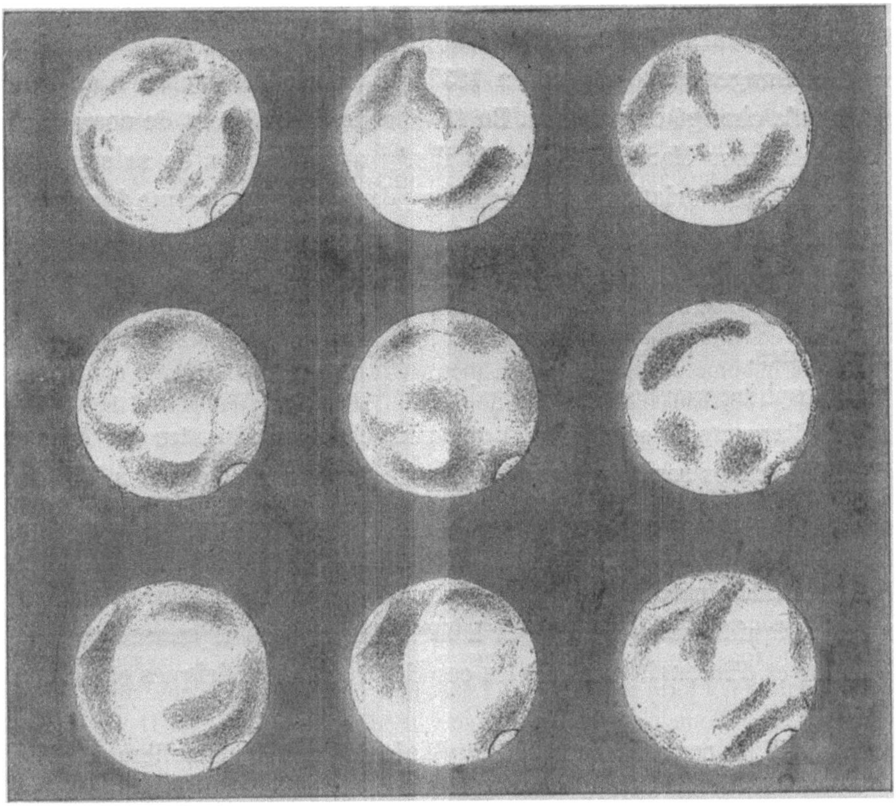

Fig. 4 Drawings of Mars, made by Mädler in 1841

The rotation period, determined with a precision superior to all previous measurements, is $24^h\,37^m\,23^s.7$.

The polar ices were studied with special care, as were the seasons in each hemisphere. Since then, we have known that the southern hemisphere has hotter summers and colder winters than the northern hemisphere, because of the greater orbital eccentricity and the inclination of the axis; the variations in the south polar ices are greater than those in the north, and they correspond to the seasons. The southern hemisphere has short, hot summers and long, rigorous winters; the northern hemisphere, on the contrary, has long, temperate summers and short, mild winters.

The measurements of the distance between the pole of cold and the geographical pole are not in accord with those of Herschel, although it is always found that the polar caps are not diametrically opposite to each other. Perhaps the poles of cold are not fixed.

The dark patches on the planet have certain fixity—certain permanence. However, there are undoubted changes. Those, which we have described in the earlier part of this work, are confirmed.

For stability, the spot *a*, the zero for longitude, is the most definite. It is the darkest and best-marked feature on the planet. (It is the Meridian Bay of my chart.) If readers will turn back to Figs. 16 and 30, Chap. 1, they will find this region to the right of the hemisphere containing the Hourglass Sea, and will note that the aspect is not the same as that shown on the chart by Beer and Mädler. The band is not clearly detached, and is less narrow; it shows that in this area *there have been incontestable changes*, perhaps of a periodical nature.

To Beer and Mädler, Herschel II. Strait appears stable, as it had also been observed by Kunowsky in 1821. Thus the serpentine arc *ac* and the patch *a*, seem to them to be definite surface features on the planet. The long broad patch *PM* of their chart is equally fixed (this is the Maraldi Sea). For the rest, despite the uncertainties and the confusion in some of their drawings, they wrote in 1832 that none of the patches observed in 1830 had changed in position. In 1837, they again recognized the Maraldi Sea with certainty. Sometimes they had the impression that considerable variations occurred in the tint, form, and extent of the dark patches. They were disposed to attribute these variations, at least in the high latitude features, to the effects of the melting of the snow, the soil becoming marshy and dark in the places where the snow had melted.

The Martian atmosphere should play an equally important role in these changes of aspect. It seems that on Mars we see two types of dark patches, those due to seas and those due to fogs or mists. Perhaps it may be that the water is not in the same state as ours, and does not form properly liquid seas, but blankets of very dense mists, which are viscous and come near the liquid state without actually being so. These aqueous blankets would show changes of extent and intensity according to the atmospheric conditions, and would follow the seasons.

We can see how our knowledge of the planet gradually advances, year by year, with the progress of the observations. We can now assert what before seemed only probable: stability, but with variations. The geographical study of Mars has become precise; Mars is a geographical globe like the Earth, not cloudy like Jupiter or Saturn; surely it has continents and seas, but these seas do not resemble our own; they show enigmatical variations which must be the object of future scientific studies.

1830.—Sir John Herschel

After having given his opinions about Mercury and Venus (*Outlines of Astronomy*), Herschel writes:

> The most natural conclusion, from the very rare appearance and want of permanence in the spots, is that we do not see, as in the Moon, the real surface of these planets, but only their atmospheres, much loaded with clouds, and which may serve to mitigate the otherwise intense glare of their sunshine.
>
> The case is very different with Mars. In this planet we frequently discern, with perfect distinctness, the outlines of what may be continents and seas. (See Fig. 5). Mars in its gibbous state, as seen on the 16th of August 1830, in the 20–ft reflector at Slough.) Of these, the former are distinguished by that ruddy colour which characterizes the light of this

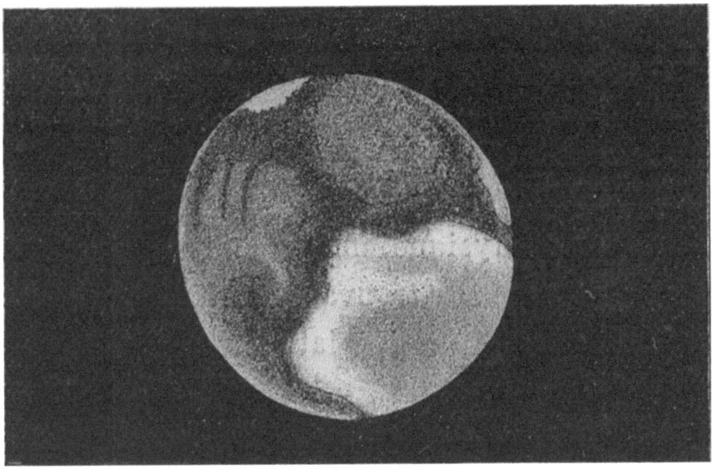

Fig. 5 View of Mars, by Sir John Herschel, on 16 August 1830

> planet—which always appears red and fiery—and indicates an ochre tinge in the general soil—like what the red sandstone districts on the Earth may possibly offer to the inhabitants of Mars, only more decided. Contrasted with this (by a general law of optics), the seas, as we may well call them, appear greenish. These spots, however, are not always to be seen equally distinct, but, when seen, they offer the appearance of forms considerably definite and highly characteristic, brought successively into view by the rotation of the planet, from the assiduous observation of which it has even been found practicable to construct a rude chart of the surface of the planet. The variety in the spots may arise from the planet not being destitute of atmosphere and clouds; and what adds greatly to the probability of this is the appearance of brilliant white spots at its poles—which have been conjectured, with some probability, to be snow; as they disappear when they have been long exposed to the Sun, and are greatest when just emerging from the long night of the polar winter.

In 1828, on 22 June, Dr. Pearson observed on the disk of Mars a dark patch, vertically elongated, not far from the left or western border, and 4 days later he again saw this patch, no longer vertical, but horizontal and elongated along its upper edge. He wrote to Sir John Herschel, who in turn communicated the observation to Smyth. Smyth described it in his work *Cycle of Celestial Objects*, and reproduced the drawing. These patches were certainly not identical, because the planet does not rotate in this sense; we are not in the direction of the pole.

We must now consider two of Sir John Herschel's views. The first is his opinion that at that time (1830) the yellow regions represented continents and the grey regions seas; this was generally accepted. The second is that the yellow tone of the continents is that of the surface of the soil. But the explanations given by Sir William Herschel's son must be treated with caution. It would mean that there could be no kind of vegetation on the Martian soil. Such a suggestion can hardly be accepted, since there is air, water and sunshine. If the surface of the soil is reddish, can it not also mean that it is covered with vegetation of this tint? Besides, the colour is not everywhere red; it is usually a warm yellow, which can best be compared with ripe wheat.

1830–1837.—Bessel[4]

From 1830 to 1837 the great astronomer-mathematician Bessel, at the Königsberg Observatory, made a series of observations of Mars which were not concerned with its physical constitution, but only with measurements of its diameter and its flattening. He found that at unit distance (the distance between the Earth and the Sun) the diameter was 9″.33. To him the polar flattening appeared insensible.

The same measurements gave the eccentricity of the southern polar patch as 6° 36′. We have seen that William Herschel had given a value of 8°.8 for this same southern polar patch in 1783, while the value given by Beer and Mädler was 8°.

In 1852 Oudemans of Leyden published a new reduction of these measurements.[5] He gave a semi diameter of 4″.664, which combined with the adopted solar parallax of 8″.571, gave a diameter of 0.544 for Mars and a volume of 0.161, relative to the Earth. For these same observations by Bessel he found:

Celestial longitude of the north polar point of Mars	349°.1[6)]
Or right ascension	317°.34
Latitude	61°.9
Or declination	50°.5

1831–1832.—Sir James South[7]

The English astronomer Sir James South, to whom we owe interesting measurements of double stars, presented to the Royal Society of London, on 16 June 1831 and subsequently on 13 December 1832, a series of observations of the *atmosphere of Mars*, showing that this atmosphere was not as extensive as had been supposed from the observations of the occultation of the star φ Aquarii by Mars on 1 October 1672. Cassini had observed at Briare: "On 1 October 1672, at 2.45 in the morning, Mars, seen with a 3–ft refractor, seemed to touch, with its western edge, the straight line drawn by joining the first and second stars of φ Aquarii, from which its distance was only six minutes. The star seemed to be so diminished and enfeebled that it could not be seen with the naked eye, or with a smaller telescope."

The star φ Aquarii is of the fifth magnitude. This observation has already been described.

The same occultation was observed from Paris by Rœmer: "Clouds did not allow me to see the beginning, and I do not know whether I would have been able to see it immediately, because three-quarters of an hour later, when the sky had cleared,

---

[4] *Königsberg Beobachtungen*, Vol. XXIII, 1847, p. 94, 95.

[5] *Astronomische Nachrichten*, no. 838, 1852, p. 351.

[6] We have seen that Herschel gave: Longitude 347°47′, Latitude 59°42 and Schröter, longitude 352°55′, latitude 60°33.

[7] *On the extensive atmosphere of Mars. Philosophical Transactions*, 1831, p. 417. —*Id.* 1833, p. 15.

I searched carefully around Mars and only found the star after 2 minutes, when it was already 2/3 of a diameter of Mars away from the eastern edge of the planet. I began to see it without difficulty only when the distance had increased to 3/4 of a diameter." (*Mem. de l'Acad.* Vol. VII, p. 359.)

Here we have a fifth-magnitude star whose brilliancy seemed to be reduced, by the influence of Mars, at a distance of 6 minutes.

There was considerable difficulty in seeing this fifth-magnitude star when it was very close to Mars, though there is no difficulty in seeing a star of the same magnitude when it is close to the limb of the Moon. From this, one can judge that Mars is surrounded by an atmosphere of some sort.

Sir James South noted that William Herschel had made a contrary observation on 27 October 1783, when he had followed a star of the 13th or 14th magnitude to a distance of 2″.56 from the planet and found it "not otherwise affected by the approach of Mars than what the brightness of its superior light might account for."

We have already noted this observation also.

On 19 February 1822, Sir James South wrote: "From Backgammon Street, London, I observed a star of the 9th to 10th magnitude whose brightness was not reduced at a distance of 143″ from the limb of the planet."

He continued:

> On the following night, the star, 42 Leonis of the 6th magnitude, approached Mars; at 4 h in the morning it was very close, and looked a beautiful blue colour. It had been occulted. I could not time the precise moment of occultation, but at emersion I recovered the star about a minute and a half from the limb; it was clear, indigo blue, and made an exquisite contrast with the colour of Mars. The planet was only 47 hours from opposition, and its diameter was 16″.6.

On 17 March 1831, South had made a similar observation with the occultation of 37 Tauri by Mars. The star showed no change in brightness or in colour. There was no colour-contrast, as in the case of 42 Leonis. No, for 37 Tauri is almost the same colour as Mars.

On 28 November 1832, Sir James South made yet another similar observation. A star of magnitude 6th or 7th preceded Mars to the south. It was of a beautiful blue colour, in striking contrast to the hue of the planet. The objective of the equatorial measured 11.85 English inches (301^{mm}), and stood a magnification of 520×. The star (RA. $=3^h\ 29^m\ 19^s$, decl. $+20°22'$) was followed right up to the edge of the planet, and neither at immersion nor at emersion did it show any change in brilliancy or colour.

The planet had passed opposition just 9 days earlier.

South concluded that the old hypothesis of a considerable atmosphere was untenable, just as Flaugergues had concluded in 1796, from a similar observation.

1837–1839.—J.-G. Galle

In 1837 and 1839 Galle, using the 9–in. refractor at the Berlin Observatory, made a series of observations and remarkable drawings. Eighteen of these drawings were reproduced by Löhse in Vol. I of the *Publications of the Astrophysical Observatory of Potsdam* (1878).

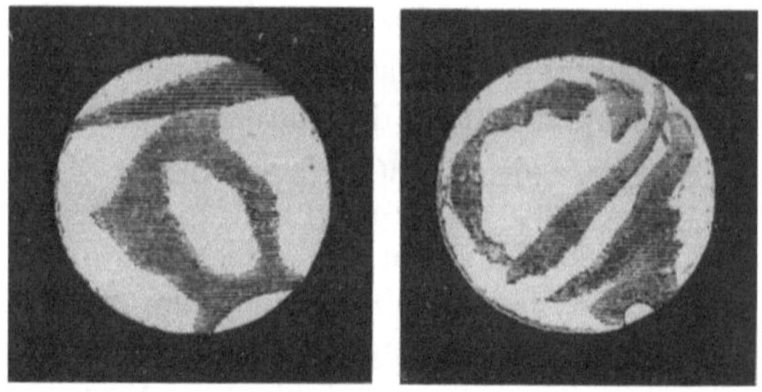

Fig. 6 Drawings of Mars by Galle, 1837–1839. A.—12 March 1837, at $10^h\ 37^m$. B.—12 March 1839, at $10^h\ 0^m$

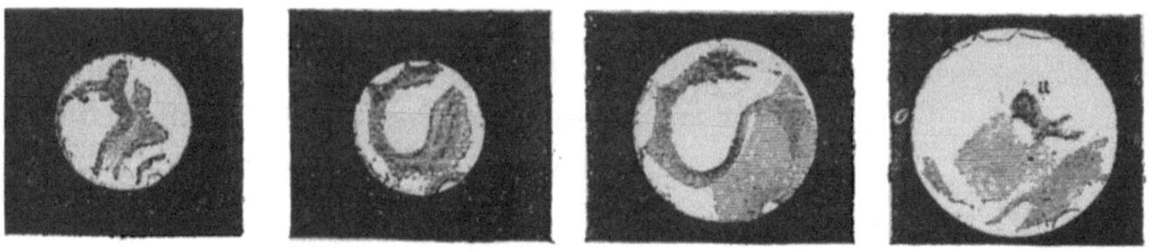

Fig. 7 Drawings of Mars by Galle, in 1839. C.—12 March, at $11^h\ 30^m$. D.—13 March. E.—14 March. F.—30 March, at $9^h\ 40^m$

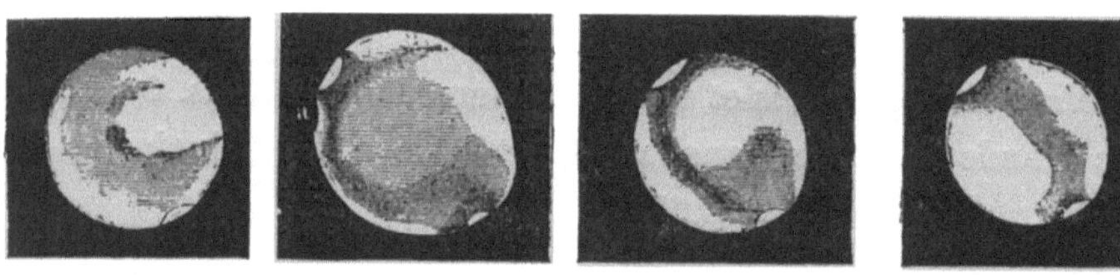

Fig. 8 Drawings of Mars by Galle, in 1839. G.—30 March, $11^h\ 10^m$. H.—31 May. I.—1 June. J.—7 June

The following sketches are reproduced here:

Figure 6A: 12 March 1837, at $10^h\ 37^m$.

Figure 6B: 12 March 1839, at $10^h\ 0^m$.—In these two views we note the north polar cap at the bottom: in the second drawing it is very small. This 1839 drawing bears a remarkable resemblance to that made by Kunowsky on 15 March 1822; the upper patch represents Herschel II Strait and the Meridian Bay.

Figure 7C: 12 March, at $11^h\ 30^m$.

Figure 7D: 13 March, at $9^h\ 41^m$.

Figure 7E: 14 March 1839, at $10^h\ 0^m$ —The sort of duck's head also indicates the Meridian Bay; the forked aspect is noteworthy, and we will return to it later.

Figure 7F: 30 March—The black patch crossed the central meridian at $10^h\ 40^m$. This drawing was made at $9^h\ 40^m$ —and the following one (Fig. 8G) at $11^h\ 10^m$.

Figure 8H: 31 May, at 14^h 30^m —The two polar patches can be distinguished not opposite to each other. A singular hollowing-out can be seen. (A similar observation by Schröter has already been described.)

Figure 8I: 1 June, at 14^h 15^m —The sombre streak, which runs from one pole to the other, seems to correspond to the Hourglass Sea, which is better identifiable on drawings 6B, and 7D, E, H.

Figure 8J: 7 June, at 14^h 22^m.

These drawings by Galle indicate *stability; but also variations in tone.*

1839.—Napoleon III

I have discovered this observation in a work which has certainly not been examined by anyone else. I include it because of its curiosity rather than its importance.

In June 1839, Prince Louis-Napoleon and M. d'Abbadie, today a member of the Institute and of the Bureau des Longitudes, who accompanied him, were visiting Sir James South's observatory in London; they observed Mars, and commented particularly on the upper polar cap, which was very prominent at the time. M. d'Abbadie made a little sketch, which it would be superfluous to reproduce, and Louis-Napoleon Bonaparte wrote a brief description *which he signed Napoleon III* (Thirteen years before he publicly began to use the name!). The planet showed a marked phase. The polar patch was so brilliant that it elongated the disk of Mars, forming a point and giving it the aspect of a pear.

This was undoubtedly a week or two after the first drawing, which preceded it.

1843–1873.—Julius Schmidt

The skilful Director of the Athens Observatory produced a collection of very numerous drawings of Mars, made in 1843, 1845, 1846, 1847, 1854, 1856, 1860, 1862, 1864, 1866, 1867, 1869, 1871, and 1873. But this beautiful series has not been published, and we know it only from the information which he gave to M. Terby. Schmidt's drawings totaled 107. The observations were made successively at Hamburg in 1843, with a magnification of 90×; at Bilk, near Dusseldorf, in 1845; at Bonn, in 1846 and 1847, with a 5–foot refractor and a heliometer; at Olmütz in 1854 and 1856, with a 5–ft refractor; and, finally, at Athens, from 1860 to 1873, with a 6–ft refractor and a magnification of 550×. In most cases it is easy to recognize the principal geographical configurations of the planet.

The four drawings below, reproduced after Terby, give an idea of Julius Schmidts observations. Here are the dates; the drawings are published in chronological order.

Figure 9A	26 September 1862 at 8^h 36^m (Athens time)	
Figure 9B	1 October 1862 at 7^h 28^m	id.
Figure 9C	16 May 1873 at 8^h 15^m	id.
Figure 9D	23 May 1873 at 7^h 41^m	id.

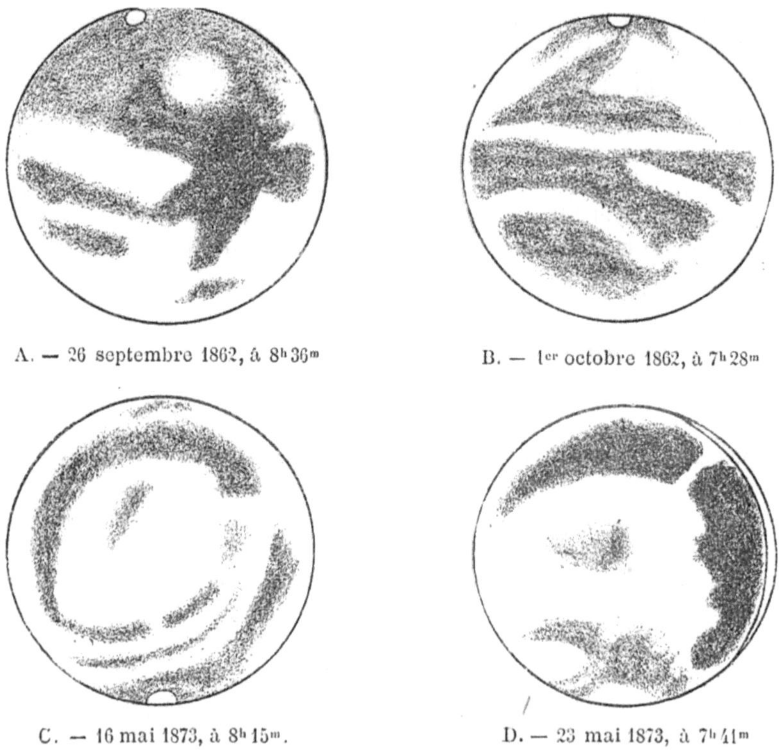

Fig. 9 Drawings of Mars, by Julius Schmidt, 1862 and 1873

Figure 9A shows the Hourglass Sea recognizably. Above, like a truly vast island, lies Lockyer Land.[8] Above again is the south polar patch, clearly away from the limb. The rest is less certain. Figure 9B shows the Maraldi Sea, and above it a sombre band which cannot be identified. Figure 9C resembles Fig. 7 of 1841 by Mädler, but none of the patches can be identified with certainty. Figure 9D appears to represent the Flammarion Sea and the Hooke Sea, which are separated by an isthmus.

These observations equally militate in favor of notable variations in the aspects of Mars.

1845–1856.—Mitchel, Grant, Warren De La Rue, Jacob, Brodie, Webb

Observations of Mars became more numerous as astronomical knowledge grew and the urge to observe increased. In this study of the planet it would be useless to give full details of all the work; this would mean a great deal of repetition, without

[8]Hellas on Schiaparelli's chart.—WS.

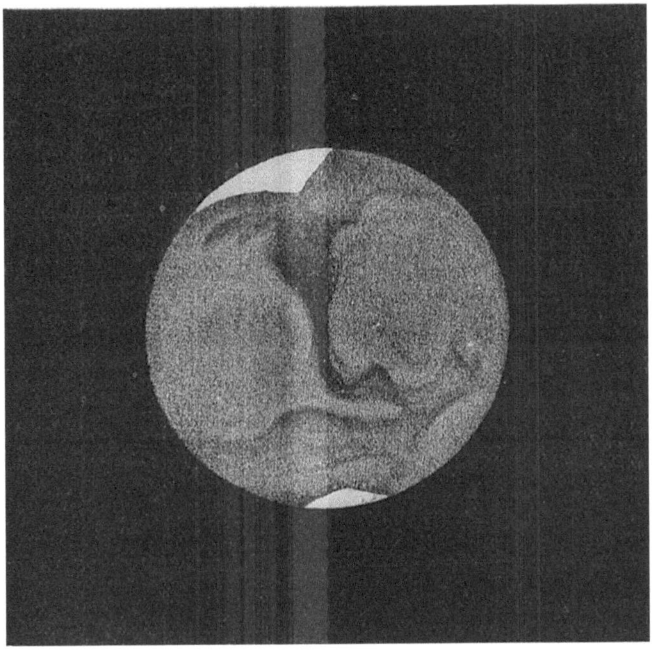

Fig. 10 Drawing of Mars, by Warren de la Rue, on 20 April 1856 at $9^h\ 40^m$

adding much, which would be new. Therefore, I propose to omit work which is of no interest, and to give only the more important observations.

In 1845 Mitchel made several observations of the planet with the great Cincinnati equatorial, concentrating particularly upon the polar snows; he believed that on 12 July 1845 he detected a black point in the snow area, together with movements at the edges of the region covered with snow. James William Grant (not to be confused with Robert Grant, author of the *History of Physical Astronomy*, London 1852) presented to the Royal Astronomical Society of London two sketches made in October 1847 and March 1854 (*Monthly Notices*, 1854, p.165). The first sketch showed the southern polar patch, and the second showed the northern. In March 1854 Jacob made two drawings upon which the principal patches are recognizable. In 1856 Warren de la Rue merits the closest attention. Of his drawings, two—those of 20 April 1856, at $9^h\ 40^m$ and $11^h\ 45^m$—are particularly notable, and are reproduced here. The first (Fig. 10) shows the Hourglass Sea as rather narrow. In the second (Fig. 11), made two hours later, the Hourglass Sea has reached the western or left edge of the disk, with Herschel II Strait occupying the upper part of the figure.

The Meridian Bay appears to the right, as a pointed tongue.

At both poles the white patches are very much in evidence. They are not diametrically opposite to each other. These two drawings are perhaps the best included in this book so far. They were obtained with the aid of an excellent Newtonian reflector of aperture 13 English inches ($0^m.33$), equatorially mounted.

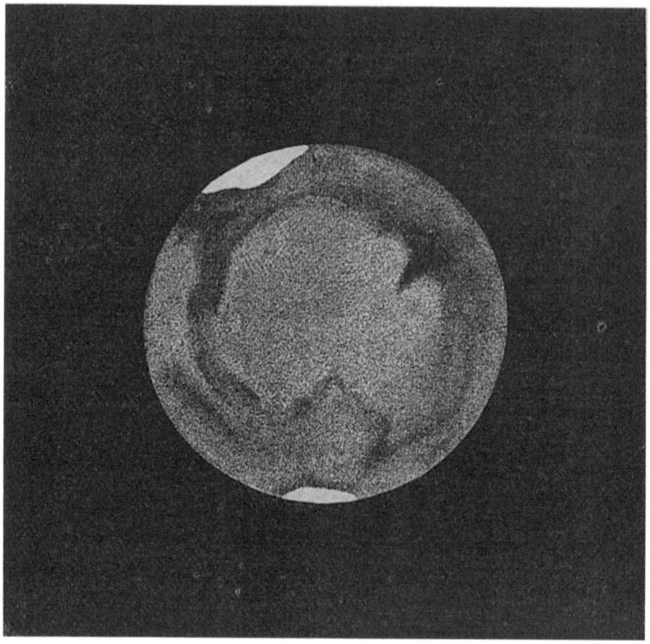

Fig. 11 Drawing made 2 hours later (at $11^h 45^m$)

To these views of the planet may be added two more sketches made at almost the same date. The first, showing considerable detail, was made on the evening of the 18 April, 1856 by Fr. Brodie; the second, a straightforward sketch, made "near the 15th" by the Rev T.W. Webb. Here are the two observations (*Monthly Notices*, XVI, 204 and 188):

18 April, $10^h 10^m$ sidereal time: Mars near the Moon; image very good; 6 1/3 in. aperture, 396× and 578. The poles are splendidly white, especially the southern. Note also two other white regions, AB and CD (Fig. 12).

Webb's sketch, although less detailed, gives a better representation than Brodie's of these four white regions (the two poles and the areas AB and CD, making up a kind of equilateral pattern). This is a sketch remarkably like Cassini's, published in the *Journal des Savants*, letter A (see Fig. 10, Chap. 1) and which was also in the front of Cassini's memoir (Fig. 11, Chap. 1). The industrious Webb also made a large number of other drawings. These are to be found in his work on practical astronomy (Fig. 13).

A physical study of the planet was published by Taylor in the *Madras Spectator* for 26 August 1845. The observations were made with a Herschel telescope. Gruithuisen referred to them in the *Astronomische Jahrbach* for 1848. The planet had a broad equatorial band, and the surface, apart from this band, was very bright. This aspect recalls that of Jupiter.

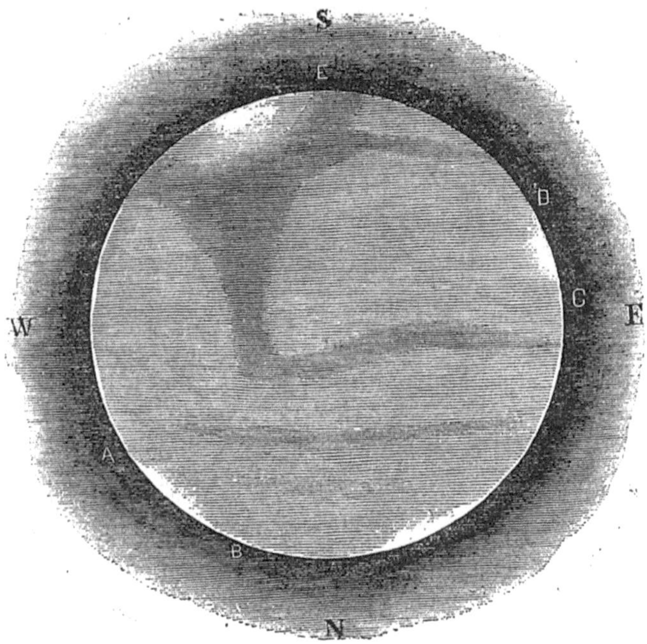

Fig. 12 Drawing of the planet Mars, 18 April 1856, by Fr. Brodie

Fig. 13 Sketch of the planet Mars by T.W. Webb, about 15 April 1856

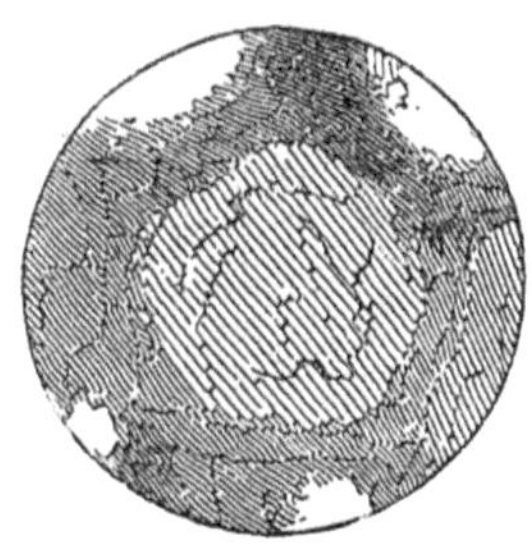

1845–1875.—Main

At Greenwich and at Oxford, Main made various observations of Mars, principally with the object of measuring the diameter. In 1845 he also made some observations of the surface.[9] On 22 August, at 11^{h} 30^{m}, at the time of opposition, he examined the Martian surface, in the company of the Astronomer Royal (George Biddell Airy):

> About 10° to the west of the apparent north point, on the planets edge, can be seen a white cap which makes a striking contrast with the sombre zone immediately above it.

[9] *Examination of the surface of the planet Mars with the telescope of the South-East equatorial.* Royal Observatory, Greenwich, 1845, p. 172.

Slightly above this dark band there is a brighter band. The most prominent sombre patch on the disk lies to the left of the great dark mass, which occupies a considerable part of the upper disk; there is also another sombre patch to the right.

The best descriptions are not accompanied by even the simplest drawing.

22 August, 11^h—The appearance of the planet has changed considerably, with the exception of the polar cap. The colour was rich red earth; the sombre patches had a very light blue tint.

Main also made observations at Oxford.[10] In particular, he made measurements of the polar flattening and the planet's diameter, of which the results were as follows:

		Flattening			Flattening
1855	9″.84	1/62	1871	9″.25	1/71
1862	9″.377	1/38	1875	9″.185	1/36
1864	9″.38	1/46			

1856.—Winnecke

Most of the present book has been devoted to studies of the physical constitution of Mars; I have not given here observations concerning the elements of the orbit, the parallax, the mass and the diameter. However, for the diameter I have included the most important measurements; those of Herschel, Schröter, Arago, and Bessel. We must also, then, note the measurements made by Winnecke in 1856, at the Bonn Observatory.[11] For unit distance, he found the diameter to be 9″.213. There was no trace of flattening; on the contrary he found a polar diameter of 9″.227 and an equatorial diameter of 9″.186.

1853.—Arago[12]

We have already noted and discussed Arago's observations of Mars given in his memoir. In Book XXIV of his *Astronomie populaire*, written in the last year of his life, when his sight—strained by overwork—was already failing, he again made studies of the seasons of Mars, together with the planet's colour and its atmosphere.

The seasons which he adopted were those given by Beer and Mädler.

There is an interesting point about the orbital eccentricity, which I have already mentioned, and which Arago himself described as follows:

MM. Mädler and Beer have followed, and largely verified by means of their observations, the explanation that the brilliant polar patches of Mars are due to the accumulation of snow.

---

[10] *Memoirs of the Royal Astronomical Society*, Vol. XXV, p. 48; Id. Vol. XXXII, p. 112. *Radcliffe Observatory Results*, Vol. XXII, XXXI, and XXXIII.

[11] *Astronomische Nachrichten*, No. 1135, 1858, p. 97.

[12] *Astronomie populaire*, published posthumously in 1854–1857. Arago had died on 3 October 1853.

> During the 668 2/3 days which make up a Martian solar year, Mädler and Beer found that in the northern hemisphere of the planet, the summer lasts for 372 days, and the winter for 296 days. The same results apply to the southern hemisphere, except that the values for summer and for winter are reversed.
>
> This inequality between the cold and warm seasons does not prevent the two hemispheres from having the same mean temperature. As for the extremes of these temperatures, it seems that the two hemispheres are very dissimilar.
>
> Thus, at the summer solstice in the southern hemisphere of Mars, the planet is actually at its least distance from the Sun, and in consequence receives the maximum amount of solar heat that it can ever do. This heat is at its least at the time of the winter solstice.
>
> It follows that if the material which produces the white patch at the south pole of Mars has properties analogous to those of our snows, then this patch ought to vary considerably more than the patch at the North Pole.

I will deal below with the theorem stating that the total quantity of solar heat received from the spring equinox to the autumn equinox is identical to that received from the autumn equinox to the spring equinox, the length of time of exposure to the Sun being exactly compensated for by the difference in distance. But even if the total quantity of heat received is the same, it is equally true that the hemisphere which is exposed to the Sun at the perihelic solstice receives at the moment more heat than the other hemisphere receives at aphelic solstice; consequently, its summer is hotter, and the polar snow will thus be more reduced.

We can imagine an orbit so elongated and an axial inclination such that the snow would never melt in the region of a pole having its winter at perihelion and its summer at aphelion.

With regard to the colour of the planet, and its atmosphere, Arago gave the following account:

> Some astronomers, physicists and geologists have written in this connection about ochre landscapes of red sandstone, from which the sunlight would be reflected. To explain the same phenomena, Lambert supposed that on Mars all vegetation was red. Alternatively, recalling that at sunrise or sunset terrestrial objects are often reddened, it may be thought that the colour of Mars is the result of modifications introduced in the rays of light by the atmosphere which surrounds the planet.
>
> But this explanation is inadmissible. If it were correct, then the colour would be most pronounced at the limb and in the polar regions, which is exactly the opposite of what we observe.
>
> We note that the red colour of Mars is more intense to the naked eye then in the telescope. In my experience, it seems that telescopically the colour is less striking when high magnifications are used.
>
> The permanent patches on Mars are never visible right up to the planet's limb, and the limb itself appears bright. These two facts are due to the presence of an atmosphere around Mars. The predominance of brilliancy of the east and west limbs is such that some observers have compared these two borders with two narrow, resplendent slices, between which the rest of the disk appears comparatively dark.
>
> Some observers have noted that the sombre patches show a light greenish tint, but this colour is not genuine. It is a contrast effect, such as is always seen when a dim white object is placed beside a bright, strongly red object.[13]

[13]This is the effect first noted by the great French chemist Michel Eugène Chevreul (1786–1889), and now referred to as simultaneous contrast.—WS

> The disposition of the permanent patches on Mars near the edge of the disk, considered as a proof of the existence of a Martian atmosphere, deserves further consideration here.
>
> Without going into details about the principles of photometry, which could be relevant in our examination, we can accept the observational result that when the sunlight shines freely down upon the uneven surface of a spherical body, the edge and the centre of its apparent disk, viewed from afar, will have almost the same intensity. This is obvious enough simply from looking at the full moon.
>
> This will not, however, be true if the rays which shine on the edges and the centre of the body do not have the same intensity.
>
> If the solar rays which illuminate the edges of the body are weaker than those striking the centre, then the edges will appear less bright than the centre.
>
> Now, if Mars is surrounded by an atmosphere which is not perfectly transparent, the rays which reach the planet's limb will be less intense than those reaching the centre, because they will have to pass through a greater thickness of atmosphere; therefore, for this reason, and even without taking into account the weakening of the light which must occur when it passes through the atmospheric layers for the second time, the solid or liquid parts of the regions near the edge should appear darker than the liquid or solid parts of the central regions.
>
> A second effect can also have important optical consequences. In the direction of each material point on the planet, we ought to see simultaneously the light sent back by the area and the light which is reflected in the same direction by the corresponding parts by way of the intervening atmosphere….
>
> Considerations of the same kind, combined with various photometric measurements of the darker and brighter areas at the centre of the disk and at different distances from the limb, lead us on to conclusions which cannot remain hidden by the optical properties of the Martian atmosphere.

From these comments of Arago—that the edges of the disk of Mars are really whiter than the interior regions, and that the patches are much less conspicuous there because of their brightness—we may conclude that the atmosphere of Mars is of considerable depth, and absorbs and reflects an appreciable part of the incoming sunlight. However, it is unquestionably more transparent than the atmosphere of the Earth and, moreover, less cloud-laden.

Arago also measured the intensity of the light reflected by the polar caps. He found it to be twice that coming from the edges of the disk.

1858.—Father A. Secchi[14]

During 1860 Mars was due to be very favourably placed for observation, and the eminent Director of the Observatory of the College of Rome prepared himself by making preliminary observations during the preceding opposition, that of 1858. He was interested as much in the physical condition of Mars as in his determination of the solar parallax. In this study he had the collaboration of his colleague P. Cappeletti, and the two astronomers made a large number of excellent drawings.

The instrument used was the fine equatorial at the Observatory, of aperture $0^{m}.244$ and focal length $4^{m}.328$. The magnifications used were 300× and 400×.

---

[14]*Osservazioni di Marte, fates durante l'opposizione del 1858. Memorie dell'Osservatorio del Collegio romano*. Roma, 1859.

The best times for observing Mars from Rome are the 2 and 3 hours after sunset, and then only during spells of consistently fine weather.

On the planet they observed patches of various colours: reds, blues, yellows; even, perhaps due to contrast, greens. Secchi commented that the drawings could not give a real idea of these tints. Copper engravings could not reproduce them and even the attempts made with chromolithography were unsatisfactory. Only pastels were successful, and forty drawings of this kind are preserved in the Observatory of the College of Rome. (I saw them personally during my visit to Rome in 1872.) It was noted that Mars appeared less red to the naked eye when, in the telescope, important azure patches were on view. This may explain the variability of Mars to some extent.

The best method of judging the forms of the patches observed is, perhaps, not to describe them, but to examine the drawings directly.

The most characteristic are those of 13, 14, 15, and 16 June (Fig. 16), which show a great blue patch in the form of a triangle and which, in their notebooks, the observers named the *Scorpion*. It did indeed remind one of the form of this creature and of this constellation. Secchi also called it the *Atlantic Canal*. This characteristic patch is none other than our famous Hourglass Sea, as it had been known for such a long time. But let me give a literal translation of the author's description:

> This Atlantic Canal is vast. Another canal,[15] small in size and joining two broader patches, is shown on the drawings of 3, 4, 5, and 7 June; we have misnamed it the *isthmus*. (This isthmus, lying about 140° to the right of the preceding sea, is known to be the narrow sea which has been named the Channel on our chart, above Christie Bay, and which Schiaparelli calls the Ganges.—C.F.) The three bays are: 1. Meridian Bay, 2. Burton Bay, or the mouth of the Indus; 3. the Channel…
>
> These two canals enclose a reddish continental area; altogether the two canals and the continent occupy about 150° of areographic longitude. The rest is covered with indefinite patches, very difficult to identify and draw.
>
> The polar patches are bounded by contours which are ash-coloured and poorly defined, but between the reddish continent and the upper polar patch there is a another very white region, which could easily be confused with the polar cap. The brilliancy of these regions is so striking that, by irradiation, they seem to project from the edge of the planet; this illusion will of course exaggerate the polar diameter.

The drawings of the polar hemisphere made by Secchi do not agree at all either with those of Beer and Mädler, or with those made later.

> With regard to the numerous questions raised by studies of the physical constitution of the planet (continued the Roman astronomer), it seems that we cannot yet give definite solutions. For example, we cannot decide whether the blue patches are really blue, or whether their colour is merely due to contrast. I incline to the view that the colour is real, because I have been able to observe small parts of the area separately by using diaphragms; however, when observed in the daytime I see the areas as almost black. The other question to be decided—whether the dark regions represent water and the reddish areas continents and the white areas clouds—is equally difficult to answer; one must first decide whether the patches are permanent or variable. If the white patches change in form, we must regard them as clouds; if not, then they could be ice-plains or continents.

---

[15]This term *canal*, found in all Secchi's descriptions, could not have been worse chosen. The mer du Sablier, for instance, does not in the least correspond to such a description.

In support of the idea that the white regions are clouds, it is notable that we can sometimes see the great patch of the Atlantic Canal as though covered with cirrus, while at other times it is not. We must see whether these aspects recur.

The reddish regions, like the bluish ones, seem too permanent for their nature to be doubted; it is probable that the former are solid, the latter liquid. The tone of the former is not uniform, but markedly *screziato*, as though filled with fine detail, about the nature of which we have no information.

A comparison of our drawings with those obtained by Mädler from 1830 to 1837 seems to establish the existence of very notable changes. However, bearing in mind the effect of the observations of differences in instrumentation and the quality of the atmosphere, we ought to reserve judgment. In particular, we have been very surprised not to recover the curious patch in the form of a bowl suspended from a thread, which was then so characteristic; and there seems to be a strong probability of change here, though perhaps this could be the lower patch of our isthmus. Was the great canal, so strong and so well-marked today, yet invisible then? But could it not be the great patch marked *pn* in the drawings made at that time? Later researchers will eventually resolve these enigma.[16]

Mars certainly seems to have an atmosphere. The brightness of the disk is much less at the limb than at the centre. Moreover, the sharpness of the outlines of the surface features decreases near the edge of the disk, which seems to demonstrate that an atmosphere exists, even if very feeble and certainly much less dense than that of Jupiter—probably even less dense than that of the Earth. It is worth drawing attention to the bright oval patch which we show on the drawings of 9 June and 10, 11, 13, 14, and 15 June, well separated from the neighbouring patch to the left. On the drawing of 8 June, however, the two patches are shown as joined together. This joining must have been only apparent, and due to a cloud in the planet's atmosphere lying about the division.[17]

From these observations, it is found that the axis of rotation is certainly not concentric with the polar patches. This conclusion had already been reached by Beer and Mädler who, however, did not regard it as definite. (The author ought to have said by Herschel, and demonstrated by him.)

Secchi was equally interested in the rotation of the planet. Using an observation made by him on 25 April 1856 at $11^h\ 20^m$ in the evening, and an identical one made on 24 July 1858 at $6^h\ 20^m$, he deduced a rotation period of $24^h\ 37^m\ 35^s$.

Here are a few extracts from his records:

1856 7 May, 11^h Rome mean time—In the middle of its disk Mars shows a large triangular patch, blue in colour, and above it a reddish patch. Seeing is bad, and it is not possible to make good observations. This patch is the Atlantic Canal, a name given for brevity to this large blue patch which seems to play the role of the Atlantic which, on Earth, separates the Old World from the New.

16 May—the disk is drawn with reddish stippling.

3 June, $9^h\ 45^m$ —Seeing good. The Isthmus is well seen. The upper polar cap is well-defined, but the lower is indefinite. The narrow canal which we call the Isthmus is well seen (Fig. 14A).

4 June, $9^h\ 30^m$ —View similar to yesterdays (Fig. 14B).

5 June, $9^h\ 40^m$ —Similar, The Isthmus is further across the disk, and so is the bright patch to the left (Fig. 14C).

---

[16] We can say that these changes definitely take place.

[17] This bright patch is called Phillips Island on my map. Its neighbour to the left is Lockyer Land, joining Dreyer Island to Kunowsky Land. We have here indications of important variations.

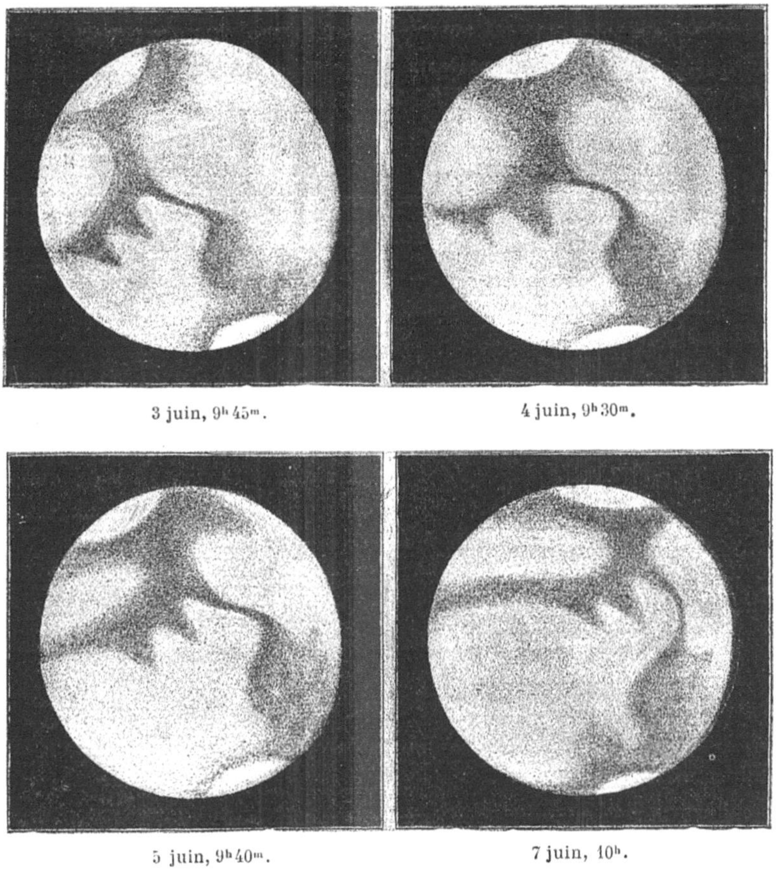

Fig. 14 Drawings of Mars made by P. Secchi. Rome, 1858

7 June, 10^h —A large reddish continent lies between the Isthmus and the Atlantic Canal (Fig. 14D).

8 June, 9^h 10^m and 9 June, 9^h 45^m —Observation of this reddish continent. In its lower part can be seen a sort of promontory, pointing toward the lower polar cap (Fig. 15A, B).

10 June, 9^h 0^m —The lower part of the figure merits special attention, because a region of bright cloud extends between the polar patch and the red continent (Fig. 15C).

11 June, 9^h 45^m —The planet presents prodigious and indescribable variety in tint. The green-blue canal is followed by a greenish fringe to the left, which extends as far as a yellow patch. At the lowermost boundary of the canal may be seen many very small white strips. These are very remarkable. Are they clouds? If they are not seen again under good conditions of seeing, one will be forced to believe that they have undergone change (Fig. 15D).

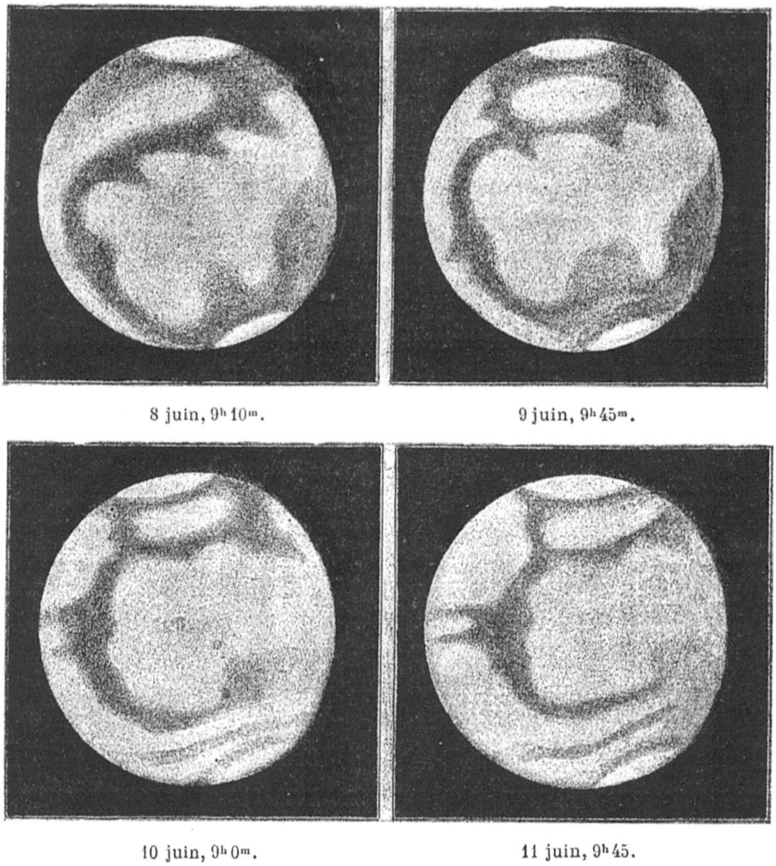

Fig. 15 Drawings of Mars made by P. Secchi. Rome, 1858

13 June, $9^h\ 30^m$ — The great blue canal (Fig. 16A) is almost in the middle of the disk; this patch is so vast that to the naked eye Mars appears less red than usual. The three principal arms of this patch make up the form of a γ or, rather a Scorpion:

Angle of the eccentric polar patches, 200°.5.
Angle of the axis of the triangular patch, 218°.
Right arm, 283°.
Left arm, 160°.5
Width of the black patch, 3″.175.
Polar diameter of the planet, 18″.371.
Distance of the patch from the upper pole, 7″.304.

These measurements were made at 9^h. The drawing was made at $9^h\ 30^m$.

14 June, $9^h\ 15^m$—Same aspect as before. Very remarkable (Fig. 16B).

15 June—Same aspect as on the preceding evenings (Fig. 16C).

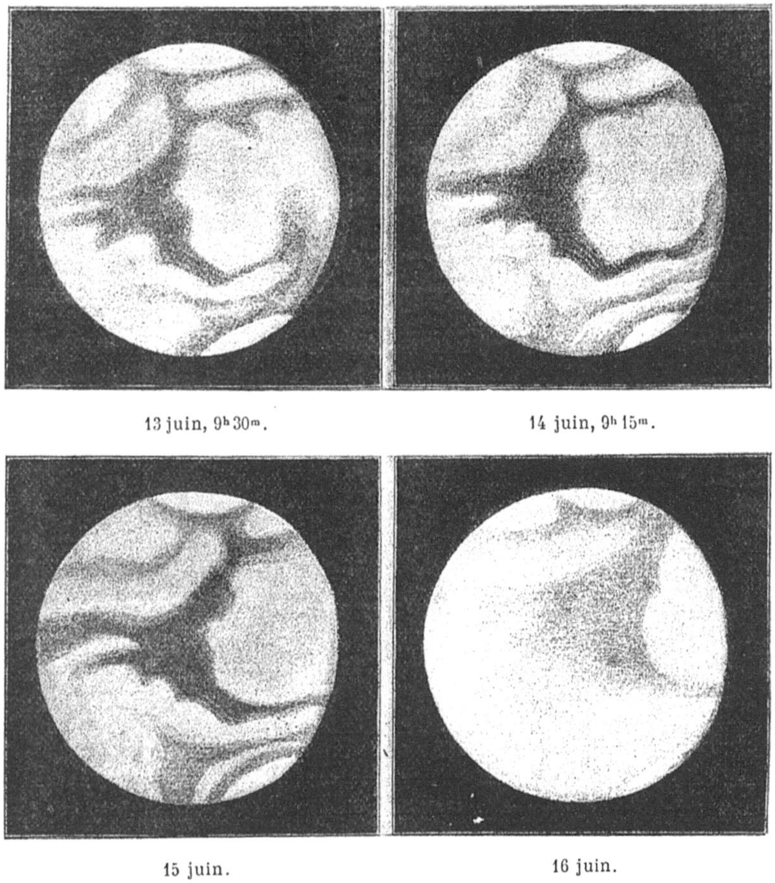

Fig. 16 Drawings of Mars made by P. Secchi. Rome, 1858

16 June—Atmosphere disturbed (Fig. 16D).

17 June, 9^h 36^m—Aspect (Fig. 17A) which is exactly like that of 25 April 1856, at 13^h 37^m sidereal time, which enables us to deduce (1) the rotation period of the planet and (2) the permanence of the Atlantic Canal, which appears on the right. The figure of the dark patch is crossed by several white veils. What are they? On the 15th they were not visible.

16 June, 9^h 20^m—Mars is particularly interesting. To the left may be seen the beginnings of a dark streak. In this area the planet is yellow, while all the rest of the disk is reddish and dappled. In this phase (Fig. 17B) the great canal tends to disappear and seems to be prolonged in the lower direction as far as the edge of the disk, but when it is seen in the middle of the disk, as from 13 to 15 June, it is obviously interrupted a long way before reaching the neighbourhood of the pole.

20 June, 9^h 40^m—The drawing was made when the canal had already almost reached the edge of the disk, and one could see only light ashy streaks on a red background (Fig. 17C).

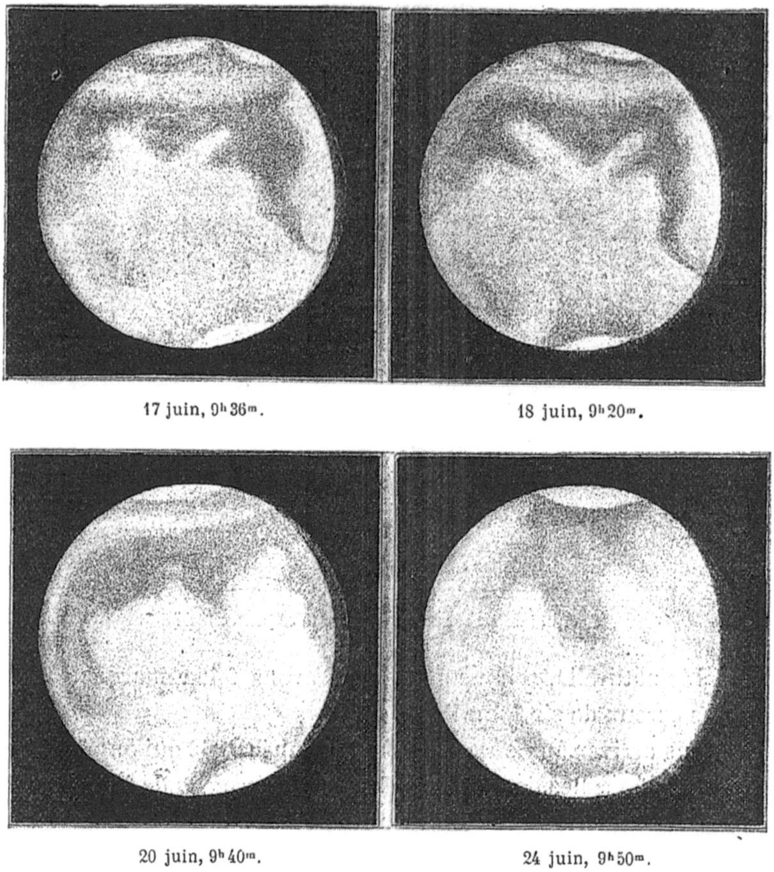

Fig. 17 Drawings of Mars made by P. Secchi. Rome, 1858

24 in June, 9^h 50^m—Thick atmosphere; the disk appeared chestnut colour (Fig. 17D).
1 and 2 July—Few patches. To the naked eye Mars appeared redder than usual.
23 July—The great blue patch strongly resembled a scorpion.
24 and 31 July, 5 and 13 August—Study of the lower polar patch; *it is certainly double*, made up of two patches in contrast with each other.

Such were the observations relating to the opposition of 1858. Father Secchi was unable to continue during the opposition of 1860, as he had intended, but he again observed in 1862 (see below). The observations just described were excellent, and among the best obtained up to that time. The later ones were even better.

1860.—Emmanuel Lias[18]

Emmanuel Liais, astronomer of the Paris Observatory and later nominated by the Emperor of Brazil to the directorship of the Observatory of Rio de Janeiro, observed Mars during the opposition of 1860, concentrating upon the physical aspect and the solar parallax. The mean of his measurements gave 25″35, or 9″.91 for unit distance. On 23 July he made the sketch of Mars which is reproduced here from his work *L'Espace cèleste*, published in 1865. In his sketch (Fig. 18), the south pole is turned toward us; while a little of the north pole can be seen below and what seems to be the Maraldi Sea.

Liais emphasized that—as Arago had shown—the reddish colour of the planet is not due to its atmosphere, but to the colour of the ground, and thought that *vegetation* was a likely cause. This latter explanation appeared to be an extremely natural one, as we have already noted when discussing Sir John Herschel's comments about the ochre lands.

We now come to the observations of 1862. Together with those of 1864, they are among the most valuable in the progress of our knowledge of Mars, because at these two oppositions, Mars was almost at its closest to the Earth. However, we had already a basic knowledge, sufficiently well-founded, to give a general idea of the constitution of our neighbour world. The following extract shows the state of opinion at this time.

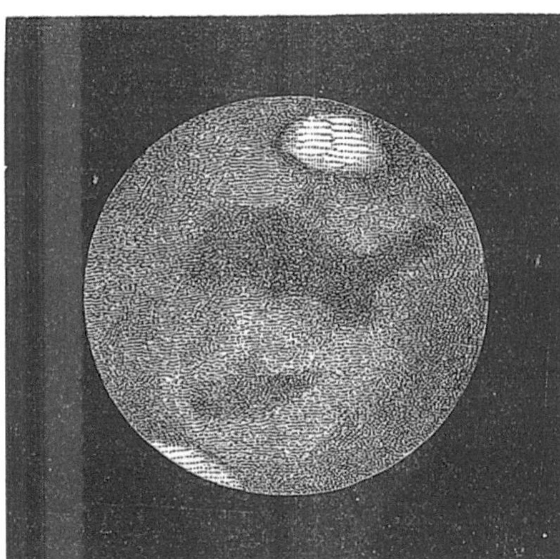

Fig. 18 Sketch of Mars, by M. Liais, on 23 July 1860

[18] *L'Espace cèleste*. Aspect and diameter of Mars.

1862.—C. Flammarion[19]

In the first edition of *La Pluralité des Mondes habités*, published in 1862, I gave the following résumé (at page 21) of the state of knowledge at the time concerning the conditions of habitability of Mars:

> Around twenty million leagues from the Earth orbits the planet Mars; it presents some striking points of resemblance to the Earth. It is 58,178,600 leagues from the central star; its year is 687 days, and its diurnal rotation period is 24^{h} 39^{m}. The atmospheric envelopes which surround it and Earth; the snows which appear periodically over the poles of both planets; the clouds which extend from time to time in their atmospheres; the geographical arrangement of their surfaces in terms of continents and seas; the seasonal variations and the climates common to these two worlds; lead us to believe that both planets are inhabited by beings whose organization is of similar character. If one of them had been born in isolation, then the other, existing under the same conditions, should follow this same course.

From the second edition of this book (1864) to the sixteenth (1871) I have published the sketch given here (Fig. 19), made at the time from a comparison of the various observations of the better-known hemisphere of the planet. The little chart includes the Hourglass Sea and the surrounding seas and shows at a glance the difference between Martian geography and our own. At the end of the seventeenth edition (1872), I gave a coloured plate of the same hemisphere, drawn from more recent observations. The existence of continents, atmosphere, clouds, snows, seas, and polar ices is regarded as definitively proved.

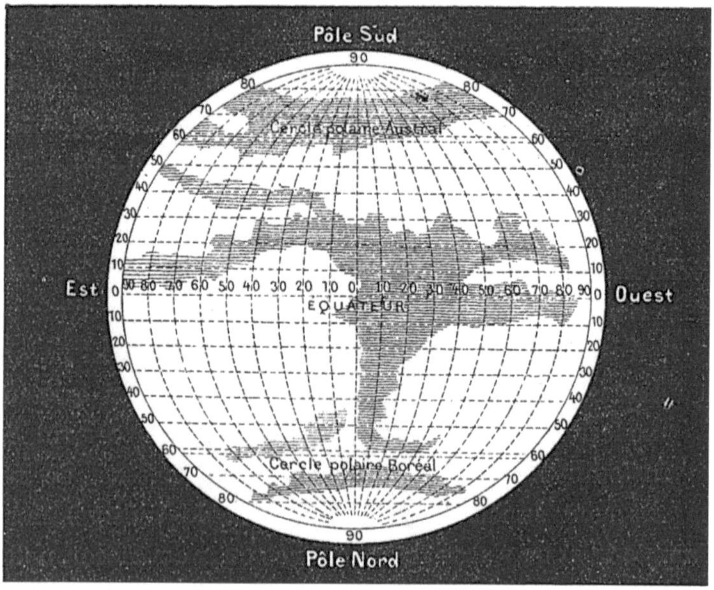

Fig. 19 The best-known hemisphere of Mars (1862–1864). (Figure reproduced from the 2nd edition of *la Pluralité des Mondes habités*)

[19] *La Pluralité des Mondes habités*.

In this figure I have indicated East and West as we use them on Earth—that is to say, as they would appear to an inhabitant of Mars.

We now come to the observations made by Father Secchi in 1862, and published in 1863.

1862.—A. Secchi[20]

The following is a translation of the résumé of the memoir published by the eminent Italian astronomer. These observations were made at the Observatory of the Collegio Romano; they form a continuation of the series made in 1858, and the same instrument was used.

Secchi wanted to take full advantage of the perihelic opposition of the planet to continue his researches into its physical constitution:

> Mars is the best-studied of all celestial bodies, except for the Moon. On it, Herschel and other astronomers have observed not only seas and continents, but also the effects of the seasons of winter and summer; however, the disagreements which exist between the modern observations and the old ones cause certain misgivings. Modern instruments ought to be able to resolve the problems, because they are superior even to those of William Herschel. Our drawings of 1858 are not in accord with those of Mädler, notably with regard to the white polar patch— Mädler showed it reduced to a small, brilliant circle, while we find it vast and complex. But at the last opposition, it reverted to the form shown by Mädler.
>
> These differences are due to two causes. The first is the angle at which Mars is seen. In 1858, the two poles were equally visible, while now the north pole is hidden and the south pole is turned toward us. On 26 September 1862, at $9^h\ 45^m$, the planet was seen in a position corresponding to that of 4 June 1858 (the one at the upper right of Fig. 14), but obliquely and foreshortened, with the upper pole tilted toward us, as in Mädler's third drawing of 1832.
>
> The second cause of variation is that the polar patches really do change constantly. The vast white fields are shown as faint and restricted in the form of Mädler's small polar cap. It is clear that the variations can be explained only by a melting of the snow or a disappearance of the clouds covering the polar regions. And indeed the pole visible during the opposition of 1862 is also the pole turned toward the Sun; this pole is having its summer, and Mars is only 15° from perihelion; therefore the temperature of the pole is at its highest, corresponding to the middle of our month of July. Note also the strong inclination of the axis of Mars to the orbit, giving very marked seasons.
>
> These aspects also proved that liquid water and seas exist on Mars; this is a natural result of the behavior of the snows. This conclusion is confirmed by the fact that the blue markings which we see in the equatorial regions do not change sensibly in form, whereas the white fields in the neighbourhood of the poles are adjacent to reddish fields which can only be continents. Thus, *the existence of seas and continents, and even the alternations of the seasons and the atmospheric variations, have been today conclusively proved.*
>
> From these observations of 1862 it is found that the very characteristic features of the planet drawn by Beer and Mädler returned in an unequivocal manner. Thus, the patch that they lettered *efh* corresponds to that which I call the Cook Sea; their *np* is my Marco Polo; their patch *a* is the Franklin Canal. I have given no names to the reddish regions, and I have confined myself to giving names to the darkest, most certain and most constant of the dark patches.

[20] *Osservazioni del pianeta Marte, Memorie dell'Osservatorio del Collegio Romano.* Nuova Serie: Vol II. Roma, 1863.

From those researches I have come to the conclusion that as well as these permanent patches, there are variable ones, which should be studied closely and consistently. The existence of an atmosphere cannot be doubted, because of the limb brightening and also because of independent spectroscopic observations.

Extract From the Observations

21 September 1862, at 20^h 50^m sidereal time—Mars showed its upper polar cap, very reduced in size and fully turned toward us. Its angle is 145° from the centre. The blue patch in the form of a Y can be seen clearly; its form resembles that of a scorpion, but its narrow part is hidden. For brevity, I will call this blue canal the Cook Canal, applying to Mars the names of celebrated navigators, and I give the name of Cabot Continent to the reddish continent which extends to the right (Fig. 20A).

26 September, 9^h 45^m —The Cook Canal lies almost exactly in the middle of the disk. But, while in the 1858 drawings its broadest part, which I call the Scorpions body, is well

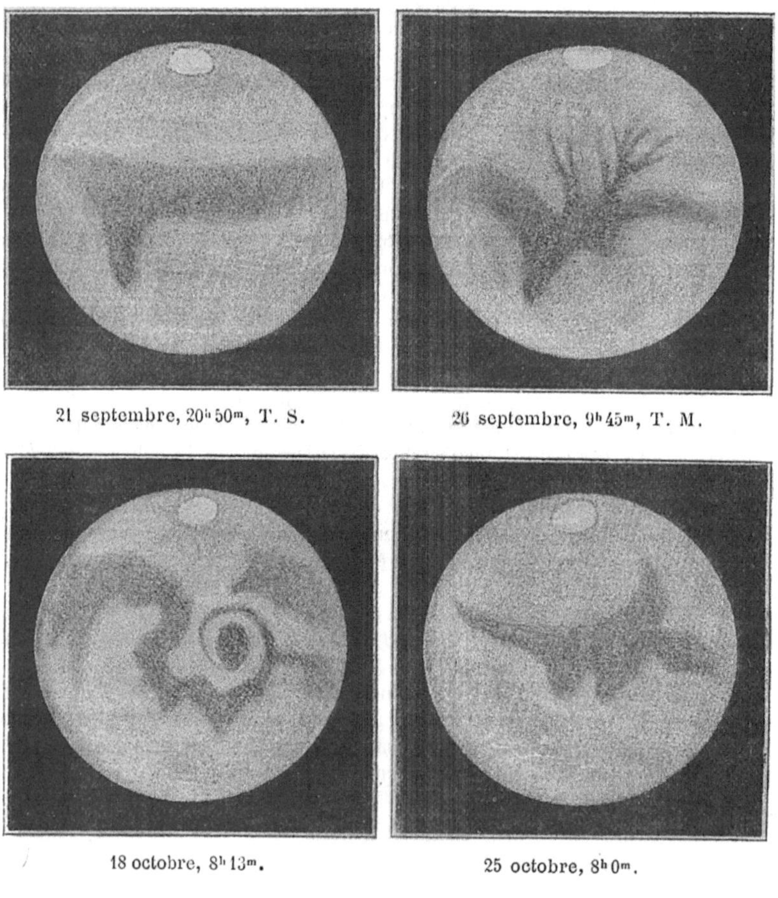

Fig. 20 Drawings of Mars made by P. Secchi. Rome, 1862

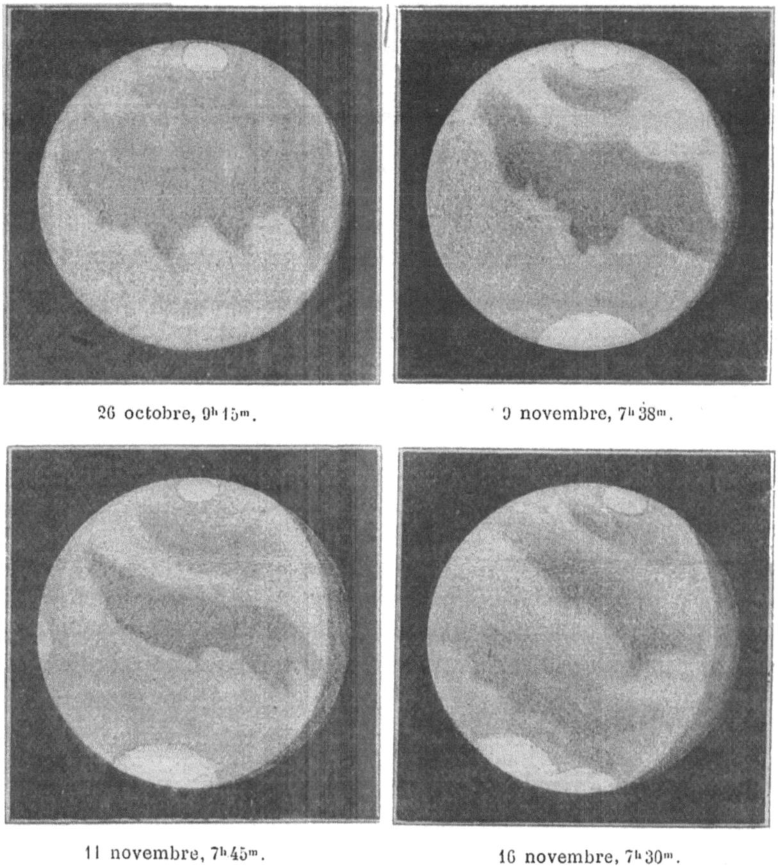

Fig. 21 Drawigns of Mars made by P. Secchi. Rome, 1862

above the centre, it is now right at the centre. (See, in particular, the drawings of 14 and 16 June 1858 on p.140.) This is due to perspective. This year, the region shows exactly the aspect which Beer and Mädler drew in their Fig. 3 of 1832. Details near the limb cannot be distinguished, which proves that *the atmosphere of Mars is absorbing*. Between the polar patch and the Cook Sea may be noticed several shades of blue, and streaks shot through with yellow and red, difficult to draw. I believe we are looking at an archipelago (Fig. 20B).

18 October, $8^h\ 13^m$ —The polar cap is well away from the edge (Fig. 20C). I can note a dark patch different in tone from those to which I am accustomed; I have never seen it before. It seems to be surrounded by a ring, or a spiral cyclone. The regions near the pole are reddish; they were certainly white during the former year. I believe we are seeing a great squall on Mars.[21]

25 October $8^h\ 0^m$ —Polar patch well marked and well detached (Fig. 20D). Between it and the canal is a large reddish region which I call *Columbie*.

26 October, $9^h\ 1^m$ —The easiest feature to recognize is the Franklin Canal to the right (Fig. 21A). Between the pole and the sea which joins the Cook Canal to the Franklin Canal I see a reddish area speckled with curved lines (Columbie).

[21]This patch, compered here with a squall, is the Terby Sea of my chart. Secchi wrote: "*La crederei una gran burrasca in Marte*."

9 November, $7^h 38^m$ —The view this evening is remarkable because of a large blue patch which I have never seen in these proportions before (Fig. 21B). It is obviously a prolongation of the Cook Sea. It was not seen in 1858, no doubt because the real variations, rather than differences in perspective, or different conditions in the atmosphere of Mars. Very nebulous in this region, notably on 18 and 20 June 1858, this patch is similar in shape to the patch which Beer and Mädler lettered *PM*.

11 November, $7^h 46^m$ —The great patch PM which I call the Marco Polo Sea appears clearer and clearer (Fig. 21C). Between it and the white pole I can see a very curious dark cloud.

16 November, $7^h 30^m$ —The great blue sea is well seen (Fig. 21D). The lower or north polar cap is double.

18 and 26 December—The continuation of the observations proves that the upper or southern polar snows have diminished considerably, and are reduced to nothing more than a small white circle.

These observations by Father Secchi are as curious as they are important. They confirm our preceding views about the continents, the seas and the atmospheric influences on Mars, as well as the *certain variations* which occur on the surface of the planet with regard to the form and extent of the seas.

We can now, more completely than before, undertake a study of the determination of Martian geography. In order to tackle this more accurately, it is essential to refer to the base chart of Mars published in this book (*fig.* 31), and to compare it with the drawings made by the industrious Roman astronomer.

In the 1858 drawings, the celebrated Hourglass Sea can be recognized with certainty on five drawings, those of 10, 11, 13, 14, and 15 June (Figs. 15C, D and 16A–C). We may reason in the following manner:

As we have noted, the Hourglass Sea was called the *Scorpio*n by the Roman astronomers, and the resemblance is indeed picturesque. The tail of the Scorpion is the Nasmyth Channel of my chart, and ends in a small sea called the Lassell Sea; the right tentacle, above the body, is the Dawes Ocean, which is prolonged toward the pole by the Lambert Sea. The first branch to the right is Herschel II Strait; the large tentacle to the left is the Flammarion Sea, which is prolonged by the Hooke Sea; the small tentacle above is probably the Main Sea, though exaggerated. This region is very variable in all the drawings. At the bottom of the figure, on all the five drawings, may be noted a white zone; the Laplace Canal, then a grey region (the Delambre Sea) and next a bright zone, followed by a dark zone surrounding the lower pole.

We again find the Hourglass Sea on the drawings of 21 and 26 September 1862 (Fig. 20A, B). Father Secchi gave no less than three names to this sea: the Scorpion, Atlantic, and Cook Sea.

In the drawings of 3, 4, 5, and 7 June 1858 (Fig. 14A–D), we see another side of the planet. This narrow elongated sea is the second characteristic aspect of Mars in these observations; the Roman astronomers called it sometimes the Isthmus, sometimes the Franklin Canal. It is almost 180° to the right of the Hourglass Sea, so that the two patches can never be seen at the same time. In my charts, I have given this sea the name of the Channel (*de Manche*, as the French refer to the *English* Channel—WS). Schiaparelli calls it the Ganges. The strait is not shown on Green's chart, which will be described further below.

If these four views of Mars are carefully studied, it seems that the first pointed tongue, going from left to right, is the Meridian Bay; the second, 20° to the right, is Burton Bay, called by Schiaparelli the Margaritifer Sinus and the mouth of the Indus; and the third, lying at the same distance below, seems to be made up of Christie Bay and the Channel. The identifications are not absolutely satisfactory. The drawings give longitude 25° for the mouth of the Indus; while the longitude of the Channel is not 50°, but 56°. However, it seems impossible to make any other identification. The Channel is absent on a great number of drawings, though we will find it again later. It is shown perfectly on two drawings made by Dawes on 12 and 14 November 1864, and on one made by Schiaparelli on 28 November 1879. The observations by Secchi and Dawes lead me to give greater emphasis to this Isthmus on my chart than it had on Schiaparelli's.

The 1862 drawings do not show it. On the eight drawings made at Rome in that year, the first two have—as we have seen—shown the Hourglass Sea. The third shows the Terby Sea, taken by Secchi to be a cyclone. The fourth enables us to identify the three bays on the 1858 drawings (Meridian, Burton, and the Channel), and it is the same on the fifth. The sixth, seventh and eighth show the Maraldi Sea, which Secchi calls the Marco Polo Sea and which Beer and Mädler lettered *PM*.

These drawings are sufficient to confirm the opinion that the Martian globe shows permanent geographical configurations, but show at the same time that these configurations are subject notable variations, which are in some measure due to the observers and to instruments, while many others—as, for example, the width of the Channel—are inherent in the physical constitution of the planet itself. This last point is of the greatest importance.

Secchi continued his observations in 1864. Among the drawings of that year, we may note, after Terby (Fig. 22), that of 1 December 1864, at 7^h, which at first appears to represent a very narrow form of the Hourglass Sea but actually shows the Maraldi Sea together with a strait going down from it in the form of an elongated triangle. This is evidence of the changes which occur on Mars in this particular region, because the elongation probably corresponds to Fig. 52B, Chap. 1, observed by Schröter on 2 November 1800, and to the left hand points of the marking shown in Figs. 51 and 53B, Chap. 1.

Fig. 22 Sketch made by P. Secchi, on 1 December 1864

1862.—Lockyer[22]

We now enter a fruitful period in the study of Martian geography. During the very favourable opposition of 1862, several astronomers devoted themselves to the work, and we will examine, among others, the results obtained by Father Secchi in Italy, Lockyer in England, and Kaiser in Holland. First let us consider the work of the English astronomer, summarizing his important memoir as completely as possible.

The doubts and difficulties in connection with the permanence of the Martian features are due chiefly to the hopeless lack of accord between drawings made at different epochs. Opinions are remarkably contradictory; thus, to cite only two examples—Cassini in 1670 recognized the patches which he had discovered in 1666 with his Campani refractor of 16½in. focal length and Maraldi in 1720 declared that it was impossible to reconcile them with drawings made in 1704, 1717, and 1719, while in our own time Secchi found in 1858 that his drawings were incompatible with those made by Beer and Mädler in 1830 and 1837.

The inclination of the planet can cause marked differences in aspect, because of the foreshortening effects; the features seem sometimes to be shifted in both latitude and longitude. It will therefore be convenient to compare drawings made under identical conditions of inclination. Thus the opposition of 5 October 1862, at heliocentric longitude 12°, is comparable with that of 19 September 1830 (heliocentric 356°) (Figs. 23 and 24).

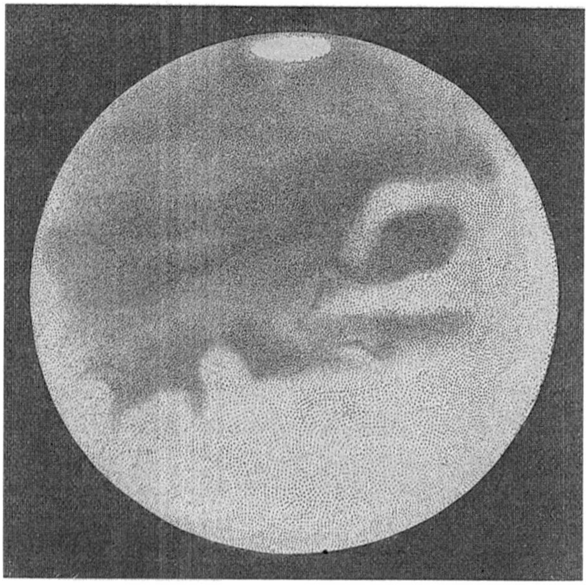

Fig. 23 Drawing by Lockyer, 17 September 1862, at $10^h\ 50^m$

[22] *Measures of the planet Mars, made at the opposition of 1862. Memoirs of the Royal Astronomical Society*, XXXII, p. 179–190.

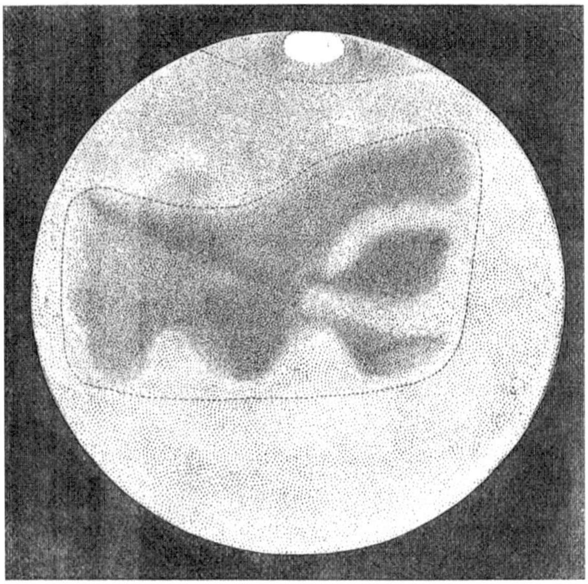

Fig. 24 Drawing by Lockyer, 17 September 1862, about $10^h\ 50^m$

Allowing for the fact that Lockyer's objective was of 6¼ in. aperture and 8½ft focal length vs. Beer and Madler's 3¾in. aperture and 4½ft focal length, Lockyer claimed that his drawings agreed perfectly with those of 1830. (The larger instrument would, of course, show a correspondingly greater amount of detail.)

These observations of 1862 confirmed, in a most satisfactory manner, the absolute permanence of the configurations of the planet. There were nevertheless inexplicable disagreements between the observations made with different instruments, even in the hands of the most skilful observers (Figs. 25, 26, 27, and 28).

While the complete fixity of the general features was not in doubt, variations in the details and tones of the bright and dark regions were observed daily—one might even say from hour to hour. Lockyer did not doubt that these changes were caused by clouds passing over the different regions. He wrote:

> These changes are, I doubt not, caused by the transit of clouds over the different features.
>
> A clear, cloudless atmosphere, here as on Mars, would have the effect of rendering the dark regions of the planet darker and more distinct; the lines and coasts, if we may so call them, are so fine and so light that it is quite impossible to represent them exactly. Beer and Mädler have already commented that generally a definite time elapses before the patches—originally vague when observing begins,—become sharp and well-defined.
>
> Clouds, on the other hand, have the effect of making the dark regions less dark, in proportion to the density of the clouds, and the bright regions less bright in the same ratio. *They can never make a bright region look dark.*[23] Thus when we see a dark patch as well-defined, we can be certain that there are no clouds above it, and that we are looking at the actual surface of the planet. However, we cannot be sure, at least in view of our information drawn from the old observations, that dark regions are not at a lower level than the bright areas.

---

[23] Is this certain?

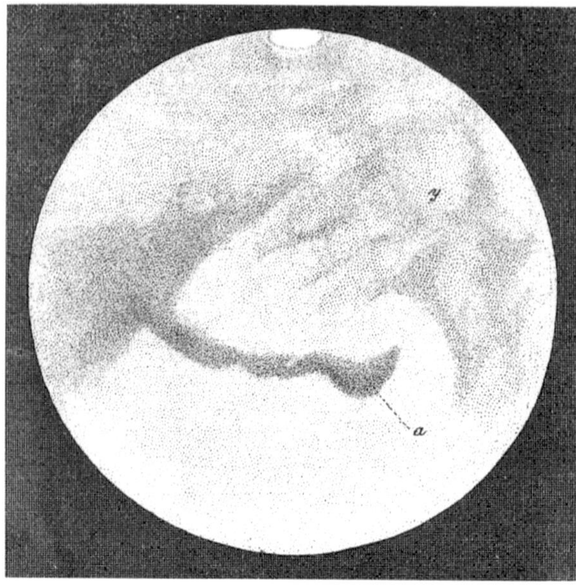

Fig. 25 Drawing by Lockyer 23 September 1862, at 9ʰ 40ᵐ

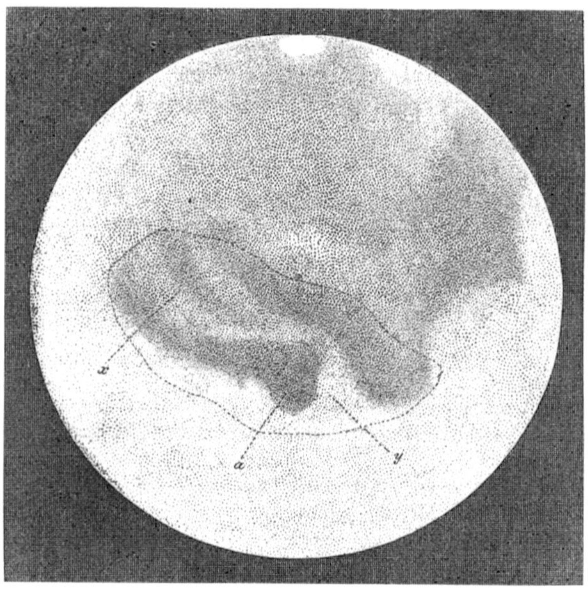

Fig. 26 Drawing by Lockyer, 23 September 1862, at 10ʰ 25ᵐ

Several examples of clouds were suspected by Father Secchi in 1858. However, Lockyer presented some unmistakable cases. In the drawing made on 3 October, at $10^h 30^m$ (Fig. 31), the space stretching from x to y was lacking in any dark feature; in the sketch made on the same evening at $11^h 23^m$ (Fig. 32) a patch was seen near y,

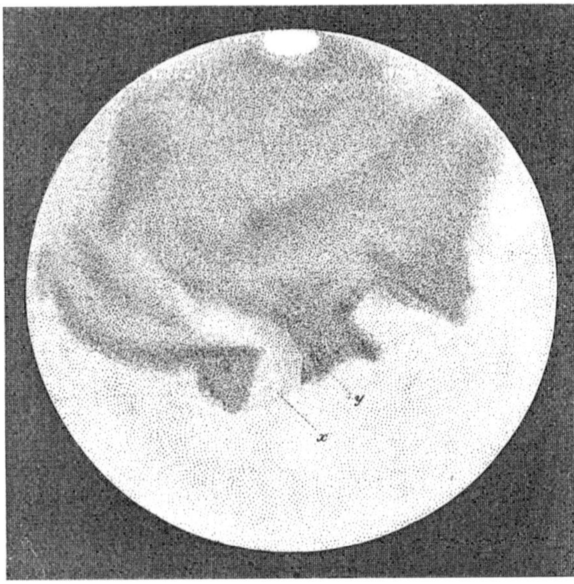

Fig. 27 Drawing by Lockyer, 23 September 1862, at $11^h\,55^m$

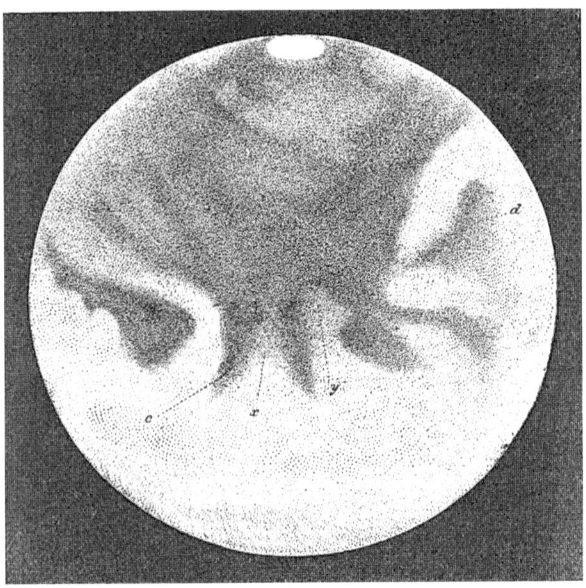

Fig. 28 Drawing by Lockyer, 23 September 1862, at $12^h\,55^m$

which spread out progressively and extended as far as x at $11^h\,51^m$, when the drawing in Fig. 33 was made. Lockyer wrote as follows:

> Now, this region is one of those which we know best, because it has been excellently observed by Warren de la Rue, Father Secchi and others, and there is no doubt that Secchi's drawing no. 8 does not represent the normal aspect of this equatorial region at longitude 28°.

> The observed changes can be easily explained by assuming that at the beginning of my observations the feature concerned, and which is persistent in Secchi's *fig.* 10, 11, 13, 14, 15, 17 and 18, was veiled by clouds which gradually dissipated up to the end of the observation; although the feature was not completely unveiled, it was much more clearly visible at the end of my observations.

He was referring to an area which our readers will know very well: that of the famous Hourglass Sea. Accepting, with Lockyer, the influence of white clouds, we can see that the zone marked *f* and *y* is very variable (further from the left border of the Hourglass Sea). Lockyer further wrote:

> To take another example, in my drawing no. 14, Beer and Mädler's patch *a* is completely invisible, while in no. 15, made several minutes later, it is absolutely evident and very notable.
>
> But quite apart from the clouds which, as we have seen, from time to time totally or partially obliterate the dark regions of the planet and give rise to variations of colour and tone which deform the appearance of the features, we must agree that an atmosphere as dense as that of Mars, with its mists and its fogs, must also play a definite role. I mention this fact particularly with the aim of establishing that although patches in the southern hemisphere can be observed with certainty in the middle of the summer, when they appear at the edge of the disk and as they are passing over the limb, we can see them even more distinctly in the northern hemisphere in the middle of the winter, blotting out the geographical features north of latitude +30° even when the features are on the central meridian. We have here new evidence of the *great intensity of the seasons* on Mars, an intensity already demonstrated by the great extent of the polar snows in winter and their rapid melting in summer. As Beer and Mädler have commented, the southern hemisphere of the planet is always the easier to study from our point of view, because it is turned toward us at the times when the planet is at its closest to us.
>
> With regard to the red and green colours so often described for the geographical features of Mars, my observations lead me to the same opinion about their nature as that expressed by Father Secchi from his studies made in 1858. To me, also, *the red regions represent continents and the green regions seas.* I do not believe that the green colours are due to contrast effects; to me, they appear real.
>
> The dark regions seem to me to be certainly green, as is agreed by all those who have observed Mars. Through my telescope, this colour is particularly marked in the feature lettered *PM* on the map by Beer and Mädler[24] (drawing made on 15 October, from $9^h\ 8^m$ to $9^h\ 20^m$). This colouring is certainly not due to the object-glass of my telescope.
>
> The patches which appeared darkest in 1862 are the same as those which had the same appearance in 1830; these seas are generally almost completely surrounded by lands.
>
> The variation of the polar snows is a most interesting subject for observation. In 1830, the summer solstice in the southern hemisphere of Mars fell on 8 September, and the minimum of the polar snows (1/20 the apparent diameter of the planet) was observed on 5 October—that is to say, 27 days after the greatest altitude of the Sun in the southern hemisphere. In 1862 the solstice fell on 30 August; on the 23rd of that month the snowy zone showed a diameter equal to 1/5 of that of the planet, but on 25 September it was reduced to 1/10, and on 11 October to 1/13, so that it could barely be distinguished. After that, the snows began to increase again.
>
> This very rapid melting of the south polar ices can be attributed to the great eccentricity of the orbit of Mars, and to the fact that summer in the southern hemisphere occurs when the planet is near perihelion. The centre of the snowy polar cap does not coincide with the pole, but lies at several degrees from the geographical pole, near longitude 20°; on the other

[24] This corresponds to Mare Sirenum and Mare Cimmerium on modern maps.—WS.

hand with the north or lower pole, visible in 1857, Father Secchi has stated that the ice-cap really is centred on the pole.

Sometimes the polar snow appears so brilliant that, like the crescent of the new moon, it seems to project beyond the planet. On one evening, as clouds passed in front of Mars, the polar snow remained visible, looking like a nebulous star. (This was noted as early as the 1700s.)

Finally, Lockyer commented that his 6¼-in. ($0^m.16$) refractor was equatorially mounted and clock-driven; the magnification generally used was 191×. The splitting of double stars such as χ Aquila, γ^2 Andromedæ and λ Cassiopeiæ was a guarantee of the power and definition of the telescope.

I have reproduced sixteen of Lockyer's drawings in facsimile. They are the most important, with regard to our knowledge of Mars, that we have yet encountered in this book. I have arranged them in chronological order.

In the first and second of these telescopic views, it is easy to recognize the circular feature which looks like an eye—which has been called the Terby Sea. Above it, and to the left, lies the De la Rue Ocean; below, a grey region subject to frequent variations; to the left, a little rounded bay, the Christie Bay; and a little further away a second, Burton Bay. The southern polar snows are brilliant and sharply-defined, as on all the drawings.

Lockyer himself calls this eye-like feature the Baltic Sea.

In the third drawing, we may note in particular Herschel II Strait and the patch *a* (the Meridian Bay). A little later in the evening this strait and this patch were slightly more advanced toward the left (Fig. 26) and an hour and a half afterwards, more advanced still (Fig. 27). There is a great resemblance between these drawings and those of Beer and Mädler. Above this elongated sea we may make out a second, the Arago Strait, and between the two there is a tongue of land which is white or rather grey. This is a variable country, which appears sometimes continental and sometimes maritime. Figure 28, still more advanced, shows the Arago Strait in the centre of the disk; it is bounded by two pointed tongues, one of which is Christie Bay; the other, somewhat deformed, is the Eye.

One senses, even at a glance, that the geographical features have been drawn with precision and certainty.

Figures 29 and 30, of 25 September, again show the Herschel II. Strait (the Meridian Bay is veiled in the first drawing; these dark clouds are presumably of atmospheric nature, as Lockyer maintains). At *e*, the Strait is joined on to the Hourglass Sea. The three drawings of 3 October (Figs. 31, 32 and 33) show, with complete certainty, the Hourglass Sea, Flammarion Sea, Hooke Sea, and below it, the Maraldi Sea. Zöllner Sea and Lockyer Land are equally easy to recognize. To the left of the Hourglass Sea, the region *xy* is misty.

The drawing made on 9 October shows the Hooke Sea and the Maraldi Sea. Above lies the Maunder Sea, between the Hooke Sea and the Maraldi Sea; also above the Niesten Isthmus; to the right, the Hooke Sea, Cassini Island and Dreyer Island; toward the edge, Zöllner Sea. Below the Maraldi Sea extends Herschel I. Continent (Fig. 34).

The drawing made on October 11 shows the same aspects, but to the left of the Maraldi Sea can be seen another—the Schiaparelli Sea, separated from the Maraldi Sea by Webb Land. (Green's chart is rather imperfect on this point.) (Fig. 35)

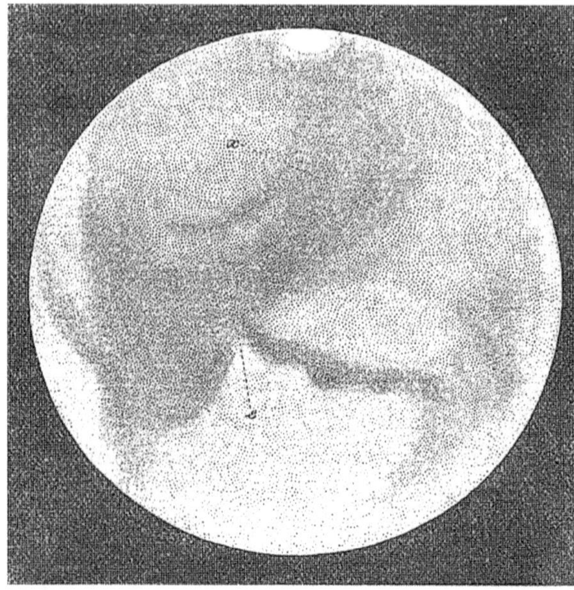

Fig. 29 Drawing by Lockyer, 25 September 1862, at 10h 44m

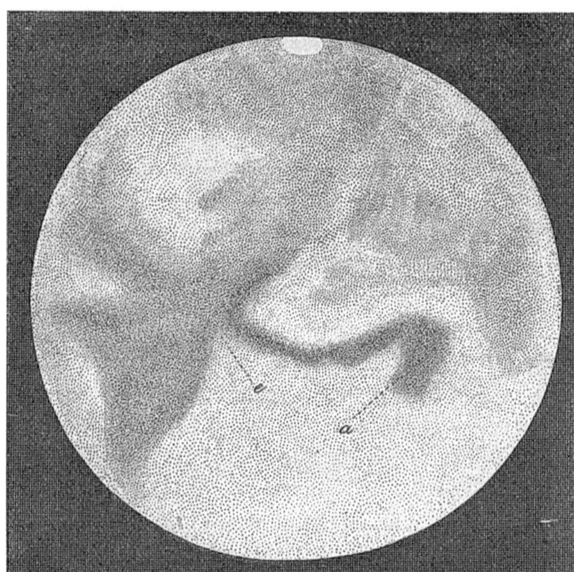

Fig. 30 Drawing by Lockyer, 25 September 1862, at 10h 50m

Finally, the two drawings of 15 October show the Schiaparelli Sea darker than the Maradli Sea to which it is joined (but this is not always the case); above, the Maunder Sea; to the left, the Eye or the Terby Sea in the middle of the disk, surrounded by the De la Rue Ocean, from which it is separated by Kepler Land. The region lying below the Terby Sea is less dark and less extended than it was on 17 September; this also is a very variable region (Figs. 36, 37, and 38).

On all the drawings, the lower or northern half of the disk is almost always devoid of features, except near the point of the Hourglass Sea.

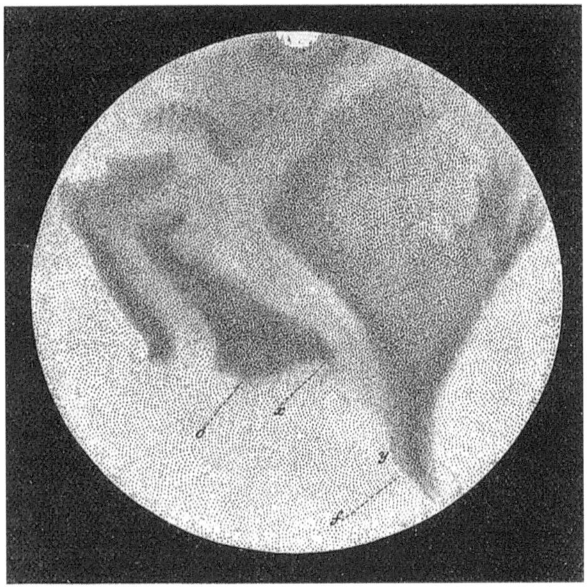

Fig. 31 Drawing by Lockyer, 3 October 1862, at 10^h 30^m

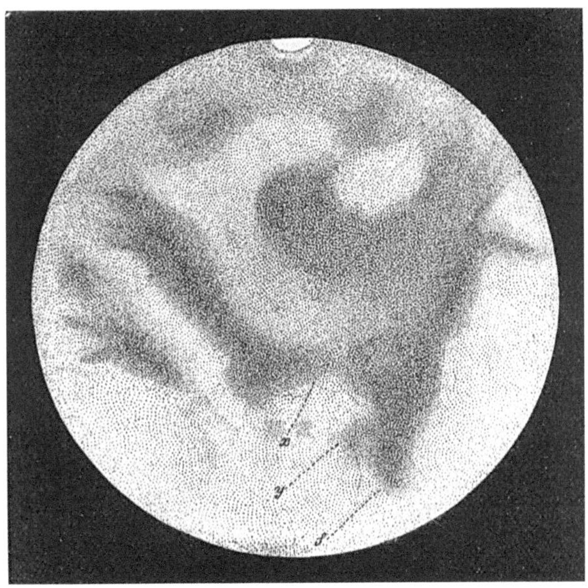

Fig. 32 Drawing by Lockyer, 3 October 1862, at 11^h 23^m

From this whole series of observations, we may conclude that the following points have been established:

1. Permanence of the position of the features;
2. Variations in the extent and darkness in tone of the features;
3. Various degrees of darkness in the seas; the Hourglass Sea, Herschel II Strait, Maraldi Sea and Schiaparelli Sea are generally the darkest.

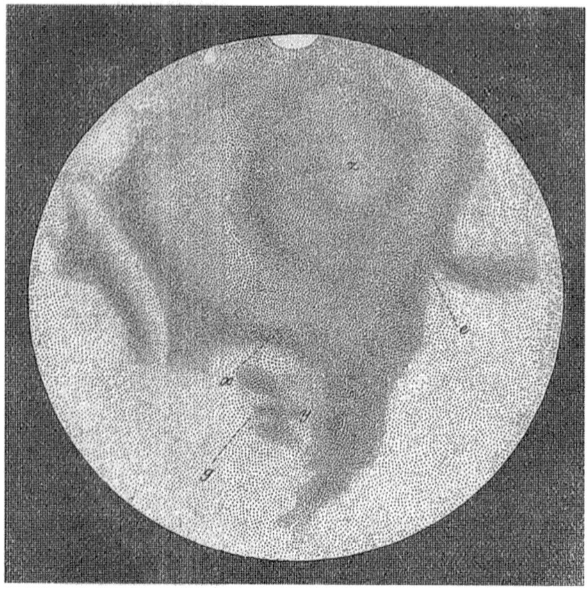

Fig. 33 Drawing by Lockyer, 3 October 1862, at $11^h\,51^m$

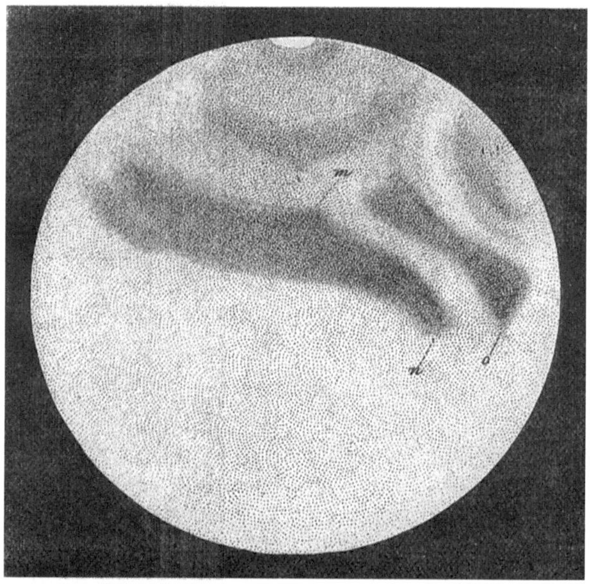

Fig. 34 Drawing by Lockyer, 9 October 1862, at $10^h\,47^m$

To the Memoir we have just summarized, Lockyer has added the following note; it concerns the observations of Mars made during the same period by Professor Phillips and the Rev. W. R. Dawes:

Mr. Phillips concludes from his observations that, against a permanent background of bright and dark features, there is on Mars a variable atmospheric envelope which condenses and fluctuates, partly modifying the aspect of the basic configurations, and even disguising them to some extent and joining bright areas to dark ones;

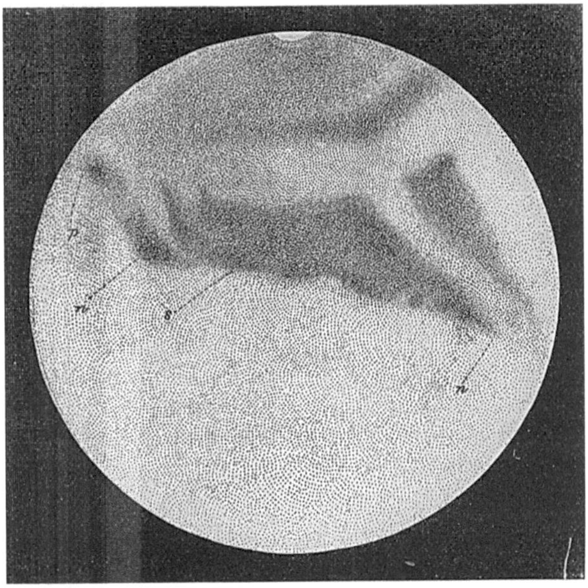

Fig. 35 Drawing by Lockyer, 11 October 1862, at 11^{h} 4^{m}

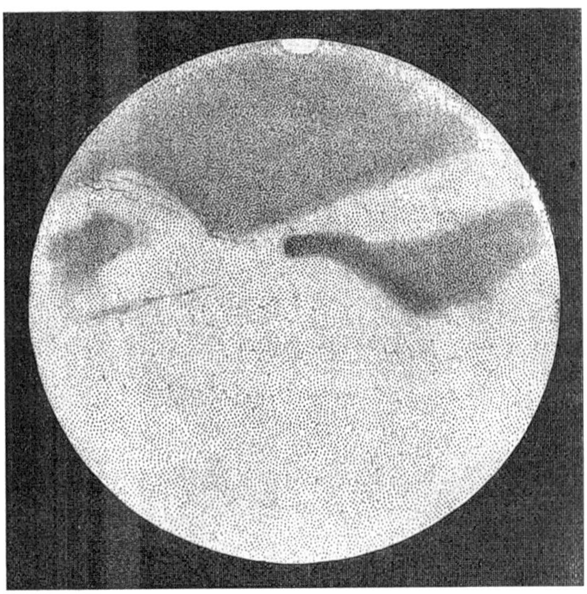

Fig. 36 Drawing by Lockyer, 15 October 1862, at 9^{h} 8^{m}

new lights and shades[25] which present no constancy, a thin, vaporous atmosphere, probably resting on a surface of lands, snow and water.

This inference is as remarkable as it is interesting, and it will be confirmed by later research.

[25] This seems more probable to me than the assertion that Lockyer made above, that we can explain the phenomena more easily by black clouds seen from above and lit by the Sun.

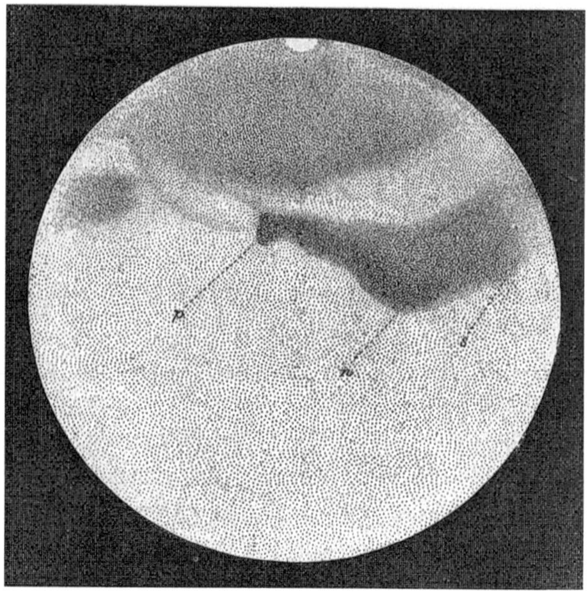

Fig. 37 Drawing by Lockyer, 15 October 1862, at $9^{h}\,20^{m}$

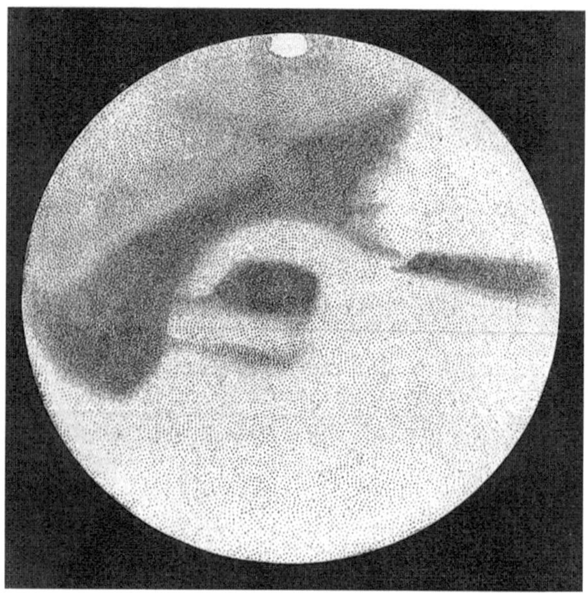

Fig. 38 Drawing by Lockyer, 18 October 1862, at $8^{h}\,0^{m}$

In reference to Dawes's observations, Lockyer remarks that they are in perfect agreement with his own. But one drawing made several minutes after his of 3 October at $11^{h}\,51^{m}$ confirmed the passage of the clouds mentioned above, as at the time of Dawes's drawing the clouds had completely disappeared and the geographical features of the region had become clearly visible.

These excellent telescopic drawings by my learned friend Mr. Lockyer, whose name during this epoch has become unforgettable in the cause of the progress of contemporary astronomy, represent a very important advance in the physical study of Mars.

Same Year, 1862.—Phillips

During the same apparition, that of 1862, Professor Phillips at Oxford made some very detailed observations of the planet, communicated to the Royal Society of London on 12 February 1863. A résumé follows.

First, Phillips commented that the various views of Mars were mutually discordant, and that in comparing them he had been very perplexed. Were the patches permanent? Were they seas? Were they lands? The assertions given above by Secchi and Lockyer were in complete contrast to Phillips's uncertainty.

Reflectors are better than refractors for colour estimates; refractors give the sharper images. Phillips made his observations with a 6–in. ($0^{m}.152$) refractor, equatorially mounted and clock-driven.

From his various drawings Phillips chose three, which are reproduced here. The first shows the Hourglass Sea, Herschel II Strait, Dawes Ocean, Lambert Sea—which goes down toward the pole—and an upper polar sea. On the second, to the left we see the Hourglass Sea, Maraldi Sea, Hooke Sea and Zöllner Sea. The third drawing shows Maraldi Sea to the right, Schiaparelli Sea in the middle and Terby Sea to the left. The first of these drawings is very similar to that made by Sir John Herschel on 16 August 1830 (Figs. 39, 40, and 41).

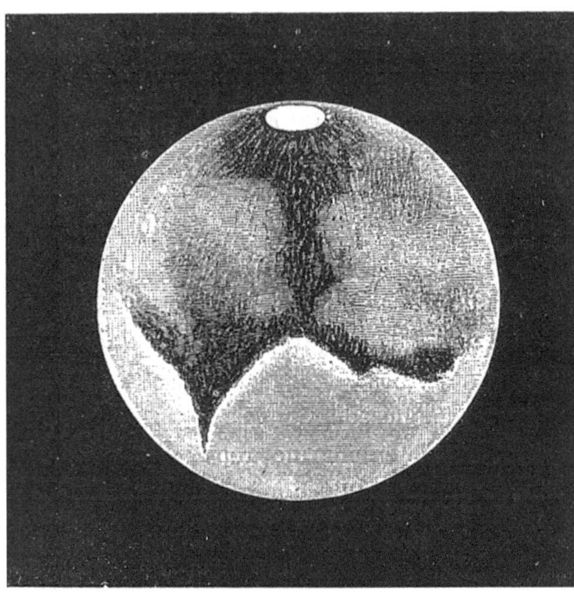

Fig. 39 Drawing of Mars by Phillips, 27 September 1862

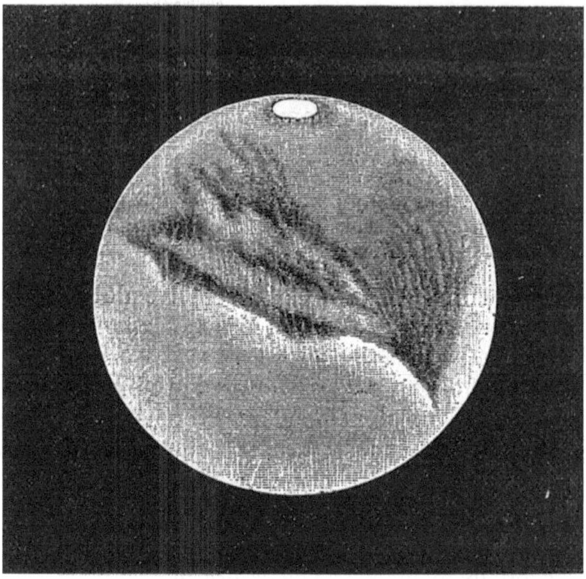

Fig. 40 Drawing of Mars by Phillips, 11 November 1862

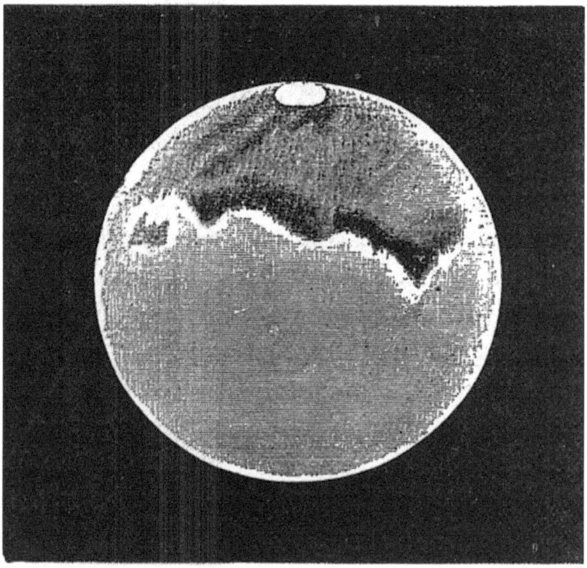

Fig. 41 Drawing of Mars by Phillips, 15 October 1862

Phillips expresses his regret that he cannot be sure whether the grey patches represent true seas, and not as with the Moon, merely grey plains. He commented that we would have positive proof in favour of the former idea if we could see there the image of the reflection of the Sun. Without taking irradiation into account, the Sun's image, reflected in the Martian seas, would be only 1/20 of a second of arc across, but even so it would appear conspicuous. A thermometer bulb 1 in. (25^{mm}) in diameter reflecting the sunlight is visible from 25 yd (22 m) as a star; the reflecting surface is scarcely 1/240 of an inch in diameter, and consequently subtends, after an adjustment for irradiation, about 1^h of arc. Using a magnification of 300× for observing Mars, 1/20 of a second becomes 15^h. This would be perceptible. Phillips therefore concludes that under certain conditions we should see the image of the Sun reflected in the waters of Mars, whether the waters are calm or whether the image is diffused by the action of waves.

The Martian atmosphere should play a major role in modifying the geographical aspects which we observe.

The polar snow lies to the side of the pole, and some way from it.

With the telescope, the continents are red and the seas green.

Phillips adds, in conclusion, that the different aspects could be due to cloudiness. There is, he says, an enormous transport of moisture from one hemisphere where it is winter to the other where it is summer, giving rise to storm systems and vast cloudy masses. These are not distributed along the parallels of latitude as with the rapidly-rotating planet Jupiter but rather their arrangement depends upon the influence of lands and waters.

Same Year, 1862.—Observatory of Lord Rosse[26]

Lord Rosse communicated to the Royal Society of London six drawings made by his assistant during the very favourable period in 1862. These drawings were made on the following dates with the great 6–ft ($1^m.83$) reflector (Fig. 42):

—22 July, at 22^h 30^m sidereal time—Definition imperfect.
—14 September, 6^h 26^m sidereal time—Definition quite good.
—16 September, at 23^h 55^m sidereal time—Very good definition. Magnification 1,200×. There was light mist, and quite possibly the clarity was the best of the season.
—6 October, 2^h 10^m sidereal time—Definition good.
—29 October, 1^h 00^m sidereal time—Definition bad.
—6 November, 1^h 40^m sidereal time—Definition very bad.

On the first and last of these drawings the Hourglass Sea may be recognized. The third clearly shows the circular Terby Sea. On the second, to the right of this lake, is Schiaparelli Sea. It seems that there are scattered clouds in each of these drawings made with Lord Rosse's great reflector.

If we compare these drawings with the preceding ones, we can see that they provide confirmation. Thus, for example, that of 16 September strongly resembles Secchi's of 18 October; also Lockyer's of 18 October and 17 September. That of 6 October gives the same aspect as Lockyer's of 3 October. All the features recognized above

[26]Observations made at the Parsonstown Observatory (Ireland).

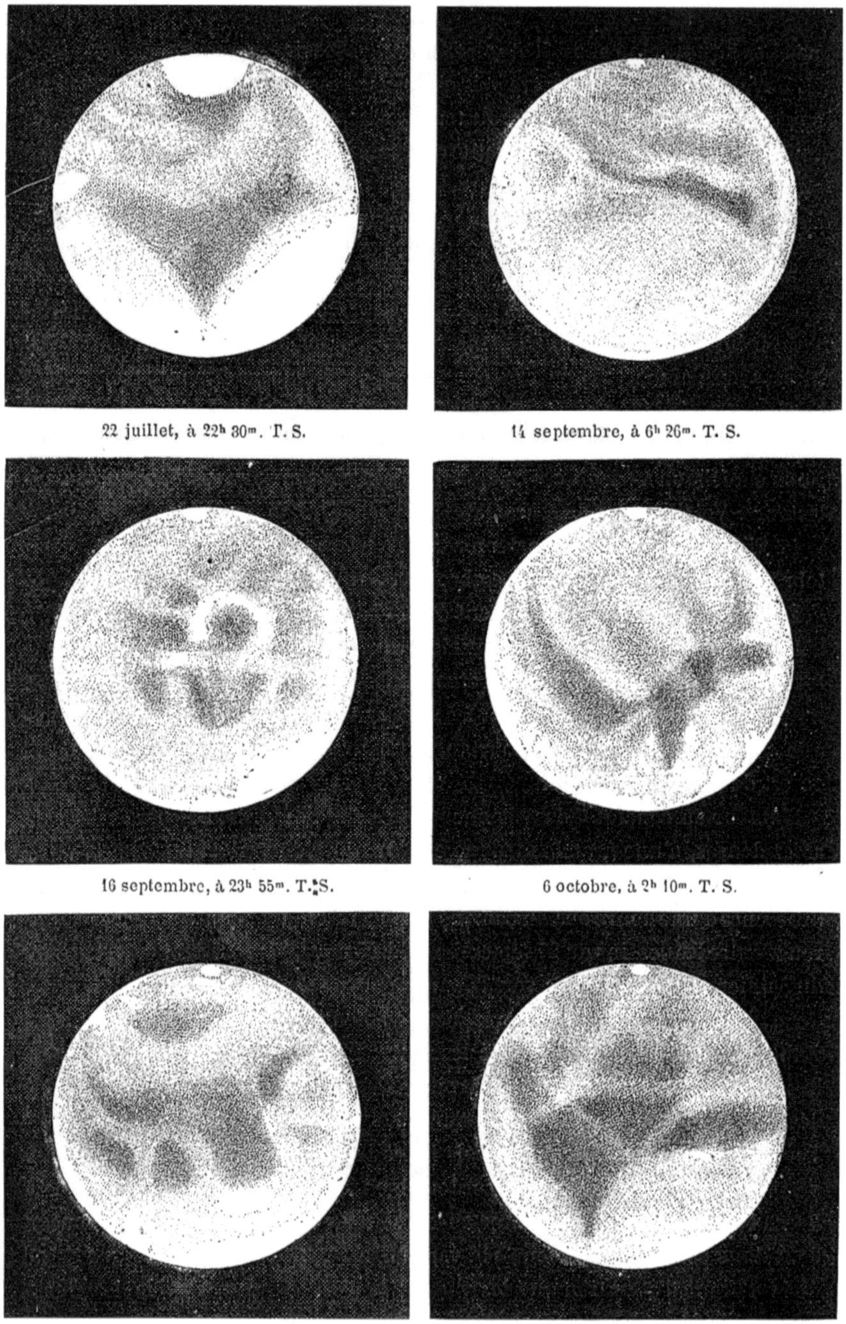

Fig. 42 Telescopic views of Mars, obtained in 1862 at the Observatory of Lord Rosse

are represented in those drawings, depending of course on the hemisphere turned toward us.

They also show that each observer sees in his own particular way, according to his eye, his experience, and his method of drawing. M. Faye told me one day, that on a fine evening at the Paris Observatory, he had begun to draw Mars in the company of one of his colleagues (Goujon); this was in Arago's time. Comparing their drawings, made with the same instrument and within a quarter of an hour of each other, little resemblance was found. More than once I have made the same comment.

Same Year, 1862.—Lassell

In the same year Lassell, with his great 4–ft ($1^{m}.20$) reflector, made a remarkable series of observations, and communicated them to the Royal Astronomical Society of London. There were 24 drawings, from 13 September to 11 December 1862. From these sketches I have chosen eight of the most interesting to offer to my readers. Here are their dates, in order:

(1) 25 September; (2) 27 September; (3) 11 October; (4) 13 October; (5) 23 October; (6) 25 October; (7) 4 November; (8) 5 November.

Magnification used ranged between 474× and 760× (Fig. 43).

The snowy cap of the upper or south pole is clearly visible on all the drawings. The observer comments that the patches have varied during the observations. Thus, he says, the face presented on 27 September is the same as that of 5 November, and yet the drawings bear little resemblance to each other; it is the same with the others.

Lassell concludes that the changes are undoubtedly produced by clouds of sufficient density, great extent, and a wide variety of forms.

This conclusion is not really so unusual as Lassell believes, because even a difference of an hour or two can sometimes produce perceptible changes. The difference between the drawings of 27 September and 5 November is a case in point, where the Hourglass Sea is further to the left in the first drawing than in the second.

On the drawings of 4 and 5 November, our readers will recognize the Hourglass Sea very distinctly, with the Flammarion and Hooke Seas to the left, the Zöllner Sea above, Herschel II Strait to the right, and the Arago Strait and Lambert Sea above it; on the sketches of 23 and 25 October, the Terby Sea, De la Rue Ocean, and the bays (Christie, Burton and Meridian) appear. These two latter drawings are in good agreement with those of Lockyer, Lord Rosse and Secchi, since each shows a very dark region above the Eye, the variability of which will be discussed later.

It is worth noting that the views of the planet obtained with the gigantic telescopes of Lord Rosse and Lassell do not always contain much more detail than those made with instruments of average power.

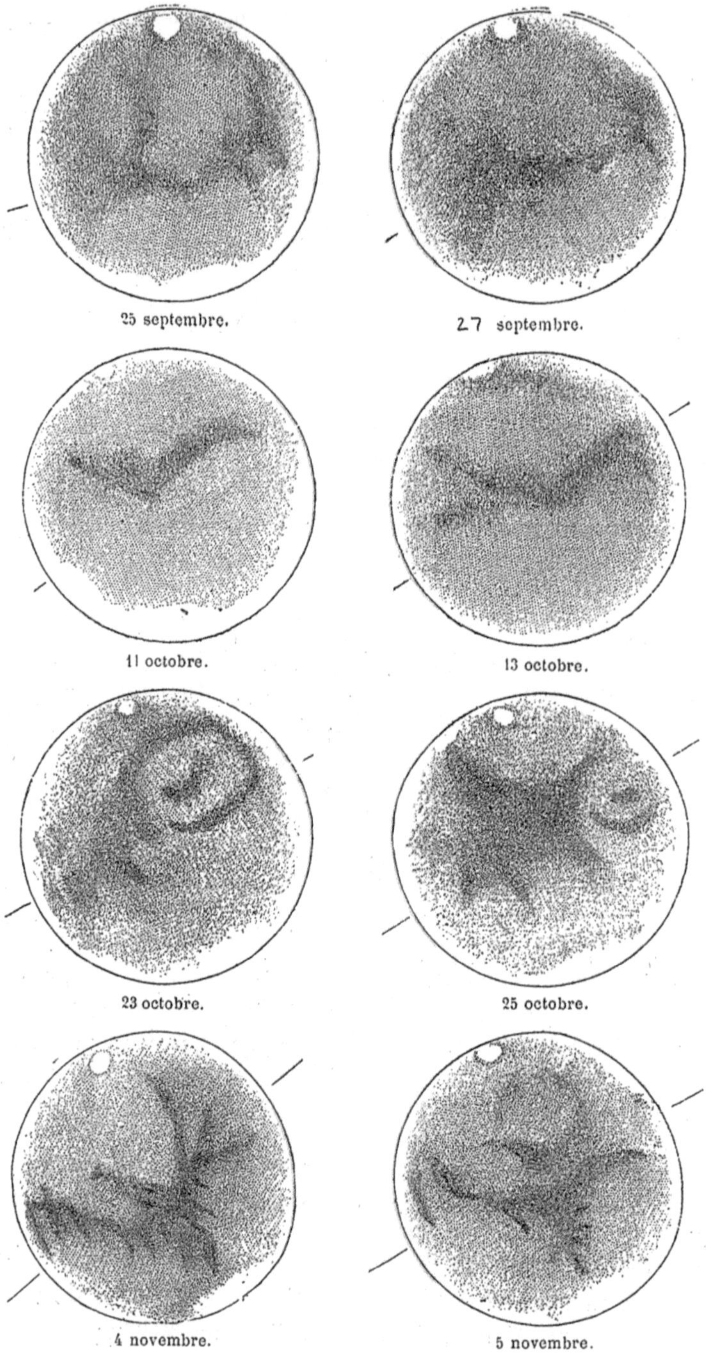

Fig. 43 Telescopic views of Mars, obtained in 1862 by Lassell

Same Year, 1862.—Main, Linsser, Nasmyth, Harkness, Grove, Knott, Ellery, Bulard, etc.

Almost all astronomers who had good instruments at their disposal made observations of the planet in the Martian year of 1862. In this book it is impossible to report all of them, or to reproduce all their drawings. Apart from the drawings made by Secchi, Lord Rosse, Lockyer, Phillip's and Lassell—already described—the observations made by Kaiser must be singled out as being particularly important; these will be discussed together with Kaiser's observations of 1864. But as a matter of principle we must also mention those of Main, Linsser, Grove, Knott, Harkness, Ellery, and Bulard.

The observations and measurements made by Main at Oxford have already been described.

Among the observations made in 1862, we must note those by Linsser at the Pulkova Observatory (Russia). In 1864, in Heis's *Wochenschrift für Astronomie*, he published an interesting paper in which he claimed that the drawings he had made agreed perfectly with those of Beer and Mädler. He asked whether the sombre patches could represent continents rather than seas, because of the various degrees of tone. He made a new calculation of the rotation period of Mars, and gave a value of

$24^h\ 37^m\ 22^s.9$.

Linsser's drawings do indeed confirm those of Beer and Mädler. He drew Herschel II Strait, which he called *Schlangen förmige Fleck* (the serpentine spot), as well as the Maraldi Sea (*pn*) and the Hourglass Sea.

The observations by Nasmyth in England must also be noted. On one drawing, made on 25 September, it is possible to recognize Herschel II Strait, with Phillips Island above. (*Memoirs of the Literary and Philosophical Society of Manchester*, 1862, 1863, page 303.)

Harkness, of the Washington Observatory, published the drawings he made on 6 and 30 September 1862 (*Annals*, 1862, p. 152). The first of these shows the Maraldi and Hooke Seas, the second the Herschel II Strait.

For 1862, note also the observations made by Knott in England and Ellery in Melbourne. These confirm the conclusions reached from earlier studies. Following Terby, I reproduce here four of Knott's drawings; they were made on 23 September at $8^h\ 20^m$, the same time on 22 October, 9^h on 3 November and $7^h\ 15^m$ on 27 November. These drawings lead to conclusions different from Lockyer's, since they indicate rapid and considerable changes in the aspect of Mars. They were made with a 7 1/3-in. refractor.

In the *Monthly Notices*, XXIII, p.75, Grove describes a series of drawings which he made in October and November 1862 with a 4½-in. (114^{mm}) refractor; they demonstrate certain variations on the Martian surface. Grove maintained that there must be clouds condensing on vast aqueous areas (Fig. 44).

As we have seen, other observers also made drawings during the 1862 and 1864 oppositions.

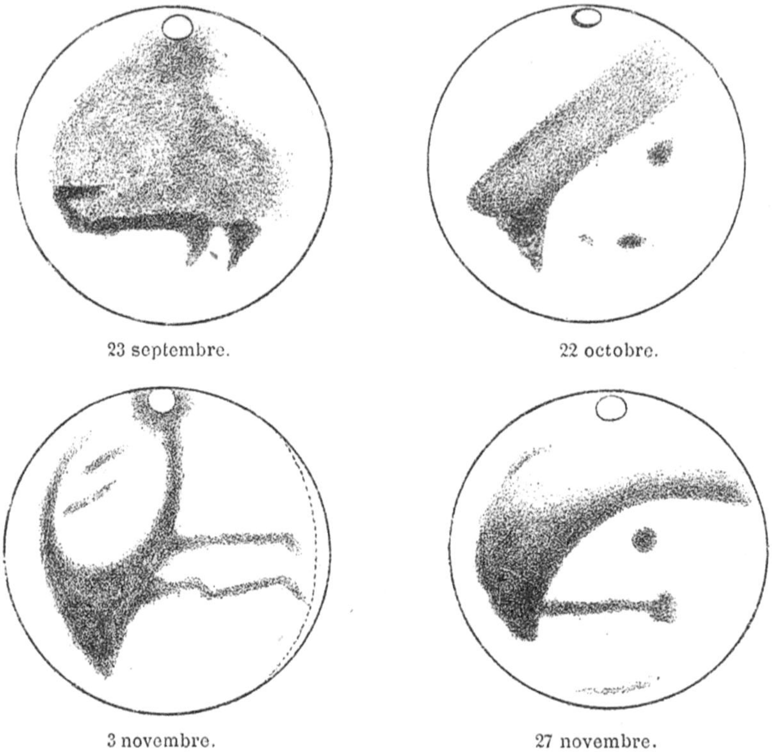

Fig. 44 Drawings of Mars, made by Knott in 1862

In France, at the meeting held on 15 December 1862, Bulard presented to the Academy of Sciences several drawings of Mars; they are not reproduced here, and of them I have no detailed information.

1862–1864.—Green and W.-L. Banks

Nathaniel Green and W.-L.Banks, both of whom were artists and amateur astronomers, lived in England; Green in St. John's Wood, Banks in Ealing. Both observed Mars during the oppositions of 1862 and 1864, the former with a French 4¼-in. (108^{mm}) refractor and eyepieces giving magnifications of 160×–240×, the latter with a 3¾-in. (95^{mm}) English refractor. They made a hundred drawings, of which 24 were published in the February 1865 issue of the old astronomical journal, the *Astronomical Register*.

These little drawings are charming, and I regret that I cannot reproduce them faithfully here. Comparisons of them bring out two incontestable facts: the first that despite the practical skill of the draughtsmen, they do not always agree in their representations of the planet; and the second, that the permanence of the Martian geographical configurations does not exclude considerable variations due, at least in part, to effects of the Martian atmosphere, which can sometimes produce bright or dark streaks in the form of equatorial bands.

In several of these drawings the Hourglass Sea is excellently shown, and on one in particular, made by Green at $10^h\ 30^m$ on 24 March 1864, there is a vague misty region shown. Later, this has been regarded by many observers as being subject to variations, due possibly to flooding.

1862–1864.—J. Joynson, Noble, Williams

During the oppositions of 1862 and 1864, Joynson, whose observatory was situated at Waterloo, near Liverpool, presented to the Royal Astronomical Society of London a series of 92 drawings of Mars, made in 1862; there were 104 others made during 1864. The Monthly Notices (10 March 1865) reproduced only two sketches, showing a grey band going round the planet. These drawings were made on 8 and 12 December 1864. The Terby Sea is recognizable, and is very black; the band is formed by the succession of seas—Schiaparelli, Maraldi, Flammarion, Herschel II and de la Rue.

In 1862, Joynson used a 3½-in. refractor, and in 1864 a 6–in refractor. In each case the magnification was 350×. He believed that the band shown on these two drawings went right round the planet without interruption. At the same meeting of the Society, Lockyer commented that the sea is not continuous, but is crossed at many points by lands. He said that the different patches on the planet were of different degrees of intensity, with some of them darker than others. These differences were shown in 1862 by Joynson, and also by Phillips and Frankland, exactly as they had been drawn by Beer and Mädler in 1830. By correlating his 1862 sketches, Joynson found a rotation period of $24^h\ 37^m\ 37^s$.

In England, Noble made some drawings with a 4–in (102^{mm}) refractor. He had begun studying the planet at the opposition of 1858, and continued until 1877. Williams obtained some drawings at the opposition of 1862, a dozen at that of 1864, and another dozen in 1867 (see *Monthly Notices*, XXV, p.170, and Terby, *Aréographie*, p.27). With his 4¼-in. refractor, he showed the principal patches on the planet, with their most striking characteristics. This period, from 1862 to 1864, was very fruitful. The most important work was carried out by Kaiser, Director of the Leyden Observatory, to whom we now turn.

1862–1864.—Kaiser[27]

At the oppositions of 1862 and 1864, this energetic observer carried out the most important Martian researches that we have yet described in the present monograph. The memoir which he published in the Annals of the Leiden Observatory was

[27] *Untersuchungen über den planeten Mars bei dessen oppositionen in der Jahren 1862 und 1864.—Annalen der Sternwarte in Leiden, Dritter band*, Haag 1872, p. 1–87.

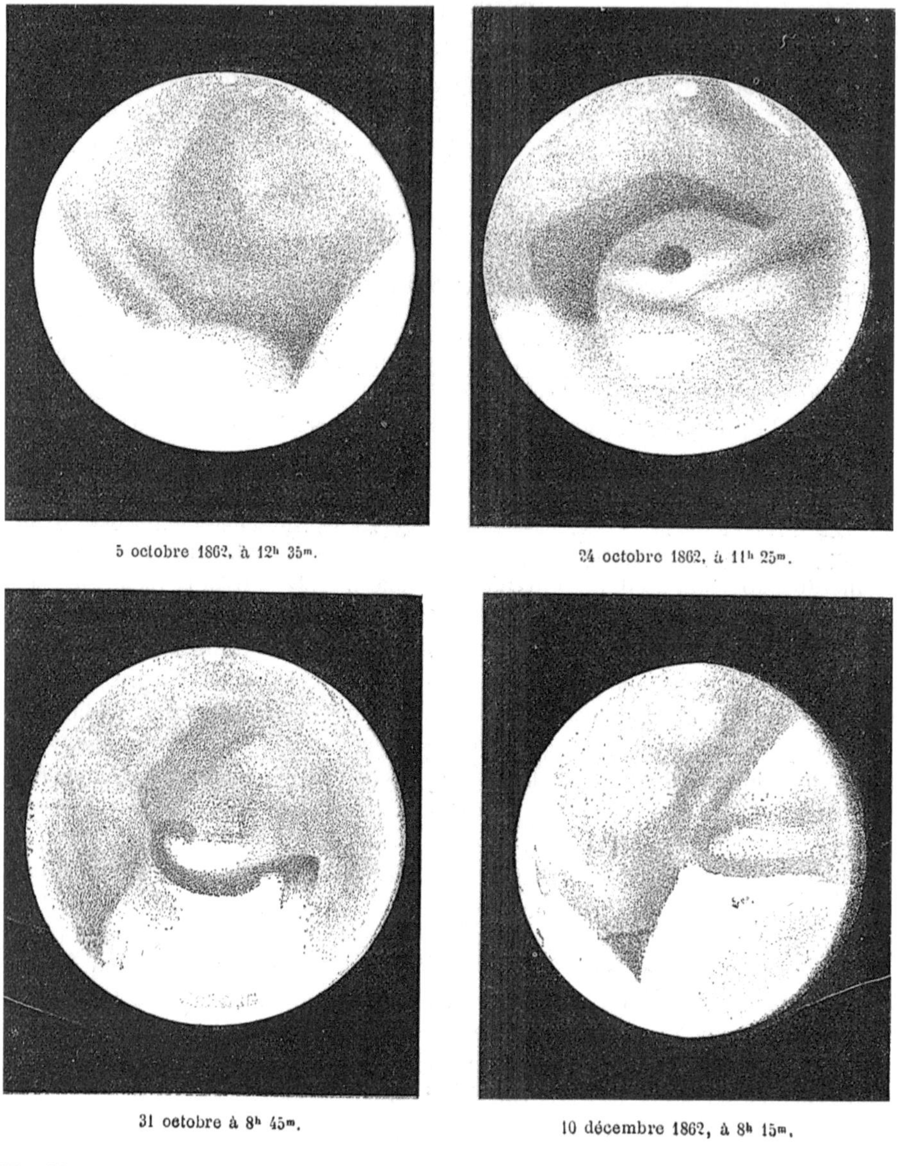

Fig. 45 Drawings of Mars, by Kaiser, in 1862

divided into several sections: the study of old drawings of Mars made from 1636 to 1864 and the observations made at Leiden with regard to the geographical configurations, the length or the rotation period, the polar caps, the flattening of the globe. The Memoir was accompanied by drawings and maps; the principal ones are reproduced here (Fig. 45).

There are four drawings made in 1862. The first shows the Hourglass, Flammarion, Hooke, and Maraldi Seas; Dawes Ocean, Zöllner Sea, Beer Continent, Herschel

Fig. 46 Telescopic view of Mars, by Kaiser, on 11 November 1864, at 10^h 30^m

I. Continent, Niesten Isthmus, and Cassini and Dreyer Lands; one can even make out Lockyer Land. The upper or southern pole is marked by a small circle or snow to the left of the Hourglass Sea; the whole area is misty. The drawing is complete, and in perfect agreement with the chart drawn from all the observations combined. It seems that at this time—5 October 1862, at 0^h 35^m —the aspect of the planet was not modified by any clouds at all. The tones themselves were noted.

The second drawing shows the Terby Sea, the De la Rue Ocean above it, and Schiaparelli Sea to the right. Instead of the Channel, a misty streak can be seen. The polar snow is detached from the edge of the disk (Figs. 46 and 47).

On the third drawing, we find Herschel II Strait and the Meridian Bay; the sombre ribbon is shown detached from the bright background, as in the time of Mädler, but instead of being circular the Bay is rectangular, ending in two points. The Bay was first seen as forked by Dawes on 22 September 1862, with an 8¼-in. Alvan Clark refractor. Above is the Lambert Sea; to the left, the Hourglass Sea. Phillips Island is very bright.

The fourth drawing seems to relate the first to the third.

It is interesting to compare Kaiser's *fig.* 1 (5 October) to Fig. 32 made by Lockyer on 3 October at 11^h 23^m, and *fig.* 4 to the drawing made by Lord Rosse on 6 October. All these agree well, allowing for the margin of error produced by different observers and their differences in drawing style. The same impression is had from comparing *fig.* 2 with Lockyer's drawing made on 17 September. Without a shadow of doubt, all these features are real, but they are shown with different degrees of certainty.

Thus we are gradually becoming more precise in our knowledge of the real geographical form of the Martian surface. The analogy with the Earth with regard to the distribution of seas and continents becomes more and more evident with observational progress.

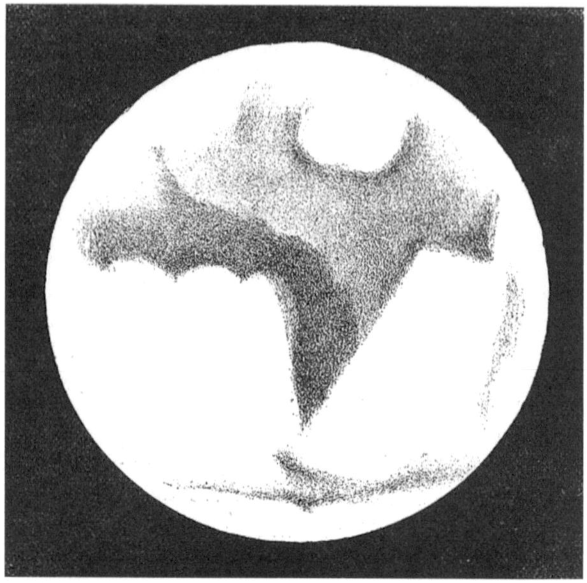

Fig. 47 Telescopic view of Mars, by Kaiser, on 22 November 1864, at $10^h\ 45^m$

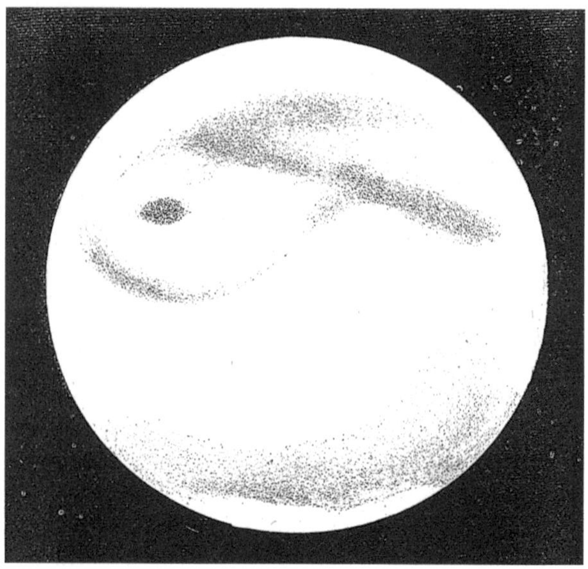

Fig. 48 Telescopic view of Mars, by Kaiser, on 10 December 1864, at $10^h\ 10^m$

This precision was increased rapidly by the six excellent drawings made in 1864, reproduced here (Figs. 46, 47, 48, 49, 50 and 51).

In the first (Fig. 46), we again see the Meridian Bay, broadened and merged with the neighbouring sea; Burton Bay, broad and double; and Christie Bay. All these appear, apparently, too broad.

In the second (Fig. 47) the Hourglass Sea is very dark; and above Dawes Ocean is Lockyer Land. Below the Hourglass Sea there is the Nasmyth Channel.

Fig. 49 Telescopic view of Mars, by Kaiser, on 18 December 1864, at $10^h\ 0^m$

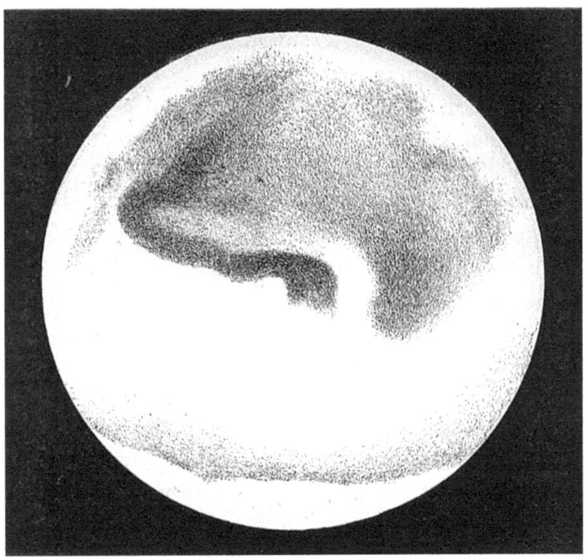

Fig. 50 Telescopic view of Mars, by Kaiser, on 23 December 1864, at $9^h\ 25^m$

On the right of the disk, there is an unidentified streak which could be Schiaparelli's Euphrates, about which more will be said later.

In the third drawing (Fig. 48), the Eye is the Terby Sea; Copernicus land, beneath which lies a dark band; and to its right, the Schiaparelli Sea.

In the fourth drawing, the Terby Sea is shown to the right, with the sombre streak below it (to be discussed later); Kepler Land, De la Rue Ocean, Christie Bay, Arago Strait and the double Burton Bay, with the Meridian Bay to the left.

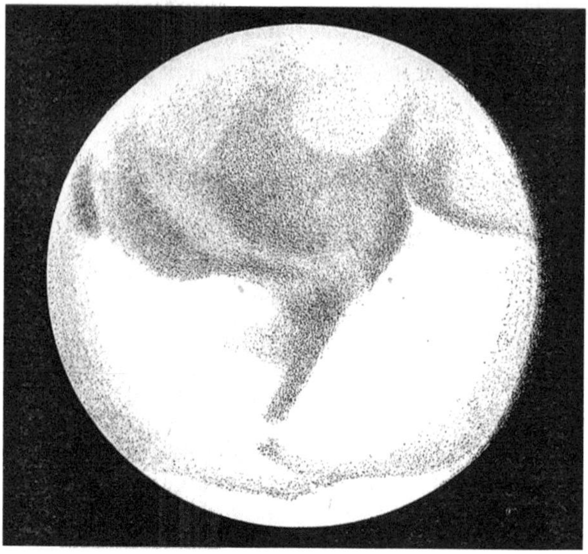

Fig. 51 Telescopic view of Mars, by Kaiser, on 28 December 1864, at $8^h\,0^m$

In the fifth sketch (Fig. 50) we see Herschel II Strait, appearing very dark and detached; Phillips Island (which is almost an actual island), and Arago Strait below the Delambre Sea.

In the sixth (Fig. 51) we see the Hourglass Sea, with its appendage; the Main Sea, diffuse; and all the features described above.

Henceforth, we can unmistakably affirm that the configurations on Mars are fixed and of a geographical nature, but are variable in aspect, no doubt because of changes in the atmosphere above them.

The Leiden astronomer compiled a geographical map of the planet by combining all his observations. This map is reproduced here in facsimile.

It is interesting to compare this map with my own (Fig. 3). Several differences are obvious at a glance. For instance, I have not separated Herschel II Strait, as Kaiser has, because in my view the sketch made by the Leiden observer does not represent a permanent configuration; generally the Strait is connected with the sea (Dawes Ocean). Generally, also, the Arago Strait is less broad at its extremity. These are two regions particularly noted for their variability, perhaps because the water of these seas retreats or evaporates, perhaps because of a change of tone—sometimes dark and blue, sometimes yellowish and bright. The more we progress in our studies, the more plausible becomes the idea of variations due to water: evaporation, flooding, and aqueous precipitation for various lengths of time.

In these observations by Kaiser, he recorded that for him the most remarkable and characteristic patch was the oval feature taken to mark the prime meridian; it lay at longitude 0°, latitude 26°S. Kaiser identified it with Beer and Mädler's spot *d* (see their map, Fig. 2). This is the circular lake which is called the Terby Sea on my map. Note its oval form on Kaiser's chart. This feature changes over the years. It is sometimes conspicuously oval, as here; sometimes perfectly circular. The area is equal to that of France.

The Hourglass Sea descends in a point below the equator, obliquely toward 150°. The round, circular spot taken by Beer and Mädler for the zero of longitude is shown as square in Kaiser's drawing, and 90° to the left of the oval patch which he takes as the zero for longitudes. It is therefore seen that Kaiser's longitudes differ by 90° from those which I have adopted, following Beer and Mädler (Fig. 52).

During 1862, 1863, 1864 and 1865, Kaiser used the 7–in. refractor at the Leiden Observatory, with Airy's double ring micrometer, to make a series of measurements of the polar and equatorial diameters of Mars.[28] The measures gave the following values:

Equatorial diameter	9″.468
Polar diameter	9″.387
Flattening	1/117

As we have seen, measurements of the flattening of Mars are not in total agreement:

W. Herschel gave a value of	1/18
Schröter	1/81
Arago	1/30
Bessel	0
Main	1/62, 1/38, 1/46, 1/7 and 1/36
Winnecke, Dawes and Johnson found a polar elongation	

Kaiser was equally concerned with the rotation period of Mars, comparing his observations with the best chosen from those of his predecessors and, justifiably, identified the vertical patch drawn by Hooke in his two observations of 3 March 1666 with the feature lettered *f* in the observations made by Beer and Mädler in 1830, and now known as the Hourglass Sea.

Hooke's observation of 3 March 1666 (old style) corresponds to 13 March in the Gregorian calendar, not adopted in England until 1752. The first sketch was made at 20^m, the second at 30^m. The patch had not yet reached the centre of the disk; it did so at $2^h\ 46^m$. In his 1862 observations Kaiser noted the transit of the same patch across the central meridian of Mars on 1 November, at $6^h\ 10^m$. He then made the following calculation:

Transit of the Hourglass Sea Across the Central Meridian of Mars:

Hooke	1666	Mar. 14	$2^h\ 56^m$
Kaiser	1862	Nov. 1	$6^h\ 10^m$
Difference:		$71{,}821^d$	$3^h\ 14^m$

During the interval, Mars rotated 70,004 times, giving a rotation period of:

$24^h\ 37^m\ 22^s.735$

[28] *Durchmesser des Planeten Mars, gemessen im Jahre 1862–1863–1864, mit Airy's Doppelbild Micrometer am 7 zölligen Refractor.—Annalen der Sternwarte in Leiden, Dritter Band.*—Haag 1872, p. 227,227, 241, 65.

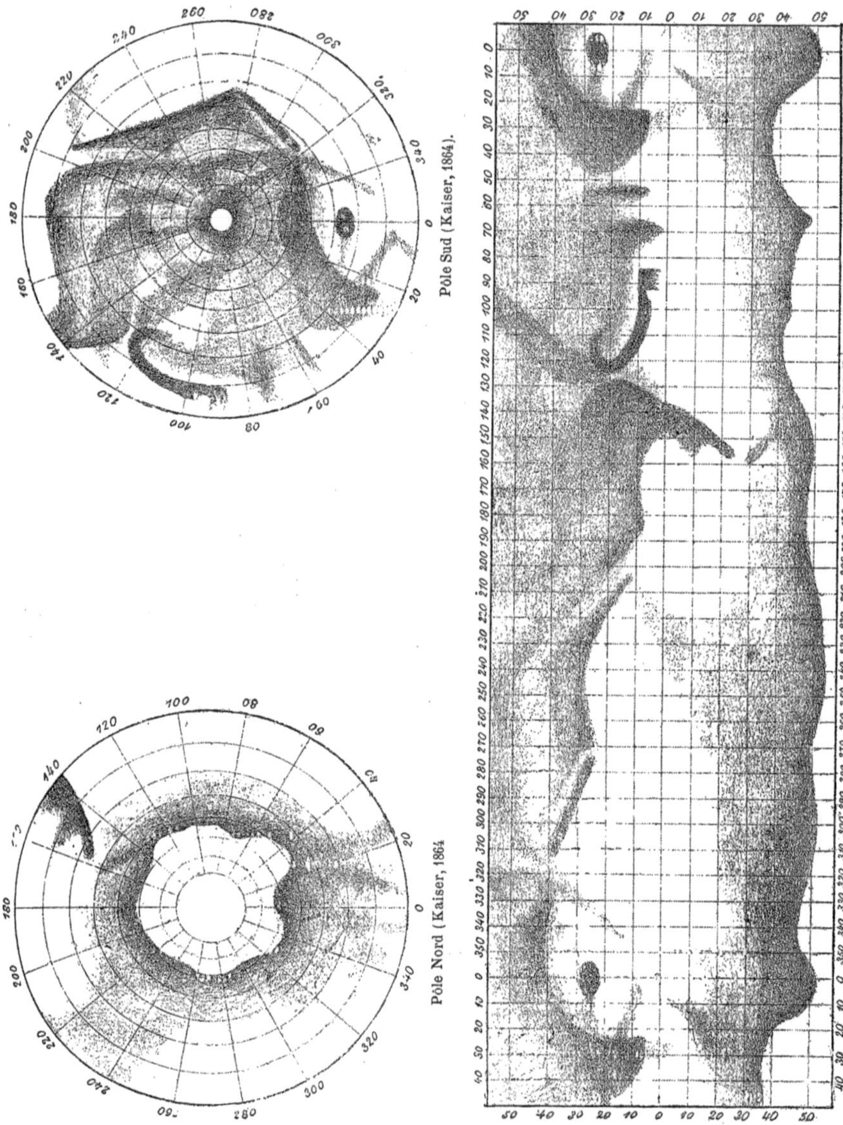

Fig. 52 Geographic chart of the planet Mars, constructed by Kaiser in 1864

From Hooke's observation and that of Beer and Mädler on 30 September 1830, at $17^h\ 22^m$, Kaiser obtained:

Difference $60{,}101^d\ 14^h\ 26^m$, or 58,581 rotations;
Rotation period: $24^h\ 37^m\ 22^s.706$

From Hooke's observation and that of Huygens on 13 August 1672, at $0^h\ 11^m$:

Difference $2{,}344^d\ 9^h\ 14^m$ or 2,285 rotations;
Rotation period: $24^h\ 37^m\ 22^s.62$.

The agreement is remarkable, despite Kaiser's doubts as to the identity of Hooke's vertical feature with the Hourglass Sea. He did not doubt the identity of the sea with the triangular patch on Huygens drawing of 13 August 1672, from which he derived:

Huygens	1672	Aug.13	$12^h\ 10^m$
Kaiser	1862	Nov.1	$6^h\ 10^m$
Difference:		$69{,}476^d\ 18^h$	

During this interval the planet made 67,719 rotations, giving a period of:

$24^h\ 37^m\ 22^s.643$

The same observation by Huygens combined with that by Beer and Mädler on 30 September 1830, at $17^h\ 22^m$, gave:

Difference $57{,}757^d\ 5^h\ 12^m$, or 56,296 rotations;
Rotation period: $24^h\ 37^m\ 22^s.595$

Summarizing, since 1864 the rotation period of Mars has been given as

$24^h\ 37^m\ 22^s.6$,

to an accuracy of about a tenth of a second.

1862–1864,—Spectral Analysis of the Atmosphere of Mars: Huggins, Miller, Rutherfurd and Vogel[29]

During 1862 William Huggins, member of the Royal Astronomical Society of London, and A. Miller, Professor of Chemistry at Kings College, made the first attempts to apply spectral analysis to studies of Venus, Jupiter, Mars and Saturn.

[29] Rutherfurd, Astronomical Observations with the Spectroscope (*American Journal of Science*, January 1863). Miller and Huggins, On the Spectrum of Mars (*Philosophical Transactions*, 1864). Vogel, Beobachtungen auf der Sternwarte zu Bothkamp, Heft. I, p. 66 (*Astronomische Nachrichten*, 1864).

The results obtained were published in the *Philosophical Transactions* for 1864. In the United States, Rutherfurd began the same kind of research at the same time.

In this study, we are not concerned with the other planets. With Mars, the spectrum was observed 6 November 1862 and 17 April 1863; the principal lines of the solar spectrum were clearly seen, but no fainter lines were found.

Huggins and Miller again examined Mars on 10 and 29 August 1864, using an improved spectroscope. In the red part of the spectrum they found no absorption lines of the type known in the spectra of Jupiter and Saturn, but at the edge of the red, toward the B and *a* lines of the solar spectrum, they detected three strong lines.

Toward the F line in the solar spectrum—that is to say, at the beginning of the blue or immediately after the green—the spectrum of Mars showed a large number of absorption lines, which considerably reduced its brilliancy. These strong bands were almost equidistant, and continued up to the violet end of the spectrum. The absorption due to these bands is presumably the cause of the predominance of the red rays from the planet. Spectroscopic apparatus with high dispersive power resolves these bands into groups of lines.

From these experiments, it is shown that Mars shines only by the light reflected from the Sun, and reflects the solar spectrum in the manner of a mirror. Secondly, the atmosphere produces the absorption lines seen by Huggins and Miller. What do the spectral lines indicate? We will soon know.

In August 1864 Huggins and Miller commented that the brilliancy of the spectrum of Mars had diminished in a remarkable manner toward the F line, because of a series of groups of lines, strong and more or less equidistant, beginning near the F line and continuing toward the most refrangible lines in the spectrum. In November 1864, these lines were much weaker and could scarcely be distinguished from the numerous lines due to the solar spectrum. Huggins's impression was that on 10 and 27 August the light of Mars was reddened; he also found that the patches were more distinct than in November:

> If this opinion is confirmed by other observations, we must agree that toward the end of the year the atmosphere of Mars was more heavily charged with mists and vapour. These mists reflected a considerable part of the incident light, and by shading and hiding the lower layers of the planets atmosphere—and hence the surface itself, which is probably responsible for the red colour— produce the absorption lines which weaken the blue and violet regions of the Martian spectrum. By a corresponding series of telescopic and spectroscopic observations, it will undoubtedly be possible to make effective studies of the meteorology of the planet.

Researches carried out in Germany by Vogel agreed well with those made in England and the United States with regard to the absorption lines in the spectrum of the atmosphere of Mars.

In reply to a question from Mr. Pritchard—whether a simple mist produces lines in the spectrum indicating the nature of the substances involved—Huggins replied that the mist could have no power of selective absorption to produce definite lines. The small particles making up the mist were large relative to the wavelength of light, thereby weakening the intensity of the blue and green rays to a greater extent than for the red. The light reflected from a mass of mist would be bluish in colour.

In this connection, Lockyer recalled that in 1862 Mars appeared redder to the naked eye than 1864. This observation is an agreement with Huggins's comments. It seems that when the Martian atmosphere is free from clouds and mists, the light from the planet is reddened, and the patches on the surface show up more distinctly. This point will be referred to again in a discussion of the 1867 observations.

We now come to the opposition of 1864, when conditions were similar to those of 1862 apart from the fact that Mars was a little less close. However, many astronomers were concentrating upon observations, and it was hoped that useful results would be obtained.

At this point, having completed our discussion of the observations by Kaiser and the spectroscoptists, we pass on the work of the English observer Dawes, the astronomer with the eye of the eagle.

1864.—Observations by Rev. W.-R. Dawes[30]

At the meeting of 9 June 1865, the skillful and energetic observer presented to the Royal Astronomical Society of London eight magnificent drawings which are reproduced here, and which represent considerable progress in our study of the planet. These observations were made from November 1864 to January 1865, with an excellent 8–in. (204^{mm}) refractor made by Thomas Cooke & Sons:

> Many curious and interesting details which I have never recognized so distinctly were manifested during that opposition (of 1864). One of the most remarkable is a long, thin streak which runs in a NE-SW direction in the northern hemisphere, and which is shown distinctly on the drawings of 12 and 14 November and more weakly on that of 10 November, as well as that of 21 January. I have already noted and drawn this streak since the year 1852, when the north pole was similarly turn toward us, but although the planet was then excellently placed for observation (its declination was +24°) I have never seen it so distinctly with the 6 1/3-in (160^{mm}). Munich instrument which I then used, as with my present 8-in.
>
> Another interesting object has been the forked shading, drawn in particular on 14 November as well as on the 20th and the 12th (less distinctly). I had previously noted it in 1852 in the form of an oval bay with a regular bank, and I did not once suspect that its contour was divided or irregular. But on 22 September 1862, I saw a forked aspect very clearly, and it has been the same all through the last opposition. This aspect gives the impression of two very broad *river mouths*, but I have never been able to recognize these rivers. The excellent drawings made by Mr. Lockyer in 1862 show the bay on several occasions, but do not show it divided into two points. It will be very interesting, at future oppositions, to find out whether this form is permanent or variable. *It is probable that the sea retreats from this part of the bank and leaves a tongue of land visible.*
>
> It is very difficult to note, with certainty, variations in the aspects of the different patches which are due to atmospheric effects over the planet itself. The difficulty can no doubt be reduced if one takes care to compare the telescopic views with the configurations already known in the regions concerned; but to me it seems preferable to abstain from all reference and all preconceived ideas, so that the drawings made are absolutely independent.

[30]Hopefield Observatory, Haddenham, Buckinghamshire.

Nevertheless, the atmosphere must play a certain role in the causes of the variations. Thus, during three consecutive nights—20, 21 and 22 January—I observed a very white patch exactly at the position marked *a* on the drawing of 21 January.[31] This white patch was certainly not visible on 10 and 12 November. It gave the impression of being an enormous mass of snow, and was as brilliant as the south polar cap in 1862. Unfortunately, a series of cloudy nights prevented me from continuing these observations.

Nothing appears more certain, than that the red tint of Mars is not produced by the atmosphere of the planet; the reddish colour is always more pronounced toward the centre of the disk, precisely where the atmospheric envelope is thinnest. Toward the edges of the disk, the grey patches are almost entirely hidden by the denser atmosphere, and the colour reflected from these edges is white or greenish white; the latter colouration may be an effect of contrast with the red of the centre.

On 1 December, several hours after opposition, I obtained several measurements of the disk with an excellent double-image micrometer. It was not possible to recognize any trace of flattening; on the contrary, I found the polar diameter rather greater than the equatorial, though by an insignificant amount 0″.02. This result recalls the measurements of Mars made by Mr. Johnson with the Oxford heliometer.

My impression is that the atmosphere of Mars is not generally very cloudy. During the last opposition, the principal configurations nearly always showed up brightly and sharply. I was not able to say, with certainty, that there were regions masked by mist and cloud. The sole exception to this permanence is found with the very white patches occasionally noticed, which gave the impression of being masses of snow or else cloudy masses whose surfaces reflected in the sunlight very efficiently. One can also note a remarkable fact about these variations by comparing the drawings of 14 November with those of the 10th and 12th of the lower pole; at the point mark a on the drawings of 10 and 12 of November there was a small grey streak, very evident at midnight on 14 November but certainly not existent on the other dates, though the neighboring details were seen perfectly.

To these extremely interesting observations, Dawes added the following postscript:

In looking back at my drawings of 1852, I saw in that year a particularly white streak along the bank marked a on the drawing of 20 November 1864. It attracted special attention because of its brilliant whiteness. It seems, therefore, that we have some effect which is either permanent or at least frequent, causing this exceptional brightness. However, since the region is in the neighborhood of the equator, it seems more reasonable to attribute the whitening to clouds rather than snow, at least if we are dealing with plateaux above the level of the sea (*Monthly Notices of the Royal Astronomical Society*, XXV, and *Memoirs of the Royal Astronomical Society*, XXXIV) (Figs. 53 and 54).

As noted above, these magnificent drawings by Dawes represent a considerable advance in our knowledge of Martian topography. The forked Meridian Bay was shown in its normal form, as well as Herschel II Strait, the banks of the Hourglass Sea, and most of the configurations shown on my map. At the Meridian Bay, which naturally gave rise to the idea of two broad river-mouths, Dawes looked for the rivers without being able to find them; Schiaparelli discovered them 13 years later, in 1877.

The white island observed on 21 January is shown on my chart, at the intersection of the 60° meridian with the south polar circle. It is not always visible, and neither are its surroundings.

Note also what Dawes said about the colouration of Mars and its atmosphere. The reddish hue of the planet is always more marked in the central region of the disk

[31] This corresponds to the feature Schiaparelli later named Argyre.—WS.

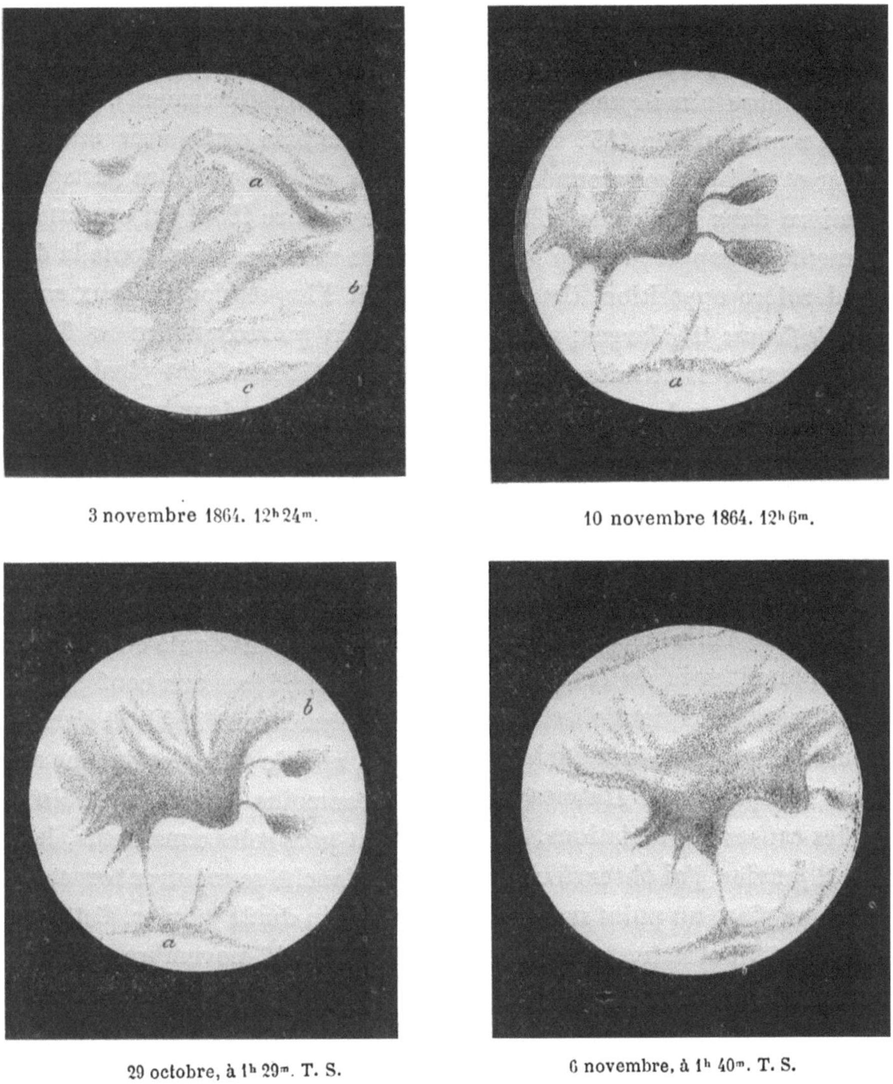

Fig. 53 Drawings of Mars, by Dawes, in 1864–1865

than at the edges. Therefore it cannot be produced by the atmosphere, because it is at the centre of the globe that the light reflected from the surface passes through the thinnest layer of the air. This had already been pointed out by Arago.

The almost constant visibility of the patches of Mars, the rarity of clouds, and the weakness of gravity at the surface of the globe lead us to believe that the atmosphere of Mars is very tenuous. That of the earth is so dense that details of the terrestrial surface would be less clearly visible, from a great distance, than those of Mars. According to researches by Langley, 40 out of every 100 solar rays which enter our

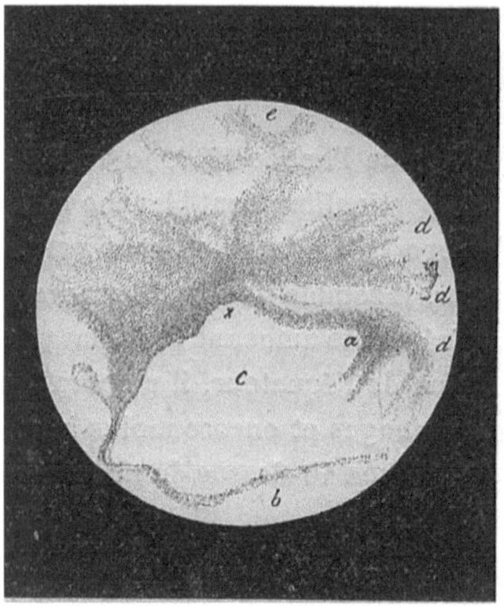

20 novembre 1864. 11h 36m.

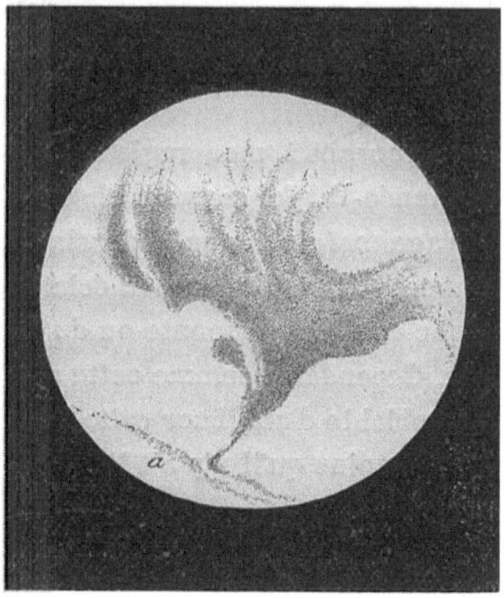

26 novembre 1864. 11h 36m.

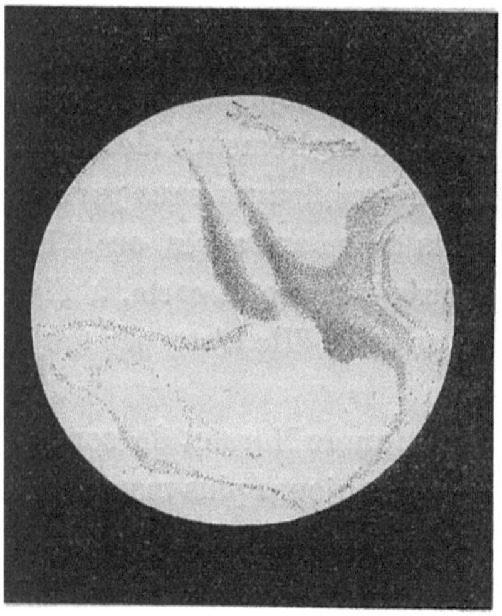

1er décembre 1864. 12h 0m.

21 janvier. 1865. 8h 0.

Fig. 54 Drawings of Mars, by Dawes, in 1864–65

atmosphere vertically are absorbed by it. Of the 60 which reach ground level, perhaps less than a quarter are reflected even by yellow sand, and this quarter must again lose 40 out of 100 in passing through the atmosphere. For the Earth, then, only 8 or 9 of each 100 light-rays would reach the eye of an observer on the Moon. From afar, the Earth should appear whitish, even with the clearest sky.[32]

[32]*American Journal of Science*, Vol. XXVIII, p. 163.

The comparison of Dawes's drawings with my chart leads to conclusions identical with those drawn from the observations by Kaiser. The Terby Sea is elongated instead of being round; it resembles the underside of a leaf. A second patch, showing the same aspect, is much larger; the Herschel II Strait is clearly detached, but the Meridian Bay is not so round and isolated as in the observations by Beer and Mädler. *Everything indicates certain variations of geographical aspects.*

1864.—John Phillips[33]

The Emeritus Professor at Oxford University, whose work for the 1862 opposition has already been described, continued his study of Mars during 1864, and presented his results to the Royal Society of London at the meeting of 12 January 1865.

First, he first he stated that the geographical aspects seen 1864 were nearly the same as those which he had drawn in 1862. He made several new drawings from 14 November to 13 December, and from them constructed the planisphere which is reproduced here. He regarded the orange regions as *land* and the greenish regions as *sea*, as is generally agreed; but in contrast to the more usual view, he noted the former as darker than the latter. A certain *fogginess* was noted under various circumstances, such as between 18 and 20 November and on other occasions. The seas were less green than in 1862, and in general everything was less sharp. However, the planet was further from the Earth in 1864 than it had been in 1862.

> White patches, undoubtedly snows, were seen between 45° and 50° latitude on the 30° meridian on the map constructed by the author, and also at latitude 50°, longitude 225°. There was less snow round the south pole than in 1862.

Phillips next asked whether the lower layers of the atmosphere played a role in the colouration of the planet, which often resembled the colour of clouds lit up by the setting sun. He remarked:

There must, at least, be considerable transport of water vapour to take the snow from one pole to the other, following the sequence of the seasons.

Phillips observed snows down to 50° or even 40° S latitude; Warren De la Rue saw them extend up to 40° N latitude in April 1856. This is a little less than with the Earth in winter. The extent of the snows is not always the same in different Martian years, as can be seen by comparing the drawing made by Sir John Herschel on 16 August 1830 with that of Phillips on 27 September 1862.

> The climates of Mars seem to be almost identical with those of our world, because there, as here, up to 50° latitude on the poles, water vapour should produce periodical snow, and from the equator to about latitude 40° the temperature remains high enough to produce normal evaporation; the atmosphere is generally clear in the equatorial and tropical regions, and the snows are variable out to a certain distance from the poles. It is undoubtedly the constitution of the atmosphere which has given rise to these quasi-terrestrial climates on a world which is more distant from the Sun than is the Earth by a ratio of 152 to 100, and on

---

[33] *Proceedings of the Royal Society*, 1865, p. 42–46.

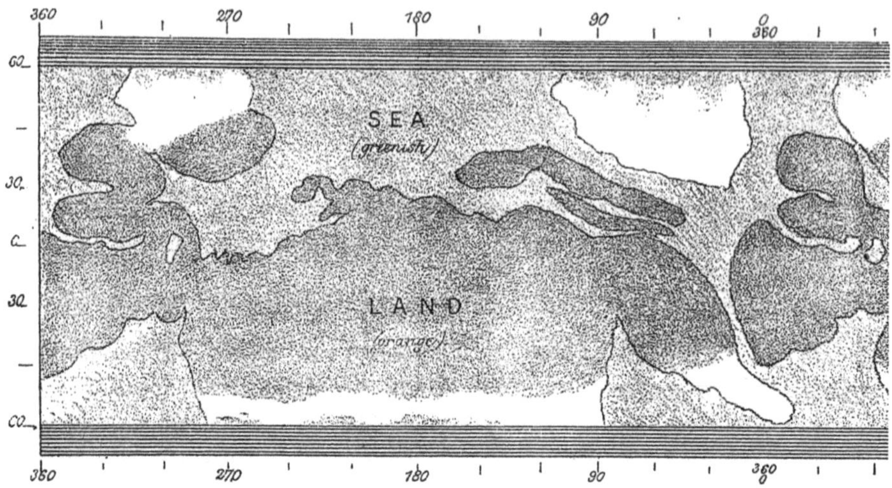

Fig. 55 Planisphere of Mars, equatorial projection, by Professor Phillips

which the heat received from the Sun is less in a ratio of 231 to 100. In preventing the radiation from escaping, and so conserving the Suns heat, the atmosphere makes the winters and nights less cold than they would otherwise be. The influence of the atmosphere on Mars seems to be the same as with us, and is even more important. The result, according to all probability, is that we should regard Mars as habitable.

Such are the results of the observations by Professor Phillips. The planisphere which accompanies his Memoir is reproduced here in facsimile (Fig. 55).

The observations seem to have been made with the same instrument as those in 1862.

It is not easy to identify features on this chart. First, it seems that pale features were seen as dark, and vise versa. It may therefore be deduced that the Hourglass Sea is the pale triangular configuration which descends almost vertically from the 20th meridian. Above the sea, the large white patch is Lockyer Land. The continent upon which the word LAND is written is Herschel I Continent. The prime meridian 0° on this planisphere corresponds almost to longitude 315° on my chart. The meridians are reckoned from right to left, from east to west, instead of from west to east. My zero lies at longitude 315° on Phillips chart, at the end of the bright ribbon which prolongs the Hourglass Sea into a long, narrow gulf to the right. There is a 45° difference between the two prime meridians.

Same Year, 1864. Félix Von Franzenau[34]

At the Vatican Observatory, Von Franzenau made a very interesting study of Mars, accompanied by six remarkable drawings which are reproduced here.

[34]Mars in November 1864. *Sitzungsberichte der K. K. Academie der Wissenschaften.* Wien, 1865, LIII Band, p. 509.

The following is a translation of the text of his short Memoir.

The favourable situation of the planet Mars during the last opposition led me to make the observations given here, which I made with the 6–in. (152^{mm}) refractor which was generously loaned to me by the Observatory. I set out to obtain representations of the planet which would be as faithful as possible with regard to the form of the patches and to the atmospheric conditions. Unfortunately, the extraordinarily bad weather, which persisted throughout, affected a large number of my observations. In all, I could make only seven drawings, and of these the last must be rejected as imperfect.

These several drawings allow me to confirm the permanence of the patches on Mars, and their striking resemblance to those shown on Mädler's drawings.

To explain the drawings, note that N S P indicate the North pole, the South pole, and Phase, respectively; by *s* I indicate the snow, or at least what resembles it, at the north pole (Figs. 56 and 57).

1. 8 November 1864, 9^h 30^m, Vienna time. *a*, *b* are the large very sombre patches separated by *v*. The blackness and sharpness of the contours of these three points of *a* is remarkable; the neighbouring regions *c*, *g*, *h*, *i*, *k* are grey shadings, scarcely perceptible, without well-defined contours; *x*, *y*, *z* are the bright red areas of the planet.
2. 10 November, 9^h 30^m —The patches remain the same, apart from a modification due to rotation. The patch *b* is considerably augmented at *d*; the band *f*, which I had previously only suspected, appeared very distinctly. As for the snow at the lower pole, it showed up as divided into two parts separated by a somber region.
3. 15 November 9^h 30^m —The progress of rotation begins to have a strong influence on the form of the patches. Patch *a* has completely lost its original aspect, and goes down well to the north; *b* is again enlarged towards *c*; as for the latter patch, it and *d* are very weakly illuminated in the region surrounding *b*; the reddish continent *x* has reached the middle of the disk, and has attained its greatest develoPMent. *s* seems almost to have disappeared, because the environs of the north pole are almost as sombre as *h* and *f*. All the northern hemisphere seems to be covered with innumerable small grey clouds.
4. 20 November, 7^h 45^m —Patch *a* has almost completely disappeared; *b* has reached centre of the disk, and in the west part *c* and *d* are more clearly visible than previously. *u* is a bright area between patches *b*, *c*, and *d*; *w* is a new bright red patch.
5. 20 November, 9^h 20^m —This drawing was made on the same evening, 2 hours later; the principal patches are closer to the central meridian of the planet, and consequently show up in more detail. *c* is even vaster and clearer; *u* can be easily distinguished, as a separation; *f* joins at *d* to the point *q*; a new patch, *p*, has appeared.
6. 22 November, 9^h —The interval between this drawing and IV covers almost two rotations of Mars. The only modification is at *r*; the snow at the north pole seems to extend further toward the south, but without well-defined limits. Note the sombre tint on the northern parts of patch b and the point *d*.

Such were the observations of von Franzenau. The most remarkable thing about them is their close agreement with the drawings of Beer and Mädler, giving extra proof of the permanence of the configurations. Another curious detail is the white isthmus above the Hourglass Sea, visible on drawings IV and VI, on 20 and 22 November. (Our readers will recognize this sea, as patch *d*.) Was it a band of clouds? This is not probable, because the same is shown on drawings on by Mädler in 1841, W. de la Rue in 1845, and Lord Rosse in 1862. Is it due to the variable depth of the sea? These topics will be discussed below. In any case, Von Franzenau's drawings lead us to a double conclusion: permanence and variations.

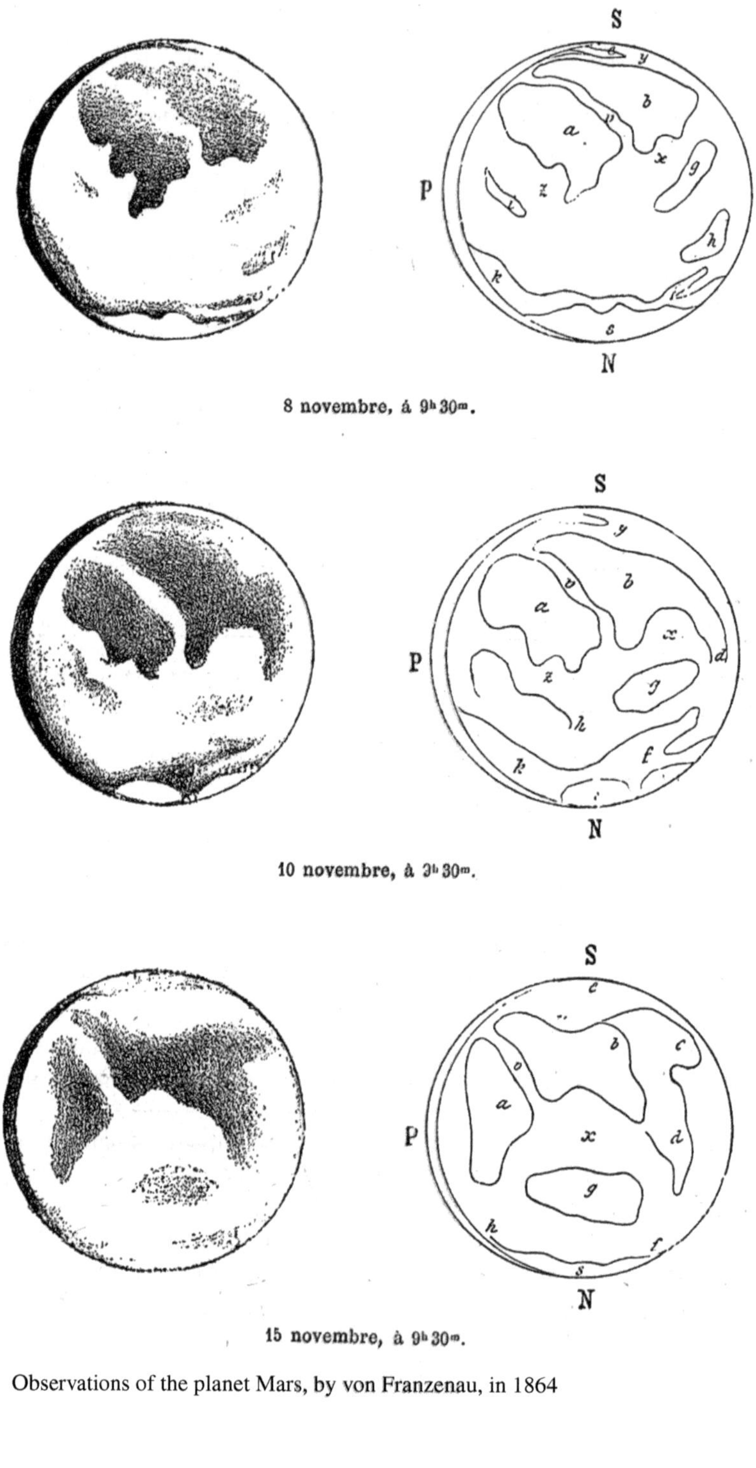

Fig. 56 Observations of the planet Mars, by von Franzenau, in 1864

Same Year, 1864.—Talmage, Secchi, and Rudolf Wolf

During the same opposition, that of the 1864, the English observer Talmage noted[35]—principally on 14 and 18 November—that the south pole seemed to project from the disk of the planet, an effect due doubtless to the irradiation caused by the brightness of the light reflected from the snow and which, measured with a micrometer, amounted to 2″.5. Talmage commented that this observation was identical with those made by William Herschel on 17 April 1777 and 20 May 1783.

On 24 November, the patches on Mars appeared more distinct than ever, although on this date our own atmosphere was disturbed. Talmage believed that from this he could conclude that the clearer our atmosphere, the less well the details on Mars would be seen. (He had observed Mars without any great success in 1862, in the excellent climate of Nice.) I take leave to question this conclusion, which would certainly be paradoxical.

Father Secchi re-observed Mars during the same opposition. His observations have already been described.

Note also, among the studies of 1864, those of R.Wolf, of Zurich.[36] In the hope of determining a new value for the rotation period of the planet, the eminent Director of the Zurich Observatory compared a drawing made by him on 19 November 1864, at $10^h\ 30^m$, with a drawing made by Secchi on 26 September 1862, at $9^h\ 45^m$ (see Fig. 20B), and found a period of

$24^h\ 37^m\ 22^s.9$.

Same Year, 1864.—J.C. Zöllner, Seidel, Schmidt: Photometry

The German physicists Zöllner and Seidel made photometric observations[37] during this opposition, finding that Mars resembles the Moon with regard to the variation of light reflected during different phases, and also with respect to the great brilliancy of parts of the edge of the disk. Also, Zöllner found that the albedo of Mars is scarcely greater than that of the Moon. Jupiter and Saturn, on the other hand, have high reflecting power. The cause seems to be that on these two planets there are clouds in their atmospheres which reflect the sunlight, while on Mars the reflection comes from the actual globe of the planet. Jupiter and Saturn had albedoes respectively four and three times greater than that of the lunar surface.

The falling-off in brilliancy of the Moon after full, particularly with regard to the bright edge, can be explained by the irregularities on its surface. Zöllner found that

---

[35]On an appearance presented by the spots on the planet Mars. *Monthly Notices of the Royal Astronomical Society*, 1865, p. 193.

[36]*Astronomische Mittheilungen*, no. 22, p. 57.

[37]*Photometrische Untersuchungen*. Leipzig, 1865.

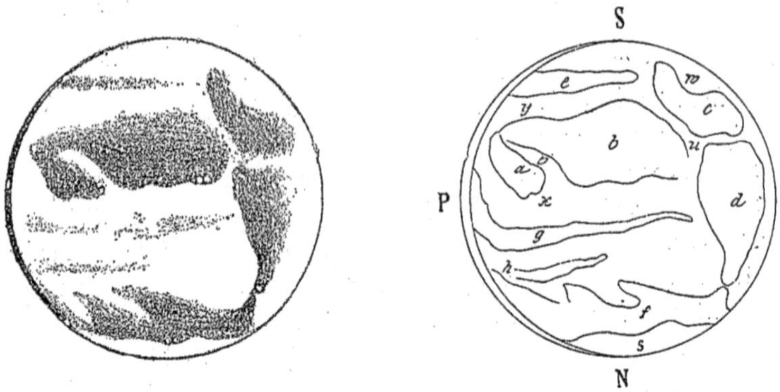

20 novembre, à 7h 45m.

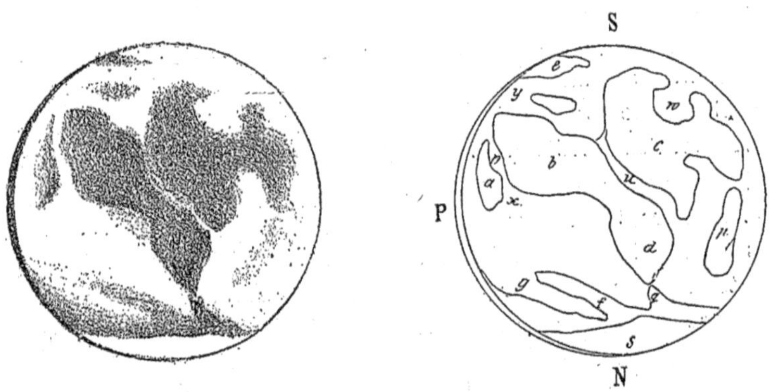

20 novembre, à 9h 20m.

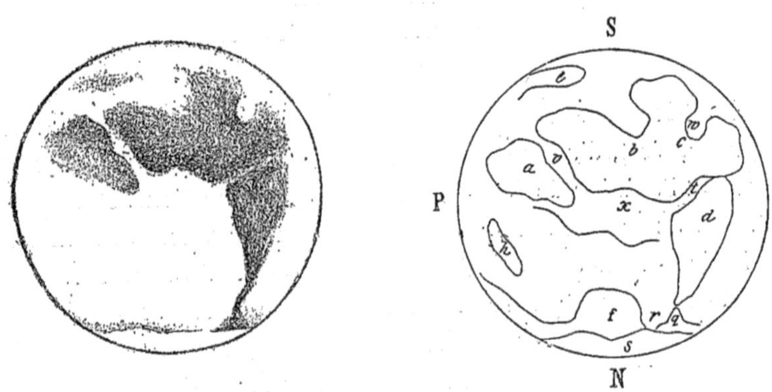

22 novembre, à 9h.

Fig. 57 Observations of the planet Mars, by von Franzenau, in 1864 for these irregularities to produce the observed changes in brightness, the mean angle of elevation of the irregularities should be 52° for the Moons surface. According to the same hypothesis, the much more rapid changes in the brilliancy of Mars demand a mean angle of 76° for the mountains there.

Zöllner gives a table to explain the albedo—or mean brightness—of each planet:

	Albedo
The Moon	0.174
White sand	0.237
Mars	0.267
Saturn	0.498
Jupiter	0.624
White paper	0.700

It can be seen that according to these data, Mars absorbs 733/1,000 or more than 7/10 of the solar light which strikes it, while Jupiter, with its cloudy atmosphere, seems to be almost as reflective as white paper, and reflects more than 6/10 of the light which it receives. Mars thus absorbs much more of the Suns radiation than does Jupiter. Seidel[38] found for the brightness of Mars, relative to the star Vega:

Mars at opposition = 2.97 × Vega,

or almost three times the brightness of Vega. The observations were made with a Steinheil objective photometer. With his photometer, Zöllner found, relative to the Sun:

Mars at opposition = 1/699,400,000 × Sun.

This determination by Zöllner corresponds to a stellar magnitude of −2.25.

From numerous observations, Julius Schmidt determined the dates when Mars should be equal to various stars of the first magnitude. Calling r the radius vector of the planet at a given moment, and D its distance from the Earth at the same moment, he found, for example, that:

Mars = Sirius, when $\log 1/\Delta^2 r^2 = 1.944$.
Mars = Aldebaran, when $\log 1/\Delta^2 r^2 = 1.258$.

1864–1875.—Dr. Terby

F. Terby, Doctor of Science at Louvain, to whose persevering and considerable work in areography we are so indebted, made a careful study of Mars from 1864 to the present time, and publishes his results in the Bulletin of the Academy of Sciences of Belgium. The first observations were made with an excellent Secrétan refractor, with an aperture at 108^{mm} and powers of 120× and 180×, and sometimes even as

[38] *Bayerische Akademie der Wissenschaften*. München, 1859.

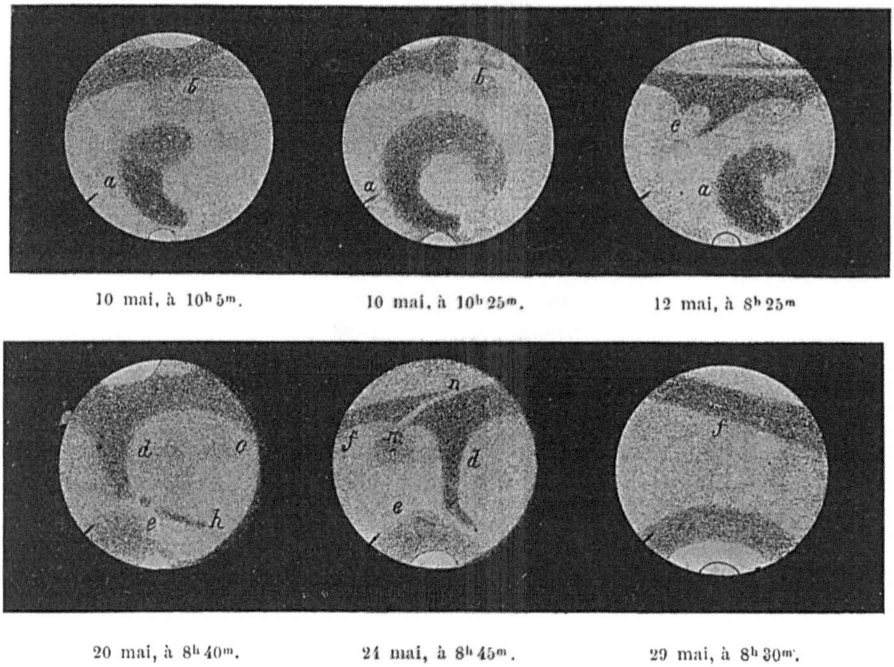

Fig. 58 Sketches of Mars, by M. Terby, in 1873

high as 240×. We will first discuss his observations made before 1877, the year which marked the start of the new cycle in our study of the planet. The Notices presented to the Belgian Academy by our eminent colleague are accompanied by 23 drawings for 1864 and 1867, 36 for 1871, 12 for 1873, and 22 for 1875. The sketches made during the two latter years are those which show the most detail. In particular, we can recognize the Hourglass Sea, Maraldi Sea, Herschel II Strait, and the north and south polar patches. From these numerous drawings, I first reproduce six particularly interesting ones made in 1873. The dates are given below each drawing (Fig. 58).

In these drawings, *a* indicates the Knobel Sea; *b*, the De la Rue Ocean; *c*, Meridian Bay and Herschel II Strait; *d*, the Hourglass Sea; *e*, the Delamare Sea and its surroundings; *f*, the Maraldi Sea. In the drawing of 24 May there is a curious separation *mn*, also seen on 22 May. This division is also shown on two drawings made in 1875, on July 20. On the drawing of 20 May 1873, the Nasmyth Channel is shown at *eh*.

In 1871 and 1873 the darkest and most easily visible patch was (as usual) the Hourglass Sea. The atmosphere of Mars sometimes seemed disturbed enough to hide the features and hinder studies of them, particularly in 1871.

Several of the 1875 drawings were of particular interest. The first, of 14 June (Fig. 59A), made at $11^h 30^m$, shows at *m* an indentation, and at *d* a very dark angular point. The long grey patch, formed by the Hooke, Maraldi and Hourglass Seas,

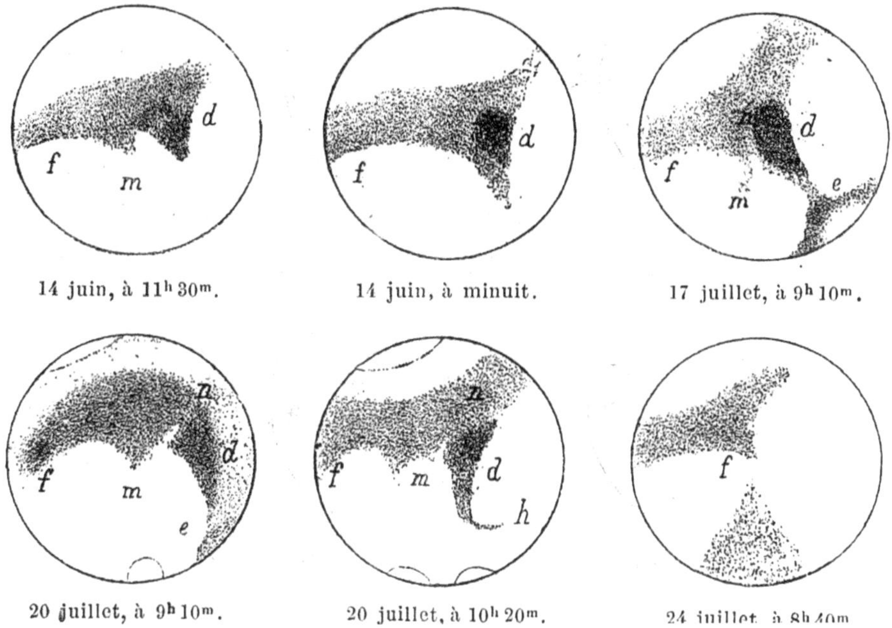

Fig. 59 Sketches of Mars, by M. Terby, in 1875

is also shown on an almost identical drawing made by Schröter on 9 September 1798, at $9^h\ 55^m$ (refer to Fig. 55, Chap. 1). The sombre angular point is the right-hand region of the Hourglass Sea, at that time appearing darker than the region below, as shown by Terby on the next drawing made half an hour later. We can also distinguish the lower part, which had not been seen earlier; this can be explained, on looking at this drawing from some distance, by assuming that the Hourglass Sea ends at this darker region, as in the first of the two drawings. This is irrefutable evidence in favour of variation in tone of the Martian seas, because the region is sometimes brighter than the vertical axis of the sea.

The third drawing shows, to the left of the Hourglass Sea *d*, an indentation *m* and a tongue of land at *n*, recalling the separation noted earlier in 1873 and which was also found in the drawings made by Von Franzenau. This same division is shown on the two drawings of 20 July.

The poles are marked by snowy patches; there are even two patches at the lower pole on 20 July at $10^h\ 20^m$, recalling a drawing by Von Franzenau on 10 November 1864 and one by Secchi on 16 November 1862. The last sketch shows the Maraldi Sea and Huggins Bay recalling an observation made by Schröter.

These observations by Terby lead us to the usual conclusion: permanence of fundamental patches, real variations in their details.

Consulting my own map, we seem to trace a bank of sand as a dark line above the Hourglass Sea and running to the left obliquely across the Flammarion Sea. All in all, we have the impression that the water here must be very deep.

1865.—C. Flammarion.—Researches with Regard to the Planet Mars[39]

The interpretation of observations of Mars is a matter of discussion. In the scientific review *Cosmos* of 26 June 1863, I discussed the observations of the polar snows and put forward the idea that the snow was chemically different from ours, and at the same time I expressed the hope that I could soon produce a complete Mercator map of Mars, succeeding the polar projections of Beer and Mädler (*Cosmos*, 1863, Vol. I, p. 751). In 1865, returning to the same subject, I claimed that according to the 1864 observations, the 0° isothermal line oscillates for both hemispheres, as with the Earth, up to latitude 45°; this indicates a mean temperature very little different from ours, despite the greater distance of Mars from the Sun. And yet it also assures us that the degree of condensation of terrestrial water and the crystallization of our snow can be duplicated on other planets. One might well think the contrary, because boiling depends upon the special relationship between the vapour of the liquid and the pressure of the atmosphere; it condenses according to the nature of the substances concerned. It is altogether too hasty to transfer terrestrial phenomena to an alien region (*Cosmos*, 1865, Vol. II, p. 315).

1867.—Huggins and Secchi: Spectral Analysis of the Atmosphere of Mars

We have already (1862) noted the first researches into the spectrum of Mars, made by Rutherfurd, Huggins, Miller, and Vogel. Huggins found that Mars shows the principal lines of the solar spectrum, and Rutherfurd detected the C, D, E, *b* and G Fraunhofer lines. At the Royal Astronomical Society meeting in London on 8 March 1867, William Huggins presented a new memoir on the subject, which will be summarized here.

Mars shines only by reflected sunlight. Its atmosphere absorbs part of this light, and its spectrum indicates which substances are present. In the blue and indigo part of the spectrum, the lines are too weak to be identified with certainty. In the red region, the Fraunhofer C line is easily visible, and its identity has been established by micrometrical measurements. Beyond this line, up to the end of the least refrangible part of the spectrum, can be seen a large number of dark lines. A very strong line has been micrometrically measured a quarter of the distance from C to B. As there is nothing analogous in this position in the solar spectrum, we must regard it as being due to absorption caused by the Martian atmosphere. The other lines in the red can be identified, at least in part, with B and *a*, the neighbouring lines in the solar spectrum.

[39] *Cosmos*: 6, 20 September and 11 October 1865.

On 14 February 1867, Huggins noticed weak lines to either side of the D line. Those to the more refrangible side were stronger than those on the other side. They occupied positions which appeared to coincide with the groups which we see when the sunlight passes through the lower layers of the atmosphere, and which are produced by absorption by gas or vapour, notably by water vapour. These lines probably indicate the existence of similar substances in the atmosphere of Mars. They are not caused by the terrestrial atmosphere, because under the same conditions they are absent from the spectrum of the Moon—even when observed at a lower altitude than Mars.

Huggins also observed the spectra of the darkest parts of the disk of Mars, that is to say the seas. Their spectra are much weaker throughout their length. The materials which form the dark regions absorb all the rays in the spectrum equally. We must conclude that in colour they are neutral, or nearly so.

The red colour of Mars cannot be attributed to selective absorption, i.e., the absorption of certain rays only, producing dark gaps in the spectrum. Moreover, it is not likely that this highly characteristic colouration has its origin in the atmosphere of the planet, because the light reflected from the polar regions remains white, even though it has passed through a greater thickness of the atmosphere than the light which comes to us from the central regions of the disk; it is in these central regions that the colour is most pronounced. Undoubtedly it originates at the surface of the planet.

The photometric observations by Seidel and Zöllner confirm this interpretation. They show that Mars resembles the Moon with regard to the abnormal value of the variation in reflected light according to the waxing or waning of the phase; also with regard to the great brightness of the regions at the edge of the disk. Moreover, Zöllner has found that the albedo of Mars, i.e., the reflecting power of different parts of the disk, is only 1½ times greater than that of the lunar surface. These optical characteristics agree with telescopic observation in that with Mars the reflected sunlight comes almost entirely *from the true surface of the planet*, not from a cloudy envelope, as with Jupiter and Saturn. In these two latter planets, the disk is less brilliant near the edges than in the central region. We have also noted that Jupiter and Saturn have albedoes four and five times greater than that of our Moon.

At the same time Huggins was investigating this problem in England, and Zöllner in Germany, Father Secchi was studying the planets Jupiter, Saturn, Uranus, Mars, and Neptune in his researches into the spectra of these bodies.[40] He wrote:

> Mars, shows terrestrial atmospheric lines weak toward the centre of the disk, but strong near the edge. These prove the existence of an atmosphere analogous to ours. He later gives two observations, of 11 February and 28 April 1869, which provide evidence of a nebulous zone near the C line and another in the extreme red. The atmosphere of Jupiter, Saturn and Uranus are very different from ours. The atmosphere of Mars seems a feeble and rarefied (*La sua atmosfera è asai piccola e sotile*).

[40] *Sugli Spettri prismatici di Corpi celesti.* 1 br. in-8; Rome, 1868. 1 br. u-4; Rome, 1872.

The same research was carried out in 1872 by Vogel in Germany, and the results confirmed those of Huggins and Secchi with regard to the existence on Mars of an atmosphere analogous to ours from the viewpoint of water vapour, which gives rise to the lines observed. We will return later (1872) to Vogel's researches into the subject.

1867–1873.—John Browning, Barnes, Johnson, Elger, Grover, Knight, Backhouse, Noble and Williams

The first of these observers, John Browning, an excellent London maker of optical instruments, published in *The Intellectual Observer* eight chromo-lithographs of Mars, from drawings made between 8 January and 24 February 1867.

He also presented to the Astronomical Society of London, on 10 May 1867, a series of thirteen coloured drawings (including those referred to above), made by him between 29 December 1866 and 24 February 1867 with an 8½-in. (216^{mm}) silvered glass made by Barnes. The colour of the disk varied from rose to ochre, the tint being redder when our own atmosphere was rather humid. The edges of the disk were very pale. The sombre patches where bluish-grey or greenish.

Light white patches have frequently been said to appear on the disk, being carried across it by rotation and becoming almost as white as the polar snows on approaching the edge of the disk. These clouds were in general poorly defined at their edges, and of circular form. They were always observed in the region of the equator.

On 31 March, at 7 hours, Browning made a final drawing, which corresponds exactly to that which he had obtained on 23 February at 9 hours. He wrote: "On these two drawings, the patch usually known under the name of the Hourglass Sea is represented as passing across the centre of the planet's disk."

At the same time Barnes, an engineer by profession, made drawings that agree well with those of Browning. In this double series, the views seem to be almost identical with those of Warren De la Rue, reproduced above; but they bear only a slight resemblance to those of Secchi, still less to those of Beer and Mädler.

In 1868 Browning constructed a globe of Mars, based on Proctors chart (to be described below), and he produced some curious stereoscopic views. Over a period of several years, Warren De la Rue had obtained excellent stereoscopic views of the Moon, using different conditions of libration at the same phase and giving an impression of surface relief.

Among the other observations of 1867, let us note those of Joynson, Elger, Grover, and Knight in England. In *The Astronomical Register*, we found observations made by these observers in 1867. The first conclusion is that there is a permanent band near the south pole. Elger comments that the colour of the disk is always stronger in the central region than toward the edges, when the patches are obscured near the limb. This is nothing new.

The southern band which Joynson believed to be continuous is the same as that seen in the drawings made by Terby in 1875. It is formed by the near-continuity of the Maraldi, Hooke, Flammarion and Hourglass Seas, Dawes Ocean, De la Rue Ocean, Cottignez Sea and Schiaparelli Sea.

Fig. 60 Sketch of Mars, by Williams, on 11 January 1867, at $11^h 40^m$

Let us also note in this period the observations by T.W. Backhouse, made during the oppositions of 1867, 1869, 1871 and 1873. They add nothing new to what has been given above.

Noble and Williams, who have already been referred to above, made new drawings in 1867, generally deficient in detail. Among those by Williams, note the one reproduced here; made on 11 January (Fig. 60) with a 4¼–in. telescope.

We can note a solution to the continuity problem in Herschel Strait, probably corresponding to that which is indicated when Kaiser's map at longitude 130°. The lower or northern snow is very extensive.

1867–1877.—R. A. Proctor[41]

We now come to Richard Anthony Proctor (1837–1888), whose work in areography was of great importance. In 1867 he began to construct a map, following the drawings by Dawes, whose principal sketches have already been given above.

We have already discussed the first attempt at Martian cartography, by Beer and Mädler, based on their drawings made in Germany between 1830 and 1837; also the planisphere of Kaiser, from his observations in Holland in 1862 and 1864, and that of Phillips, drawn from his observations made in England, in 1864. We note also the chart by Secchi, made from his Rome observations in 1858, which was reproduced in the second edition of my book *La Pluralité des Mondes habités* (1864).

The drawings by the English astronomer Dawes brought a new precision to studies of the Martian world. His contemporary, Proctor, wished to use them for a map which would be as complete as possible, and using them exclusively, drew up the chart reproduced here (Fig. 61), and which was the first chart to be published with a properly worked-out system of nomenclature.

A nomenclature, with settled names, was highly desirable, though from representations of the planet in the earliest drawings there were but a few objects shown, and a

[41]*Chart of Mars from 27 drawings* by Mr. DAWES.—*Half-hours with the telescope*, London, 1869, pl. VI.—*Other Worlds than Ours*, London, 1870, p. 93.—*The orbs around us*, London, 1872, frontispiece.—*Essays on Astronomy*, London, 1872, p. 62.—*Flowers of the Sky*, p. 167.

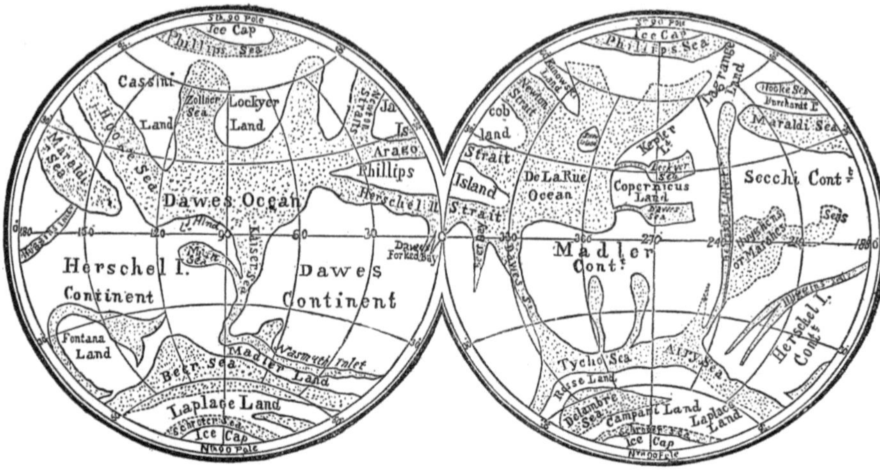

Fig. 61 Chart of the planet Mars, by R.A. Proctor, in 1867

few letters sufficed to identify them. We can see patches *a*, *b*, *c*, etc. But as the details became more numerous, such designations became inadequate, and unsuitable for comparisons. Fixed names were far better.

The same thing has happened in geography. The indication of a country by a letter or a figure, or even by its precise position, is so inadequate mentally that from the moment of its discovery the most insignificant island has been given a name which will distinguish it from all the others. Names are indispensable, above all, in the human species. One cannot easily picture men existing without names.

Unfortunately, many geographical names—as with others—are apt to be arbitrary. In the case of Mars, it seemed very natural to follow the system which had been adopted for the nomenclature of the Moon. The Moon was the first globe on which geographical features were recognized; Mars is the second. Indeed, after the Moon, Mars is the only planet whose geographical configurations are known accurately enough to be put on to a map (those suggested for Venus in the eighteenth century are extremely uncertain). It seems that when giving names to the Martian continents and seas whose exact forms can be distinguished, we should choose the names of those celebrated astronomers who have been most concerned in the study of the planet, naturally with complete impartiality as to their terrestrial nationalities.

In respect to this last, Proctor may be criticized for giving too much preference to astronomers of his own country. He is also guilty of often repeating the same names. It is inconvenient when certain names are repeated several times—and leads to confusion avoidable in a more precise nomenclature. Thus Dawes's name is used six times (Dawes Ocean, Dawes Continent, Dawes Sea, Dawes Strait, Dawes Isle, and Dawes Bay); Beer twice (Beer Sea and Beer Bay), Lockyer twice (Lockyer Land and Lockyer Sea), Phillips twice (Phillips Sea and Phillips Island), etc. Instead of this duplication, it would have been better to use the names of others, even if their work may have been of less value—astronomers such as Galileo, Halley, Lalande,

Le Verrier, or observers of Mars, such as Galle, Schmidt, Lassell, Knott, Green, Von Franzenau, Vogel, etc. These defects in the nomenclature explain why some astronomers have been disposed to modify it.

But this consideration is unimportant from the point of view of the intrinsic geography of the planet, otherwise termed areography. Proctor rendered a signal service to science by constructing the first well-defined chart of Mars, and thereby laying the foundations of areography; and for this—together with his other very considerable work—Proctor's name is written indelibly in the history of astronomy.

In his original work on Mars, Proctor was concerned with obtaining a value for the rotation period which would be as accurate as possible. At the meeting of the Royal Astronomical Society of London on 14 June 1867, he presented his first paper on the subject. Making comparisons analogous to those described above by Cassini, Maraldi, Herschel, Schröter, Beer and Mädler, Kaiser, etc., and making use of Dawes's drawings, he gave the rotation period of

88,643 seconds, or $24^h\ 37^m\ 23^s$.

Confident that this value was near the truth, he compared Dawes's observations with those of Herschel and Hooke, and found:

$24^h\ 37^m\ 22^s.745 \pm 0\ ^s.005$.

Within an accuracy of 2 days, 79 rotations of the Earth are equal to 42 of Mars.

Proctor returned to the subject at the meeting of 10 January 1868. Comparing the drawings made by Browning in January and February 1867 with those of Dawes in 1864 and 1856, he selected three of the clearest and most precise (Dawes, 24 April 1856 and 26 November 1864, Browning, 23 February 1867), when the Hourglass Sea was near the central meridian. Comparing these drawings with those made by Hooke on 12 March 1666 (see Fig. 15, Chap. 1), he found a mean period of $88{,}642{,}735^s$ or:

$24^h\ 37^m\ 22^s.735$.

The probable error does not exceed $0^s.005$.

Returning to the same question in 1869, Proctor found, with the help of a drawing made specially for the purpose by Browning on 4 February 1869, the value of $24^h\ 37^m\ 22^s.736$. He concluded that the first value should be adopted.

Kaiser had found $24^h\ 37^m\ 22^s.62$, but the difference of $0^s.115$ per rotation would produce a difference of $2^h\ 20^m$ going back all the way to the year 1666, and would put the Hourglass Sea 50° from the centre of the disk whereas Hooke's drawing showed it only 18° away; this sea was not visible on all of Hooke's drawings, and it could be lost in the fogginess toward the edges of the disk.

Again Proctor came back to the question in 1873. This difference of 1/10 of a second was the subject of new research. He found that it was due to a slight error in calculation. In counting the number of days which had elapsed between 13 August 1672 and 1 November 1862, the Director of the Leyden Observatory had given 69,476 days; this is two days too many because he had forgotten that 1700 and 1800 were not Leap Years.

Moreover, Kaiser had taken Hooke's observation as having been made on 14 March instead of 13 March. He reproduced Huygens drawing, as we have seen, with the date (13 August 1672) and also those of Hooke (12 and 13 March 1666, at 20^m and 40^m). These new considerations showed that the Hourglass Sea could give a reliable value for the rotation period, and Proctor concluded that the true period must be between

$24^h\ 37^m\ 22^s.71$ and $24^h\ 37^m\ 22^s.72$.

Therefore we adopt, as the best available period and to an accuracy of 1/10 of a second, a value of

$24^h\ 37^m\ 22^s.7$.

This is the sidereal day: it is the real period of rotation, not the solar day. Since the Earth's tropical sidereal day is $23^h\ 56^m\ 4^s.09$, the Martian day is $41^m\ 18^s.6$ longer than ours.

Proctor concentrated largely upon Mars in his books—apart from the very last, which was left incomplete at the time of his untimely death. More will be said of this later.

1871–1873.—Lehardelay, Crosley, Gledhill, Burton, Denning, Wilson, Guyon, Lowdon, Joynson, Spear

In 1871 Ledhardelay, observer at Fontenay (Normandy), made some observations with a 162^{mm} aperture Steinheil refractor.[42] These observations were made on 2, 11, 13, 23, and 24 March and 23 April. The northern polar snow was well seen. The observation on 23 March was one of the best; the southern polar snow was also seen. The planet appeared to be covered with patches in the form of rounded lobes, some slightly festooned on their common borders, yellowish in colour and represented in separation by a very thin grey line. The drawing made on this occasion is reproduced here; it was published by Terby (*Aréographie*, p.111), and was made with a magnification of 547×. A curious feature is the river, or rather canal, which is called Burton Bay on my map or the *Isthmus* on Secchi's drawings (Figs. 62 and 63).

Fig. 62 Sketch of Mars by M. Lehardelay, 23 March 1871, $10^h\ 30^m$

[42] *Bulletin de l'Association, Scientifique de France*, 31 December 1871, p. 219.

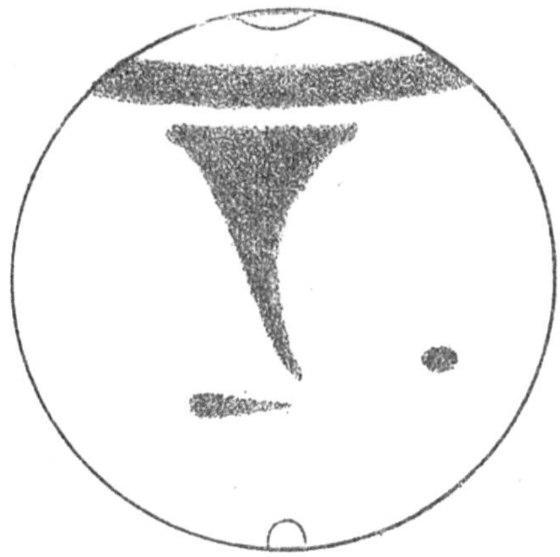

Fig. 63 Mars on 4 April 1871, at 11^h. Drawing by M. Gledhill

At the Halifax Observatory, Gledhill made observations on of Mars during the less favourable opposition of 1871, and published six sketches in *The Astronomical Register*.[43] The lower or northern polar patch was turned toward the Earth, and showed up as round and brilliant. The Hourglass Sea can be identified, and near latitude–30° the south circumpolar region shows a dark band, from which thin tongues extend northward. On two drawings, there may be seen, adjacent to the northern polar cap a dark patch in the form of a balloon, whose point touches the polar snow. (This drawing has a slight resemblance to Burton's of 23 March 1871, already described; Gledhill's feature was seen on the same day at the same time, and identification is therefore certain.) These six sketches are not reproduced here, but can be found in lithographic reproduction in the publication cited. I give only that of 4 April 1871, at 11^h (Fig. 63). The two polar caps are on view, and the Hourglass Sea is at the central meridian.

The summer solstice in the northern hemisphere of Mars fell on 2 March, opposition on 19 March.

At Loughlinstone in Ireland, C.E. Burton, using a 12–in. (305^{mm}) Newtonian reflector, made excellent the drawings in 1871 and 1873, and continued the series during the opposition of 1879.[44] In our third period of observation (1877–1892) the 1879 work will be described, together with the chart made from it. For the moment, let us discuss the drawings of 1871 and 1873.

The main point about Burton's observations is that he concluded that considerable changes take place on the surface of the planet. Three drawings in 1871 and four in 1873—or, in general, all of those which represent this side of Mars—show an immense dark patch in the form of a pear or a balloon, corresponding to the Tycho Sea. It is in the neighbourhood of the north pole, and is contained inside the Martian Arctic Circle.

[43]October 1871, p. 233.

[44]*Transactions of the Royal Irish Academy*, XXVI, p. 427.

Fig. 64 Drawing of Mars, by Burton. The Tycho Sea in 1871

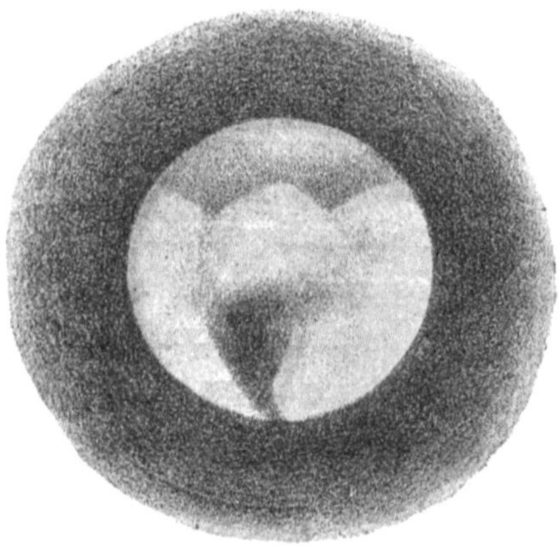

Fig. 65 Drawing of Mars, by Burton. The Tycho Sea in 1873

At the time when these observations were made, the Martian north pole was turned toward us. The balloon-shaped patch appeared very sombre, and of bluish green hue. Burton wrote:

If it is a sea, its decrease considerably surpasses, both in extent and speed, anything analogous which has occurred on the surface of the Earth in historical times.

This curious observation thus confirms our previous deductions.

I reproduce here (Figs. 64 and 65) two of Burton's drawings representing this sea, made on 23 March 1871 and 7 April 1873. Several other observers saw the same feature, notably Terby at Louvain on 12 May 1873, and Burton again saw it consistently during his 1873 observations. This patch, he commented, was as obvious and as distinctive as the celebrated Hourglass Sea.

Burton also drew the Hourglass Sea under excellent conditions; for example, see his sketches of 7 April and 4 May 1871, reproduced here (Fig. 66 and 67). In the first, to the left of the Hourglass Sea, may be noted a variable region which will be discussed later; to the right, a pointed cape (Banks Cape), which is generally shown as it is drawn in my chart.

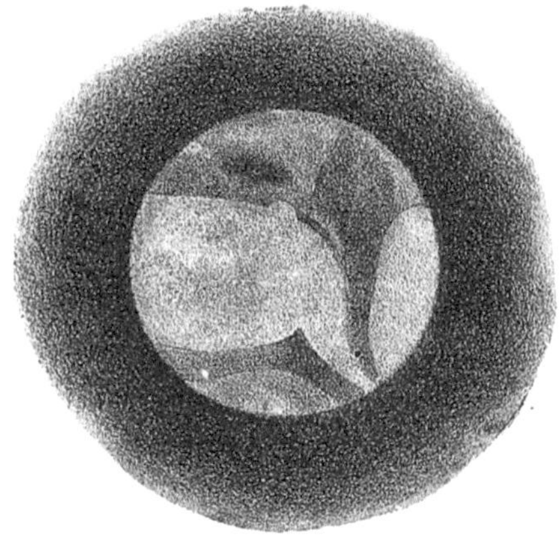

Fig. 66 Drawings of Mars, by Burton. The Hourglass Sea in 1871

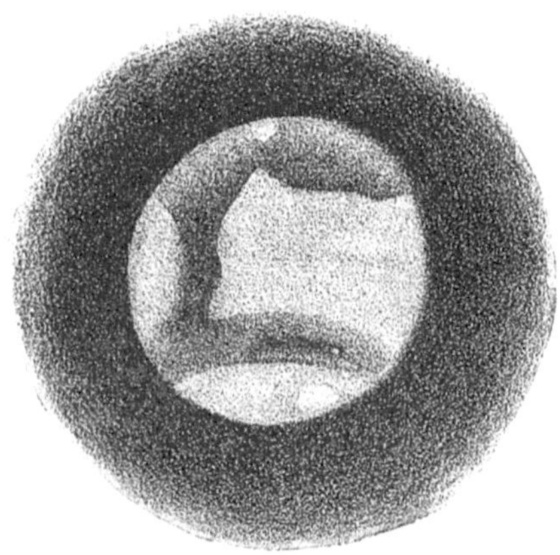

Fig. 67 Drawings of Mars, by Burton. The Hourglass Sea in 1871

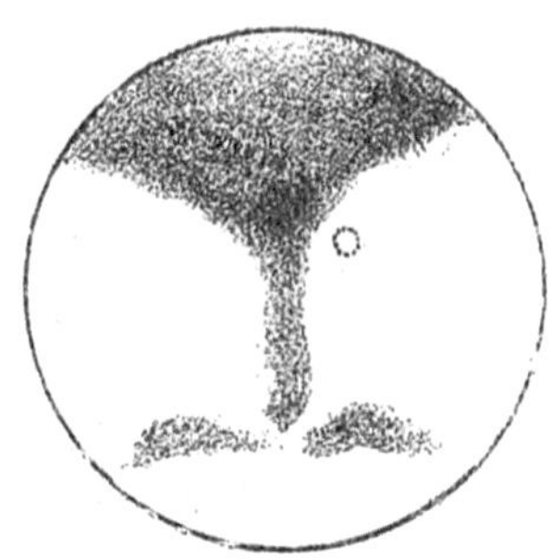

Fig. 68 Sketch of Mars, by Burton, 24 May 1873, indicating the position of a snowy plateau in the tropics

Often, to the right of the shore of the Hourglass Sea, may be seen a point indicated by the dotted circle in Fig. 68—an extremely brilliant white patch. Burton considered that this indicated the presence of a very lofty plateau, lying not far from the tropics, yet covered with snow: *the summits of a cluster of lofty mountains, or a high table-land*. This Alpine plateau lies inside the tropical zone.

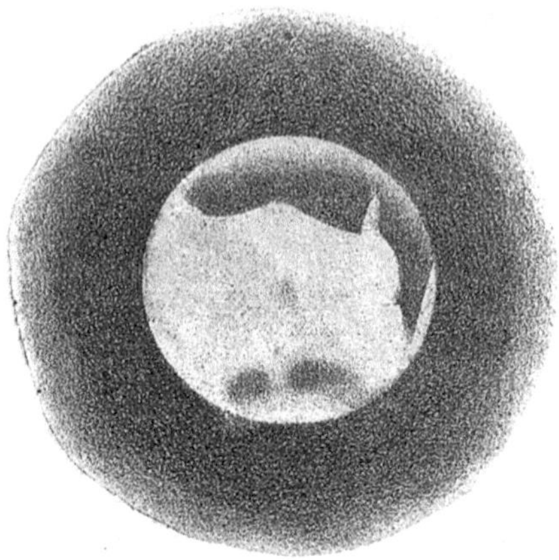

Fig. 69 View of Mars, on 29 May 1873 (Burton, showing the sand bank above the Hourglass Sea

Fig. 70 Mars on 4 May 1871, at $9^h 30^m$. Drawing by M. Wilson

Let us look now at the drawing made on 29 May 1873 (Fig. 69), on which may be seen a line separating the Hourglass Sea from the Hooke Sea, across the Flammarion Sea. This line has already been noted in connection with the drawings by Von Franzeneau and Terby, and can even be recognized on earlier drawings. It is certainly a sandbank, as it sometimes clears away.

In 1871, at the Rugby Observatory, Wilson made a certain number of very interesting drawings. In Fig. 70, one of these is reproduced; it was made on 4 May 1871, at $9^h 30^m$. An 8¼-in. (210^{mm}) refractor was used, magnification 300×. During this opposition, and also in 1877, Wilson made many drawings of Mars which are reminiscent of those of Beer and Mädler.

To these studies made in 1871 and 1873, we must add those of Denning, Guyon, and Lowdon in England. In particular, Guyon made six drawings in 1871 and ten in 1873. The sketches add nothing really remarkable to the data already given.

During the same opposition, we must also note the observation made by John Joynson at Waterloo, near Liverpool, and one by J.Spear, at Churkrato, Bengal.[45]

[45] *Monthly Notices of the Royal Astronomical Society*, 1871, p. 208 and 262.

Joynson commented that the north polar snow was much less extensive than in 1867, and was somewhat like the aspect of the south pole in 1862. He wrote: "The wine-glass-shaped channel is certainly permanent, as is the sea which dominates it."

At Bengal, on 19 November 1870, Spear noted: "The snow at the north pole is of remarkably intense brilliancy."

1872–1873.—Dr. Vogel. Spectral Analysis of the Atmosphere of Mars[46]

The skilful Bothkamp astrophysicist observed Mars on 19 November 1872, 2, 20, and 22 April, and 3 June 1873 with the object of continuing his spectroscopic researches, described earlier. In his memoir, he gives in detail the positions and wavelengths of 25 spectral lines. Here is a summary of the results obtained:

> In the spectrum of Mars, we find a large number of the lines of the solar spectrum. In the less refrangible parts of the spectrum we find some bands which do not appear in the solar spectrum, but which coincide with lines due to absorption in our own atmosphere. We can conclude, with certainty, that Mars possesses an atmosphere which, in composition, does not differ essentially from ours, and which, in particular, is rich in water vapour. The red colour of Mars seems to be the result of absorption of the blue and violet rays; at least in this part of the spectrum it has not been possible to make out absorption bands. In the red, between C and B, we may deduce the presence of lines due to the actual spectrum of Mars, but it is not possible to fix their precise positions, because their intensities are too feeble.

Vogel gives a list identifying lines in the Martian spectrum with solar and telluric lines:

Wavelength		
570.0	}	Near Brewsters δ
580.0		
592.1	}	Telluric lines near D
594.9		
628.0		α
648.8		Fairly dark line
655.6		Telluric lines near C
687.8		B

This description will be continued later, with the spectroscopic researches into the atmosphere of Mars made in 1877. These 1877 results are more complete and more precise.

[46] *Untersuchungen ueber die Spectra der Planeten, verfasst von Dr. H.C. Vogel. Sternwarte zu Bothkamp*, Leipzig, 1874.

1873.—C. Flammarion: Observations of the Planet Mars

During the opposition of spring 1873 the planet was well placed for observation. Here are the results of the studies I made of its surface with the Secrétan 108 mm refractor. The usual magnification was 202×, occasionally increased to 288 and often reduced to 150 because of the planet's low altitude over the horizon.

The résumé given here is that which was presented to the Academy of Sciences.[47]

Around the time of opposition, Mars presented its northern hemisphere to us, which is less well known than the southern. The north pole, strongly tilted toward us, betrayed itself by a very brilliant white patch which under certain conditions of atmospheric transparency seemed to protrude from the contour of the disk.

The cap is not actually very extended; to the eye it sometimes gives the impression of a white pear which scintillates on the limb below the disk. Its position indicates that the pole lies about 40° from the lower limit of the vertical diameter, toward the east (image inverted, as in an astronomical telescope). The north polar snows do not actually extend beyond aerographic latitude 80°. However we know that at times the snows can be considerably more extensive, reaching down beyond 60°, while the variations in the southern snows are greater still.

There is probably a polar sea around the north pole, because a dark patch is constantly visible there, whichever hemisphere is turned toward us by the planets rotation. This polar sea seems to extend to latitude 45°, and even beyond in certain points, but it is divided in two by the tongue of land which extends from the 65th to the 75th degree. Whatever may be this intermediate land (which can scarcely be distinguished), the sea extends in one direction as far as the ice, that is to say to at least latitude 80°, and in the other direction as far as 45° (Fig. 71).

A long, narrow Mediterranean runs from north to south, and rejoins a vast sea which extends from beyond the equator into the southern hemisphere; between the western extremity of this Mediterranean and the northern sea referred to above, there is another enigmatical feature. Normally this Mediterranean channel seems to join the two patches. Sometimes, at the western end, there may be distinguished a sharply turned, even right-angled, feature, which does not, however, interfere with the general physical details already described: *a north pole* marked by a very white patch; a north sea extending from these latitudes; *a broad network of water* reaching in an east–west direction; and a considerable *southern sea.*

It was actually autumn in the northern hemisphere of Mars. The greater part of the north polar snows had melted, while snow was accumulating around the south pole, invisible to us. The southern region was marked by a white streak near its edge. Was this snow, extending down to latitude 40°? More probably it was due to clouds.

The detailed study of the planet showed that the surface was decidedly different from the Earths with regard to lands and seas. On Earth, three-quarters of the globe are covered with water; with Mars, on the other hand, there is more continental than maritime surface.

---

[47] *Comptes rendus des séances de l'Academie des Sciences*, Vol. LXXVII, p. 278, *séance*, 28 July 1873.

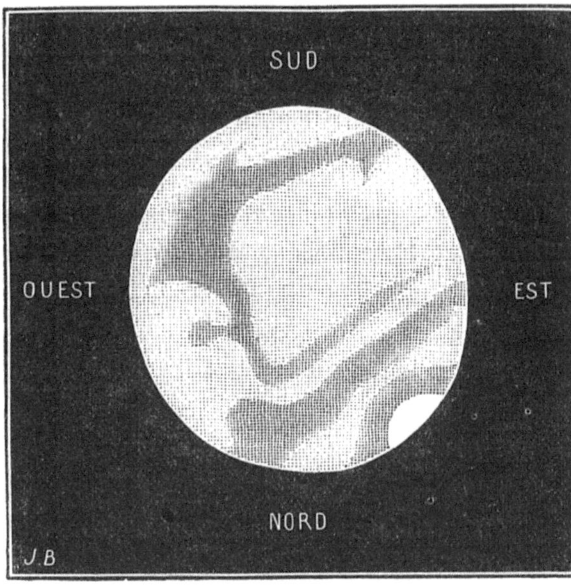

Fig. 71 View of the planet Mars, on 29 June 1873, at 10^h in the evening

Nevertheless, evaporation on Mars produces effects analogous to those in terrestrial meteorology, and spectral analysis shows that the atmosphere of Mars is charged with *water vapour*, as is ours; also that the seas, snows and clouds are composed of the same water as our own seas.

It seemed to me that the red colour of the continents was less intense than usual this year. The cause of this colouration has often been discussed, and was at first attributed to the atmosphere; but this explanation has been rejected, because the edges of the disk of the planet are less strongly coloured than the centre; they are almost white. The opposite would be true if the colouration were due to the atmosphere; the redness would be greater at the edge, because of the greater thickness of atmosphere through which the reflected light-rays have passed. Is it due to the colour of the materials making up the planet? We can accept this if the Martian continents do not remain in the state of sterile deserts, but that, under the influence of the atmosphere, rains, the life-giving heat of the sun and the elements which on Earth lead to the *development of vegetation*, produce the same type of vegetation, suited to the physical and chemical state of Mars. Now, since it is the surface which we see, not the planets interior, the red colour ought to be that of the Martian vegetation, since it is this species of vegetation which is produced there. It is true that while the seasons on Mars are of almost the same intensity as ours, we do not see variations in tone corresponding to those which we observe during the different seasons in terrestrial latitudes; but the vegetation which covers the surface of Mars is presumably very different from ours and less subject to variations over the course of the year.

However this may be, the studies made of our neighbour planet are numerous enough to allow us to form a general idea of its geography, and even of its meteorology. We can give a résumé of the facts which we have acquired with regard to the astronomical physics of the planet:

1. The polar regions are alternately covered with snow according to the seasons, and show variations due to the marked eccentricity of the orbit; the ices of the north pole do not extend beyond latitude 80 N.
2. Clouds and atmospheric currents exist there, as on the Earth; the atmosphere is cloudier in winter than in summer.
3. The geographical surface of Mars is more equally divided up into continents and seas than is that of the Earth; there is slightly more land than sea.
4. The meteorology of Mars is a little like that of the Earth; water exists in the same state as on our globe, but, without doubt, under different temperature conditions.
5. The continents seem to be covered with reddish vegetation.
6. Finally, reasoning analogous to that applicable to earth shows, without doubt, that organic conditions do not differ much from those which led to the appearance of the Earth.

Same Year, 1873.—F. Hoefer, Stan. Meunier

Some time after the presentation of these results to the Academy of Sciences, our learned friend Dr. Hoefer objected to the preceding explanation, maintaining that the colour of Mars is not due to vegetation because it does not vary with the seasons. According to Hoefer, it is simply the colour of the soil.

The colour of the soil? But then this soil would be bare! The sun, the rain, the air would allow it to remain sterile over the centuries! Dr. Hoefer, who is a fervent partisan of the doctrine of the plurality of worlds, cannot accept this sterility, which is contrary to all known effects of the forces of Nature. There must be *something* on the lands, whether it be moss or even less.

The objection about the invariability of the colour throughout the Martian year is not well-founded, and a broader examination will show its inadequacy. On Mars why restrict Nature to producing vegetation of the same type as ours? The mean conditions of temperature, density, gravity are opposed to this idea; therefore the marked difference which exists between the Martian and terrestrial vegetation can account for the variation in colour. But in addition, even on Earth we find that Nature meets this objection and shows us species of vegetation which do not change. In the tropics, olive-trees and orange-trees are as green in winter as in summer. In the North the fir, the cypress, the laurel, the spindle-tree, the box tree, the holly, rhododendron, etc., keep their greenness even in the middle of the snow. Even in our latitudes, grasses and almost a 1,000 species of vegetation scarcely vary at all. Why, then, reject an explanation which is so simple—when even on Earth we have the same examples, and when the differences in conditions of life on Mars and the Earth cannot have led to the same kind of vegetation on the two worlds?

A second objection has been put forward; that on Earth the continents are vegetation-covered only in limited areas, and that its colour dominates those of the lands, so that Mars should be ochre; but the deserts are exceptions. Only water will suffice to produce vegetation, and the sterile regions of Mars are those over which no rain ever falls. The same agents which led to the appearance of the first vegetation on Earth—the natural forces of fertility—exist on Mars as well as on our planet. There, too, we can see actual clouds and rain. Therefore it is probable that the dominant colour of Mars is due to some kind of vegetation which covers the soil,

In one of the meetings following that in which I presented the preceding observations, M. St. Meunier made the following remarks concerning the forms of the Martian seas as compared with those of terrestrial oceans:

> At the moment when the attention of observers is directed toward the planet Mars, I believe that it is of interest to submit to the academy a comment concerning this body—a comment which confirms the theory of cosmical evolution which has already been developed.
>
> We know that, from this point of view, Mars as a globe is older than the terrestrial globe, and now offers conditions from which we are far removed. Many considerations support this idea, among them the thinness of the atmosphere and the smallness of the oceans relative to ours.
>
> The fact which I want to discuss today concerns the forms of the Martian seas compared with the seas of Earth. I see here a new indication of the relative oldness, because it seems evident that our seas have taken up approximately the same proportions of those on Mars, but on Mars they have considerably diminished in size because of progressive absorption by the solid core.
>
> One of the most remarkable characteristics of Mars is the presence of many long, narrow channels, and seas in the shape of *bottle-necks*. This disposition differs essentially from anything known on earth.
>
> Now, if we take a marine chart such as that of the North Atlantic Ocean, and draw successive horizontal curves for greater and greater depths, we recognize that these curves tend progressively to delimit zones whose forms are more and more elongated. At 4000 metres, for example, we obtain forms which are comparable in every respect with those of the seas of Mars.

From this, it follows that we suppose the Atlantic to be absorbed by deep-seated masses on the way to solidification, so that the level of the ocean would drop by 4,000 m, we would have a much smaller area covered with water-in the form, that is to say, of a narrow elongated sea. These are precisely the conditions shown on Mars.

This comment by our colleague is ingenious, and must be noticed; but it is not certain that the geological form of Mars is analogous to that of the Earth, or that a diminution in the amount of water would lead to this configuration. On the Moon, for example, the orography indicates nothing of the kind, and all the deep plains are circular; this absence of water leads to another type of landscape. The orography of these two neighbour worlds is sharply contrasted. It is probable that Mars is rather flatter, because of its greater age, so that the sea-bottoms have been raised relative to the mountains, which have been reduced by rain, ice, winds and various atmospheric agents.

1873.—Nathaniel Green

This artist—already known to our readers—published in *The Astronomical Register* for 1873, p.179, a selection of six of his drawings of the planet made during this opposition, and a planisphere sketched according to those drawings.

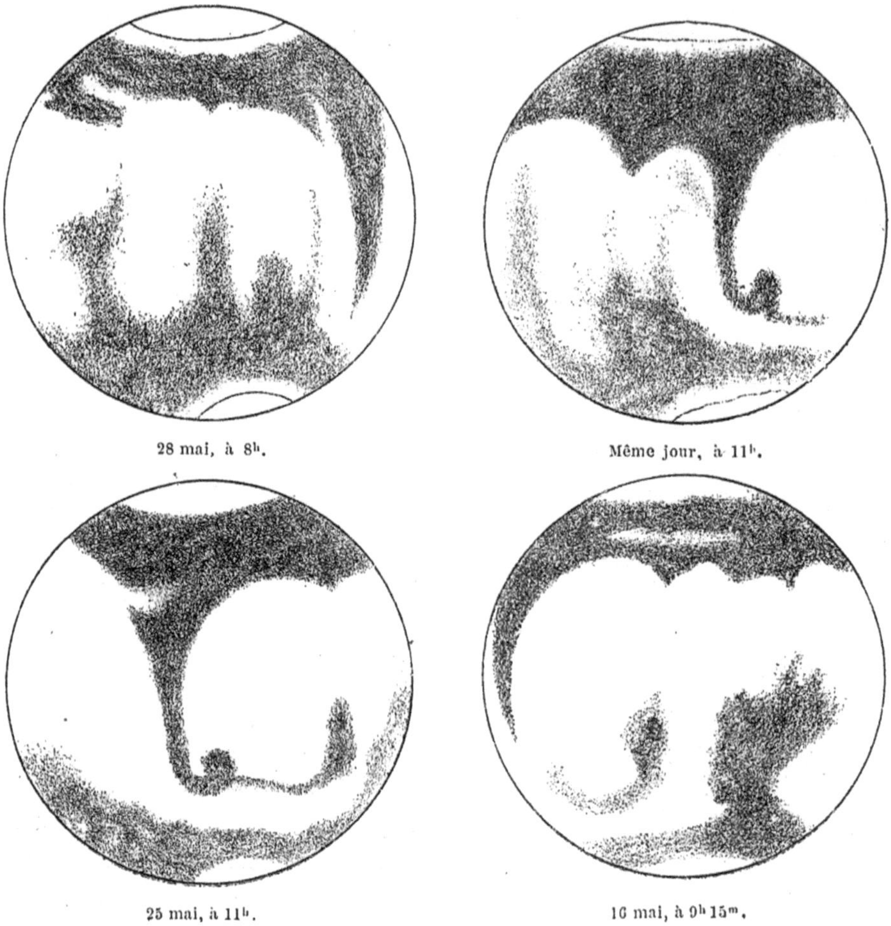

Fig. 72 Drawings of Mars, by M. Green, in 1873

The drawings themselves were made in London from 16 to 30 May 1873. The planet was low over the horizon, because of the southerly declination.

Four of the drawings are reproduced here (Fig. 72), together with the data and times for each. Our readers will recognize the advance of the Hourglass Sea from right to left, because of the planet's rotation.

The lower or north polar snow has been very marked; in December 1872 it was much more extensive than during spring at the time of opposition. After opposition, on the other hand, the southern snow increased considerably. Opposition took place on 27 April.

In the small planisphere reproduced below (Fig. 73), Green represented the features which he had been able to observe with certainty; each of the six views which we will discuss has, for its central meridian, the points given on the figure below. If we compare this planisphere with my map, we will find that point A.

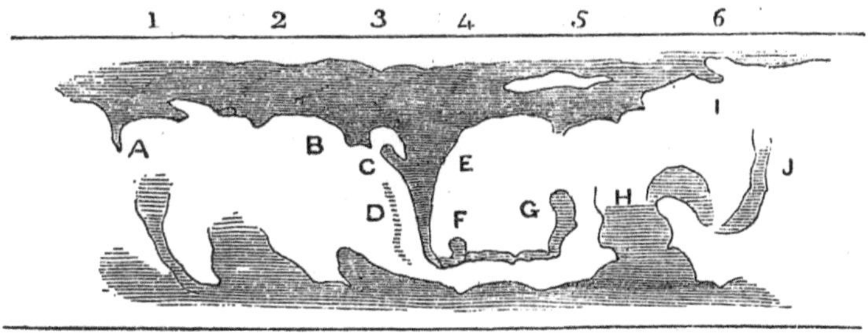

Fig. 73 Planisphere of Mars, by M. Green, from his observations of 1873

is the Maraldi Sea; B. the Flammarion sea; C. the Main Sea; E. the Hourglass sea, and the G. the Lassell Sea; H. is the upper Knobel Sea, I. the Christie Bay, and J. the Tycho Sea.

At D there is a grey streak parallel to the Hourglass Sea which does not appear on my map. "This was the most delicate streak observed in any of the drawings,"

wrote Green; he seemed to be certain of its existence. This aspect cannot surely be long-lived. Also note the feature F, which is hardly ever visible. Green wrote:

The Hourglass Sea is very dark near E. Generally, the dark areas are regarded as seas; but in this case should not one see a reflection of sunlight when the sun is on the meridian?

This question had already been asked, in 1862, by Phillips. In other words, he asked why we never see an image of the sun reflected from the Martian seas, and he could give no answer. In 1878 Schiaparelli asked the same question; he found that the diameter of the solar image should be 1/24 of a second at closest approach, as on 5 September 1877 (Phillips had given a similar figure: 1/20), and an intensity or luminosity 2,100,000,000 times less than that of the sun at unit distance, i.e. the distance between the Earth and the Sun. In his photometric researches, Zöllner (1865) had determined the amount of sunlight effectively reflected by the disk of Mars at mean opposition, and had found it is to be 1/6,994,000,000, of sunlight at the distance of the Earth. From this it is found that the total light at a perihelic opposition, such as that of 1877, is 1/2,990,000,000 of that of the Sun at unit distance. Therefore, the luminous image of the Sun reflected in the Martian seas at such an opposition should give more light than that from all the rest of the planet's disk.

This result is based on the assumption of total reflection of all these rays of the Sun. But in reality a transparent liquid, such as water, has a refractive index of 4/3, and reflects only 1/49 of the incident light. We must also take into account the absorption produced by the double passage of the rays through the atmosphere, which reduces their mean intensity. Instead of 1/49, therefore, we have, in round numbers, 1/100, the intensity of the solar image seen by reflection in the Martian seas should be:

$1/21 \times 10^{10}$ that of the Sun.

In the work already cited, Zöllner gave for Alpha Aurigae (Capella) a value of

$1/5.57 \times 10^{10}$ that of the Sun.

Therefore, at the best opposition, the solar image should appear in the spherical surface of the Martian seas with a brilliancy equal to ¼ that of Alpha, that is to say as a beautiful star of the third magnitude.

Such an image ought to be visible against the sombre background of the Martian seas, despite the brightness of the disk. But this assumes a calm, mirror-like sea surface. Now, the observations of moving clouds, cloudy streaks, polar snows formed by vapours which are driven there—all these prove that there is wind across the surface of the planet. The surface of the water must, therefore, be normally distorted to a greater or lesser extent, and even the slightest waves will hinder the formation of a single solar image, giving instead a number of broken, smaller images. It is true that the total luminous intensity of these images would be the same as that of a single image, but they would be spread over a wide area of variable extent, and would become blurred, particularly if the crests of the waves were high. This nebulous light would become imperceptible to the observer.

Summarizing, it is not therefore impossible, under the best conditions, to detect the image of the Sun reflected from a sea surface on Mars, but this can be done only under exceptional circumstances.

Such is the answer to the question put forward by Green. We will return to this observer when discussing the work carried out in the year 1877.

Same Year, 1873.—E.B. Knobel, Webb, Grover

During the same period a skilful English observer, Knobel, at his observatory at Burton-on-Trent, made a series of observations, published by the Astronomical Society of London,[48] accompanied by 17 drawings. The observations were made with a reflector with a silver-on-glass mirror 8½in. (210^{mm}) aperture, of excellent quality, and with magnifications of 250× and 300×.

In general, the drawings agree perfectly with those of Dawes and with Proctors map constructed from them. However, certain exceptions are worthy of attention. Thus eight drawings, made from 11 to 22 May, very clearly show a dark circular patch in the lower or northern hemisphere, below the Meridian bay, which corresponds to Le Verrier Land or the Knobel Sea curved toward the left and continued upward after a sort of separating bulge. This separation may be traced obliquely from the SW to the NE, while on my chart (at longitude 30°) it runs from east to west; moreover, to the right of this sea, from 8 to 22 May, Knobel saw a patch as white as snow on 22 May, when the area lay on the terminator, brighter than the disk, so that it could be taken for polar snow. This snow sat at longitude 25°, latitude 50° N.

The Knobel Sea is the feature seen to the right in the fourth drawing by Green, reproduced above.

In these sketches, the oblique continuation of the Hourglass Sea—the Nasmyth channel—is also very marked, but the Lassell Sea is not, though it is very pronounced in Green's drawings.

[48] *Monthly Notices of the Royal Astronomical Society*, 1873, p. 476.

During these observations, the coastline of the Meridian Bay was always seen with admirable clarity. The northern hemisphere of the planet was always clearer than the southern hemisphere; the north polar snow was better seen than that in the south. The Hourglass Sea was always very dark; the Main Sea was visible, but less dark.

The lamented English astronomer T.W. Webb, author of the *Celestial Objects for Common Telescopes*, made 85 drawings of Mars between 1839 and 1873, of which he communicated the principal ones to me; they have already been discussed (1856 opposition). Webb had keen eyesight, and was a skilful observer; his drawings, though of small size, are very precise with regard to many details. Also in England, Grover made five drawings in 1873, adding to the dozen he had made in 1867 and which have been referred to above.

Same Year, 1873.—Julius Schmidt: Rotation Period of Mars[49]

In November 1873 Julius Schmidt, Director of the Athens Observatory, published a mathematical memoir about the rotation of Mars. He used his own drawings, extending from 1843 to 1873 (four of which have been reproduced here), and compared them with those of Kaiser, Mädler, Herschel and Huygens. The general result of this work gave a rotation period of:

$24^h\ 37^m\ 2^s.6027$.

Let us put aside, as of purely mathematical interest, the 10–millionths of a second and even the thousandths and hundredths, and give simply: $24^h\ 37^m\ 22^s.6$.

We have already seen that the Martian rotation period had been given by Proctor as $24^h\ 37^m\ 22^s.7$. Therefore we may now be confident that the period is known to an accuracy of 1/10 of a second.

These two series, by Proctor and Schmidt, were made with equal care, and are of equal value. Taking the approximation to 1/100 of a second, we can now give $22^s.65$ as being very close to the truth, if not absolutely precise.

This is the sidereal rotation. The Martian year, which is made up of 669 2/3 rotations, has in corresponding Martian days a value of 668 2/3—namely, a rotation less, due to the orbital motion, which is in the same sense as the axial rotation. Therefore, the solar day of Mars is:

$24^h\ 37^m\ 35^s$.

During the opposition of 1873, Julius Schmidt, using the 9–in. (229^{mm}) refractor at Berlin Observatory, made an important series of observations and drawings, which appeared in Volume I of the *Publicationen des Astrophysikalischen Observatorium zu Potsdam* (1878). These six drawings of 1873 are not easy to interpret, except that of 25 May ($10^h\ 5^m$), which shows the Hourglass Sea. The others seem to disclose very large variations.

We will return to Schmidt's work in 1877 and 1879.

---

[49]*Astronomische Nachrichten*, No. 1965.

Same Year, 1873.—Trouvelot: Drawings of Mars[50]

In Vol. VIII of the *Annals of Harvard College Observatory* (1876), Trouvelot published the four drawings of Mars which are reproduced here, made with the observatory's 15–in. equatorial. The first was made on 23 May, at $11^h\ 30^m$; the second on the next day, at $9^h\ 30^m$; the third on 26 May at $8^h\ 30^m$, and the fourth on 29 May at $9^h\ 8^m$ (Figs. 74, 75, 76, and 77).

On the first two drawings, the Herschel II Strait is shown, and on the last two engravings the Meridian Bay and the Burton Bay; below, the Knobel Sea and the Tycho Sea. The Hourglass Sea, and its lower extensions toward the right (the Nasmyth Channel), are visible on the third and fourth drawings. The snows of the lower or northern pole are very evident. In a short note attached to these drawings, Trouvelot—a very skilful observer – was moved to write that the patches on Mars appeared to be surface features rather than clouds in the atmosphere. The whitish borders to the continents shown in his drawings do however suggest clouds.

In 1882 Trouvelot had an excellent general summary regarding his main astronomical drawings,[51] in which he commented on the observations he had made of the different planets in our system. He summarized his studies of Mars as follows:

> The sombre patterns showed different tones, from pale grey to deep black. The writer has never noticed green or blue coloration, and believes that these effects are due to complementary colour against the rosy tone of the continents.

Fig. 74 Drawing of Mars, by M. Trouvelot, in 1873

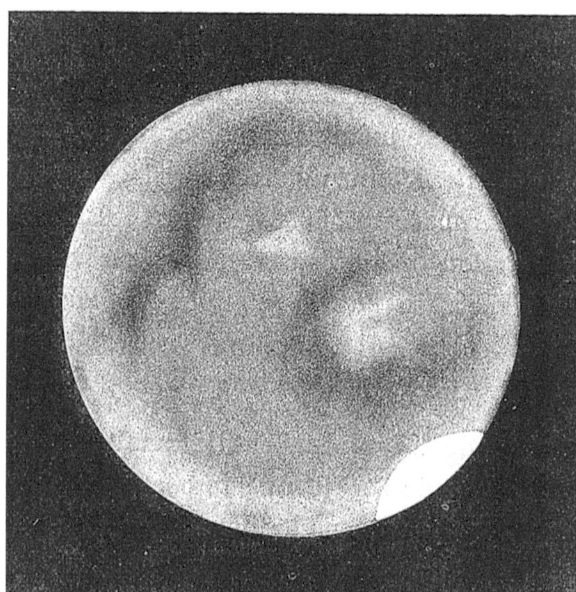

[50] *Annals of the Astronomical Observatory of Harvard College*, Cambridge, Vol. VIII, 1876.

[51] *The Trouvelot Astronomical Drawings Manual*, New York 1882.

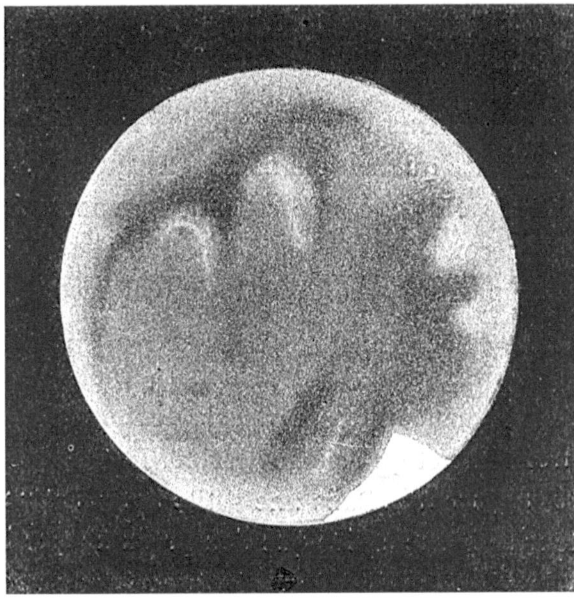

Fig. 75 Drawing of Mars, by M. Trouvelot, in 1873

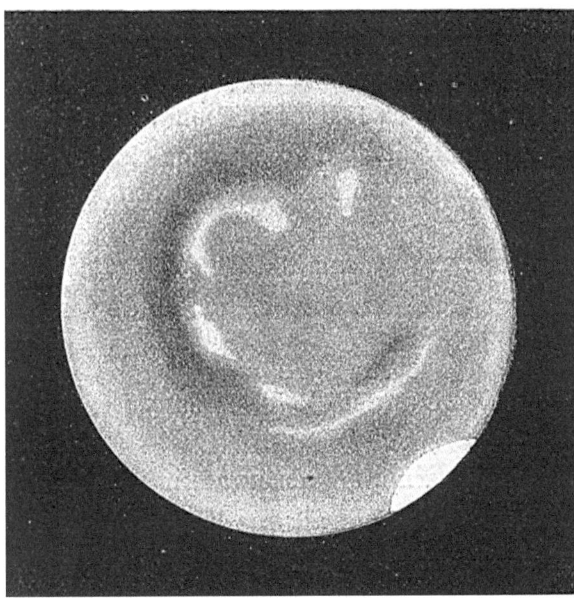

Fig. 76 Drawing of Mars, by M. Trouvelot, in 1873

Several changes are certain, notably that of the sea represented by Beer and Mädler on the southern hemisphere at 230° longitude, below the circular sea d. This is the lake shown on my map, below the Terby Sea, at longitude 90°.

Trouvelot wrote in 1882:

In 1877, during one of the most favourable oppositions of the planet, I saw a strong dark patch there. There is not the least doubt about this change.

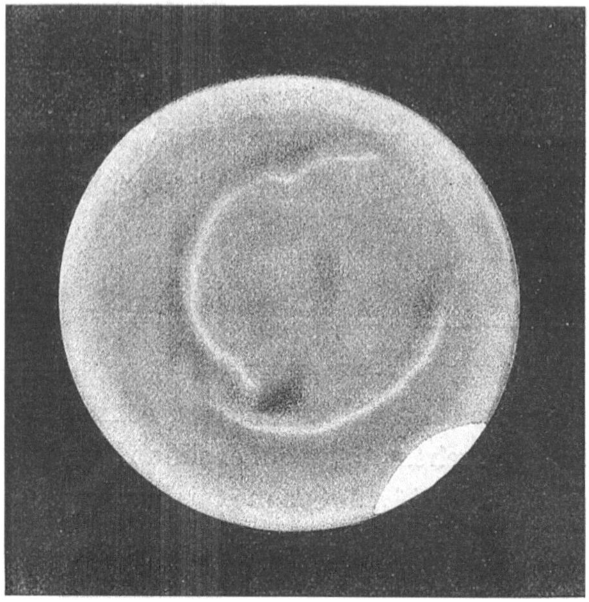

Fig. 77 Drawing of Mars, by M. Trouvelot, in 1873

This confirmed the opinion he had expressed in 1876 in *Les Terres du Ciel*.

More will be said about Trouvelot's work, and his deductions, in an account of the observations in 1882–1884.

Same Year, 1873.—Lohse[52]

At the Bothkamp Observatory, Lohse made a series of studies and drawings, from which he concluded that Mars showed considerable variations in appearance. Six drawings were presented, which do not show details identifiable with known features. These are reproduced here, with a schematic tracing explaining the first.

The first of these drawings (Fig. 78) was made on 9 May 1873, at $10^h\,10^m$. The tracing which accompanies it (Fig. 79) was made to indicate the tones: g = deep grey, dr = dark red, r = pure red, h = white, hh = very white. The south polar patch hh is not diametrically opposed to the northern.

The second (Fig. 80) was made on 25 May at $10^h\,5^m$. I believe we can identify the Hourglass Sea, near which is a white elongated patch giving some impression of the Moon hiding behind a cloud.

The third (Fig. 81) was made on 2 June, at $9^h\,45^m$. Again shown is a sombre circle which recalls the ring in the first drawing. What can we conclude from these representations of the planet, except that each observer has his own personal manner of seeing, and that Lohse exaggerated features which were really vague and uncertain?

[52] *Publicationen des astrophysikaleschen Observatorium zu Potsdam*, 1878.

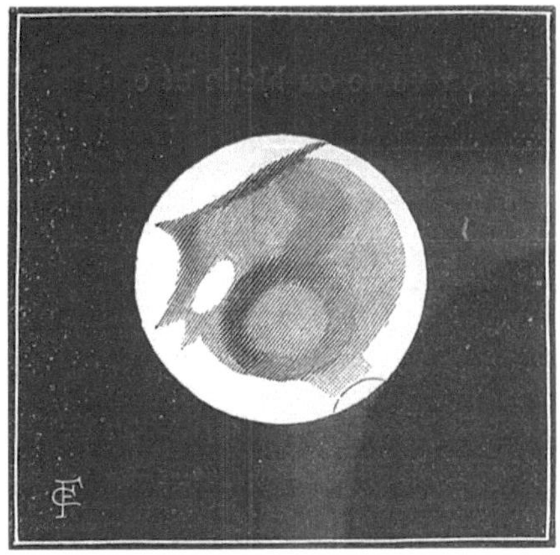

Fig. 78 Drawing of Mars, by O. Lohse, 9 May 1873

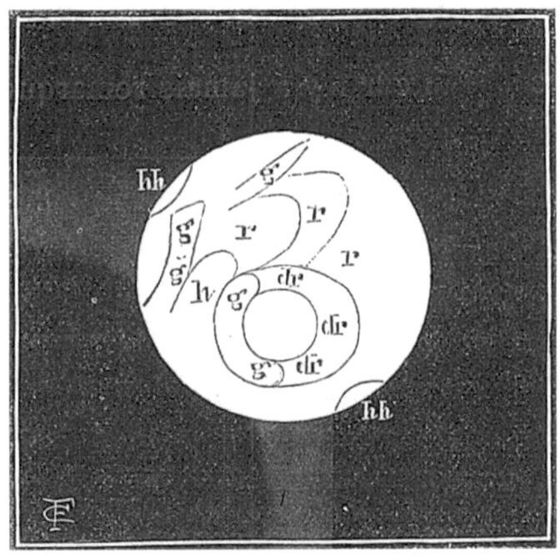

Fig. 79 Lohse's annotated outline of the above

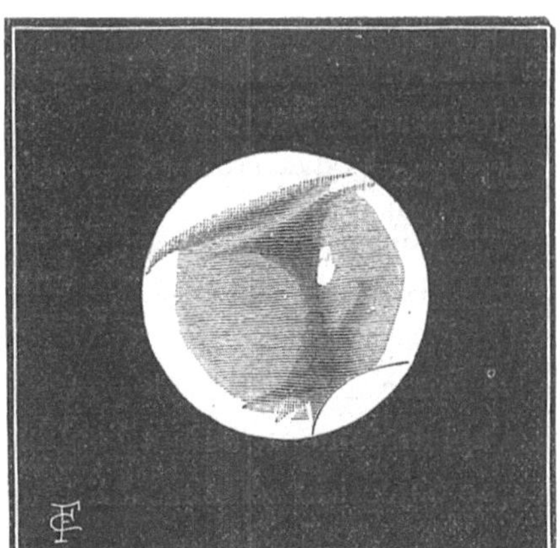

Fig. 80 Drawing of Mars, by O. Lohse, on 25 May 1873

Fig. 81 Drawing of Mars, by O. Lohse, on 2 June 1873

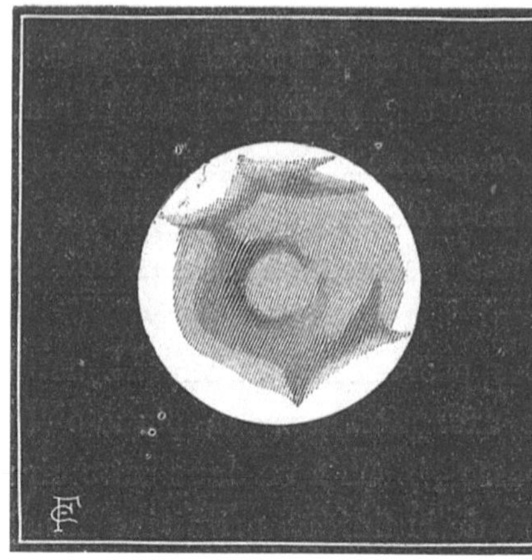

1872–1880.—Amigues, Hennesey, G.H. Darwin and Flammarion: Form of the Planet Mars

Most of the values found for the flattening of Mars seemed to be too great to be explained theoretically. The globe of Mars, smaller than that of the Earth and spinning less quickly, could develop an equatorial centrifugal force which is feeble compared with that of the Earth, and the polar flattening should be less than that of our globe, which amounts to 1/292.

Laplace took this discrepancy into account by assuming that local irregularities, analogous to those whose effects may be seen in various parts of the Earths globe, had a relatively greater influence upon the smaller globe of Mars. Arago challenged the validity of this explanation, and maintained that the shape of Mars was very regular; everything seemed similar at the poles and the centre of the equatorial region. His measurements of diameters at 45° gave him values intermediate between those of the poles and the equator, as exist in an elliptical body; however, Schröter, from his observations, had calculated that in the southern hemisphere there were mountains higher than those in the north. Amigues proposed, to the Academy of Sciences,[53] a different and highly original explanation, based upon a geometrical analysis of the problem.

Imagine a body placed on the equator of the planet. Let F be the attraction of the planet on the body, F' the centrifugal force due to the rotation. We know that the ratio F'/F is the same for all bodies on the equator of the same planet; Laplace represented it by the letter φ, which changes in value from one planet to another but is always small.

Geometers, assuming that the material of the Solar System was originally fluid, have concluded that for each planet approximating to a sphere, the flattening ought to lie between 1/2 φ and 5/4 φ.

[53] *Comptes Rendus des séance de l'Académie des Sciences*, 1874, Vol. I, p. 1557.

These provisions are justified by the observations. However, Mars is an exception, since Amigues concludes that the flattening exceeds 5/4 φ. This raises a serious objection to the hypothesis that the bodies were originally fluid.

However, it may be that the geometers have not considered the problem of spheroids with sufficient generality.

In effect, they all admit in their theories that the density of the layers diminished steadily from the centre of the spheroid up to its surface. Now, there is no proof that all the planets obey these conditions. Let us imagine, for example, that a planet cooled and solidified, taking up a certain form, and that later, because of circumstances impossible to define, a mass of cosmical materials passed in the neighbourhood of the planet and was disrupted by it, spreading out across the surface like a torrent of lava. We would be left with a spheroid in which the outer layers would be denser than the central layers. Amigues presented the general problem of spheroids as follows:

A spheroidal mass whose outer particles are fluid turns about an axis passing through its centre of gravity. The movement is slow; that is to say, the value of j is small. We assume a sphere having as its centre the centre of gravity of the spheroid, a sphere almost as large, but not exceeding it at any point on its surface. The material in the interior of the sphere is of mean density ρ (the mean density is the density of an homogeneous body of the same volume and the same mass). As for the material which is situated beyond the sphere, and which is spread out on the surface in a thin layer, it will presumably remain fluid, homogeneous, and of density φ'. Under these conditions, assuming that we have an equilibrium figure very little different from that of a sphere, we can attempt to find out what this figure will be.

The problem is evidently indeterminate, and it is easy to see that the figure we are looking for, will depend upon the distribution of the material in the interior of the sphere. We can remove this indeterminacy either partially or completely. It is the latter which will be discussed here.

We assume that the sphere described above is made up of spherical layers, concentric with the sphere, and homogeneous. This hypothesis has several advantages:

1. It seems to depart only slightly from the present physical conditions:
2. It leads on to definite problems which will make up the law regulating to densities of the layers:
3. It allows for an easy calculation.

This calculation, by ordinary methods, that is to say, by applying Laplace's functions and neglecting quantities of the second order, gives the following result.

The mass takes up the form of an ellipsoid, whose flattening is given by the following formula:

$$\varphi / 2\left(1 - 3\rho' / 5\rho\right).$$

Do not forget that our calculation applies only to a spheroid and that, consequently, the formula is valid only if the flattening which it yields is positive and small. This means that the value of ρ'/ρ must not be too big.

Discussion:

1. for $\rho' = \rho$, we obtain 5/4 φ, Newton's result.
2. for $\rho' = 0$, we obtain 1/2 φ, Huygens's result.
3. when $o < \rho'/\rho < 1$, the flattening lies between ½φ and ¾φ; this is what happens in the case discussed by Laplace and most geometers.
4. When $\rho'/\rho > 1$, the flattening exceeds 5/4 φ; this is the case which has not yet been examined.

Let us apply the formula to Mars. Its probable flattening is 1/33; this is low enough for us to use our formula. The value of φ relative to Mars being 0.0045866, we obtain the following relation:

$$1/33 = 0.00045866/2(1 - 3\rho'/5\rho).$$

We therefore have an equation of the first degree, which gives, without difficulty:

$$\rho'/\rho = 1.54.$$

Amigues' conclusions about the form of Mars are as follows:

1. The planet was formed in two or several stages;
2. The mean density of the outer layers is 1.54 times the mean density of the core, that is to say, of the planet.

The essential is to decide whether these premises are accurate—if the flattening of Mars really is 1/33.

But this flattening is difficult to measure. It seems to be below /33. The measurements made up to 1877 are:

1784	Herschel	1/16	1862	Main	1/38
1797	Schröter	1/81	1864	Main	1/46
1798	Köhler	1/81	1871	Main	1/71
1811–1847	Arago	1/30	1875	Main	1/36
1830–1837	Bessel Insensible		1856	Winnecke Insensible	
1852	Oudemans	Insensible (after Bessel)	1864	Kaiser	1/117
			1864	Dawes Insensible	
1855 Main		1/62	1877	Young	11/219

(Of these, the last last value is probably the best.)

An English geometer, M. Hennessey, has replied to Amigues' paper.[54] He comments that the results obtained by Amigues seem to be a complete verification of those which Hennessey had reached much earlier. He wrote:

> M. Amigues proposes to raise the important objection—the objection to the hypothesis of the original fluidity of the planets, because of the exceptionally large value for the flattening of Mars by saying: "The geometers have not discussed the problem of spheroids in a sufficiently generalized way".

[54] *Comptes Rendus des séances de l'Academie des Sciences*, 1878, Vol. II, p. 590.

And after having indicated the method concerned, he adds:

This calculation, made by ordinary methods—that is to say, by using Laplace's functions and neglecting the quantities of the second order—leads to the results which follow...

Hennessey comments that a very long time ago he had undertaken research into this very problem of spheroidal attractions (see *The Proceedings of the Royal Irish Academy*, Vol. IV, p.333). In the first case, he applied the results of his solutions to the question of the figure of the Earth, with the object of studying the basic parts of the theory which attempted to explain the spheroidal form by means of surface friction.

This theory was first proposed by Playfair in his *Commentaries on the System of Newton*, and was again put forward by Sir John Herschel in his *Outlines of Astronomy*. It was also cited by Sir Charles Lyell, serving as the basis for his views expressed in his *Principles of Geology*.

The results obtained by Hennessey did not confirm the theory, because the greatest compression of the Earth on the basis of surface friction would have been 1/404, which is in sharp disagreement with the results of observation. Hennessey wrote:

In 1864, for the first time, I applied my calculations to the problem of Mars, in a communication to the British Association, and a short extract from my work was published.

In February 1870, I published a memoir in Atlantis on the figure of Mars, and I applied to Mars the mathematical results of the preceeding researches. I found an equation giving the ellipticity as a function of the mean density D_1, and also the density D of the surface of the plane

$$e = 5q/(10 - 6D_1/D) = q/2(1 - 3D_1/5D).$$

In the equation, q is the ratio of centrifugal force to gravity.

Now if we use Arago's notation, replacing q by φ, D' by ρ' and D by ρ, we have

$$e = 5\varphi/(10 - 6\rho'/\rho) = \varphi/2(1 - 3\rho'/5\rho),$$

after which we have a formula exactly the same as that of Amigues.

From my formula, I have also deduced that if we accept the greater flattening attributed to Mars, we must conclude that the outer density is greater than the planets interior. But since such a situation appears contrary to the laws of physics, if the constitution of Mars resembles that of the Earth, I prefer to accept the conclusion of Bessel, Oudemans and Winnecke, who, after making very careful observations, are unanimous that the flattening of Mars is almost insensible.

An extract from my first researches into the theory of the form of the Earth, according to friction, has appeared in several scientific journals of some years ago. I am however convinced that the results obtained by Amigues with regard to Mars have been entirely independent, and made without any prior knowledge of my researches.

The complete agreement between the calculations of Hennessey and Amigues therefore confirms the opinion expressed above, in opposition to the theory of Playfair, Herschel and Lyell with regard to the form and structure of the Earth. (There is also a reference to these difficulties in Aragos memoir on Mars). Hennessey returned to the same question in 1880.[55]

---

[55] *Comptes Rendus*, 1880, Vol. I, p. 1419.

C.A. Young, of the United States, has published a series of observations on the equatorial and polar diameter of Mars. These measurements seem to have been made with the greatest care and under the most favourable circumstances; the observations were reduced and corrected for the slight effects of aberration, and the final value for e or the polar flattening was 1/219.

Hennessey comments that it is easy to see that this value agrees better with the hypothesis of the former fluidity of the planet than with the hypothesis of surface erosion by a liquid ocean having the same density as water.

If Mars were originally in a hot, fluid state, its mass would have been distributed in spheroidal surfaces of equal density, the density increasing from the surface to the centre.

The ellipticity would depend upon this law, and upon the rotation period of the planet, as is the case with the Earth. In such a liquid spheroid

$$e' = 5Q' / 2F(a'),$$

where Q' is the ratio of the centrifugal force to the gravity at the equator, and $F(a')$ is a function of the radius whose form depends on the law which regulates the variations of density going from the surface to the centre.

If T' is the rotation period of the planet, a' its mean radius, M' its mass and g' the intensity of the gravitational force at the surface, then

$$Q' = 4\pi^2 a' / T'^2 g', \quad g' = M' / a'^2$$

and consequently,

$$Q' = 4\pi^2 / T'^2 a'^3 / M';$$

For the Earth we have

$$Q = 4\pi^2 a / T^2 g \text{ and } \quad g = M / a^2;$$

and thence

$$g' = gM' / M(a / a')^2$$

and consequently,

$$Q' = Q(T / T')^2 (a' / a)^3 M / M'.$$

Astronomers generally agree that

$$a' / a = 54.$$

$T = 86164^s$, $T' = 24^h\ 37^m\ 22^s.7$ or 886427^s. If we accept that the mass values for the Earth and Mars given by Le Verrier, we have

$$M = 1/324439 \quad M' = 1/2812526Q = 1/289,$$

and it follows that

$$Q' = 1/224.07.$$

For the Earth $e = 5/2$ QF (a), and if F (a) has the same value for Mars, or more accurately the density varies from surface to centre in the same way as with the Earth, then

$$e'/e = (Q'/Q)e.$$

But, as the last determination of e gave $e = 1/293.46$, the calculation leads to $e' = 1/227.61$.

As Mars shows evidence on its surface of an aqueous fluid, we may go back to a theory previously advanced to explain the figure of Mars. It is suggested that erosion of the surface, combined with the centrifugal force which is produced by the planets rotation on its axis, will yield the observed result. This theory was suggested by Sir Charles Lyell.

Considering the theory of erosion by a moving liquid on the surface of a planet, I have found, for the ellipticity of the enveloping liquid,

$$e = 5QD + 6(D' - 1)\varepsilon / Q(5D - 3),$$

ε being the ellipticity of the solid surface, D the mean density, and D′ the density of the solid material at the surface. The maximum possible value of e corresponds to $e = \varepsilon$, and then

$$e = 5QD / Q(5D - 3) - 6(D' - 1)s$$

With the Earth, the values generally agreed for the mean density of the globe and the density of the solid crust are in round figures $D = 5.6$ and $D' = 2.6$. With these numbers, it is evident that e cannot exceed 1/417.

The smallest possible value which we can give D in the present state of our knowledge is about equal to twice D′, and it follows that

$$e = 5/7Q = 1/404.6.$$

Hennessey concludes that the erosion theory cannot explain the figure of the Earth so satisfactorily as the theory of complete original fluidity.

If Mars were an homogeneous solid, the erosion theory could give as good an agreement with the observed ellipticity as the homogeneous fluid theory, because in either case e would be 5/4 Q′, whence $e = 1/179.24$, a value which is sensibly greater than the result obtained by observation.

The researches carried out by various astronomers have recently shown that the surface of Mars presents a well-defined distribution of solid material and liquid material. The lands appear to form groups of islands rather than large continents.

If the figure of the planet differs from that deduced from the theory of original fluidity, if its flattening is slightly or considerably greater, such a distribution of land

and water could not exist. With a strong flattening, the lands would form a great girdle near the equator; with maximum flattening or a spheroidal figure, the lands would form two continents round the poles with an intermediate equatorial ocean. All recent observers agree that the planet has a distribution different from that which would occur in the latter case.

To me, it seems more likely that the old determinations of the flattening of Mars (upon which Laplace's reasoning was based) were too large, and that the real figure is close to the value given by Young, in accord with the rotation period and a gradual downward increase in density,

On the same question, the skilful mathematician G.H. Darwin[56] discussed Laplace's formulae concerning the densities rotations and flattening of the planets. Calling φ the ratio of the centrifugal force produced by the rotation (at the extremity of the mean radius of the planet) to the gravity: Mars spins in $24^h\ 37^m\ 22^s.6$ or 1.025956 sidereal days, while the density adopted by Darwin is 0.948 that of the Earth, and the sidereal Martian day is 0.997270. We have for the Earth

$$\varphi = 1/289.66$$

and for Mars

add equation

The measurements of the flattening must be affected by observational errors. Agreeing that the law of the interior density should be the same for Mars as for the Earth, the resulting flattening should be 1/298. But Darwin certainly adopted too great a value for the density of Mars; actually it is barely 0.70.

In 1872,[57] I examined the ratio of gravity to centrifugal force at the equator of Mars. Adopting a value of $24^h\ 37^m\ 22^s.7$ or 8843^s for the rotation period, we have:

Speed	$\omega = 2\pi/88643 = 0.0000709.$
	$\omega^2 = 0.0000000050239.$
	$a = 6371000\,m \times 0.53 = 3376630.$
Centrifugal force	$\omega^2 a = 0.1696$
Gravity	$g = 9\,m8088 \times 0.376 = 3\,m688.$
	$g/\omega^2 a = 217.5.$
On the Earth	$g/\omega^2 a = 9.8088/0.033858 = 289.$
Flattening therefore = 1/292.	

The ratio of centrifugal force to gravity, which is 1/289 at the terrestrial equator, is 1/217.5 at the equator of Mars. The flattening should not differ much from this value, if, as is likely, the density of the globe increases from the surface to the centre, as with the Earth. It would then be in the region of 1/226.

If Mars rotated on itself solely by virtue of its own gravitational force, in the manner of a satellite over the equator moving around the mass of a planet which is

[56] *Monthly Notices of the Royal Astronomical Society*, Dec.1876, p. 77.

[57] *Études sur l'Astronomie*, vol. III, 1872.

condensed toward the centre, the rotation period would be only $1^h\,40^m$. This figure must be multiplied by 14.77 to give the real rotation period of the planet. This number is also the square root of the number 217.5 found above, representing the ratio of centrifugal force to gravity at the equator of Mars.

We have the relationship

$$T / p = \left(g / \omega^2 a\right)^{-1/2},$$

where T = the length of the real rotation, p = the theoretical gravitational period, g = the surface velocity, ω = the angular velocity and a = the radius. But as

$$\omega = 2\pi / T, \qquad \omega^2 = 2\pi^2 / T^2.$$

Therefore, for all the planets we have the equation

$$g / \left(2\pi^2 / T^2 a\right) = T^2 / p^2,$$

$$2\pi^2 / T^2 \mathrm{x} T^2 = gp^2,$$

and, finally,

$$2\pi^2 a = gp^2,$$

which links the radius of the planet to the satellite period.

1874. —Terby: Aréographie

On 6 June 1874, the eminent Louvain astronomer presented to the Belgian Academy of Sciences "A Comparative Study of Observations made of the physical aspect of the planet Mars between the time of Fontana (1636) and the present day (1873)."[58] This extremely important work begins by an account of all the observations, and continues with a comparison of the various representations given of each region of the planet. This is a meticulous study of areography, and a detailed, careful study of the most important drawings. The principal questions concerning the geography and meteorology of the planet are discussed. Terby's aim is above all to help the observers. In conclusion he writes:

> Directed to debatable points, they do not neglect to elucidate a great many questions asked in this book, and the accuracy of Martian maps is also essential. I will be happy if the points raised in this memoir, touching upon such matters, contribute something towards our knowledge of the physical state of Mars.

[58] *Mémoires de l'Académie royale des Sciences de Belgique*, Vol. XXXIX, 1875.

This monograph on Mars has been of the utmost value, not only to observers but to all students of the planet; and science is indebted to M. Terby for one of the best documents on this subject—one of those which has made the most progress toward our general knowledge of Mars.

Note also, in the same year (1874) an excellent study by the Rev. T.W. Webb,[59] summarizing the work of Kaiser, who died on 28 July 1872—a study accompanied by two drawings made by this skilful Dutch observer. One of the particular points made in this study is Kaiser's assertion that the differences in tone which characterize the various dark patches, and the lack of sharpness of their contours, indicate that the seas do not resemble our own. Webb concludes that their blue-green colour is real, not due merely to contrast with the yellow continents.

1875.—Holden, Bernaerts, Ellery, Flammarion

During the 1875 opposition Holden, at the National Observatory in Washington, used the great 26–in. ($0^m.66$) equatorial—the most powerful telescope then in existence—to make a number of drawings, of which six were communicated to the Royal Astronomical Society of London.[60] They were made on 14, 16, 21 and 23 June and 2 and 5 August; magnification used, 400×. Despite the size of the instrument, Holden's drawings do not agree with the known appearance of the planet, as can be seen from his sketches of 16 June, from $10^h\ 40^m$ to $11^h\ 15^m$, and 23 June, from $10^h\ 20^m$ to $11^h\ 07^m$ (Figs. 82 and 83), which are the best of the series. This does not give much encouragement for the use of large instruments.

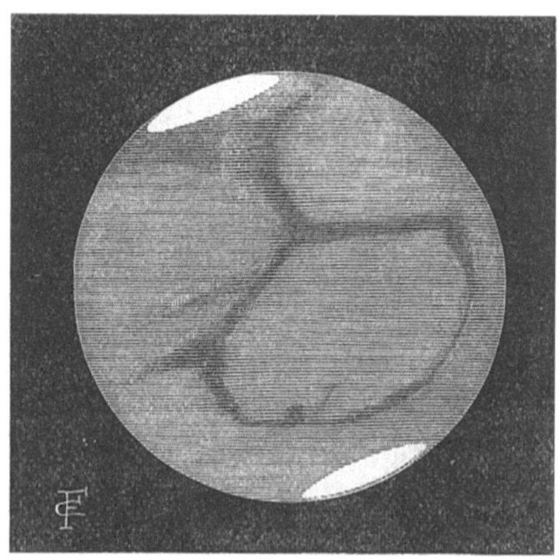

Fig. 82 Mars, by M. Holden, at Washington, on 16 June 1875

[59] *Nature*, 12 and 19 February 1874.

[60] *Monthly Notices of the Royal Astronomical Society*, November 1875.

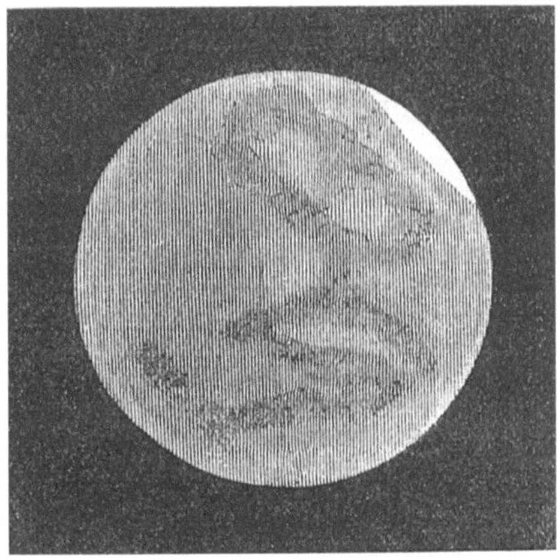

Fig. 83 Mars, by M. Holden, at Washington, on 23 June 1875

Holden made his drawings in pastels, stating that the colour best matching that of the Martian continents was salmon-red, and commenting that the principal belt on Jupiter was similar in hue; it had been drawn at the same time, and the same crayon had served for both.

On 12 August 1875, Mars was occulted by the Moon, and the phenomenon was observed at $2^h\ 58^m$ at the Windsor Observatory (New South Wales) by John Tebbutt. Nothing remarkable was seen.

During the same opposition Bernaerts, at Malines, made a series of observations and sketches,which add nothing of importance.[61]

Since 1875 we have been concerned with various comparisons between the different planets, stars, and luminous gas, with the help of a movable sextant mounted on a fixed pier, and bringing together the images of the different stars, or one star and a gas-jet.[62] The stars were, as often as possible, at 40°–50° above the horizon, and about 1 km. to the south of the Paris Observatory. There is one essential provision: the thickness of the atmosphere tends to increase the rays at the red end of the spectrum, to the detriment of those at the blue end. The experiments which I conducted gave for the colours and the contrasts

Sirius	Bluish white
Moon	Clear yellow
Jupiter	Milky yellow
Mars	Orange yellow
Antarès	Orange
Gas	Reddish orange

[61] *Bulletin de l'Academie Belgique*, 2nd series, Vol. XLV, p. 39.

[62] *Bulletin de la Société Astronomique de France*, 1st year, 1887, p. 50.

There were some very curious contrasts:

Mars and the Moon	Vivid orange and pale blue
Mars and Jupiter	Orange and pale sea green
Mars and Saturn	Orange and green
Mars and Véga	Red and blue
Gas and Mars	Orange and citron
Gas and Moon	Bright cherry red and glittering silver

Thus Mars, which to the naked eye appears as red as Antarès, its historic rival, is less red than a gas-jet seen from a distance of 1 km.

In the same year I also made many drawings of the planet.

1876.—C. Flammarion: *Les Terres du Ciel*

The first edition of this book, published in November 1876, was devoted to the planet Mars.[63]

I reproduce here the following figure (Fig. 84) which gives the opposition positions for the cycle prior to 1877, and prepares us for the perihelic opposition of 1877. C is the centre of the orbit of Mars, P the perihelion point of Mars, *a* the aphelion of the Earth, *p* the perihelion of the Earth, Ω the line of intersection of the orbits. It is seen that since 1869, each opposition has brought Mars nearer to the Earth, and that the closest approach was due in 1877.

On this orbit I have marked the solstices and equinoxes of Mars; and with regard to Martian climatology, I wrote:

> This world, as with ours, presents three very distinct zones; the torrid, temperate, and glacial. The first extends 28°42′ to either side of the equator; the temperate zone extends to latitude 61°18′, and the glacial zone from this latitude to each pole.
>
> As with the Earth, the planet rotates in the plane of the Zodiac; similarly the Sun seems to move throughout the year in the Zodiacal constellations. However, the summer solstice in the northern hemisphere does not occur with the Sun in Cancer, as with us, but in Virgo while the winter solstice is in Leo, not Capricornus. On Mars, we should therefore refer to the Tropics of Virgo and Leo.
>
> The existence of a Martian atmosphere is proved. When the surface patches are near the centre of the disk, they are clearly seen, but when they are carried by the planets rotation to near the edge of the disk, not only do they appear foreshortened due to the geometrical perspective of their position on the rotating sphere, but they also lose their sharpness, becoming pale and even invisible before reaching the limb. This effect is caused by the Martian atmosphere, which absorbs the light-rays, and interposes a veil of greater and greater thickness as the light comes from nearer and nearer the edge of the disk. Moreover, the edge of the planet is, overall, paler than the central region, because of the same atmospheric absorption.
>
> In addition, the snows, the clouds, and the researches of spectral analysis prove that water vapour exists in the Martian atmosphere.

[63] *Librarie Academique, Didier et Cie*. Vol. I, Book VI, pp. 307–440.

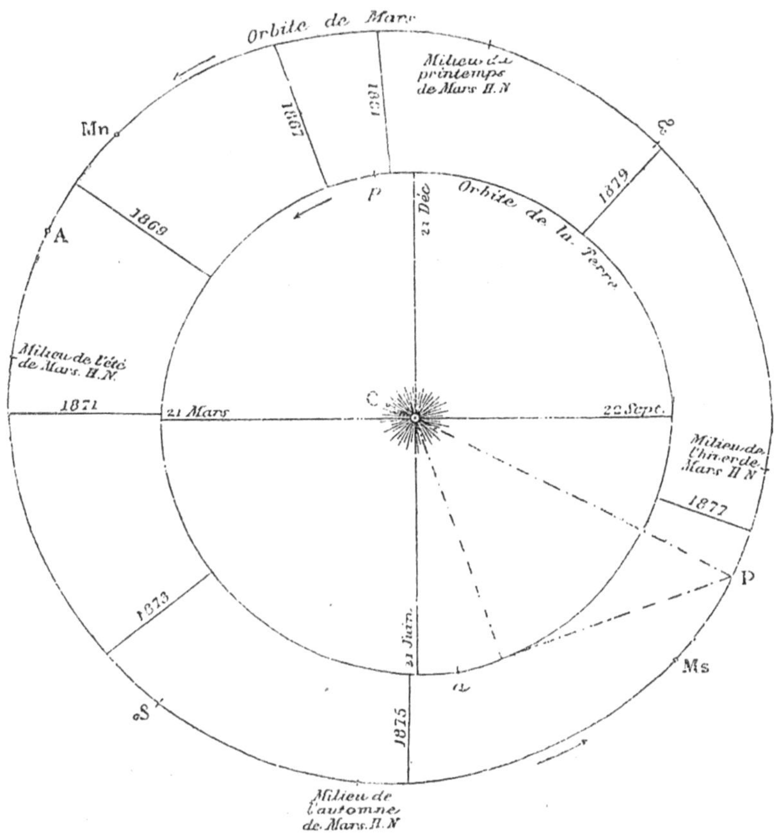

Fig. 84 Relation of the orbit of Mars to that of the Earth, from 1867 to 1877

The geography of Mars is dealt with in another chapter, where I consider all the observations made since 1636, and which are completed by a map showing the present extent of our knowledge. This map will be found below, as a preliminary to the opposition of 1877. It may be summarized as follows:

> The examination of this planisphere shows us first that the geography of Mars does not resemble that of the Earth. While three-quarters of our globe is covered with water, the distribution of land and sea on Mars is practically equal. Instead of being islands emerging from the heart of the liquid element, the continents seem rather to reduce the oceans to simple inland seas, and to make them become veritable Mediterraneans. There is no Atlantic, no Pacific on Mars, and one could walk almost right round the globe on dry land. The seas are cut up by gulfs, variously prolonged in a large number of arms stretching out in the manner of our Red Sea over the closed land. Such is the essential character of areography. The second characteristic, which suffices for recognition of Mars from a great distance, is exemplified by the Hourglass Sea and the Channel.

We may also conclude that the dark patches do indeed represent stretches of water, while the bright areas represent continents—though this interpretation has been discussed and challenged by more than one observer. (Liais, Cruls, Brett, Trouvelot, etc.).

In my own words:

> It is evident that there is water on Mars, since we can see the state of the polar ices, the variable snows, and also the clouds floating in the atmosphere; moreover, the presence of water has been proved by the spectroscope. The seas, seen from afar, appear darker than the lands, because water absorbs a great part of the light and reflects very little.
>
> We must, however, note that the seas of Mars are not equally dark; some are particularly dark (Hourglass Sea, Kaiser Gulf, Lockyer Sea, Maraldi Sea; see my chart). It may be thought that the less dark seas are sprinkled with islands, which we cannot distinguish because of their small size, and that some areas of the water are not very deep, as is the cause with, for example, our Zuyder Zee. These differences are surprising; they cannot be explained by changes in the transparency of the atmosphere of Mars, but are real. So why is not the same thing evident in terrestrial waters? Yet the colour of sea-water is far from uniform. The Marne is yellow, the Seine pale green, the Rhine dark green, etc.
>
> Moreover, it seems that the Martian seas are not changeless, because since 1830 there have been certain unquestionable variations. For instance, there is Kaiser Gulf, which since the end of the last century has presented the aspect of a thread ending in a disk, but after 1862 has become broader, ending not in an isolated black circle but in a forked bay. On the planet there may possibly be displacements of water, and variations in the colour of the water, which do not occur on Earth.

Such is a résumé of the knowledge derived from physical observations. For the first time (1876), the variations in the seas, in tone and in extent, has been established by adequate observational data.

To the long series of observations which makes up our second period, we must add the work of Capocci (1862), Schultz (1862), Vada (1863), Michez (1865), Folque (1867), Fabritius (1873), etc. But these detached contributions add nothing to our knowledge.

All the observations which we have examined have, assuredly, their intrinsic value, but on arriving at the close of our second period we must comment that the work of these last few years was essentially in preparation for the exceptionally favourable opposition of 1877. Astronomers prepared well ahead, as they had done in 1862; the planet was becoming better and better known, and the progress in optics was most encouraging.

Before coming to the third and last section of our examination, let us summarize the progress made during the second period, 1830–1877.

Conclusions of the Second Period 1830–1877

Refer back to an earlier section of this work, and recall the conclusions which had been reached in the first period.

The 13 points made in the first list have been confirmed. Many have been developed. New information has been obtained:

14. The length of the rotation period has now been fixed accurately at $24^h\ 37^m\ 22^s.65$, to within a few hundredths of a second; the true value certainly lies between $22^s.6$ and $22^s.7$. In 1830, we had still been far short of this accuracy.

15. The geography of the planet has been sketched in its principal characteristics. Several charts have been drawn up, first by Beer and Mädler in 1840, then by Kaiser in 1864, Phillips in the same year, Proctor in 1867, Green in 1873. To these geographical contributions must be added the map I published in 1864, showing the better-known hemisphere of the planet, that with the Hourglass Sea at the centre. The dark patches are essentially permanent, and we cannot agree with Schröter that they are atmospheric in nature. However, our first conclusion (point no. 8) is retained; the forms and aspects of the patches are variable. Two hundred new views of Mars have passed under our eyes during this second period. Joined to the earlier 191 of the first period, this makes a total of 391 different drawings of the planet, made by all the observers. Their study, by comparing them, establishes that each observer sees according to his eyes, his skill and his instruments; and he draws accordingly.
16. Therefore, for each drawing there is what we may call a personal equation, an individual interpretation, and as the details of a globe seen from the distance of Mars through two atmospheres are always vague and exceedingly delicate, there can be no single drawing which gives a rigorous and exact representation of Mars as it would appear to an observer close to the surface.
17. Nevertheless, despite all this variety, there is a definite basis represented in my general map. Moreover, the discrepancies between different observers cannot account for other kinds of changes, which must therefore be regarded as real. The Hourglass Sea certainly changes in breadth and tone; its left bank, particularly at the Hind Peninsula, seems to indicate lands which are sometimes dry and sometimes inundated; the circular Terby Sea has around it—particularly below it—regions which are sometimes bright and sometimes dark; the Flammarion Sea is sometimes crossed by a sort of sandbank; the Meridian Bay appears sometimes round, sometimes square, at still other times elongated and forked, and so on.
18. These aspects and their variations confirm the interpretation already put forward from the first period: the dark patches represent stretches of liquid, seas and lakes, and the bright areas are solid expanses, continents and islands.
19. The variations in the polar snows confirm that the water has the same properties as that on our own planet—susceptible to conversion into snow, ice, or clouds.
20. Spectral analysis, created during the second period, establishes that the waters are analogous to ours in chemical composition.
21. However, these aqueous stretches appear to be in a different physical state from our own seas: they may be less dense, less liquid, or covered with viscous fogs.
22. The atmosphere is less disturbed than ours, less charged with clouds and fogs, less productive of rain, more rarefied, more transparent. The water must evaporate and condense more easily than on Earth. Contrary to the views of Father Secchi, cyclones cannot be seen, but we can sometimes observe very extensive clouds (see the various drawings), at great distances from the poles—particularly in Lockyer Land, which is sometimes mistaken for a pole.
23. There is less water on Mars than on the Earth, both in extent (scarcely half the globe, instead of three-quarters, as with us) and also, no doubt, in depth, because the variation in tone of the seas seems to be due to the fact that the bottom is

sometimes visible; there is also frequent flooding across the vast plains, which, we assume, must be very flat.

24. The south or (as seen in the telescope) the upper hemisphere of Mars is primarily marine; the northern hemisphere is primarily continental. The level of the latter is therefore presumably higher than that of the former. The geological events leading from the formation of the planet have led to the develoPMent of an elevated northern hemisphere and a depressed southern. Let us note that it has been almost the same on Earth, where the great continents, Asia and Europe, North America and half of Africa, lie in the northern hemisphere; the southern hemisphere has South America, South Africa and Australia, but the level of the surface is much lower. The difference could be due to the Suns attraction on the hemisphere of Mars which is closest to the Sun during the half-period of the revolution of the line of the apsides at the critical time when the planets crust was solidifying; this attraction could have had the effect of lifting up the northern hemisphere slightly and obliquely. The continental centre seems to be in Huygens Continent, at longitude 150° latitude 20°; the marine centre almost antipodal in Dawes Ocean, at 330° longitude, 30° latitude. For Earth, the analogous points are the Carpathians and their antipodes.
25. The polar flattening of Mars is certainly less than was believed by Herschel, Laplace, and Arago; objections to the theory are without foundation. The geometrical figure of the globe of Mars does not differ much from ours, if we accept greater flattening. The ratio of centrifugal force to gravity is !/217.5. The polar flattening should differ little from this value.
26. Rivers should exist on Mars, since there are clouds, seas and rains. The Meridian Bay seems to be the mouth of two great rivers.
27. Though the globe of Mars seems to be less irregular than ours in orographic relief, there certainly seem to be mountains of some height, and some elevated plateaux. Thus the two islands drawn on my chart at 47° and 297° longitude, are sometimes visible and sometimes not; they are undoubtedly mountains sometimes covered with snow. It also seems that there is a high plateau near the equator to the right of the Hourglass Sea, and another at the intersection of 185° longitude with 65° south latitude.

To sum up: the analogies between Mars and the Earth have been established by a series of observations. The climatic conditions seem remarkably similar to our own; the temperature on Mars is almost the same as ours, though the physical conditions of atmospheric pressure and density, and the lower gravity, produce effects analogous to a temperature difference.

Third Period.—The Martian Cycle 1877 to 1892

Our third period includes the Martian cycle from 1877 to 1892, which has been more fruitful than any of those preceding it.

Recall first that the combination of movements of the Earth and Mars round the Sun produces a 15 year cycle, over which Mars is presented to us in all its aspects. The planet completes one revolution round the Sun in $686^d\ 23^h\ 30^m\ 41^s$; the Earth takes $365^d\ 6^h\ 9^m\ 11^s$. The result is that the Earth and Mars lie along the same line, relative to the Sun, about every two years: 780 days on average. The interval is not constant, because neither planet moves at uniform velocity, since their orbits are elliptical, not circular. We can consider the opposition dates of Mars in the cycle which will now be discussed. This can be done by means of a table.

Oppositions of Mars, 1877–1892

		Distance from Earth		
	Date	a.u.	km.	Maximum diameter
1877	5 September	0.3767	55,532,000	25″
1879	12 November	0.4824	71,877,000	19″
1881	26 December	0.6028	89,817,000	16″
1884	31 January	0.6691	99,696,000	14″
1886	6 March	0.6699	99,830,000	14″
1888	11 April	0.6050	90,145,000	15″
1890	27 May	0.4849	72,265,000	19″
1892	4 August	0.3773	56,217,000	25″

Because of the ellipticity of the two orbits, the oppositions differ markedly with regard to the minimum distance separating the two planets. When opposition takes place at the time when Mars is at perihelion, at longitude 334°, on August 27 on Earth, the distance is reduced to its minimum. These favourable oppositions recur every 15 years. Perihelic oppositions took place in 1877 and 1892. I have calculated the distance in kilometres, using solar parallax 8″.81, and find that the average distance from the Earth is 149,000,000 km.

C. Flammarion, *Camille Flammarion's The Planet Mars*, Astrophysics and Space Science Library 409, DOI 10.1007/978-3-319-09641-4_3

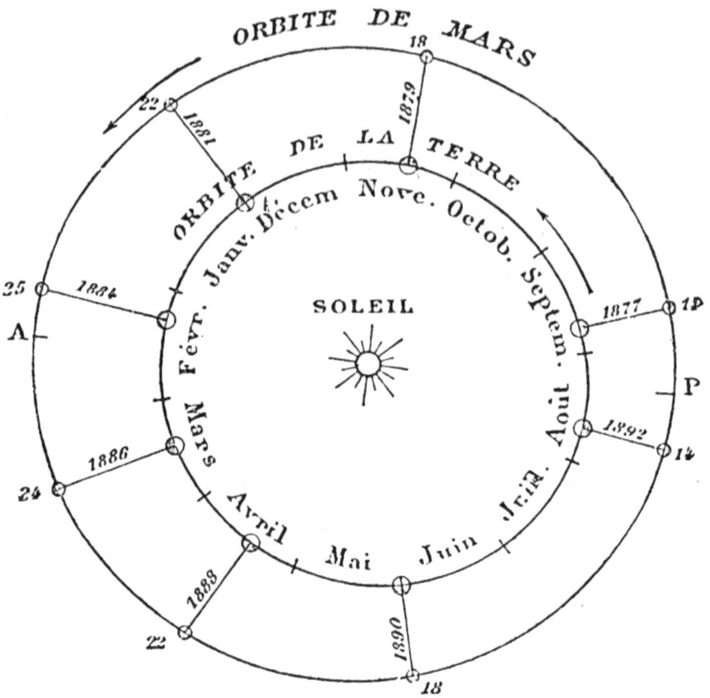

Fig. 1 Cycle of the oppositions of Mars

Minimum distance does not coincide exactly with the date of opposition, because opposition represents a difference of 180° in longitude between the Sun and the planet, whereas Mars may cross the plane of the Earth's orbit shortly before or after opposition. But the difference is never more than a few days. The figures given above indicate the closest approach for each year.

If we express these distances in kilometres or in leagues, we find a range for these seven oppositions of between 55 and almost 100 million km. or 14 to 25 million leagues, which is considerable. The cycle is shown in the diagram (Fig. 1), which is self-explanatory.

This cycle, from 1877 to 1892, which forms the third section of our historical study, is also a naturally-defined period in the observation of Mars, and if this monograph is continued in the future it will give the cycles in the years ahead: 1892–1907, 1907–1922, etc.

The year 1877 produced magnificent results. If the opposition of 1830 truly opened our geographical and climatological studies of Mars, that of 1877 ushered in an era of analysis exceeding anything for which one could have dared to hope.

We will here review the work accomplished. Now that the observations are becoming more detailed and more precise, and we can trace the effects of the seasons upon the surface of the planet, we can give, as headings, the dates of the solstices and the equinoxes corresponding to the times when the observations were made.

1877
Opposition: 5 September
South Pole Inclined to The Earth

Calendar of Mars		
	Upper (S) hemisphere	Lower (N) hemisphere
1 May 1877	Spring equinox	Autumnal equinox
27 Sept. 1877	Summer solstice	Winter solstice
6 Mar. 1878	Autumnal equinox	Spring equinox

The opposition of September 1877 inspired me to draw up a standard planisphere based upon all the information available. This chart of Mars was published in April by the Academy of Science, with a long note accompanying it and explaining the reasons for producing it. It seems that the earlier charts, even Proctor's left much to be desired, and I tried to produce something a little better, so as to help observers in identifying features shown on their drawings.

1877.—C. Flammarion: Chart and Observations of the Planet Mars

The chart is shown in Fig. 2, presented here as provisional and destined to be greatly improved in the future.[1]

Such is the point at which we resume our aerographical studies. At the same time, with a $0^{m}.20$ reflector and magnifications of from 210× to 300×, I made a large number of drawings of the planet. Four of them are reproduced here (Fig. 3), reduced to a scale of 1″ of arc to 1″, showing accurately the size of the disk on the dates indicated. The first was made on 30 July at $11^{h}\ 0^{m}$; the Hourglass Sea was nearly at the centre of the disk, slightly past the central meridian (longitude of central meridian, 312°). The second was made on 23 August, at $11^{h}\ 0^{m}$; the circular sea or the Eye was appreciably past the central meridian, whose longitude was then 105°. The third drawing was made on 14 September, at $10^{h}\ 0^{m}$; it shows the Maraldi Sea and the Hooke Sea (longitude of the central meridian, 240°). The fourth drawing was made on 26 October, at $7^{h}\ 55^{m}$, and vaguely sketches the form of two open wings of a large sail, formed by the union of the Schiaparelli and Maraldi Seas (longitude of the central meridian 190°). By this latter date, Mars was already a long way from the Earth.

These four drawings also show, in a striking manner, the diminution of the polar snow at the upper or south pole. The summer solstice of the southern hemisphere of Mars occurred on 27 September 1877.

[1] *Comptes rendus de l'Académie des Sciences*, meeting of 27 April 1877, p. 476.

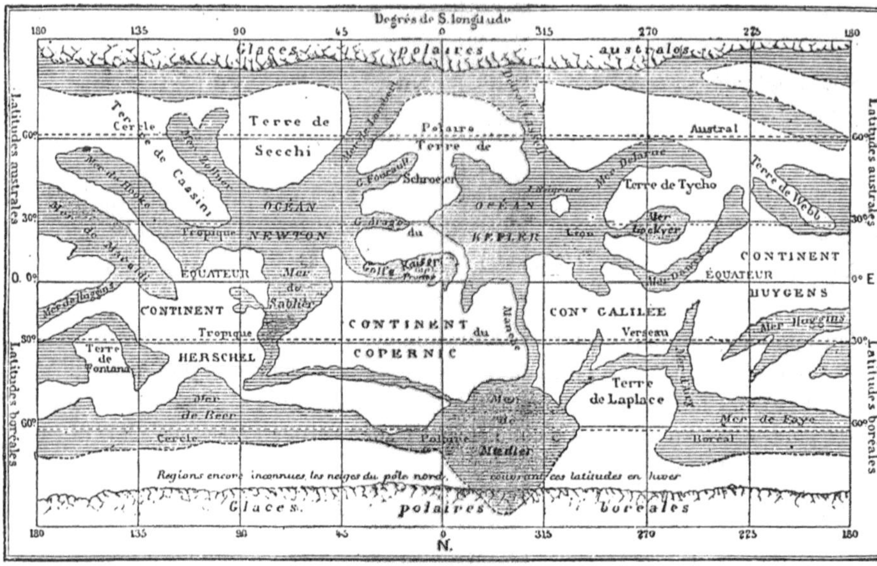

Fig. 2 Provisional geographical chart of the planet Mars, by M. Flammarion, in 1876. This chart is the same as that published in the first edition of *Terres du Ciel* (November 1876)

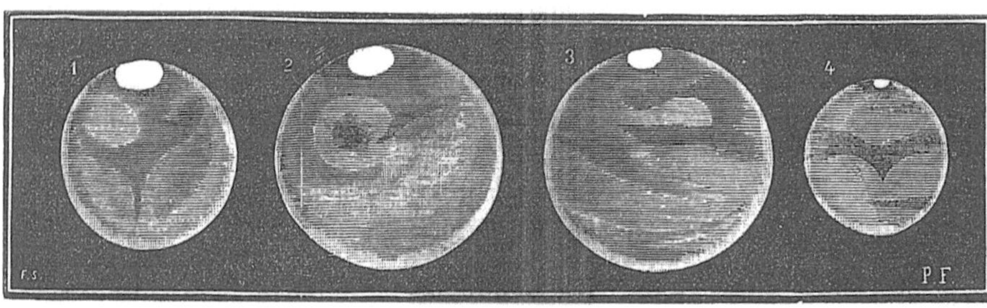

Fig. 3 Sketches of Mars, on 30 July, 23 August, 14 September and 26 October 1877

To these sketches, which represent the total circumference of the planet, we can add a certain number of others made during this opposition. Also reproduced here is the drawing of 27 September, at 8^h 35^m (Fig.4), in which the circular lake is detached as a very clear black feature. The sea which surrounds it does not seem to continue above, as on the drawing of 23 August. The De la Rue Ocean and Christie Bay are well shown.

One point worthy of attention during the period from July to October 1877 is that the visible hemisphere of Mars has been almost constantly clear and cloudless.

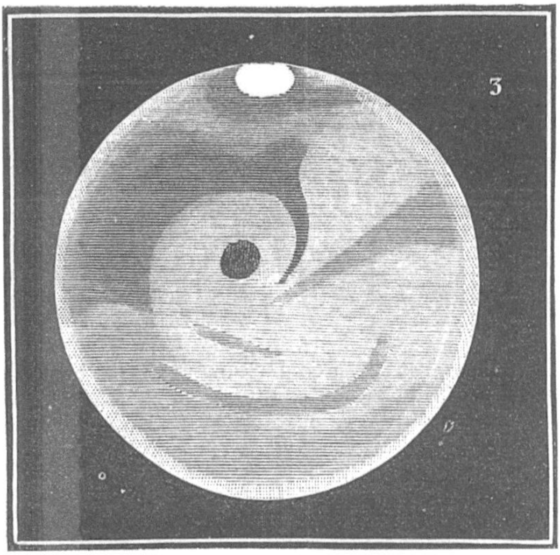

Fig. 4 Mars, 27 September 1877

1877.—Paul and Prosper Henry: Drawings

At the Observatory of Paris, the Henry brothers made a certain number of drawings of the planet during the same opposition, using the $0^{m}.24$ equatorial. These drawings agree remarkably closely with my chart. Some particular points should be noted. Thus in several drawings the circular sea appears to be almost connected with a light streak, as if the Schiaparelli Sea were continued as a thin thread up to it. Such an aspect is, for example, shown on the drawing of 22 August, at 11^{h} 50^{m}. A beautiful drawing made on 5 September, at midnight, clearly shows the Meridian Bay and Burton Bay, but the Dawes Ocean seems to be limited to the south by a stream which corresponds to the Arago Strait. Even more remarkably, the observers saw that on several occasions the Hooke Sea was crossed by a white line, giving the impression of a gigantic rectilinear bridge of clouds or a sandbank. Such an aspect was, for example, shown on the drawing of 10 September, at 9^{h} 50^{m}. The right bank of the Hourglass Sea was very dark. Two hours later, this sea, carried round by the rotation of the planet, was shown in its full amplitude. These four very curious drawings are reproduced here (Fig. 5). It is interesting to compare them with my complete chart (Fig. 31).

To these drawings may be added that of 27 August (Fig. 6), by the same observers, which represents the first observation made from France of the satellites of Mars. These satellites were discovered at the Washington Observatory by Asaph Hall, the more remote on 11 August 1877 and the closer on 17 August. The news was telegraphed to Europe, notably to Le Verrier, Director of the Paris Observatory—this was near the end of his life; he died on the following 23 September—and attempts were made to confirm the discovery, as curious as it was unexpected. On 27 August the Henry brothers succeeded; they masked the planet and saw the outer satellite. At this time the hemisphere of Mars with the Eye near its centre was turned toward the Earth.

Before continuing with our study of the planet, this is the right place to relate the astonishing discovery of the satellites.

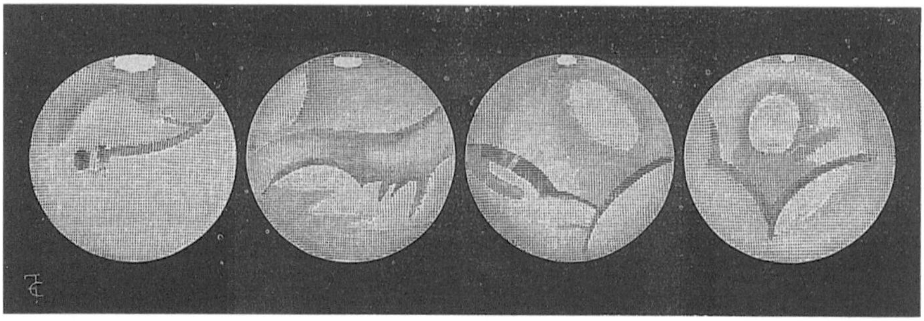

Fig. 5 Sketches of Mars, on 22 August 1877, 5 September, 10 September at $9^h\ 50^m$, and the same day at $11^h\ 45^m$ by the Henry brothers

Fig. 6 Mars, on 27 August 1877. The first observation made in France of a satellite of Mars

1877.—Asaph Hall: Discovery of The Satellites of Mars

The discovery of the two satellites of Mars was certainly one of the most curious and most interesting of modern times. It may be said that it was due—as is generally the case with discoveries—to the result of the most laudable persistence. We have already seen that 1877 was particularly notable because Mars came almost to its closest to the Earth, and that opposition occurred on 5 September. Professor Asaph Hall, astronomer at the Observatory of Washington, decided that this would be an extremely favourable opportunity to study the neighbourhood of Mars, using the great equatorial at the observatory. He said, rightly, that even if many observers had already been disappointed in their hopes of finding a satellite of the planet, this was not sufficient reason for saying definitely that no satellite existed, particularly since observations in 1877 were so exceptionally favourable for a search.

He therefore began work in the first evenings of August, examining the regions round the planet with minute care, and, to avoid being hindered by the great brilliancy of the planet, masking it or putting it out of the field of the telescope, so that he would be able to glimpse the slightest trace of any satellite in the neighbourhood.

The first nights were fruitless, tiring and discouraging, and the astronomer had decided to give up the search when Mrs. Hall, secretary to her husband, strongly insisted that he should try for one more evening. This was on 11 August. Hall went to the equatorial, and 3 hours later believed that he had caught sight of a luminous point which made his heart pound. But he had scarcely made sure of its existence when a thick mist rising from the Potomac River interrupted his observations. The sky remained obstinately covered with clouds during the next few nights. Finally, 5 days later, on the 16th, the sky cleared, the astronomer made haste to go to his telescope, recovered the tiny point, did not lose it again, and after 2 hours observation satisfied himself that it was moving across the sky with the planet. Therefore, it could not be a fixed star. But perhaps-a real hazard!—it could be one of the innumerable small planets which move between Mars and Jupiter, and which happened to be passing Mars exactly at this moment? He consulted his ephemerides, and found that on this date the asteroid Europa should be passing exactly behind Mars.

A preliminary calculation showed that if the tiny point he had observed were a satellite, it should be hidden by the planet for part of the night following, the 17th; but should reappear before dawn close to its original position. If it were the asteroid Europa, it would be found on the same evening a little to the south-east of Mars.

The night of the 17th was marvelously clear, and Mars had scarcely risen above the horizon mists when Hall impatiently pointed his telescope toward it. No satellite was visible, which was a good sign. At 4 o'clock in the morning the astronomer gleefully saw the tiny speck of light come quietly out from the radiance of the planet, just as he had calculated; it really was a satellite of Mars.

This was not all. In observing the satellite and following its movement, Hall slightly later, found a second, even fainter and closer to the planet!

The news was telegraphed to the principal astronomers all over the world, and in spite of the initial scepticism it was not long before the discoveries were confirmed in full.

These two small satellites were followed, with large telescopes, during the months of September and October 1877; then they were lost to view as Mars receded from the Earth. They were recovered in 1879, when the planet came back to our neighbourhood, and were even seen by observers with less powerful instruments, because they are more easily seen when their existence is known than when they are unsuspected. They were again recovered during the opposition of 1881, and have been seen at all subsequent oppositions.

These two little moons were named by their discoverer Deimos (Terror) and Phobos (Flight), recalling two of the verses of Homer's *Iliad* (Book XV) which tells how Mars descended to Earth in order to avenge the death of his son Æsculapius:

> He ordered Terror and Flight to put on their shields, and himself put on his shining armour.

Fig. 7 Apparent orbits of the satellites of Mars, as seen in an astronomical telescope

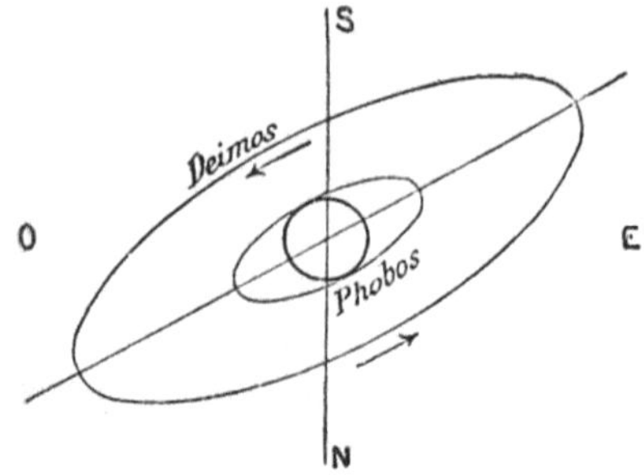

Phobos is the first and closer in, Deimos the second. The elements of their orbits are as follows:

The radius of Mars:	= 3,364 km
Distance of Phobos:	= 2.77 radii of Mars
	= 9,321 km
Distance of Deimos:	= 6.92 radii of Mars
	= 23,281 km

These distances are reckoned from the centre of the planet. If we subtract the radius of Mars, it is found that the distances from the Martian surface are 6,000 km for Phobos and less than 20,000 for Deimos.

As the angular diameter of Mars is 9″.57, the greatest elongations are only 13″ for Phobos and 32″ for Deimos.

The revolution period of Phobos is strangely rapid: $7^h\ 39^m\ 15^s$. That of Deimos is also very rapid: $30^h\ 17^m\ 54^s$; this is almost equal to four revolutions of Phobos, which indicates that both satellites are of similar origin. Both orbits are practically circular, almost in the plane of the Martian equator, and inclined at about 26° to the ecliptic. The system is represented in Fig. 7; it is seen that the satellites move in the equatorial plane of Mars.

Because of the smallness of the satellites and their closeness to the planet, they cannot be seen without excellent instruments. However, it is easier to see an object already known to exist than to discover it; instruments much less powerful than the Washington equatorial are sufficient for the observer to see these two luminous points, and even measure their positions.

The existence of the satellites had been suspected by analogies, and several astronomers—W. Herschel, d'Arrest, etc.—had even spent many hours in searching for them. It had been seen that the Earth has one satellite, Jupiter four, Saturn eight; therefore Mars, which is between the Earth and Jupiter, ought logically to have two. Kepler himself was the first to reason this out, in 1610, and Voltaire followed the tradition in his *Micromégas*.

These two celestial bodies are so small that it is impossible to see them as appreciable disks; all we can do is to estimate their probable volume by carefully measuring the amount of light which they reflect. This was done at the Harvard College Observatory by Professor Pickering, whose photometric measurements confirmed estimates made by other observers. Assuming that the satellite surfaces are similar to that of the planet itself, their diameters cannot be greater than 10–12 km. The first, Phobos, is brighter and probably the larger of the two; yet it shines only as a faint star of the 10th magnitude, while Deimos is only of the 12th magnitude. However, Deimos is the easier to see, because it is further from the planet and is less overpowered by the light of Mars. It is nevertheless remarkable that these two luminous points, *whose diameters are scarcely greater than that of Paris*, can be seen across a distance of 15–20 million leagues by man-made instruments. [2]

The apparent movements of these satellites in the Martian sky are particularly curious. As has been noted, Deimos revolves round Mars in $30^h\ 17^m\ 54^s$, while the planet itself rotates in $24^h\ 37^m\ 23^s$. It follows that Deimos moves very slowly from east to west in the sky of Mars.

Since the difference between the revolution period of Deimos and the rotation period of Mars is only $5^h\ 41^m$, the satellite seems to take 131 h to go right round the Martian sky; this is equivalent to 5 Martian days and 8 hours, and this will be the short month upon which the inhabitants could base their calculations.

The movement of Phobos is quite different. As it completes its revolution from west to east in $7^h\ 39^m$, and as the planet turns on itself in $24^h\ 37^m$, Phobos seems to rise in the west and set in the east, after having crossed the sky at a speed corresponding to the difference between the two movements; that is to say, about 11 hours. This is a case which is unique in the Solar System.[3]

What are the apparent sizes of these two satellites as seen from the planet?

We know that for an object lying at a distance equal to 57 times its own diameter, the apparent diameter will be 1″, while an object at 570 times its own diameter will subtend an angle ten times smaller: that is to say, 6 minutes. Since Phobos is 6,000 km. from the surface of the planet, and is probably about 12 km. in diameter, it will lie at a distance of 500 times its own diameter, and will consequently show a disk of about 7 minutes.

This is little less than one-quarter the diameter of our full moon, which is 31 minutes. It is also one-third the mean diameter of the Sun, which, as seen from Mars, is 21 minutes.

Deimos, at 20,000 km from the Martian surface, is reduced to a small disk 2 1/2 in. in diameter.

The light received from these two satellites by the inhabitants of the planet is extremely feeble. Even at the zenith, Deimos will have an apparent diameter equal to only 1/15 that of our full moon, so that its surface area will be 225 times smaller.

---

[2] In giving for Deimos an apparent diameter of 0″.031, Mr. Hall himself writes (*Monthly Notices*, Feb. 1878, P. 207) that this angle corresponds, at the distance of our Moon, to a circle 187 ft (57 m) in diameter, so that the idea of establishing a system of luminous signals to communicate with the inhabitants of Mars *is by no means a chimerical project.*

[3] On observing the transit of this satellite across the meridian (*Monthly Notices*, *Id.*, p. 208), the Martian astronomers have a very exact method of determining their longitude, not to a factor of 29, as with our Moon as a multiple of errors of observation, but only to an error of below one-half. However, we cannot doubt that the Martian astronomers have their own difficulties. These are perhaps due to a dense atmosphere and strong refractor; moreover, it is no sinecure to observe three transits of the meridian in 1 day!

Also, the light received from the Sun varies, according to the position of Mars, between 1/2 and 1/3 of that received by our Moon. It follows that the brightness of Deimos should be somewhere between 1/450 and 1/675 of that of our moonlight. Phobos is three times larger, showing a disk 6′ to 7′ in diameter, and giving ten times as much light, i.e. between 1/45 and 1/67 of the intensity of our moonlight. Such are the two tiny satellites.

The discovery of the satellites made it possible to determine accurately the mass of the planet, which had previously been rather uncertain. Hall found that relative to the Sun, the mass of Mars is 1/3,093,500, which is 0.105 that of the Earth.[4]

The spectacle of Mars seen from each satellite would be admirable, and the brilliancy of the planet would be marvellous. Seen from Phobos, it would cover almost a quarter of the whole sky, and from Deimos about 1/11 of the sky. From Phobos its surface area would be greater than 6,400 times that of our full moon; from Deimos over 1,000 times; brilliance 2,500 and 400 times respectively.

Some cosmological comments may be made about the satellites of Mars.

The best hypothesis of the formation of the celestial bodies is that which regards them as condensations from primordial material (Kant and Laplace). An immense nebula originated with the Sun, and the planets condensed out of this nebulosity, their revolution around the central fire being due to the original rotational movement of the nebula.

It will be the same with satellites relative to their primary planets; the Moon originated from the terrestrial nebula, or was thrown off from the equator; the satellites of Mars, Jupiter, Saturn, etc., are of analogous origin. According to this hypothesis, each satellite should circle round its primary in a period greater that the axial rotation of the primary, because, since the separation of the satellite, the planet would have continued to condense and to spin more and more quickly, by virtue of the principle of the law of areas.

It is the very rapid movement of the inner satellite of Mars which is a puzzle. Does it contradict the nebular theory?

No, it does not. In the Saturnian system, the particles which make up the inner ring have revolution periods which are shorter than the rotation period of the planet. The periods for which centrifugal force equals gravity are for the distances of the transparent ring—1.36 to 1.57—5^h 50^m and 7^h 11^m, and for the boundary of the broad central ring, 7^h 12^m.[5] The rotation period of Saturn is 10^h 15^m.

---

[4] The principal determinations of the mass of Mars, before the accurate results made possible by studying the movements of the satellites, were:

1802—Delambre, from the perturbations of the Earth (value adopted by Laplace): 1/2,546,320.

1813—Burckhart, by the same procedure: 1/2,680,337.

1828—Airy, correcting Delambre on the basis of observations at Greenwich: 1/3,734,602.

1853—Hansen and Olufsen, again by the perturbations of the Earth: 1/3,200,900.

1858—Le Verrier, by the same procedure: 1/2,994,790.

1876—Le Verrier, by the perturbations of Jupiter: 1/2,812,526.

[5] *Études sur l'Astronomie*, volume III, 1872, p. 30.

In the equatorial zone of Mars, as with that of Saturn, an atmosphere remained after the separation of the satellite and the isolation of the ring respectively; this upper atmosphere was very rarefied, but nevertheless provided resistance to the movement of the satellite, and caused it to make a gradual approach to the planet. This approach was increased by an increasing velocity of motion. It is likely that the period would have been stable only in perfectly empty space—in the pure æther. A satellite moving in the equatorial plane of Mars, close to the surface, in empty space would have a revolution period of $1^h\ 40^m$, as has already been noted.

But let us now continue with our comparative study of the observations of Mars.

1877.—Niesten: Observations and Drawings

L. Niesten, astronomer at the Brussels Observatory, made a large number of observations between 21 August and 10 November 1877, with the $0^m.152$ equatorial, using powers ranging from 90× to 450×—mostly between 180× and 270×. These observations include 42 drawings, from which four are reproduced here: (1) the circular Terby Sea; (2) The Hourglass Sea; (3) Maraldi Sea and Hooke Sea, joined by the Niesten Isthmus; and (4) the Maraldi and Hooke Seas, separated. The author himself has summarized his observations as follows:

> *Polar patch.*—As the northern hemisphere of Mars was having its summer, the polar cap was shown all through the series of observations. Oval on 21 August, the patch becomes rounded; it became very small after 14 September, and became flattened again as from 20 October. Its colour was fresh white. Its brilliancy varied sensibly from one day to another, and even during the course of the same evening (21 September $8^h\ 15^m$ and $11^h\ 15^m$). Note the dull appearance on 21 August and 18 October, and the exceptional brilliance on 22 September, 26 August and 2 November.
>
> *Seas.*—In all the observations, the polar patch is seen to be surrounded by a sea, less well-marked at longitude 150°, darker toward longitude 0°, 90° and 180°.
>
> The Zöllner Sea was more conspicuous than the Lambert Sea on 11 September. They were connected (see Fig. 31). Lockyer Land was very bright on 13 October.
>
> The drawings of 14 and 15 September and 18 and 20 October (Fig. 8) represent the eastern extremity of the Maraldi Sea, to which the Hooke Sea is joined. Burckhard Land was consistently oval in shape.

Niesten thinks that the sea which, on the 1876 map, ends at longitude 180°, latitude 80°S., is in fact prolonged further. This is perfectly correct, and the chart shows the prolongation (Maunder Sea). The Belgian scholar writes:

> On the chart, I have kept the name of the Lockyer Sea for the circular lake which has since received the name of the Terby Sea; which had been given to another sea on the first plate of the preceding map, *Terres du Ciel*. The appearance of my drawings, greatly resembles the corresponding region shown on M. Flammarion's map. Indeed, Proctor's map is certainly very far from the truth with regard to the Dawes Sea, shown on this map as a long ribbon of sea extending along the 240° meridian.

Herschel II Strait is shown very much as it is on my chart, but with no trace of Phillips Island.

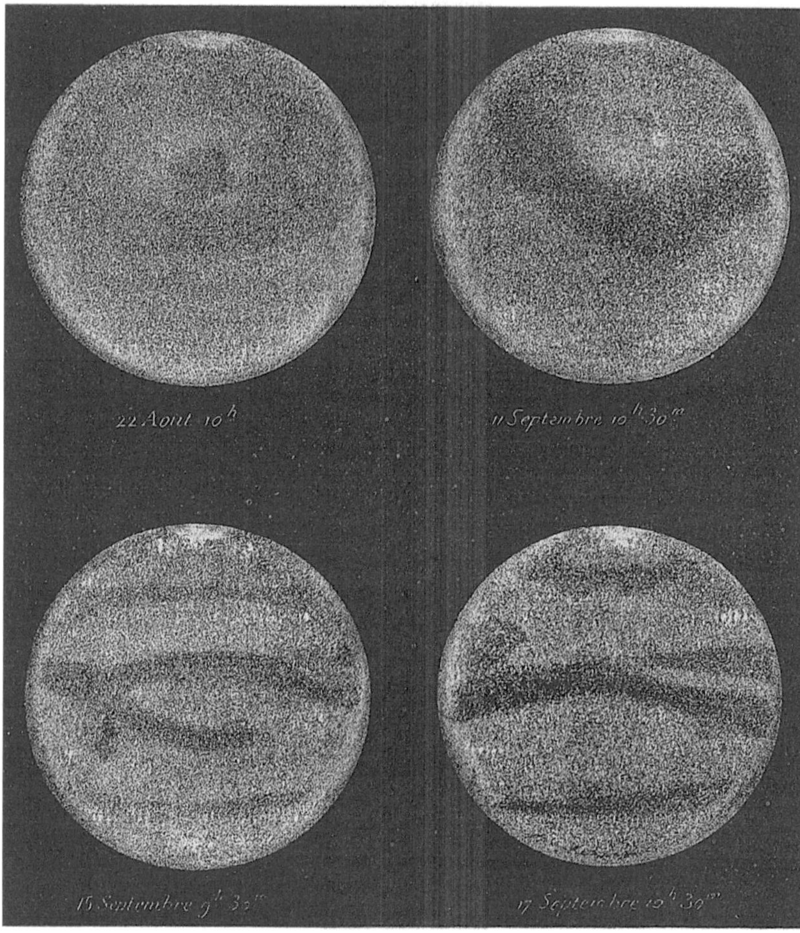

Fig. 8 Observations of Mars, by M. Niesten in 1877

The patches show a bluish grey colour, of varying darkness; the darkest have been the circular lake, the north and central parts of the De la Rue Ocean and the easternmost extremity of the Maraldi Sea. The disk of the planet was generally pale yellow, sometimes orange yellow, ochrey on 21 August. Note also the reddish reflection along the Maraldi Sea and sometimes at the south of Dawes Ocean, as well as in the environs of the circular lake and Herschel II Strait. They are not evident on Burckhard Land.

1877.—F. Terby: Studies of the Planet Mars[6]

The eminent Louvain astronomer communicated a summary of his observations to the Belgian Academy. I give here the most important points (9^{mm} refractor, magnification 120× to 140×).

[The figure no. below refer to those in our Fig. 9.—WS.]

The south polar patch has been constantly visible, as it is necessary to emphasise; its somewhat oval form could often be noted when turned toward the Earth-facing hemisphere. This patch was most brilliant, whitish and most extended after the end of August, while the Herschel II. Strait appeared on the disk (*fig.* 1). It has been weaker and less extended during my observations made during the middle of September (*figs.* 2, 3, 4, and 5), while I observed the Hooke and Maraldi Seas. After 21 September it again became very white and very brilliant, while I observed the easternmost extension of the Maraldi Sea *pf,* De la Rue Ocean,

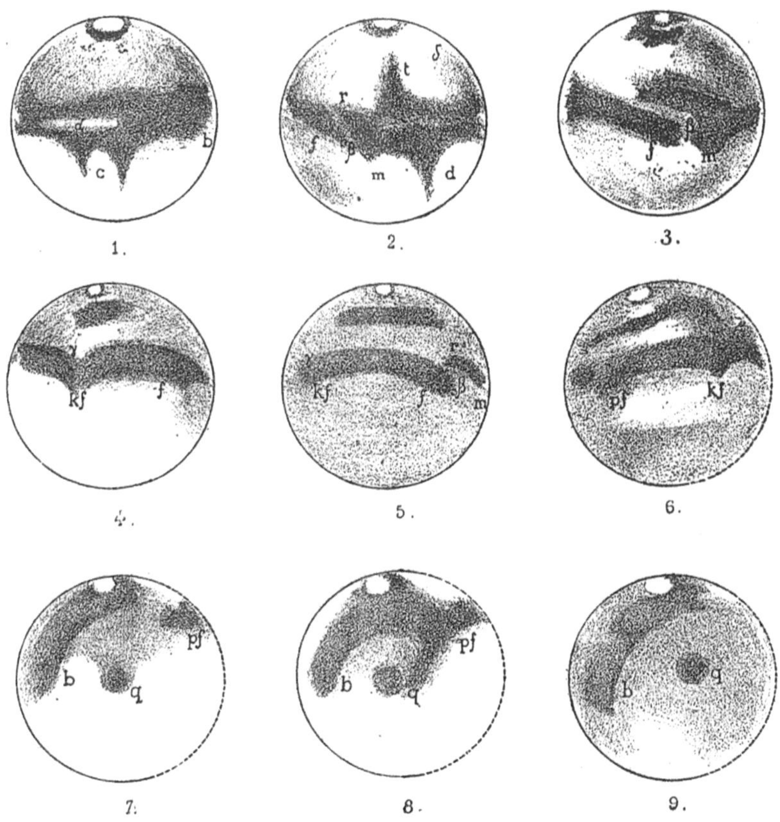

Fig. 9 Drawings of Mars, by M. Terby, in 1877

[6] *Bulletin de l'Académie des Sciences de Belgique*, 1878, t. I, p. 33.

Lockyer Sea, Herschel II. Strait and the Hourglass Sea. During the period covered by these observations, from 30 August to 20 October, I observed no northern snows.

Fig. 1. 30 August 1877, $10^h 30^m$ to $10^h 45^m$. The south polar patch is very brilliant, very white and rounded. The dark border which surrounds it (Phillips Sea) is the blackest area on the disk. The region observed is that of Herschel II Strait. In *c*, two bays are visible; in *b*, the De la Rue Ocean. With a magnification of 120x a brilliant region could be seen at *a* this corresponds to Phillips and Jacob Islands. Momentarily this zone gave to the patch the aspect of two parallel bands. The northern part of the disk, situated under the sombre zone, was much more brilliant than the region lying between the sombre zone and the polar cap. The sombre zone is darker to each side, in the neighbourhood of the edge of the planet.

Fig. 2. 11 September, from $10^h 5^m$ to $10^h 30^m$. The polar patch is much smaller and dimmer than on 30 August. The dark patches are themselves very feeble. However, it can easily be noticed that the various regions are of unequal darkness. The Hourglass Sea *d*, Hooke Sea *mr*, Maraldi Sea *f* and Burckhard Land *b* were seen; Zöllner Sea *t* was doubtful. Summarizing, this observation was unsatisfactory because of the astonishing feebleness of the patches, which are normally so easily visible.

Fig. 3. 14 September, 10^h to $10^h 25^m$. The image is admirably sharp, thanks to light mist. The snowy patch seems to be turned entirely toward us, but it is neither fresh white nor brilliant. Hooke Sea *mr* appears to have a form different from that shown in Proctor's chart. The side along Burckhard was seen with confidence, and the map should be corrected. Burckhard Land is broader to the north than to the south, and shows a curvature which has never before been suspected. I also see that Hooke Sea does not extend as far as the western extremity of the Maraldi Sea. I see a dark band between the small dark zone which surrounds the polar patch, and the Hooke and Maraldi Seas.

Fig. 4. 17 September, from $8^h 30^m$ to $8^h 35^m$. I see the Maraldi Sea *f*, and the dark zone making up a bay *kf*. At *g* to the east of this bay, I have seen the tongue of land which I propose to name Webb Land, and which is so well-marked in several of Webb's drawings. This solution of the continuity of the band remains dubious, from observations with my refractor. Between the Maraldi Sea and the polar patch a sombre band can be distinctly seen. The polar patch is always smaller and less brilliant than on 30 August.

Fig. 5. 17 September, at $10^h 15^m$. The polar patch seems even smaller, but became more brilliant by $8^h 30^m$.

Fig. 6. 21 September, from $08^h 15^m$ to $8^h 30^m$. The polar patch has again become whiter and more brilliant, without however equalling its aspect on 30 August. It can be seen that the extremity *kf* to *pf* of the Maraldi Sea is not indicated on Proctor's map. The north part of the disk was definitely greyish.

Fig. 7. 27 September at $8^h 15^m$. The polar patch is white and brilliant; the De la Rue Ocean *b* can be seen, a small sea *q* is the darkest area on the disk, rounded like the shadow of a satellite of Jupiter; it corresponds to the region occupied by the Lockyer and Dawes Seas on Proctor's chart. These two small seas are merged here, so do they really exist separately? Also seen is the extremity *pf* of the Maraldi Sea.

Fig. 8. 27 September from $8^h 40^m$ to $8^h 55^m$. This observation shows a connection between the De la Rue Ocean and the Maraldi Sea.

Fig. 9. 28 September, from $8^h 5^m$ to $8^h 15^m$. Polar patch white and brilliant. The little sea *q* is again the darkest region.

The author continued his observations up to 23 October, and published six more drawings. These results are remarkable when it is remembered that the instrument used was only 90^{mm} in aperture. The excellent object-glass was by Secrétan. Without doubt the seeing and methods of observation were excellent.

1877.—O. Van Ertborn: Observations and Drawings

These observations were also made in Belgium, at Aertselaer, near Antwerp. The Baron Octave van Ertborn began his studies of Mars in 1860, and made several drawings around the time of each opposition. In 1877, in the Mémoires de l'Académie de Belgique,[7] he published 26 drawings made between 15 August and 3 November, with the aid of a 108^{mm} refractor on an equatorial mount. Magnifications 125×, 205× and 255×.

The southern hemisphere of the planet was shown, with the contours of the continents and seas clearly defined, while those of the northern hemisphere were seldom visible, and remained veiled by mists.

The principal geographical configurations are recognizable on these drawings. The author thought that he had made out the Bessel Pass, but he found it impossible to see the Dawes Sea. He also believed that he had noted a thread linking the circular sea with the neighbouring ocean. The Hourglass Sea was somewhat short. The polar caps were very brilliant on all his drawings, which were very remarkable in view of the instrument employed.

After having obtained a perfect temperature equilibrium, the dome was opened several hours before the beginning of the observation. The equatorial was driven by a perfectly regular clock movement, which—the observer added—was very important, because to see the smallest details it was necessary to watch for a long time; only continuous sensation could affect the retina. A country site was also very important for the steadiness and sharpness of the images. The author, who was helped in his observations by several people, wrote:

> The eye of the observer plays a vital role in the observation of very small or very faint objects. To ensure that nothing is lost from view, and that nothing is neglected, delicate observations should be made by people whose eyes are not tired. It is the retina which gives the sensitivity and the definition; the role is more important than the provision of large instruments. My nephew can see 15 Pleiades with the naked eye; to him the stars and planets show no rays. At 92 metres distance, he can see objects 1 centimetre square against a dark background. On 5 September 1877, he saw the companion of μ Andromedæ with the help of my 4-in.; he had never heard of it, and it is a very severe test even for 8-in. objectives.

Two of van Ertborn's best drawings, made under excellent atmospheric conditions, are reproduced here. The first, made on 5 September at 11^{h} 50^{m} (Fig. 10) shows the Dawes and the De la Rue Oceans, the Hourglass Sea, and Herschel II. Strait. The second, made on 29 September, at 7^{h} 45^{m} (Fig. 11) shows the De la Rue Ocean, undoubtedly Burton Bay in the ribbon below, the circular Terby Sea attached by a thread, and Bessel Pass.

Unquestionably van Ertborn tended to accentuate the outlines of vague, indefinite and doubtful contours. This is the principle reason for the diversity of his drawings.

---

[7] *Observations de la planète Mars pendant l'opposition de 1877.* (*Mémoires des Savants étrangers*, vol. XLII, 1879.

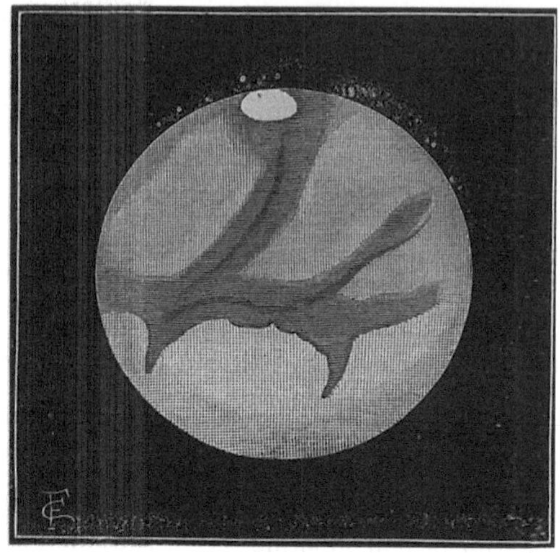

Fig. 10 Drawing by M. Van Ertborn, 5 September 1877

Fig. 11 Drawing by M. Van Ertborn, 29 September 1877

1877.—Cruls: Observations, Drawings and Rotation Period

To L. Cruls, then Astronomer at the Observatory of Rio de Janeiro and now Director there, we owe a beautiful series of observations and photographic drawings. These observations were made with the $0^{m}.25$ equatorial, generally with a magnification of 240×; once on 13 October, when the planet was near the zenith, with 340° and 580°.

The observations extended from 16 August to 13 October. During the whole of the period the south pole was consistently shown as intensely white. On 13 October it did not touch the limb of the planet, but was fully on the disk, isolated and much reduced in size.

A rare condition for the majority of these observations, thanks to the position of Rio de Janeiro, was that the meridian distance of Mars from the zenith during this opposition did not exceed 12°.

Cruls cited an opinion of Liais, then Director of the Rio Observatory, that the sombre patches were not seas, but lands more sombre than the rest of the planet, and showing a range of darkness. These tones varied according to droughts and rains. The lands were regarded as more different from each other than was the case with the seas, the waters showing a tint intermediate between that of the bright lands, which appeared like sand, and the dark lands, giving the impression of forests or prairies.

According to this point of view, the sombre patches are due not to stretches of water, but to lands covered with vegetation.

The sombre patches in the polar regions appear vague; those between latitudes 50°S. and 40°N. appear more distinct.

The presence in our atmosphere of a light layer of vapour, tempering the excessive light of the planet, increased the clarity of the patches.

A beautiful series of 21 photographic drawings accompanies this Memoir. One of them is reproduced here; that of September 16 (Fig. 12). One can immediately recognize the Maraldi Sea, Hooke Sea, Flammarion Sea and Zöllner Sea; also Herschel Continent, Burckhard Land, Dreyer Island, Webb and Cassini Lands, etc.

Cruls used his observations to make a direct determination of the rotation period, from the return to the meridian of the circular sea on 24 August, 3 September and 3 October; also by the return of the western part of the Maraldi Sea, on 16 August and 27 September. Corrected for the position of the Earth, he found the rotation period to be $24^h\ 37^m\ 34^s$.

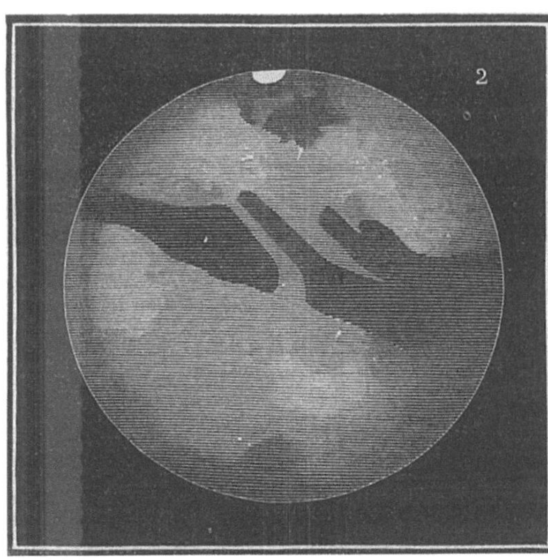

Fig. 12 Drawing of Mars, by M. Cruls, on 16 September 1877

These observations by the eminent Rio astronomer have added to the valuable documents concerned with studies of the planet. However, I cannot agree with the opinion of my illustrious friend concerning the dark patches. In my view, the greatest probability is in favour of seas. First, the presence of water vapour in the Martian atmosphere is demonstrated by four distinct facts; (1) spectral analysis; (2) the polar snows, which vary with the seasons; (3) the clouds of vapour sometimes extending over vast countries; (4) the clouds—rare, but they exist. Now, this vapour can be provided only by stretches of water; these stretches of water ought to be darker than the continents, since they absorb more of the incident light, and this is indicated by the darker regions. Following this, their mobility is an indication of their nature; we have already shown, and will again recognize later, even more surely, that several of the sombre patches vary in size; now, is not water the mobile element par excellence? It seems, then, that the dark patches represent the stretches of water which unquestionably exist on the surface of the planet.

Affirm what you like—perhaps forests, prairies, etc., as indicated by Liais and Cruls—without exceeding the limits of reason. It is very possible, and even plausible. In my balloon voyages I have often noted that prairies, and swamps covered with reeds, seem darker than rivers. But in them, water plays the fundamental role.

1877.—J.-L.-E. Dreyer: Observations and Drawings

At Lord Rosse's observatory at Parsonstown, J. L. E. Dreyer used the 3-ft ($0^{m}.91$) reflector to observe the planet. The 6-ft ($1^{m}.83$) reflector did not give better images.

This observer published 12 drawings.[8] The two most interesting are reproduced here, the first on 7 September at 11^{h} 50^{m} (Fig. 13) and the second on 3 October at 11^{h} 10^{m} (Fig. 14). In the first, the Herschel Strait and the Meridian Bay can be seen in the form of a detached ribbon, recalling *fig.* 96 by Lockyer on 25 September 1862, and that of Knott on 23 September 1862, as well as that by Kaiser on 31 October 1862. All these drawings by Dreyer represent this part of the planet, showing the same aspect. Herschel Strait is sometimes very sombre.

In the second drawing the Terby Sea is shown, also clearly detached from the adjacent ocean, clearly circular or rather slightly elongated east-west. The lake below is not well marked. It is shown feebly on the drawings of 28 September and 1 October.

The lower or northern hemisphere is very pale, and almost devoid of patches, in all the drawings-except for the Hourglass Sea.

The long vertical band named Bessel's Inlet on Proctor's chart was not seen on a single occasion.

[8] Notes on the physical appearance of the planet Mars. *Scientific Transactions of the Royal Dublin Society*, 1878, Vol. I, p. 64.

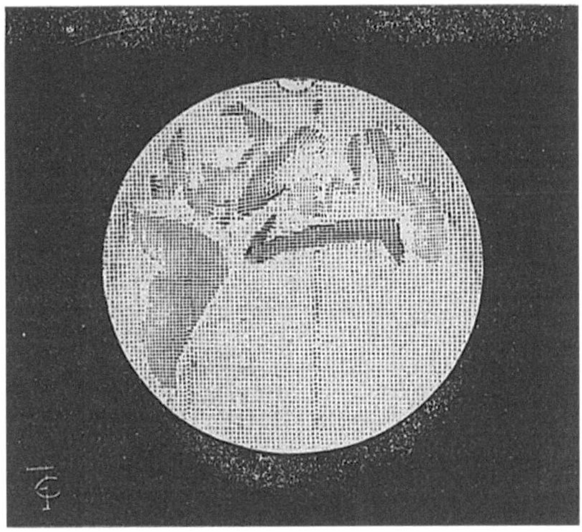

Fig. 13 Drawing of Mars by M. Dreyer, 7 September 1877

Fig. 14 Drawing of Mars by M. Dreyer 3 October 1877

1877.—O. Lohse: Observations and Drawings[9]

This observer, whose studies made in 1873 have already been described, made a new series of observations and drawings during the opposition of 1877. These resemble the aspects previously known; however, the differences are worthy of attention. They can be judged from the four drawings reproduced here (Fig. 15).

[9] *Beobachtungen und Untersuchungen über die physische Beschaffenheit des Jupiter und Beobachtungen des Planeten Mars. Observatorium zu Potsdam*, 1878.

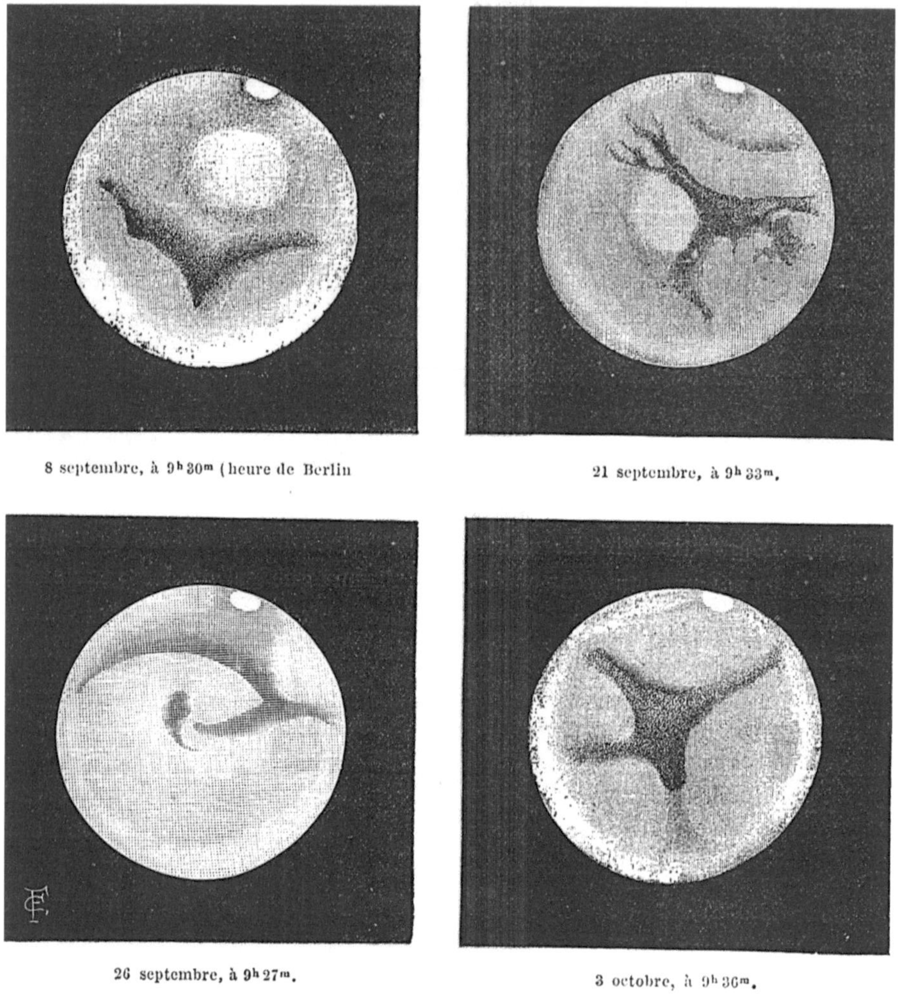

Fig. 15 Drawings of Mars, by M. Lohse, in 1877

The first was made on 8 September, at 9^h 30^m (Berlin time). The Hourglass Sea can be seen near the centre, and above it the vast space of Lockyer Land.

The second was made on 21 September, at 9^h 33^m. The longitude of the central meridian had then become 158°, and the Maraldi Sea was crossing the meridian; it is scarcely recognizable.

The third was made on 26 September, at 9^h 27^m. The Terby Lake was about to cross the central meridian; it is elongated, and to the right the neighbouring sea is strangely curved back and turned toward it. Compare this drawing with that of Schiaparelli, which will be described later, made on the same day,

The fourth drawing was made on 3 October, at $9^h 36^m$. The longitude of the central meridian was then 51°, and Christie Bay is on view.

These drawings were made with a 5 ½-in. equatorial.

From these observations, the author drew the following conclusions:

The disk of Mars becomes white, and brighter progressively toward the edge, in all the drawings the patches are effaced by this ring of brightness. The disk of Jupiter, on the other hand, decreases in brightness from the centre toward the edges, as is constantly emphasized by the transits of the satellites, which seem brilliant when crossing the peripheral zone and dark when they reach the central regions. Therefore, the atmosphere of Mars differs essentially from that of Jupiter.

The great differences observed between the drawings of Mars are caused, in part, by the fact that the atmosphere of the planet is not absolutely transparent. It seems that the transparency is subject to oscillations, as is indicated by the variations in the red colour, seen through this atmosphere. There seems to be vapours there: mists which, for one reason or another, do not combine in clouds analogous to our own. This light mist, unequal in distribution, is more or less transparent, and allows us to see the geographical configurations, except toward the edges of the globe, where the thickness is greater and the gaps or bright intervals are masked by the angle of projection; therefore sunlight is better reflected at the edge than at the centre.

As for the reason which stops the clouds from becoming as dense as those of the earth, Lohse states that he considers as reasonable the opinion of an English observer, Brett, who will be referred to later; the planet should be very warm, and this warmth hinders the condensation of clouds, except in the polar regions.

Like the English observer, the German astronomer thinks that the polar caps are due not to snows, but to very high-altitude clouds in the upper regions of the Martian atmosphere. This opinion is contrary to that generally held, but the brilliant whiteness of the polar caps, which sometimes appear to project beyond the edge of the disk, is favourable to it.

I will return to these assertions when discussing the observations made by John Brett.

1877.—N. E. Green: Drawings and Map

Nathaniel Green, the English artist and painter, who has already been mentioned, went to the island of Madeira and set up a 13-in. Newtonian reflector, first at an altitude of 1200 ft and then at 2,200, with the aim of obtaining the best possible views of the planet. The eyepiece most frequently used gave a magnification of 250×, though 400× was sometimes employed.

Under these very advantageous conditions, Green studied Mars with the greatest attention, and made a large numbers of drawings. He noted, with sound reason, that a pencil or brush gave too great an intensity to the very delicate and sometimes very vague features, which could be recognized only with difficulty and even then only with the most experienced eye.

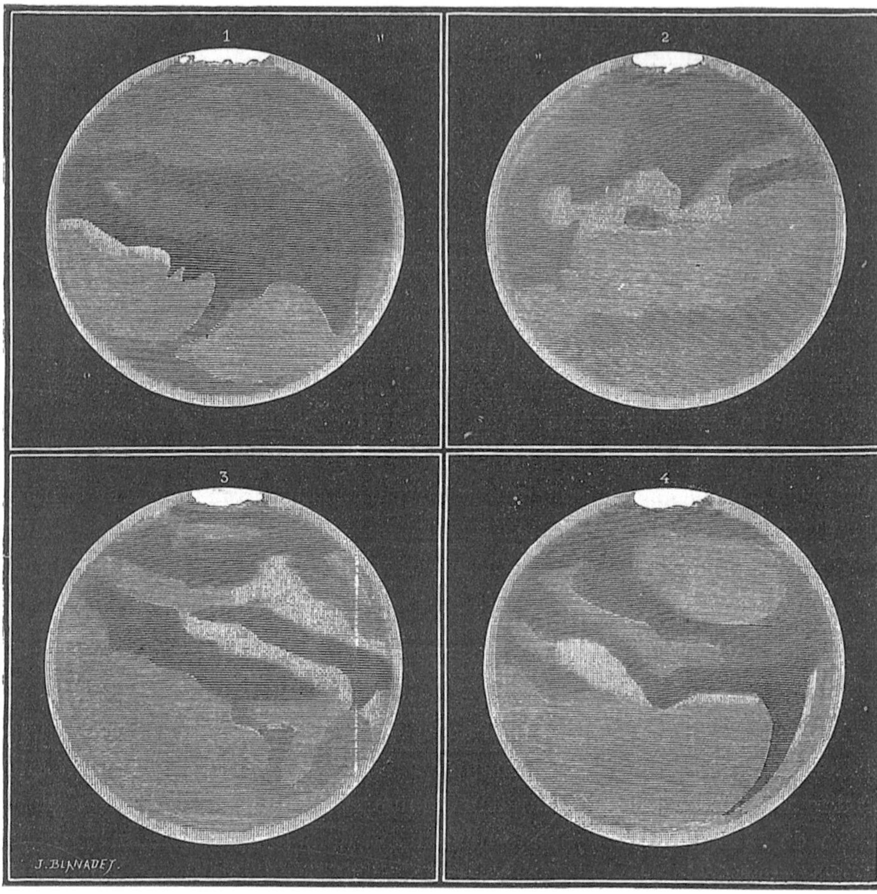

Fig. 16 Drawings of Mars, by M. Green, at Madera in 1877

Green presented his work to the Royal Astronomical Society of London, and published 12 of his drawings in lithograph form, plus two drawings of the south pole. From all these he compiled a general map, more complete than any previously published. This map has become a classic.

The first four of the drawings are reproduced here (Fig. 16), representing the whole of the planet, made (1) 1 September, $10^h\,40^m$ Greenwich time; longitude 7°; (2) 29 September, $9^h\,0^m$, longitude 232°; (3) 18 September, $11^h\,45^m$, longitude 232°; (4) 15 September, $11^h\,10^m$, longitude 250°. In the first drawing the central meridian crosses Burton Bay; in the second, to the right of the Meridian Bay; in the third, the Hooke Sea and Maraldi Sea; in the fourth Gruithuisen Bay, with the Hourglass Sea coming up to the right.

As has been said, the total of drawings made between 19 August and 5 October—26 favourable nights, 21 useless—allowed this skilful observer to construct a general map, which is reproduced here (Fig. 17). It deserves study in close detail.

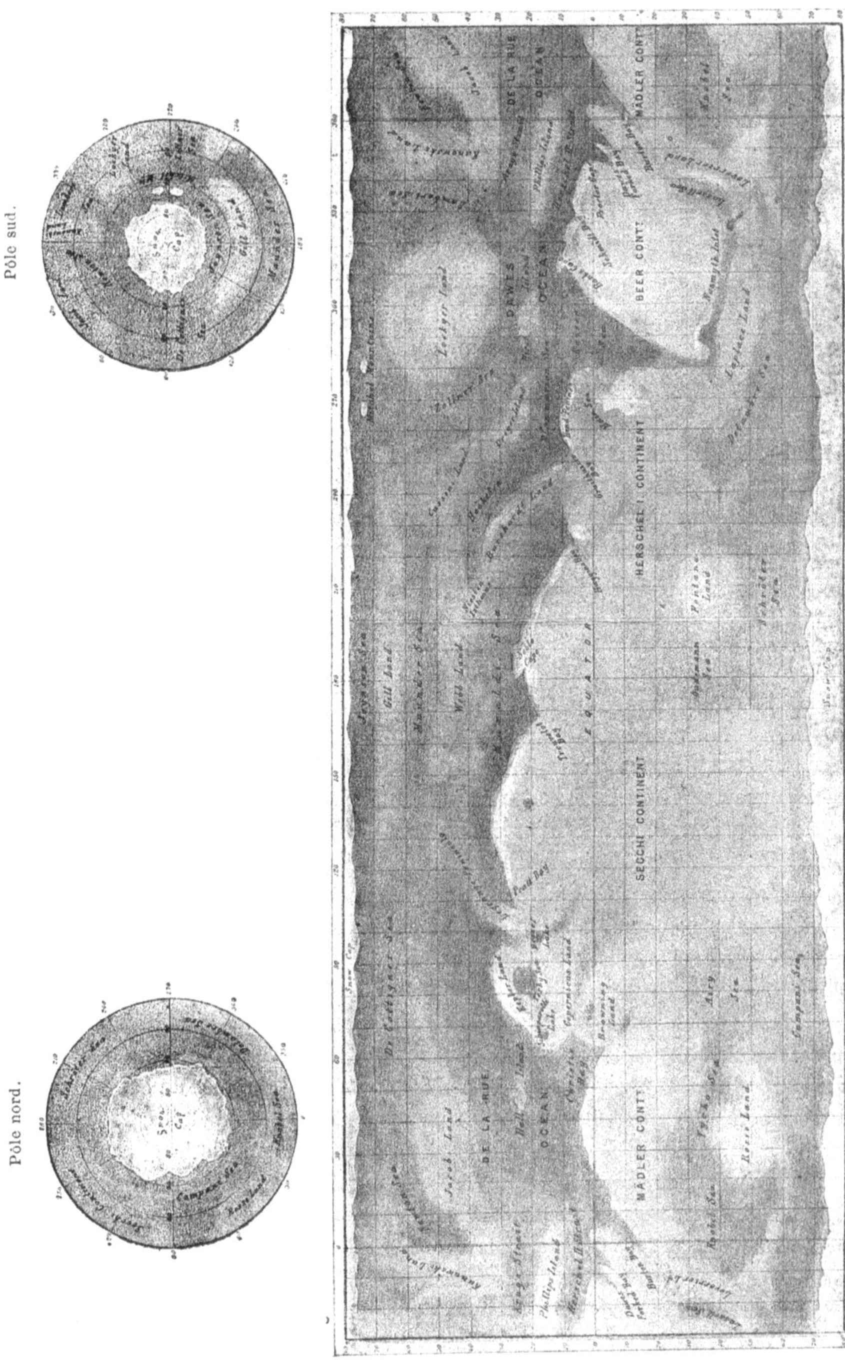

Fig. 17 Geographical chart of the planet Mars, by M. Green, from his observations of 1877. Published by the Royal Astronomical Society of London

The regions round the north and south poles are shown above the planisphere. On his map, Green has shown no details which he has not himself verified. It differs considerably, in several ways, from that of Proctor. Thus, despite the most careful attention, it has been impossible to prove the existence of the long pass called Bessel Inlet. Moreover, I had already deleted it from my map, even before the 1877 observations.

Green thinks that an appearance of this kind can sometimes be produced from an atmospheric current from north to south.

The environs of the circular lake, the Terby Sea, are also very different; the circular lake is a clearly-defined, sombre patch; Green observed it 18 times from Madeira without seeing the Dawes Sea, which showed rather as a vague grey shading. The chart is particularly interesting to study in this connection. Note especially the little Schiaparelli Lake and the snowy Hall Island.

The Maunder Sea, shown on Green's map above the Maraldi Sea, between Webb Land and Gill Land, is visible on the drawings made on 18 September and other dates. It was shown for the first time on my 1877 map, extending from longitude 247° to 180°—that is to say, reckoning west-east, from 112° to 180°. On Green's map, this sea is shown as a little further right and a little longer, from 130° to 220°. It was observed at the same time by Maunder at the Greenwich Observatory, as a band intermediate between the Maraldi Sea and the Joynson Sea.

The oceans' shores were on several occasions shown with the appearance of the whiteness of snow, as the author himself wrote:

> *Snows.* —Near the polar snows, snow was also seen on various continents. In the drawing of 20 April 1856, according to Warren de la Rue, the whole land south of the Hourglass Sea was evidently covered with snow, because at this time the pole was out of view and the shape of the white region was precisely that of Lockyer Land, seen foreshortened. This aspect was also clearly seen on *figs.* 3 and 4 published by the author in 1873 in *The Astronomical Register* [not reproduced by Flammarion —WS]; there also the south polar cap was out of view, but the snow covered the whole of Lockyer Land. There is also a very marked indication of the presence of snow on the white line which forms the boundaries of Beer Continent, near the equator, and it is probable that this continent was protected by chains of mountains of considerable height, as with the continents of North and South America, which are protected on their west side by the Andes. These bright lines are not confined to the edges of Beer Continent, and are also seen in other regions, such as Kepler Land, to the south of the Terby Sea; Hind Peninsula, and the snowy Dawes Island.
>
> *Clouds.*—The clouds of Mars (writes Green) are evidently much less dense than those of the Earth, to the point where no proper clouds can exist near the equatorial regions. Brett has been led to regard these polar caps as cloudy formations, but this hypothesis is contradicted by the forms of the caps and especially by the clearly-defined fixed points where the clouds gather and form, as we see, in the same places year after year. But if they are not clouds in the true sense of the term, they are certainly vapours sufficient to veil completely vast continental stretches. In the drawing of 29 September, for example, not only the De la Rue Ocean is veiled, but even its western part, clearly-defined on previous occasions, is on this occasion hidden by clouds. In the drawing of 18 September, the zone which surrounds the pole has become very clear, while it is indistinct on other drawings. In 1877, at Greenwich, and in 1862 in Lockyer's observations, parts of Dawes Ocean are temporarily masked by white clouds, These indicate white masses, rivaling in brilliancy those of the pole observed on the limb near the north pole and particularly on the eastern side; one of these is shown on a drawing published by the author in 1886. These white patches are

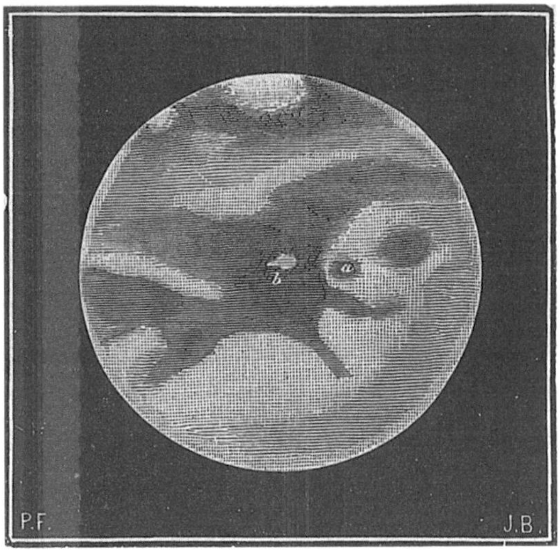

Fig. 18 Aspect of Mars on 2 September 1877. Drawing by M. Green

observed only near the limb, and do not advance with it; they cannot therefore be attached to the surface, and the only explanation is that they are clouds or vapours, which are dissipated by the rising sun, as often happens in our own climates.

Atmosphere.—The principal characteristic of the atmosphere of Mars is that it consistently enfeebles the geographical features and the colourations when near the limb. Noble has commented that this enfeeblement is more pronounced near the western limb than near the eastern, which indicates that sunrise is generally clearer than sunset. The most pronounced evidence of this atmospheric ring concentric with the centre of the disk was presented on 20 September; the bluish-white of this ring showed a great contrast with the orange tone of the southern region, a contrast which was considerably increased when clouds passed across the planet. The northern hemisphere showed no geographical features; the atmosphere there was of low transparency. It seems that the environs of the north pole had become charged with vapours which soon condensed to form the polar caps.

Seas.—The seas were of greenish-grey, due perhaps in part to contrast with the orange-yellow continents, but perhaps in part real, so far as can be judged from the variations of the tones observed. If the darkness of these tones corresponds to the depths of the water, the Hourglass Sea and Terby Sea are the deepest. This latter sea has well-defined contours, and is slightly elongated in an east-west sense. This sea is generally shown as attached to the De la Rue Ocean by a dark channel, as is best seen when under oblique viewing. But when looked at directly, the aspect is not of a channel but of a small lake, which can be clearly seen, as shown here (Fig. 18) on the drawing made by the author on 2 September, at $1^h\ 10^m$ in the morning.

This little lake is lettered *a* in the drawing, *b* indicates the position of the snowy Hall Island. In this connection Green commented that when an elongated patch is imperfectly seen, there is a tendency to picture it pointed at the end, and this tendency is common to all observers; it could explain certain straight lines traced by Dawes and Schiaparelli. This is the case, Green thinks, with the apparent joining of the Terby Sea to the De la Rue Ocean.

Polar Snows.—On this subject, Green makes some interesting observations. First, it may be noted that on the first of the drawings reproduced here, that of

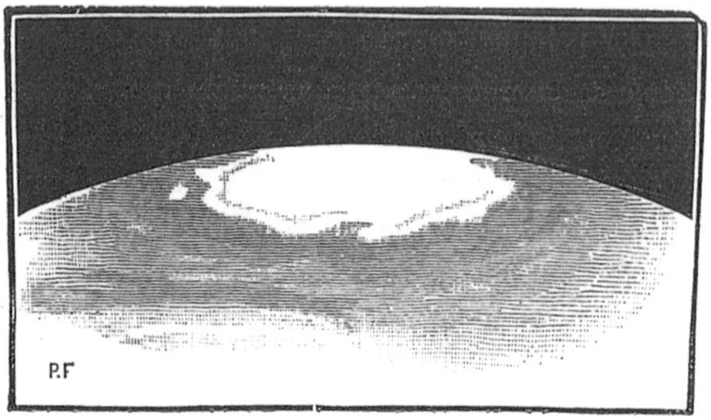

Fig. 19 The snows of the south pole of Mars, on 1 September 1877

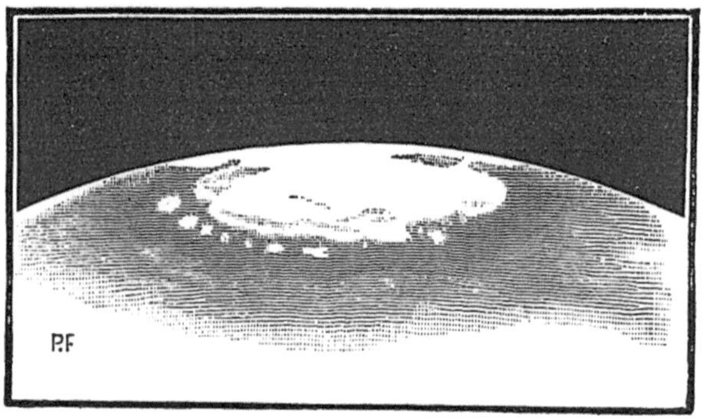

Fig. 20 The same pole, on 8 September

1 September, there is a white point to the west of the polar cap, which in all probability represents snow. It is better seen in the drawing in Fig. 19. But let as again quote the author:

> Without doubt, this is snow remaining on an elevated region, while the snow on the lower ground around has melted. This point is as brilliant as a star, and cannot possibly escape notice. On 8 September, at $0^h\ 30^m$, I observed it again; but on this occasion I could perfectly distinguish two separated points, and two days later, from 10^h to $11^h\ 30^m$, I could make out two concentric zones of snow, as shown here (Fig. 20). These alterations in form were undoubtedly due to effects of perspective, the various patches being presumably of the same form as on 1 September. I never saw them to the east of the polar cap, and this is a matter of particular interest. Indeed, their great brilliance to the west of the pole; their decrease after crossing the central meridian; and their invisibility on reaching the eastern side; can be easily explained by assuming that the slopes of the mountains keeping the snow, are inclined to the south-west; in this case they will be shielded from the rays of the Sun for the greater part of a rotation, but they are fully exposed to the sunlight, and consequently better seen when they reach the region of the western limb. It is interesting to note that this luminous point was observed and drawn in the same fashion on 30 August 1845, at Cincinnati, by Mitchel; it must certainly be associated with a local feature on the planet. I have named it after Mitchel in memory of an enthusiastic friend of astronomy.

Another observer, Brett, examining Mars on the night of 1 September, described this point near the pole as an *auxiliary patch*. This provides confirmation of the preceding observation.

The decrease of the snowy polar zone is obvious. During July, the zone had occupied an area twice as large as at the end of September.

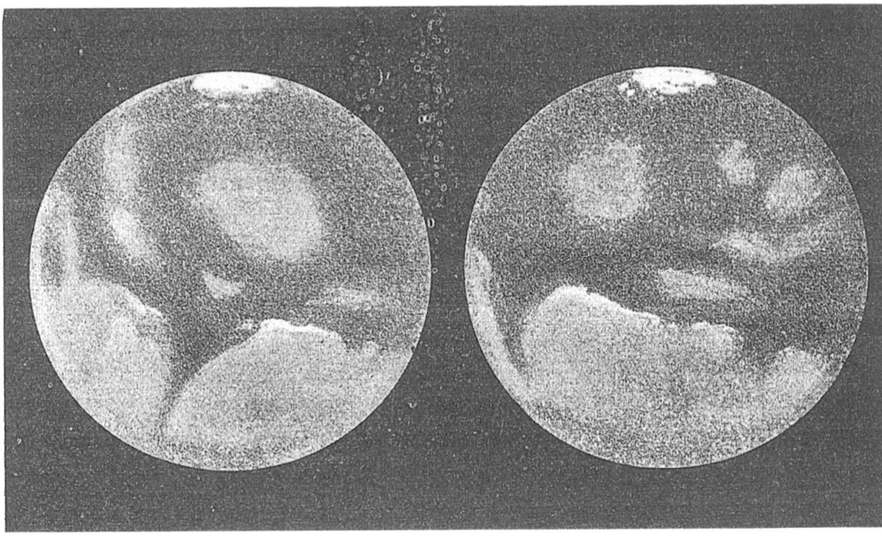

Fig. 21 Telescopic views of Mars, by M. Green, on 10 September at $11^h\,20^m$, and on 8 September at half past midnight

Fig. 22 Explanatory diagram

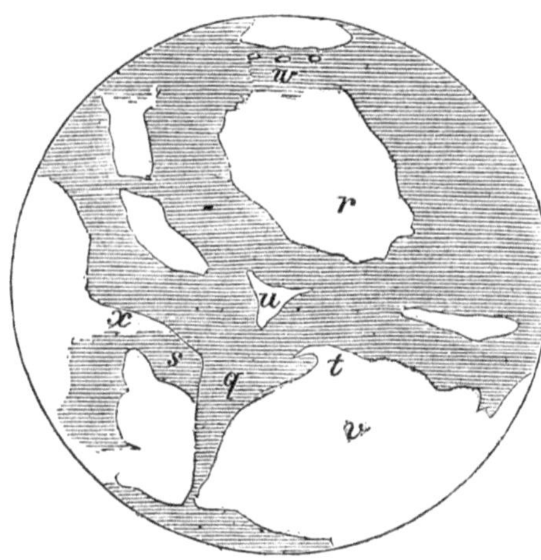

Such are the observations of the skilful English painter, who therefore devoted part of his career to representing the curiosities of the sky. An examination of his general map sums up all the facts revealed by his meticulous researches. I have kept to a photogravure reproduction, so that no modification can be introduced by the author's hand; but this means that the continents are less bright than they ought to be, because they are too yellow photographically. The reader must therefore tolerate this inevitable photographic effect.

Among Green's drawings, two more are presented here (Fig. 21); they deserve the closest attention and are particularly remarkable. They were obtained under exceptional atmospheric conditions—particularly the first (Fig. 22).

To the attentive observer, they will be a veritable treasure to the eye and the spirit. The first was obtained on 10 September, at $11^h 20^m$. The atmosphere was so transparent and so calm that the observer could distinguish the tiniest details on the surface of the planet. He used an eyepiece made especially by Browning for the observation of Mars, finished with great care, giving a magnification of 400×. Around the polar cap may be seen several patches of isolated snow. The region west of the Hourglass Sea, in the position of Main Sea and Hind Peninsula, shows a half-tone which is neither continental nor maritime, and which gives the impression of flooded terrains or swamps. This has already been noted on the drawings by Dawes, and I will return later to these apparent inundations—which may be real.

On the rivers east of the Hourglass Sea, at the angle of Herschel Strait, may be distinguish a bay or small sea, almost separated by a sort of near-island. Above and to the east, there is a cape already observed and drawn in 1862 by Banks at Ealing (see *The Astronomical Register*). Between the Hourglass Sea and Lockyer Land, the author has also distinguished a sort of triangular island, slightly different from the background which surrounds it; this island was seen and drawn on 3 August of the same year by Hirst at Sydney, and on 16 September by Trouvelot at Cambridge.

The second view, obtained on 8 September at $12^h 30^m$, completes the preceding one, especially for all the region east of the Hourglass Sea up to the Meridian Bay.

Finally, note that Green saw no trace of the canals recorded by Schiaparelli, and which will be discussed below. But it is only right to add that these features were not discovered until February and March, four months after Green's last drawing.

These drawings by Green take, perhaps, pride of place in all the drawings of the planet so far discussed in this history.

1877.—Harkness, Noble, Pratt, John Brett, G.D. Hirst, Bredichin, Bernaerts, Hartwig, Schur, Ellery, De Konkoly, Bœddiker, Weinek, Klein, Duval, etc. Various Observations

Before coming to the most important observations of this priceless opposition of 1877, those of Schiaparelli at Milan, we must complete the preceding notes by reviewing all the other observers who obtained results of different degrees of success.

These observers of Mars are a little like days—they follow each other, but do not resemble each other! One may perhaps become slightly disillusioned.

In 1877, the most powerful instrument in the world was the 26-in. ($0^m.66$) equatorial at the Washington Observatory, with which Hall discovered the satellites of Mars. Professor William Harkness observed Mars with it on several occasions, from 18 August to 18 October 1877, but could not obtain good images with a magnification of 400×; he had to be content with using a power of 175×. He made eight drawings, and after each session Hall stated that he had never had better views.

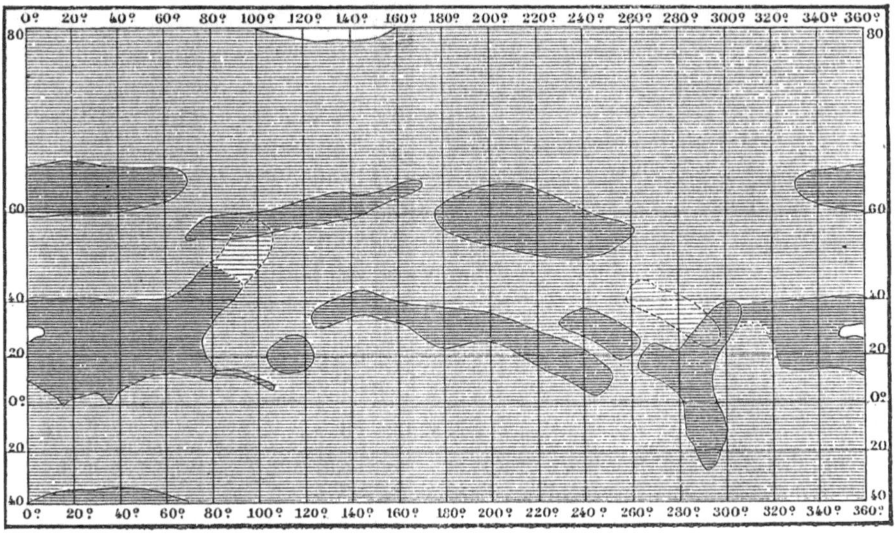

Fig. 23 A map of Mars produced in 1877 at the Observatory in Washington

These eight drawings were used to draw up the planisphere given here (Fig. 23), drawn on Mercator's projection.[10]

As the observer himself said, it was a meager reward. One can scarcely recognize the Hourglass Sea, the Maraldi Sea, Terby Sea and de la Rue Ocean. This was all that could be seen with an 108^{mm} refractor!

The only interesting result of these observations was a determination of the position of the south polar snow, by Hall. He gave the position as longitude 20°.66, distance from the pole 5°11′.[11]

At the Royal Astronomical Society of London, at the meeting of 9 November 1877, Captain Noble presented a series of drawings made at his observatory at Forest Lodge, Maresfield (Uckfield) with a 4.2-in. equatorial refractor and 61 in. in focal length, giving good images with a magnification of 255×. Noble, who had studied Mars over a long period—he had begun his observations in 1858—noted that the patches became invisible toward the edges of the disk; in particular he commented, as Green had done, that the markings could be distinguished closer to the following or eastern limb than to the preceding or western. He concluded that on Mars, sunrise is brighter than sunset. The mornings are clearer than the evenings. It is the same on Earth, at least in our climates; sunshine is more frequent in the morning than in the evening, and all photographs confirm that morning light is better than afternoon light.

[10] On the Physical Configuration of Mars. *Monthly Notices of the Royal Astronomical Society*, November 1879, p. 13.

[11] *Astronomische Nachrichten*, Vol. XCI, 1878, p. 223.

At the following meeting of the same Society, H. Pratt presented a series of drawings made with an 8.15-in. Newtonian equatorial and a magnification of 400×. The weather was not favourable, and in general the images were not good. The drawings were made with extreme difficulty. Details seen during favourable moments disappeared moments afterwards. The reddish tint of the planet appeared paler than in the preceding oppositions. Pratt confirmed the comments of Captain Noble with regard to the better visibility of the patches nearer the sunrise border than near the sunset border.

The persistence with which a great number of well-known patches are seen again year after year, and seen for nights and nights during each opposition, surely proves that the features belong to the globe and not the atmosphere. However, the differences in the details observed at the same instant, by the same person, with the same instrument, under the same atmospheric conditions prove that there are definite variations in the transparency of the Martian atmosphere. The idea of *local obscurations*—proving that there can be cloudy conditions in the atmosphere—seems sufficient to explain the discrepancies though it is not easy to decide why certain forms should be visible at certain epochs and obliterated or greatly modified at other times.

In general the atmosphere of Mars is very transparent, but from time to time the configurations become invisible, undoubtedly because of the temporary opaqueness of the planet's atmosphere—though the opaqueness is never comparable with that of the masses of the clouds of Jupiter. The effects are not due to a lack of transparency in our own atmosphere, because at times when Mars is misty, Saturn, at a lower altitude, is very sharp. This has been recorded several times, notably on 14 November.

At the same meeting another English observer, John Brett,[12] presented a series of observations made between 2 August and 8 October, with a 9-in. Browning telescope; he observed from the extreme south of England, near the Lizard. His results were not encouraging, and were even contradictory.

The disk of Mars was always less clear than those of Jupiter and Saturn; it was a bad telescopic object.

Brett believed that the atmosphere of the planet was so opaque that it hindered any feature from being seen clearly unless it were near the centre of the disk. He compared it with that of Jupiter, and thought that the latter was not an atmosphere in the usual sense of the term:

> The disk of Mars is very white at the limbs, proving the thickness of the atmosphere. On the contrary, Jupiter is more brilliant at its centre than at its limbs, proving the opposite. It could be liquid and semi-transparent to a great depth below its surface; Mars, on the other hand, is a solid body, its general topography is permanent, and it has a considerable atmosphere. However, it is not cloudy. From 2 August to 8 October, the author observed the planet without detecting a single one. The principal patches were identified. These cannot be seas, because these would produce evaporation and hence clouds. Since one hemisphere was totally without clouds over a period of two months, the idea of seas cannot be accepted. Nobody can suggest that the Martian atmosphere could not be dense enough to support clouds; one has only to look.

---

[12] *Monthly Notices of the Royal Astronomical Society*, Vol. XXXVIII, p. 58.

Thus wrote John Brett. And the polar snows?

> The white patches at the poles are generally regarded as snows, but there are one or two objections to this idea, notably the absence of clouds. First, the north polar cap, now in view, is surrounded by a sombre shading, which is the same as the tone of the assumed seas extending toward the equator, and which is not separated by any strait. Therefore, if the white patch is snow, it lies upon the sea or upon a polar island.

Brett cannot accept this, and at the same time comments that this white polar patch is very often seen not confined to the globe, but extending above it. He attributes this effect to irradiation, but the distance is too great for this explanation to be correct; and, moreover, the white suspension produces a shadow to the east when the planet has passed opposition, as I saw notably on 28 September at 9^h—the best evening of the year. Brett continues:

> This is not snow, but an enormous cloud, which is formed on the planet at the only place where it can form—at the pole. This is the only region cold enough for vapour to condense because the rest of the planet is very warm.

Here, certainly, is something new, and we have already seen that several observers of the planet accept these conclusions. But all the assertions are dubious, and it is not even difficult to refute them. First, it is incorrect to claim that the atmosphere of Mars is as opaque as this, because, on the contrary, almost all observers agree in recognizing its transparency. It is incomparably more limpid than our own. At the distance of Mars, and under the same conditions, it would be impossible to distinguish on Earth as much detail as on Mars, even on the clearest days.

The complete absence of clouds is certainly an error. Undoubtedly clouds are very rare; but it will suffice to compare the excellent drawings of Lockyer and Green, from 1862, to recognize their existence and their movements. Water vapour, whose presence in the Martian atmosphere is shown by spectral analysis, does not condense into clouds as readily as on Earth, but it sometimes throws up a veil, which makes it hard to distinguish vast countries; and there is no doubt that the polar caps are so produced, because, as Brett says, they do not lie above the level of the globe, but sometimes appear by irradiation to project from the disk, because they are as white as snow.

What Brett takes to be the density of the Martian atmosphere is in fact the effect of the presence of water vapour, which exerts a very marked absorbing action at the greatest density of the atmosphere over the whole of the planet.

Brett adds these facts—taken together—oppose the opinion that Mars could be inhabited.

While these preceding observations were being made in Europe, another zealous observer, Hirst, was studying the planet from Sydney (New South Wales), and made a very painstaking drawing with a 10¼-in. (260^{mm}) refractor. Hirst commented that it was only toward the middle of August that the geographical configurations became very clear, but was unsure whether the cause lay in our atmosphere or in that of Mars.[13]

[13] *Monthly Notices of the Royal Astronomical Society*, Vol. XXXVIII, p. 58.

At the Moscow Observatory, Bredichin observed the opposition of Mars in connection with the solar parallax. This is not the place to give his positional measurements, but on 6 September Bredichin made a drawing[14] which showed, in particular, the brilliant whiteness of the polar cap, and also made out the Maraldi Sea in the form of outstretched wings.

We have already described the observations made in Belgium by Terby. To these must be added those made at Malines by Bernaerts[15] with a 9-in. refractor, and which were accompanied by drawings. The most interesting feature of these sketches is that they connect the Zöllner and Lambert Seas with the south polar sea, as shown on my chart.

The diameter of the planet was again measured, notably at the Strasbourg Observatory by Hartwig.[16] He found:

Equatorial diameter	9″.421 ± 0″.012
Polar diameter	9″.300 ± 0″.022
Flattening	1/78

At Breslau Observatory, at the same opposition, Schur found:

Equatorial diameter	9″.262 ± 0″.016
Polar diameter	9″.168±0″.018
Flattening	1/99

The combination of these measurements made by Arago, Bessel, Kaiser and Main, together with these new measurements, gives for the mean diameter at unit distance 9″.352.

During the same opposition, at the Melbourne Observatory in Australia, Ellery[17] made a series of measurements of the polar and equatorial diameters of Mars. The results were bizarre; sometimes the former diameter was greater than the latter (which should be constant), sometimes smaller, For instance:

27 Aug.	Polar diameter	24″.185	Equatorial diameter	24″.550
29 Aug.	"	24″.918	"	25″.488
30 Aug.	"	25″.172	"	25″.082
6 Sept.	"	25″.602	"	25″.287

At his observatory at O Gyalla in Hungary, de Konkoly made a series of interesting observations[18] and, in particular, published 15 drawings made between 19 October and 16 November, a study which he continued during the following oppositions. The principal geographical configurations were recognizable except for some

[14] *Annals of the Moscow Observatory*, Vol. IV, 1878.

[15] *Bulletin de l'Académie de Belgique*, Vol. I, p. 33.

[16] *Untersuchungen über durchmesser des planeten Venus und Mars. Publ. der Ast. Gesellschaft*, Leipzig 1879.

[17] *Monthly Notices*. Vol. XXXVIII 1878, p. 400.

[18] *Beobachtungen angestltt an astrophysikalischen Observatorium in O Gyalla in Ungari, I Bond* 1878.

variations in aspect, due undoubtedly to the differences between observers—to which we have become accustomed.

At the Prague Observatory, Weinek,[19] to whom we owe those charming drawings of lunar craters, obtained views of Mars on 8, 21 and 29 September which—strange to say—show no interesting detail, even though the instrument was an 8-in. equatorial armed with a magnification of 192×, and Weinek is a very skilful observer.

At the Göttingen Observatory, Bœddicker, who subsequently followed up his studies from the observatory at Birr Castle in Ireland, observed during the 1877 apparition, and obtained ten views of the planet.[20] I regret that I cannot publish all the drawings. Each has its own value, without question, but it is impossible to give every observation in a single monograph on Mars, and already the limit has been reached! All the work, all the documents known to me have been taken into account, even if they have not been used to the full. Certainly we can say that during the 1877 opposition, which was so fruitful, some of the drawings are very similar to each other.

Note, for instance, the two drawings by Klein at Cologne, made on 27 September and 24 October, published in the German astronomical review Sirius. In France, several amateur observers have sent me a great number of drawings, among which should be particularly noted those of E. Duval, a farmer at Saint-Jouin (Seine-Inférieure).[21]

We must also note the work of Dreyer[22] and Grover[23] with six drawings in September and October 1877, Lamey,[24] Fergola,[25] Lindstedt,[26] etc. These are judged as being of lesser importance than the preceding observations. The last two were concerned solely with the positions of Mars on the meridian, using stars for comparison.

1877.—Schiaparelli: Observations, Map and General Study

We have now reached the greatest work which has been carried out with regard to Mars.

The illustrious Director of the Milan Observatory—as skilful in observation as in calculation—to whom Science owes more than one brilliant discovery, notably that of the orbits of meteors and their association with cometary orbits, undertook studies of Mars which were more successful and more fruitful than any previously undertaken, and which eclipsed those of all his predecessors.

Each opposition period since that famous year, 1877, has been marked by considerable research on the part of this eminent astronomer. Here, I propose to deal

[19] *Beritchte der K. Sächs, Gesellschaft der Wissenschaften*, 15 Dec. 1877.

[20] *Veröffentlichungen von der Königl. Sternwarte zu Göttingen*, 1877.

[21] *Ils ont été publié dans le journal hiebdomadiare, La Nature*, Dec. 1877, p. 80.

[22] The aspect of Mars in 1877. *Astronomische Nachrichten*, Vol. XCIII, 1878.

[23] *English Mechanic*, Vol. XXVI, 1878.

[24] *Considération sur un essaim d'asteroides autour de Mars*. Autun, 1877.

[25] *Observazioni di Marti*. Naples. 1879.

[26] *Beobachtungen des Mars*. Lund, 1878.

with the work carried out in 1877, which the author has himself described in a special work.[27]

These observations were made with an excellent equatorial constructed by Merz of Munich, of $0^m.218$ aperture and $3^m.25$ focal length. The magnification used was 322×; only in January, February and March, when the apparent diameter of the planet was greatly reduced by extra distance, from 25″ to 5″, was a power of 468× used.

When beginning his observations, Schiaparelli did not wait for the planet to become close; but the results obtained were so encouraging, and the atmospheric conditions remained so favourable, that he threw himself into his main work with the greatest pleasure.

Schiaparelli's work in 1877 was divided into five sections. (1) a new determination of the direction of the axis of rotation. (2) a topographical triangulation of fundamental points on the Martian surface. (3) description of different regions of the southern hemisphere and part of the northern. (4) the south polar cap. (5) the Martian atmosphere. All these will be carefully examined.

It is important to begin by knowing the exact direction of the planet's axis of rotation. The author took as his basis the direction determined by Oudemans, from the observations of the north and south polar caps made by Bessel in 1830, 1835 and 1837.[28] For 1834, these give a direction of:

R.A. 317° 34′; declination +50° 5′.

The annual variation due to terrestrial precession is +0′.485, and +0′.247. The co-ordinates for 1877 therefore became:

R.A. 317°55′, declination +50°16′.

The zero for longitude was taken as *a* in the chart by Beer and Mädler, at the Meridian Bay, as had been adopted, notably by Marth, who, at each opposition since 1875, had calculated the ephemerides of the daily aspect of Mars.[29] Longitudes are reckoned from left to right—for a telescopic view, which gives an inverted image with south at the top, the longitude reckoning is therefore from west to east for the observer looking at Mars, or east to west for an observer actually on Mars. In other words, longitude increases from the preceding to the following limb.

To calculate the areographic longitude of the central meridian, Marth adopted a value of 88,642.7 seconds of mean solar time for one complete rotation of Mars, relative to the stars. Schiaparelli also adopted this value,

Sixty six observations of the positions of the snowy cap (*macchia nevosa*) were given for the position of the south point of the axis of Mars as seen from the Earth, 164°.90 for 27 September, at 0^h GMT, which corresponded to the mean of the observations.

Adopting the polar diameter as measured by Kaiser (9″.387) and 8″.80 for the horizontal equatorial solar parallax, we find that a degree of arc on a great circle on

---

[27] *Osservazioni astronomiche e fisiche sull' asse di rotatzione e sulla topografia del pianete Marte. Reale Accademia dei Lincei. First Vol., 136 pages and planisphere*. Rome, 1878.

[28] *Astronomische Nachrichten*, no. 838.

[29] See *Monthly Notices*, 1875, p. 305; 1877, p. 301, etc.

the globe of Mars will be equivalent to 0°.533 on the terrestrial equator—that is to say, 59 km. The probable error of the position obtained from the polar snow is about 7 km. The author concluded that the position angles of the polar cap, taken at a single opposition of the planet, are not sufficient for a precise determination of the axis, and he returned to the problem at the following opposition, that of 1879.

Areographic Triangulation of Fundamental Points

Observers had formerly declared that it was impossible to make micrometrical measurements of the patches on the globe of Mars. Such was not the opinion of Schiaparelli. He thought that when the diameter of the planet was not less than 20″, it would be quite possible to measure positions with a micrometer, and that the probable error would not exceed 1° of the arc of a great circle.

Since he wanted to establish Martian topography on an accurate basis, the Milan astronomer followed the principles of terrestrial topography. He chose a certain number of points which were both distinct and easy to recognize, distributed all over the planet, and used them as a fundamental réseau for the determinations of the positions of all other features.

The determination of the areocentric position of a point on the surface is obtained by noting the moment when the point crosses the central meridian, and, at the same time, measuring micrometrically its distance from the centre of the disk. We can then easily convert to longitude and latitude.

Sixty-two points were measured in this way. They are listed in the following table, together with the names given to them by Schiaparelli.

No.	Name	Longitude°	Latitude°
1	Vertice d'Aryn	0.00	+4.56
2	Secondo corno del golfo Sabeo	003.54	−02.37
3	Istmo della Terra di Deucalione	017.82	−02.52
4	Ombra dell'isimo stesso	017.83	+04.56
5	Golfo delle Perle, bocca dell'Indo	023.59	−04.90
6	Bocca dell'Idaspe	027.38	+04.41
7	Capo degli Aromi	038.40	+08.30
8	Capo delle Ore in Argyre	039.78	+39.38
9	Capo delle Grazie in Argyre	051.86	+53.84
10	Golfo del'Aurora, bocca del Gange	055.74	−02.32
11	Punta dell'Aurea Cherso	061.49	+25.26
12	Primo punto di Thaumasia	066.36	+23.79
13	Confluente del Chrysorraos col Nilo	084.16	−18.88
14	Lago del Sole, centro	090.24	+25.22
15	Lago della Fenice, centro	106.45	+19.42
16	Bocca del Fasi	106.93	+44.88
17	Colonne d'Ercole, bocca esterna	119.81	−44.88
18	Centro d'Icaria	119.92	+37.86

(continued)

(continued)

No.	Name	Longitude°	Latitude°
19	Primo punto del Mare delle Sirene	131.37	+31.32
20	Primo punto di Thyle I	134.12	+65.08
21	Colonne d'Ercole, bocca interna	138.02	–
22	Centro di Thyle I	151.86	+65.08
23	Base australe d'Atlantide I	159.80	+37.54
24	Primo punto del Mare Cimmerio	165.80	+37.49
25	Golfo del Titani	174.24	+18.17
26	Ultimo punto del Marre delle Sirene	176.52	+25.34
27	Stretto d'Ulisse, mezzo	187.08	+74.08
28	Punto della riva australe dell'Oceano	188.15	–07.12
29	Fiume dei Lestrigoni, bocca sull'Oceano	200.19	–04.50
30	Golfo dei Lestrigoni, ultimo seno	201.79	+18.01
31	Scamandro, bocca sul Mare Cronio	202.52	+55.41
32	Scamandro, punro di mezzo	202.57	+48.98
33	Fiume dei Ciclopi, bocca sull'Oceano	205.05	–15.77
34	Base australe d'Esperia	211.10	–
35	Capo boreale di Thyle II	221.61	+62.28
36	Centro di Thyle II	223.53	+69.93
37	Golfo di Ciclopi	224.98	+12.43
38	Primo punto del Mare Tirreno	226.41	+37.81
39	Centro d'Esperia	231.62	+22.79
40	Bocca australe delle Xanto	234.11	+51.13
41	Ultimo punto del Mare Cimmerio	238.87	+09.85
42	Esperia, base settentrionale	250.28	+13.22
43	Piccola Sirte	256.94	+06.24
44	Capo Circeo, in Ausonia	266.59	+15.68
45	Punto della costa d'Ausonia	266.79	+22.70
46	Lago Tritone	267.15	–20.38
47	Primo punto dell'Ellade	270.74	+49.49
48	Lago Meride	277.09	–
49	Biforcazione d'Ausonia	282.32	+13.33
50	Congiunzione del Nepente col Nilo	286.25	–26.26
51	Gran Sirte et bocca del Nilo	290.45	–17.09
52	Punto più australe dell'Ellade	–	+57.99
53	Centro dell'Ellade	294.12	+46.30
54	Punto più boreale dell'Ellade	–	+30.38
55	Ultimo punto del Mare Tirreno	296.09	–00.67
56	Ultimo punto dell'Ellade	315.07	+44.08
57	Corno d'Ammone	318.32	+10.40
58	Scilla e Carridi	324.17	+20.31
59	Ellesponto, punto di mezzo	326.11	+48.22
60	Primo punto della Noachide	334.82	+48.40
61	Bocca del Phison, nel golfo Sabeo	338.85	+05.05
62	Primo corno del golfo Sabeo	357.27	–02.37

As can be seen, these names are drawn from ancient geography and even, on occasions, from mythology. Many of them sound pleasing to the ear. Schiaparelli stated that the names in Proctor's chart were too few to be conveniently transferred to his own map, so that he had introduced a new nomenclature for his personal use. The prime meridian has been named the *Point of Aryn*, recalling a legend of the Middle Ages. The so-called town of d'Aryne was supposed to be the *dome of the world* in Middle Ages charts, lying at an equal distance from the north, the south, the east and the west; it was therefore taken to lie on the equator and to mark the central meridian.[30]

This is our Meridian Bay, whose points are named the first and second horns of the Sinus Sabæus—the first being that which first passes across the central meridian—as seen by the observer, because of the planet's rotation. This is also in the sense of the numbering of the degrees of longitude.

Because of atmospheric circumstances, this zero point for Martian longitudes cannot be the only object which is measured; and as it is the origin of the longitude system, there must be a consistent error in the values given in degrees, even though there is no alteration in the relative positions. The author promised to verify this zero point at a later date.

If we compare this zero meridian with that on Green's chart, we find that there is a difference of 7°; Green's zero passes to the right of the Meridian Bay. This difference applies all over the map; for example at 90°, 290° etc.

The Mercator projection chart of Mars given by Schiaparelli in his Memoir is reproduced here (Fig. 24). On it we find the 62 points given in the Table. It is a truly remarkable piece of work, and one showing features which the old observers of Mars could never have suspected. It depended for its successful completion on an unflagging persistence, an excellent eye, a rigorous method of observation and a good instrument.

If we compare this planisphere with my map, it is easy to identify the geographical configurations. The Hourglass Sea becomes Syrtis Major; a little less accentuated on Schiaparelli's map, no doubt because in 1877 it was less broad and less dark than usual. Herschel II Strait becomes Sinus Sabæus; the circular Terby Sea becomes Solis Lacus; Kepler Land, Thaumasia Fœlix; Huygens Continent, Memnonia; Maraldi Sea, Mare Cimmerium; Hooke Sea, Mare Tyrrhenum, etc. The map does not extend beyond latitude 40°N., because in 1877 this part of the planet was not seen to advantage. The Italian astronomer completed the map during the following oppositions.

Note that the author places west at the right and east at the left, instead of the other way round, which is in the sense of all celestial images as seen in an astronomical telescope. This system does not apply to an Earth-based observer, but to an observer who is actually on Mars. On Mars, as seen from Earth, one point is east of another if it passes the meridian first; thus Vienna is east of Paris and passes the meridian ahead of it. This system is very logical, if only because it is hard and fast.

[30] See SANTAREN, *Essai sur l'histoire de la Cosmographie au moyen âge*, Vol. 1, pp. 94, 368, and Vol. III, p. 310.

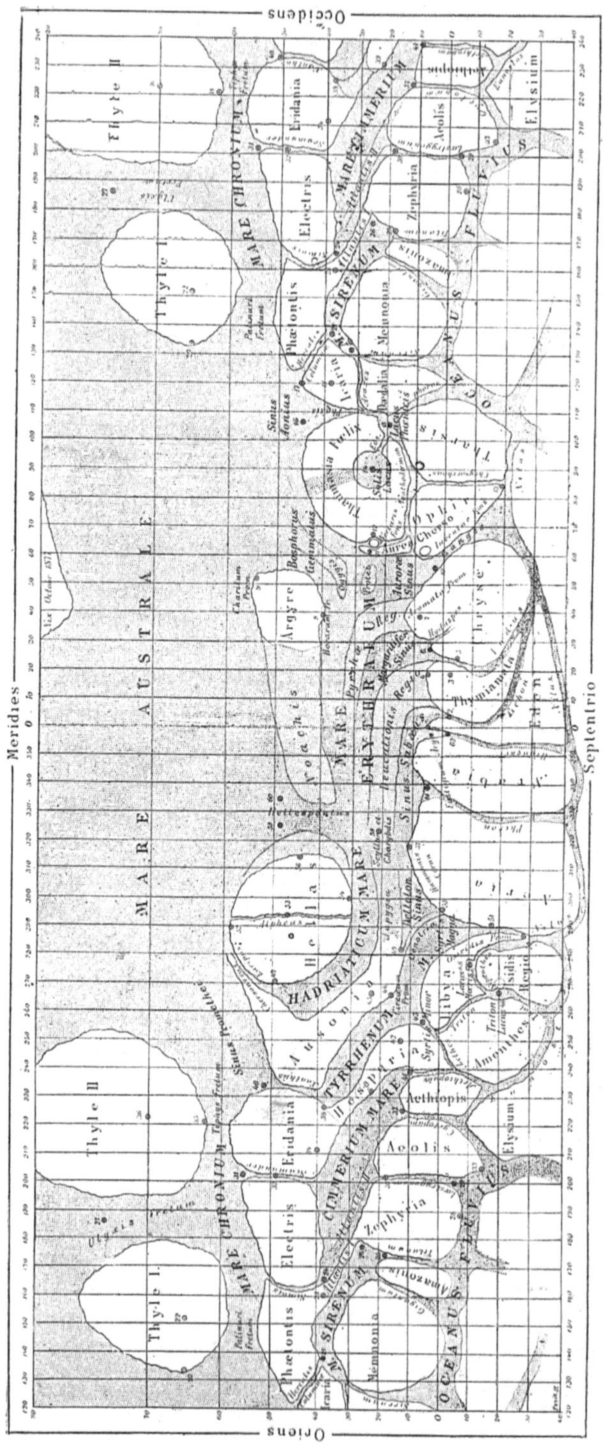

Fig. 24 Triangulation and areography, by Schiaparelli, in 1877, from the positions of 62 measured points, a new map

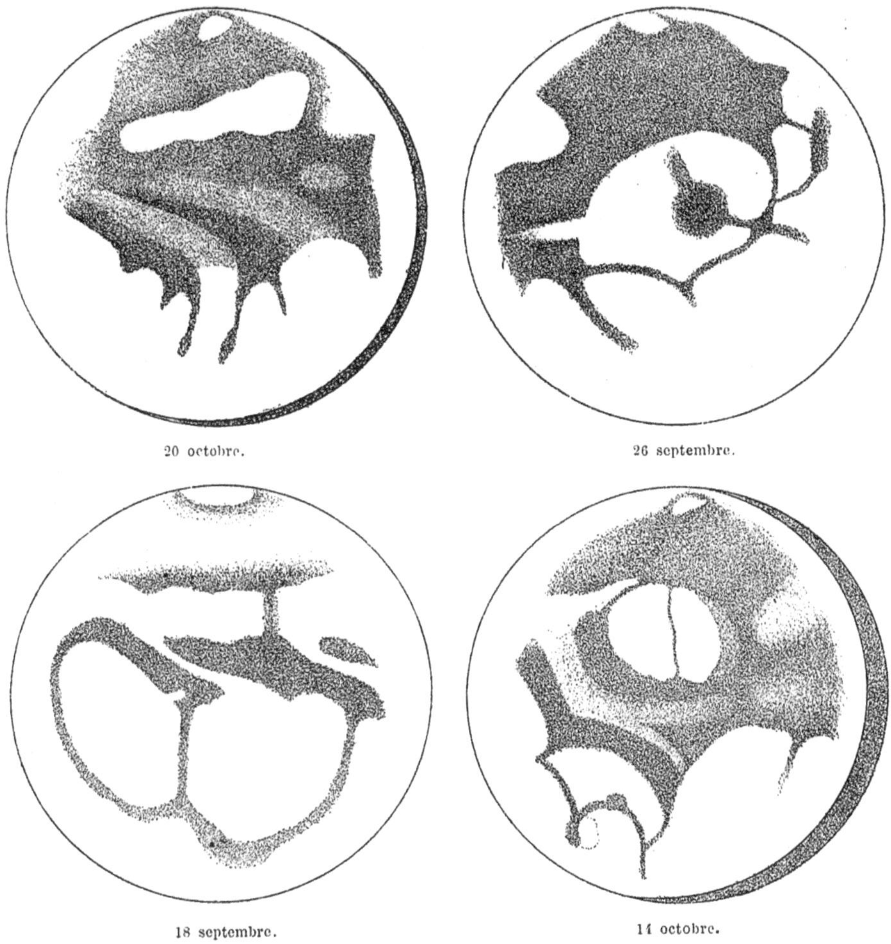

Fig. 25 Drawings of Mars, by M. Schiaparelli, in 1877

The arm of sea called the Channel on my map, at the extremity of Christie Bay, is very broad, and has been given the name of Ganges.

The two points of the Meridian Bay are prolonged, as far as a southern sea, by two streams which are called Hydaspes and Gehon. We have noted above that in 1864 Dawes, convinced that in this position he had seen the mouths of large rivers, had looked for these rivers without success.

Not far from here is another large canal, the Phison.

Further away, we again see these curious streaks—the famous "canals."

On these numerous drawings, most of which show specific details and not many of which show the entire disk, the author published the four which are reproduced here (Fig. 25), covering the whole of the planet. They were made on 20 October, 26 September, 18 September and 14 October, the longitudes of the central meridian, in degrees, being respectively 18°, 85°, 181° and 298°. The latitude of the centre of the disk is, on average, −24°.

The second of these drawings seems to contradict the maps, showing a white area to the left of Kepler Land which surrounds the circular lake. This, wrote Schiaparelli, was due to a mass of bright clouds, *una massa di nubi illuminate* (it is interesting to note that Green had observed an analogous effect on 29 September). In the first drawing, the same disagreement is shown for the large Argyre Island, and the explanation is the same.

Schiaparelli then gave the following general description of the planet.

Description of the Surface of Mars

Schiaparelli begins by commenting that since it is important to record, rapidly, what is seen through the telescope, one must never lose time, and that the drawings made quickly, are the best:

Up to here, Schiaparelli has not committed himself. But he is, with very good reason, more explicit later. What is his precise opinion? It is as follows:

> With regard to the nature of the sombre patches, it is possible to put forward an infinite number of more or less arbitrary proposals. However, there are two which can be supported by a sufficient analysis, and of these two there can be only one which gives a plausible explanation of all the observed facts.
>
> The first, which compares the patches on Mars with those on the Moon, assumes the surface of the planet to be entirely solid, the variety in the tones being due to the materials making up this surface. While not entirely impossible, such an hypothesis does not succeed in explaining the observed facts, at least if we do not complicate it by other subsidiary hypotheses, more or less bizarre, The existence of the polar snows, which is certain rather than only probable, together with the fogs and clouds, proves that in the atmosphere of Mars there is a meteorological circulation in which vapours rise in certain regions and condense in others. It does not seem that this could occur unless the surface of the planet plays a part. If the vapours on Mars condense into crystals in certain places, and then condense in other places into liquid form, the explanation is adequate. These liquid condensations, at least if we assume that the surface of the planet is an exactly equipotential surface, should unite in low-lying areas and produce seas or lakes of greater or lesser extent. The tracks along which these liquid condensations reach their reservoirs can only be streams or rivers, either regular or intermittent in flow. It is true that the entire system might be hidden or subterranean, as with the circulation of water in the deserts of Africa; alternatively the lakes in question could be very small, and invisible from Earth, so that the mechanism of the circulation of the atmospheric vapours would be unobservable. All this is possible, but the suppositions will become useless until the moment when we see on the planet appearances precisely similar to those which would be presented to an observer on Mars looking at the condensation of vapours in the atmosphere of the Earth.

This reasoning, published by Schiaparelli in 1878, is of the same kind as that which may be found in the first edition of my *Terres du Ciel* (1876, p.429). And how could it be otherwise? The analogy is too obvious not to be our guide, though it must not lead on to conclusions which are too narrow and too geomorphic, as might be said. I ask my readers to follow this passage:

> Martian meteorology is a very similar reproduction of that of our own planet. On Mars, as on the Earth, the Sun is the supreme agent of movement and of life, and its action there determines results analogous to those which prevail here. The heat vapourizes the water of the seas and makes these vapours rise into the upper atmosphere; the water vapour assumes a visible form in the same way as with the production of our clouds, that is to say by temperature-difference and saturation. Winds are caused by the same temperature-differences. We can follow the clouds, carried along by the air-currents, over the seas and the continents, and, as I have said, many photographic observations have already been made of these meteorological variations. If we cannot yet precisely see rain falling on the countries of Mars, we can at least infer it, since the clouds dissipate and re-form. If we cannot yet see snow falling, we can infer this also, because, as with us the winter solstice there is a time of frost. Thus, there is on Mars as on the Earth, an atmospheric circulation and a drop of water which the Sun steals from the sea and is returned there after having fallen from the cloud which received it. There is more; when we should be on our guard against any tendency to create imaginary worlds in the image of our own; when we are presented with such an organic similarity—as in a mirror—it is not difficult to avoid going still a little further in our deception. Indeed, the existence of continents and seas shows us that Mars, like our world, has been the scene of interior geological movements which have caused uplifts and depressions. There have been collapses and upthrusts modifying the primitive uniformity of the crust. In consequence there are mountains and valleys, plateaux and basins, steep ravines and cliffs. How is this rainwater returned to the sea? By springs, streams, riverlets and rivers. The drop of water which falls from a cloud crosses permeable lands, as with us; it slides over impermeable terrain, again sees daylight in a liquid form; prattles in the streams; flows in the rivers and descends majestically toward the river-mouth. Thus it is difficult not to see on Mars, scenes analogous to those which make up our Earthly landscapes; streams running in their pebbly beds gilded by the Sun; rivers crossing the plains and falling in cascades to the floors of the valleys; great rivers descending slowly to the sea across vast countries. The sea-shores receive there, as here, the tribute of the canals, and the sea is sometimes calm like a mirror, sometimes whipped up by tempest.

I reproduce this passage here to show the agreement between the two lines of reasoning. For my part, I continue to believe that the dark patches on the Martian globe represent seas. We will see later (1879) that Schiaparelli has changed his mind and has become very sceptical in this respect.

But let us continue to describe the work of this eminent observer. We next come to an examination of the seas, and to the variations in their tones and depths, the deeper ones absorbing the sunlight more strongly and appearing darker to us, these shallower seas allowing the light to reach their floors. The nature of the liquid, and that of the material, held in suspension, must also exert an influence. The Milan astronomer added:

> Without making any special hypothesis with regard to the nature of those liquids, the variations in their tones can be explained simply by difference in depth, transparency and chemical composition. The difference in salinity between terrestrial seas causes the great differences between the colours of these seas. The saltier the water, the darker it appears. In general, the salinity of terrestrial seas decreases with latitude, because of the lesser evaporation and the greater precipitation, which explains why the polar seas are brighter than the equatorial ones. This has been shown by Maury with regard to the contrast between the waters of the Gulf Stream with the Atlantic; the bright green of the North Sea and the polar seas, and the sombre azures of tropical seas and the Indian Ocean. It is the same on Mars. There, also, the polar sea is less dark than those of the torrid zone, and the seas of the temperate zone are of an intermediate tint. All this leads us to regard the Martian seas as similar to the terrestrial ones.

> The complicated réseau of dark lines which links the patches which we regard as seas forms an extra argument in favour of the same hypothesis. These lines owe their colour to the same cause as with the seas, and can only be canals or communicating straits. Their broadening and their mouths are to be expected, according to this explanation. Nothing analogous to the réseau can be seen on the Moon. If there were material there of different colours, we would have to ask ourselves how such a reticular distribution could have been produced.
>
> We see, therefore, that the hypothesis of a maritime and continental constitution of the surface of Mars is a very probable one. It will become almost a certainty it we can manage to affirm, in an indisputable manner, the real disappearance and eastern emergenge of the Lake of the Sun. This canal, which was seen by Mädler in 1830, by Kaiser, Lockyer, Rosse and Lassell in 1862, then by Kaiser and Dawes in 1864, could not be recovered in 1877, despite the most careful searches which I carried out, and which showed details of much smaller size. If the change is confirmed in the future, it will be difficult to find an explanation simpler and more natural than that of a change in the amount of water in this region, analogous to what happens in China along the course of the Yellow River.

The reader will certainly realize that what we have called the pear's tail in the following drawings—Beer and Mädler, at point *d* of the left-hand hemisphere; Lockyer, Figs. 23, 24 and 38; Lord Rosse, Fig. 42, no. 3; Kaiser Fig. 49; Dawes Figs. 53 and 54—is the appendage of the Terby Sea, which is not shown on my map because I regard it as essentially variable.

In 1879 Schiaparelli recovered it; it had again become visible. He gave it the name of the Nectar Canal.

In 1877, Green noted that there in a small lake (designated *a* in Fig. 18) which he called Schiaparelli Lake. He believed that this little lake, forming an intermediate point between the De la Rue Ocean and the Terby Sea, could, because of the indecisive images usually found, be taken for a canal joining the two seas. For my part, I maintain that there are certain changes there from one year to another, and that the basic explanation of liquid variations is justified. But let us once more go back to the work of the Italian astronomer:

> On Mars there are regions of a tint intermediate between those of the sombre seas and the bright continents. What do they represent? If we regard the patches on Mars as simple colourations of a solid surface, these variations in tone do not require any other special explanation. The predominant mineral and, equally, the predominant vegetation can offer all possible gradations in tone, and all possible colours. But if we attribute the bright-dark distribution to liquid surfaces, we have a more natural explanation, better able to fit the observed facts, It will then suffice to regard the tone as proportional to the absorption of the Sun's rays by the liquid, and in this case the grey regions under discussion will be submarine banks or damp depressions. On the Mare Erythraem we can see clouds condensing in preference to other areas, which agrees with a lower temperature, due precisely to these depressions or banks. The Isthmus of Hesperia, at the place where the Mare Tyrrhenum and the Mare Cimmerium are closest together, should therefore be a place of possible communication between the two seas.

He here refers to the Niesten Isthmus, at the point shaded in my map. Probably, there is almost always a light layer of water there, and when seen obliquely it appears darker than when seen face-on.

As for the depths of the seas, Schiaparelli recalls that according to Father Secchi's experiences in the Mediterranean, even a very white object becomes invisible at a depth of 60 m. However, according to M. de Tessan, the southern extremity of Africa, the bank of the Aiguilla, seems to reduce the darkness of the waters, although it is 200 m below the sea-surface. The depth of water in the Martian coastal seas is probably slight, as it is with the canals.

The atmosphere is sometimes hazy; it is generally clearest when the Sun reaches its greatest altitude above the region concerned. Sometimes the veiling is so marked that nothing can be made out in the area.

The effect of the clouds is to whiten the regions below them. If therefore we have a region which first appears sombre and then bright, we may assume that in the first case we are seeing the actual surface of the planet, and in the second that we are seeing a layer of cloud or mist.

Let us now discuss details of the geography of Mars.

Beer Continent=*Great Diaphragm*, containing *Aeria*, *Arabia* & *Hammonis Cornu.*

(It is necessary, to follow those geographical details, to have both my chart, Fig. 31, and Schiaparelli's, *fig*. 174, before you.)

This is the greatest stretch of bright continent on the Martian globe; throughout the observations, from September 1877 to March 1878, it was impossible to discover even one patch on the whole continent. Hammonis Cornu (no. 57 in the triangulation) corresponds to Banks Cape. The banks of the Hourglass Sea and Herschel II, Strait are clear-cut, and without indentations; the seas are sombre, probably deep.

Herschel II Strait=*Sinus Sabæus*, *Phison*, *Sinus Meridiani*, *Hiddekel* & *Gehon.*

Opposite Beer Continent, the region which extends to the south of the Strait is not luminous, but grey. This is why the Strait should not be separated, as it often is on the old drawings. A small gulf, Schmidt Bay, receives the course of a river, the Phison, already seen by Kaiser on 22 November 1864. Schiaparelli saw it extending as far as the Nile. Next, at the Meridian Bay, are seen two rivers, the Hiddekel and the Gehon, the first parallel to the Phison and the second curved. The Hiddekel was seen only on 28 February 1878, when the planet was very small and the two points of the Bay could no longer be distinguished; therefore it was only the observer's judgement which placed the mouth of the river at the first point. The course of the river is uncertain. The Deucalion peninsula seems to be a submerged land; it lacks the brightness and clarity of the continents.

Arago Strait=*Margaritifer Sinus*. **Burton Bay**=*Mouth of the Indus*. **Mädler Continent**=*Chryse* & *Hydaspes*.

Arago Strait is a decidedly sombre sea. When the images are blurred or one is not certain of one's orientation, it is easy to mistake it for the Hourglass Sea, as has happened more than once. The Indus, a broad river, merges after coming from the Nile and forming a curve. Its course could not be followed as far as the Nile until after 24 February 1878, because previously the continent was covered with clouds, but then it was well seen, although the diameter of Mars was reduced to 5″.7. The peninsula of Pyrrha Regio seemed, like the Deucalion, to be a submerged land.

Channel=*Ganges*. **Christie Bay**=*Aurorae Sinus*.

It must be admitted that the nomenclature of the celebrated Milan astronomer is euphonious and charming, even without taking into account its erudition and ancient connections. Gulf of Pearls, Aurora Gulf, Phoenix Lake, Icarie, Elysian Fields, lands of Deucalion and Pyrrha; one can imagine nothing more gracious. Personally, I hope with all my heart that this ingenious areographical nomenclature will replace all preceding systems. But perhaps a large number of these lightly-marked features are essentially variable, sometimes even fading to complete invisibility.

The Aurora Gulf is vast and sombre, it has been thus represented by the majority of observers. From here emerges the Ganges, *one of the broadest and most easily visible canals on the whole of the planet*. Schiaparelli saw it under all conditions between 28 August and 25 February. It went as far as the Nile. This is the Channel Manche on my map. In 1858, Secchi drew it admirably; he gave it the name of Franklin Isthmus or Canal. To the right of this isthmus, there can be seen a narrow, vertical canal—that is to say, running in a north-south direction—which has received the name of Chrysorrhoas, and which joins a watercourse which is no less light—and probably no less variable—running from east to west below the circular lake or Terby Sea.

Terby Sea = *Solis Lacus*. **Kepler Land and Copernicus Land** = *Thaumasia*.

We have now come to the circular lake, which has been compared with an Eye of which it forms the iris. Schiaparelli wrote in 1877 that it was round, though perhaps a little elongated in the vertical sense. On 30 September, the diameter of the planet being 21″.79 and that of the lake was 2″ or 10°.5. It was very dark, particularly at the centre. The darkness decreases from the centre toward the edges, but in steps, not gradually. This is one of the most circular geographical configurations on the whole of the planet. A small canal joins it, to the right, to Lacus Phoenicis. Another streak, less dark but broader, runs to the south. It has been impossible to see anything else there.

(This region is the scene of considerable variations. I will return to this subject later. The streaks shown by Schiaparelli are called Nectaris Fons, Aurea Cherso, Agathodæmon, Eosphoros, Chrysorrhoas and Lacus Phoenicis, and do not appear to be stable.)

Jacob Land = *Noachis & Argyre.*

This is an island which appears bright in its right-hand portion, and sombre in its left-hand portion, as though it remained constantly submerged beneath a shallow layer of water. However, it sometimes seems wholly bright, as is shown in the drawing made on 20 October 1877. But in general it shows the aspect represented on my map. Schiaparelli thinks that snow and clouds are often found in this white region. Dawes' drawing of 21 January 1865 shows this island as white, and it has also been called Dawes' Snowy Island.

Phillips Island = *Deucalionis Regio.*

This island has already been mentioned. It gives the impression of a submerged peninsula. Schiaparelli wrote:

> *La terra di Deucalione, e tutti le altre simili, siano continenti sottimarini*, according to all appearances. In nearly all the old drawings we have seen that this semi-island is as white as the adjacent continent, and the Herschel Strait, ending at the Meridian Bay, is clearly detached showing black against the bright background. On Proctorís map, made from drawings by Dawes, the Meridian Bay is in communication with the Arago Strait, and makes the region seem like a complete island (Phillips Island). It is the same in Secchi's drawings of 1858. Terby inclines to the view that this is the true configuration, and that the attachment of the Island to the continent is an illusion caused by clouds which make the channel look white.

Pyrrha Regio is a similar case.

(Can it be that the water assumes a state intermediate between the liquid and the cloudy, and condenses above the surface in the form of layers of viscous mists, dark and very dense?)

Schiaparelli thinks that all those lands surrounded by water should produce vapours which become dense to a greater or lesser degree, and to a distant observer show up in various shapes according to their own actual shape; and also to the wind. However, nothing of the kind was observed to 1877 anywhere in the southern hemisphere of Mars, except for clouds over Jacob Land.

Hall Island = *Protei Regio.*

This is an island isolated in the De la Rue Ocean, near the same latitude as the Terby Sea. It has been seen on Green's map. It is closer to the equator than Dawes' Snowy Island, and is rarely observed. Green drew it as very white on 2 September at 1h10m and 2h20m. Schiaparelli saw it on 2 October and 4 November; he thought that on 26 September and 4 October he had observed not the Island itself, but its meteorological image, a white cloud indicating its form.

Schiaparelli Sea = *Mare Sirenum, Herculis Columnae, Araxes & Lacus Phoenicis.*

It is to the skilful Milan astronomer that we owe a new clarification of this curious Martian region. Previously, the sea bearing his name had been confused with the Maraldi Sea. His observations have defined it much more accurately.

This sea is prolonged by two straight arms, one which descends toward the left as far as Bessel Lake, the other which goes up toward south as far as the Cottignez Sea. The first of these two is called Araxes on Schiaparelli's chart, and instead of being rectilinear, it is sinuous. "*Alla sua curvature, dit-il,che e molto evidente, e costituice un caso piutosto raro nei canali onde è sparso il pianeta, ho posto, particulare attenzione*," Schiaparelli wrote. (There have been changes here; see below for later observations.) The second arm is called Herculis Columnae, while the lake is named Lacus Phoenicis. The Lagrange Peninsula is called Icaria. This region was shown on the sketch made by Kaiser on 10 December 1864, but the phase is invisible. The region was also shown on a drawing by Lockyer made on 18 October 1862.

The Sirenum canal was seen on 18 September, but the lower part was broader, pale and badly defined. The author attributed this aspect to turbulence in the atmosphere of Mars, which seemed to have been cloud-laden over long periods. On 6 January 1878 the diameter of this region of the planet was reduced to 8″.2. The Sirenum canal was clearer than ever, and this clarity lasted until 21 March. Schiaparelli thought that when the Sun reached the equator—on 22 February—it dispersed the mists. This is possible. But it is equally possible that the canals change with the seasons. We have already noted that there were several other canals which did not become evident until February.

The broadening of this canal, as with the Eosphoros, the canal of the Giants, that of the Titans, etc., is attributed by Schiaparelli to a division of the mouths into deltas, as we see on the Rhône, the Rhine etc. In this case the water will flow from the south toward the north; from the Mare Sirenum toward the Ocean, as there will be a gradient from south to north,

Maraldi Sea=*Mare Cimmerium, Huggins Bay & Cyclops River.*

Readers will have been familiar with this sea for a long time. The greatest peculiarity of the region is the existence of a canal which is called Huggins Bay on my map and the Cyclops River on Schiaparelli's. This seems also to be one of the most variable configurations on the planet. During September, October and November only an indistinct grey shading could be made out; Schiaparelli attributed this aspect to the atmosphere of Mars which, he said, was then as cloudy there it was over the equatorial lands. But on 25, 28 and 30 December the canal was seen very clearly, although the disk was reduced to 9″. This clarity persisted until the end of the observations. The canal descended vertically from the Maraldi Sea along the 223rd degree. To me it is more oblique, and more like the view shown on Dawes' drawing of 1 December 1864; perhaps its course is subject to certain changes.

Hooke Sea and Flammarion Sea=*Mare Tyrrhenum & Syrtis Minor.*

These seas succeed the preceding one, and lead on to the Hourglass Sea. The Hooke Sea is darker in the north than in the south. Between the Hooke Sea and the Flammarion Sea, a pointed gulf penetrates the lands; it has been given the name of Gruithuisen Bay, while Schiaparelli calls it the Syrtis Minor. It is in fact the joining of two rivers, the Lethe and the Triton. Its course is curious and doubtful; it is very difficult to distinguish. Not far away can be seen the river of the Ethiopians.

Hourglass Sea=*Syrtis Major, Nile, Libya, Nepenthes, Tritonis Lacus & Moeris Lacus,*

This is the region of Mars which has been known for the longest period; as we have seen, it was first drawn in 1659. Schiaparelli has given the Hourglass Sea the name of the Syrtis Major, which to me seems to be less happy than his other names. He gave the name of the Nile to the lower region which turns to the right in a long canal, and which is called Nasmyth Channel on my map.

To the left of the Hourglass Sea, toward the little Main Sea, Schiaparelli drew a canal, the Nepenthes, which abutted on to a lake, Tritonis Lacus; and in his centre lay another lake, the tiny Moeris Lacus. Another river very easy to distinguish, the Triton, comes from the lake to the Syrtis Minor, describing a graceful curve. This continental region, surrounded by the waters, has been given the name of Libya. To me this region seems to be subject to frequent, considerable variations, and it seems that there is extensive flooding, particularly to the north. Note the shaded edges of these areas on the drawings made by Dawes, 26 November 1864; Kaiser, 28 December 1864; Kaiser, 22 November 1864; and Lockyer, 3 October 1862.

The Hourglass Sea is much narrower on Schiaparelli's map than on the drawings made by Mädler, Secchi, Lockyer, Kaiser, von Franzenau, etc. One must conclude that its banks, too, are variable.

Above the Nepenthes, in Isidis Regio—which is itself very white—Schiaparelli observed a point as brilliant as the polar cap on 14 September. To him it seemed to form a square of side 1″.5—about 8°, or 480 km. If it were snow, we must conclude that there is a group of high mountains on the shores west of the Triton Lake.

Cassini Land and Dreyer Island=*Ausonia & Iapygia.*

From his observations, the Milan astronomer concluded that the clouds have a marked tendency to form over lands surrounded by water. This is undoubtedly why the Herschel Strait and the Meridian Bay are so frequently detached and isolated as a dark ribbon, while the Deucalion peninsula is often whitened by clouds. Ausonia becomes sombre, toward Iapygia. Under normal conditions, when the atmosphere of Mars is clear, the regions appear as shown on the drawing made on 14 October 1877.

Lockyer Land=*Hellas*, *River Alpheus.*

Also a singular region. A round island, slightly elongated S.E.-N.W., whose diameter is not less than 30° or 1,800 km. In colour yellow, like the continents, but sometimes as white as snow. In 1877 the island was shown divided from south to north by a canal which Schiaparelli named the Alpheus.

South Polar Sea

Apart from the two Thyle islands, the polar cap of Mars is limited by the 60th degree, parallel south, and is entirely maritime. The opposition of 1877 was extremely favourable for the examination of this regions; in October the axis was inclined by only 65° to our line of sight, so that all the polar snow remained constantly in view, surrounded by the sombre sea. The atmosphere appeared cloud-free.

Such is the résumé of the geographical observations. A few words more, about the polar caps.

The position of the centre of the south polar snow was found to be: Longitude. 29°47. Distance from the pole, 6°15.

At the same time, Asaph Hall had found a longitude of 20°66 and 5°11. The mean of these two determinations gave **25°06** and **5°63**. The average of the measurements made by Kaiser, Lockyer and Linsser in 1862 gave 15°51 and 4°26. In 1830 Bessel found 21°55 and 6°59. Schiaparelli concluded that at different Martian solstices the south polar cap, when reduced to its minimum extent, always occupies almost the same position on the planet. Undoubtedly there is a sea-floor there.

Schiaparelli's map of the southern hemisphere of Mars (Fig. 26) excellently shows the position of the triangular snow-cap remaining near the pole.

These polar snows vary in size according to the season. Here are the observations:

Date/Days before or after solstice/Diameter of the polar snow, degrees/Date/ Days before or after solstice/Diameter of the polar snow, degrees

23 August		–34	28.6	22 September		–4	14.7
28	"	–29	23.9	24	"	–2	3.8
3 September	–23	26.0	25	"	–1	11.5	
10	"	–16	23.9	26	"	0	11.5
10	"	–16	18.5	30	"	+4	12.5
11	"	–15	20.2	1 October	+5	13.7	
12	"	–14	17.4	2	"	+6	11.8
13	"	–13	16.9	4	"	+8	12.7
14	"	–12	17.4	10	"	+14	10.4
15	"	–11	14.1	12	"	+16	9.5
15	"	–11	16.1	13	"	+17	9.3
16	"	–10	16.1	14	"	+18	7.6
18	"	–8	19.1	27	"	+31	7.0
20	"	–6	18.5	4 Nov.	+39	7.0	

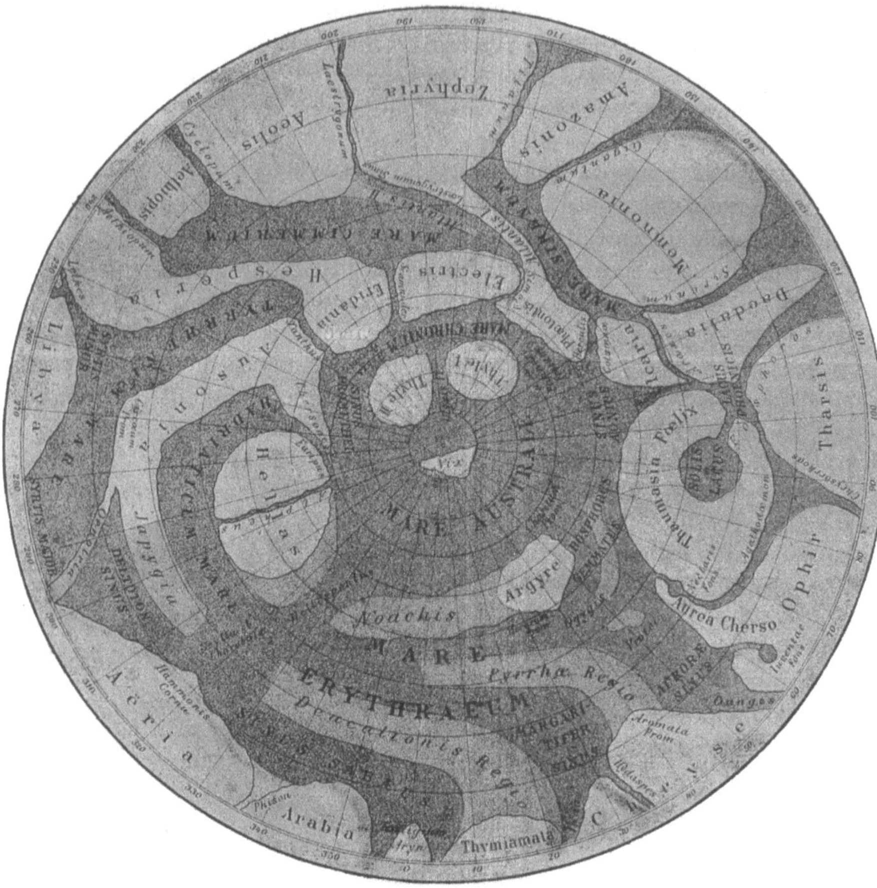

Fig. 26 The southern hemisphere of Mars, in 1877. Drawn by M. Schiaparelli

Disregarding the fluctuations which are inevitable in measurements which are so difficult, we can see that there is a rapid decrease of the polar snows from 28° to 7°. Having seen this great reduction, Schiaparelli waited to see the snow disappear altogether. He did not see this, however.

The observations could be continued only with difficulty after the latter date, because of the great obliquity of view, the gradual advance of the shadow and the formation of clouds in these regions. At the beginning of December the snows appeared to grow again. The summer solstice occurred on 26 September. The minimum of the snows therefore followed 2 months later.

And at this minimum the snowy area presented a triangular form.

Schiaparelli ended his admirable work with considerations of Martian meteorology and geology. These questions will be discussed later. Can we doubt, after this description, that the progress made by the illustrious Milan astronomer surpassed, at a bound, all previous work? We will again find his results during the oppositions which will follow.

Same Year, 1877.—Maunder: Spectral Analysis of Mars[31]

While the beautiful physical observations described above were being made by skilful observers, other investigators continued the spectral analysis research which had begun in 1862 and 1864 by Huggins, Miller, Rutherfurd and Vogel. It was continued in 1867 by the same investigators, together with Secchi, and in 1872 by Vogel. At the Greenwich Observatory, Maunder found in the spectrum of Mars lines at the following wavelengths, expressed in decimeters:

		23 Aug	21 Sept.	26 Sept
Band δ of Brewster's spectrum	{1st edge		5640	5639
	{2nd edge		5661	5717
Group of lines near D	{lst edge		5889	5887
	{2nd edge		5907	5897
Weak band, middle		6019	6022	
Band, α middle		6287	6287	6298
Very weak band, middle				6511
Group of lines near C	{1st edge	6544	6537	6544
	{2nd edge	6572	6587	6575
Very weak band, middle				6695
Very weak band (B?), middle			6852	6895

If we compare these absorption lines with those found in 1867, we find at least a reasonable agreement.

Of these bands, the three which are most marked, band α and the groups near D and C of the solar spectrum, were observed at the same time in the spectrum of the Moon, which was at the same altitude above the horizon; but they were narrower on the Moon than with Mars.

We must try to detect differences between the spectrum of the sombre patches and that of the bright regions. The dark patches yield a spectrum much fainter than that of the rest of the disk; the contrast is very marked in the red and yellow and less in the violet. No specific line or band has been found. All the absorption bands appear fainter toward the edges than in the inner part of the disk.

The bright areas appear orange, varying very gradually from whitish yellow to red over a period of time.

(Also, during this precious opposition of 1877, a certain number of observations of the positions of Mars compared with adjacent stars were made in connection with the determination of the solar parallax. This work will not be discussed here.)

Finally, note that the first photograph of Mars was taken in this year by Professor Gould, Director of the Cordoba Observatory; he gave a lecture in connection with the Philadelphia Exhibition in which he announced that he had succeeded in obtaining photographs of 84 celestial objects, among them Mars, Jupiter and Saturn.

[31] *Monthly Notices*, Vol. XXXVIII, November 1877, p. 34–38.

The principal bright and dark areas are recognizable, but these photographs cannot withstand enlargement.

Opposition of 1879

The opposition of 1879 was but a little inferior to that of 1877, with regard to the importance of the observations. True, the planet did not approach as close to the Earth, but for Northern Hemisphere observers it was higher above the horizon, and the extra clarity of the images furnished more than ample compensation. Moreover, the recent discoveries gave strong encouragement to all the observers.

Situation of the Planet				
Opposition: 12 November. Diameter 19″.3.				
Pole inclined toward the Earth: the southern, but less so than in 1877.				
Dates	Latitude of centre°	Apparent diameter″	Phase (zone deficient)″	Angle Sun/Earth°
12 Aug.	−15.2	11.4	1.7	46
12 Sept.	−10.5	14.2	1.7	41
12 Oct.	−9.8	17.8	0.9	26
12 Nov. (opp.)	−14.5	19.3	0.0	0
12 Dec.	−18.2	15.3	0.6	23
12 Jan.	−17.2	11.0	1.0	35
12 Feb.	−12.7	8.1	0.9	38

Calendar of Mars		
	Upper (S) hemisphere	*Lower (N) hemisphere*
14 Aug.1879	Summer solstice	Winter solstice
21 Jan.1880	Autumnal equinox	Spring equinox

1879.—N.E. Green: Observations[32]

This author first explained that the aim of his observations had been just the same as in 1877, and in addition he aimed to see whether any change had taken place during the interval.

The atmosphere in England was not favourable, and the best views were obtained when Saturn was completely hidden by mist, so that the brightness of Mars was very much reduced.

[32] On some changes in the markings of Mars (*Monthly Notices*, March 1880, p. 331). See also *The Astronomical Register*, Dec. 1879, p. 295.

Apart from a few details, all the configurations can be identified on his chart. Niesten at Brussels, Burton near Dublin and Denning at Southampton made drawings giving added confirmation.

Certain variations in aspect have been observed. One of the principal is a white band at latitude 20°S., extending from 260° to 360° in longitude, joining up Dreyer, Hirst and Phillips Islands in a long line. To the east of Phillips Island, the bright band turns toward the equator, and passes between the Meridian Bay and Burton Bay, going on to join Beer Continent. Now, this was precisely the aspect seen and drawn by Beer and Mädler in 1830, Lockyer in 1862 and Kaiser in 1864, while at Madeira, in 1877, this part of the globe of Mars was marked by a half-tone on which the islands were seen only indistinctly; the space between the Meridian Bay and Burton Bay always stayed dark enough to maintain the continuity of the equatorial band.

To the north of the Terby Sea, the sombre patch named the Dawes Sea on Proctor's map was drawn by Dawes, Lockyer and Kaiser; but it was certainly invisible in 1877. Green found it impossible to recover. It had therefore disappeared, or nearly so. Yet it returned in 1879, much as it had appeared in earlier years.

As for Schiaparelli's canals, Green believed that he had glimpsed some of them, but wrote that for his own part he

> did not think that they could be permanent geographical markings, because if all the dark lines seen by observers were combined on the same map, the greatest confusion would result.
>
> It is possible that some of these lines are the limits of very faint patches, so faint as to be invisible, or spaces between veilings in the atmosphere. In these two cases, the positions would vary.
>
> These observations lead us to regard the large dark patches as the most permanent configurations, though subject to partial obliteration or even physical disappearances due to the interposition of some sort of brighter atmospheric veil.

At the meeting of the Astronomical Society of London on 12 March 1880[33] there was an interesting discussion about the difficulty of making accurate drawings of some of the more doubtful details on Mars. Those who took part were Green, Brett, Knobel and Christie. The conclusion was that it was impossible to be sure, and that the atmosphere of Mars as well as our own air could produce variations of different degrees. As for real changes, Green did not believe in them:

> The changes of which I speak I do not for a moment suppose to be actual changes on the planet, but are doubtless due in great measure to its atmosphere.

Such is not my own opinion. To me, it seems that real changes occur on Mars, and that they are pronounced enough to be visible from Earth.

At the same meeting, Mr. Brewin showed some drawings made during the opposition of 1879.

---

[33] *The Observatory*, April 1880, p. 369.

1879 Dr. Terby: Observations and Drawings

As before, Terby made his observations with an excellent 108^{mm} refractor by Secretan, and also with the 108^{mm} equatorial at the Brussels Observatory.[34] They extended from 28 September to 18 December, and 23 drawings were made.

The principal ones were:

27 October, from $9^h\ 45^m$ to 10^h (Fig. 27A). It is clear that this drawing is incomplete, and it is impossible to find much detail. At ζ I recognize Schiaparelli's Œnotria. In this sketch

[34] *Aspect de la planète Mars pendant l'Opposition de 1879* (*Bulletin de l'Academie de Belgique*, March 1880).

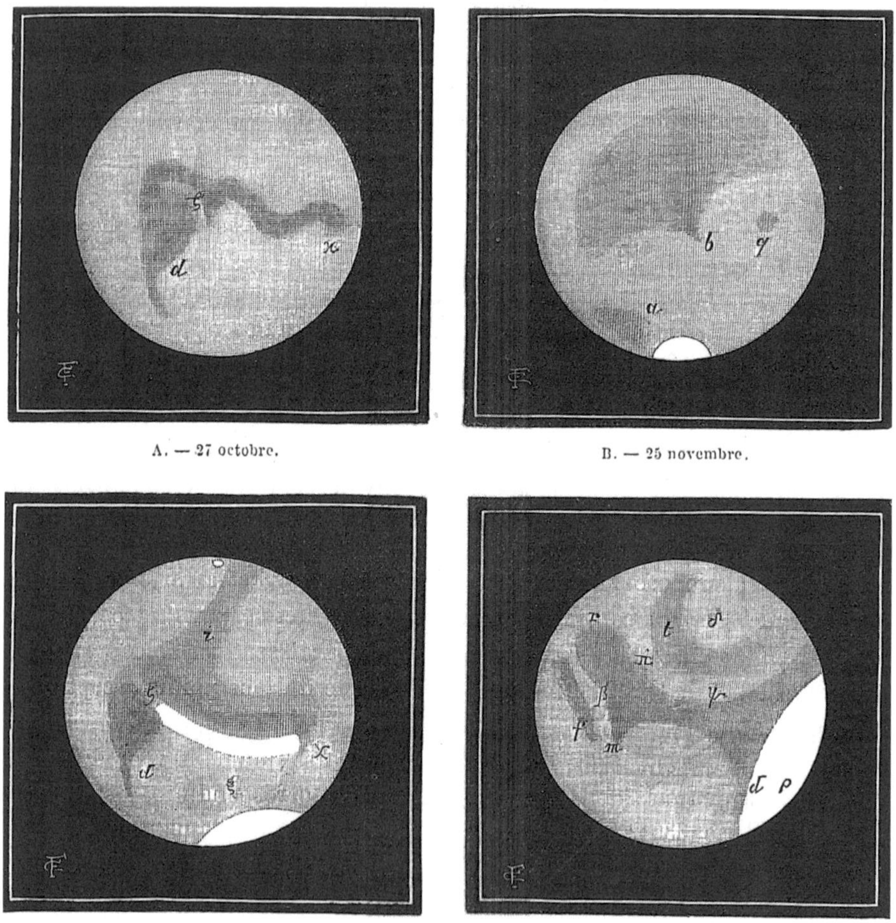

Fig. 27 Drawings of Mars, by M. Terby, on 27 October, 25 and 29 November and 6 December 1879

Herschel Strait fully justifies the epithet *Schlangenförmig* bestowed upon it earlier. This observation is very good and accurate.

25 November, from $8^h\ 45^m$ to $9^h\ 15^m$ (Fig. 27B). Image very good. The large patch is very feeble (De la Rue Ocean) Patch *a* is the darkest area—darker than on 16 October (Tycho Sea) then comes the Circular Sea, less dark, though on 15 and 16 October it was darker. At $9^h\ 15^m$, polar whiteness; the patch *a* seemed to disappear. The positions of patches *a*, *b* and *q* were drawn with great care, and related to the mean time of the observation, 9^h.

25 November.—It is interesting to note that the Tycho Sea was much blacker than it had been on 16 October. This effect cannot be attributed to the position of the patch on the disk, because on the contrary the south pole was more tilted to the Earth on 25 November.

29 November, from 6^h to $6^h\ 30^m$ (Fig. 27C). Very good image. Schiaparelli's Œnotria is seen at ζ; *i* is the Lambert Sea, very feeble. The Sinus Sabaeus $\zeta\ \kappa$ is bordered with white up to the broken line π. It was the same at $8^h\ 40^m$. A very light shading was seen at ζ. The two polar patches were certainly seen on this occasion; the upper was extremely small, as in 1877, but less brilliant. The progressive inclination of the south pole toward the Earth therefore brought this small cap into view; it was reduced to its minimum size, and even lost its brilliancy, following the summer solstice.[35] It is difficult to indicate the boundary of the great dark patch in the south. The image, at first excellent, became very unsteady with the approach of a cloud which rose above the horizon and put an end to the observation.

6 December, from $5^h\ 41^m$ to $6^h\ 1^m$ (Fig. 27D) and at $6^h\ 11^m$. The image was very unsteady; I used a diaphragm which stopped the aperture down to $0^m.077$ and, as I have often found, I could immediately continue observing under the best conditions. Burckhardt Land β, Cassini Land π, Iapygia ψ, and the Zöllner Sea τ were seen with certainty. δ is Lockyer Land. The continent Agria, ρ, very brilliant and very white. The close attention with which I examined these details introduced some doubt as to the precise positions on the disk, Nevertheless, I could say that the Zöllner Sea τ was recorded at about the mean time of the observation $5^h\ 51^m$, that of the Hourglass Sea at the beginning and that of Burckhardt Land at the end. This latter feature is therefore a little too near the western edge with respect to the mean time of the observation, and the general configuration is slightly distorted.

In these observations by Terby, the northern or lower polar cap appears double. He was able to see the upper polar cap three times, though it was very small. Note, in particular, the white border of the Herschel Strait, the Hourglass Sea and the Meridian Bay (*clouds or snows?*) and the black sea *a* on 25 November, corresponding to the Tycho Sea.

In the same year, 6 December 1879, the skilful Louvain astronomer presented a memoir to the Belgian Academy, with the aim of establishing that the canals discovered by Schiaparelli in 1877—upon which doubts had been cast by a certain number of astronomers—had already been shown on some earlier drawings, notably on those by Holden in 1875 made with the great 26-in. ($0^m.66$) equatorial at the Observatory of Washington.

In that year Holden made six drawings, of which the best have been reproduced here. I cannot, alas, share Terby's optimism with regard to the correspondence of these views with Schiaparelli's.

Terby also believed that confirmation of the canals could be obtained from the drawings by Knott and Schmidt in 1862, Gledhill, Lehardelay, Vogel and Lohse in

[35] Summer solstice for the southern hemisphere fell on 14 August 1879, and the autumnal equinox on 21 January 1880.

1871. Knobel, Lohse and Trouvelot in 1873, and Cruls and Niesten in 1877. To me, however, this correspondence does not seem absolutely certain… Far from it!

1879.—Niesten: Observations and Drawings[36]

The observations by Niesten during this opposition extended from 3 October to 26 January, and included no less than 40 drawings, made with the 6-in. ($0^m.15$) equatorial in the east turret of the Brussels Observatory. Magnification ranged from 90× to 450×. Niesten was assisted by Stuyvaert; two sketches were made in succession, and no detail was finally fixed on the drawing without having been confirmed by both the observers.

Before making their observations, the two astronomers worked out the position of the globe of Mars and what they ought to find, according to the maps by Green and Schiaparelli. They recognized several canals; others were not clearly defined, but seemed rather to be due to the demarcation of different tints covering adjacent regions. In noting the contours of certain configurations, lightly tinted with grey or grey-orange, they could identify these contours with some of the canals.

Niesten proposed to use Green's nomenclature for the seas and continents, which are stable and certain, and Schiaparelli's for the rivers and canals, which appear to be variable and uncertain. It seems to me that this proposal is entirely acceptable, and may be recommended.

Among the drawings, first note the four which are reproduced here in heliogravure, and in which the central meridian corresponds to longitudes in degrees of 67°, 150°, 250° and 330°; they therefore cover the whole of the planet.

In the first drawing, made at midnight on 29 October (Fig. 28A), note, at the centre, the Christie Bay with Hall Island. Phillips Island is not joined to the continent. Note the Terby Sea or circular lake, above. This lake is shown on the drawings made on 15 and 29 October, 25, 26 and 27 November and 19 December; Hall Island on those of 15 and 29 October, and 26 and 27 November. The north part of the shading denoted the confluence of the Nile, the Jamuna and the Ganges.

In the second drawing (Fig. 28B), of 19 December at 6^h in the evening, the Schiaparelli Sea has developed to its full extent. Its union with the Maraldi Sea produces the well-known aspect of two open wings for flight.

In the third drawing (Fig. 28C), made at 35 minutes past midnight on 9 November, Burckhardt Land, which separates the Maraldi Sea from the Hooke Sea, is at the centre. Niesten Isthmus, Webb Land, Cassini Land, and Dreyer Island are seen; a grey shading bordering the north polar cap; Œnostos and Astapus.

In the fourth drawing (Fig. 28D), of 27 October at $10^h\ 15^m$, may be recognized as the Hourglass Sea and Herschel Strait, very sinuous and bordered with white; clouds

[36] *Observations sur l'aspect physique de la planète Mars pendant l'opposition de 1879–80 (Annales de Observatoire de Bruxelles*, Vol. VIII, 1880).

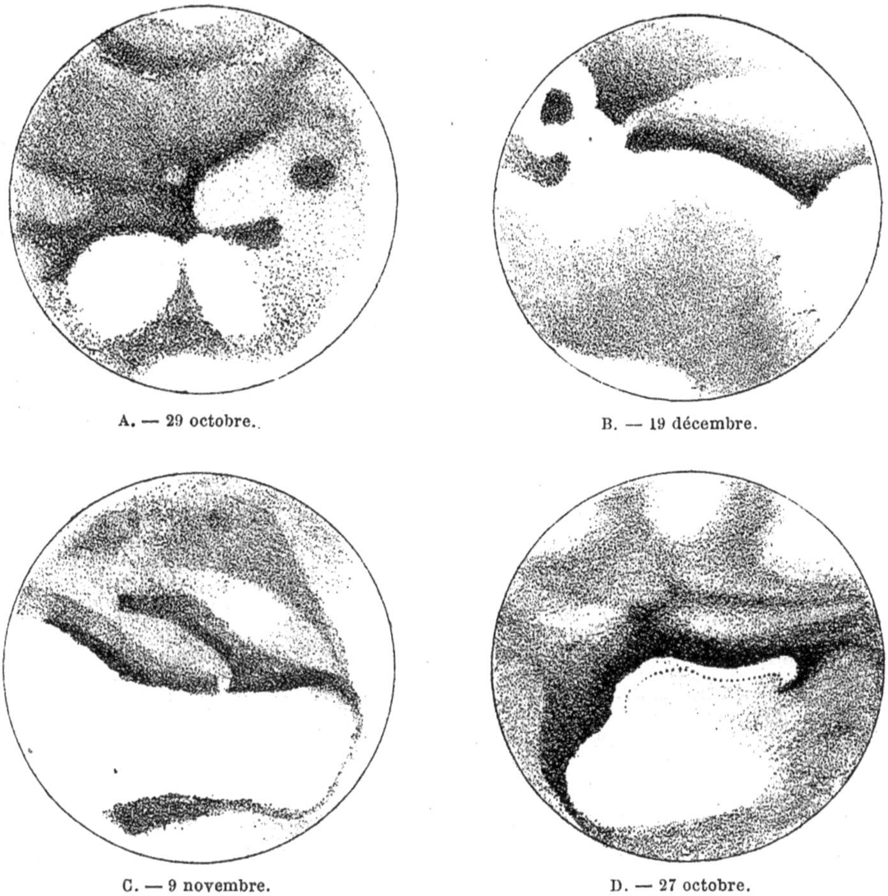

Fig. 28 Views of Mars, by M.Niesten, on 29 Ocrober, 19 December, 9 November and 27 Ocrober 1879

or snows? The same white border is visible on Terby's drawing of 29 November. The southern lands again stand out against the grey background. Between the grey shadings may be seen the trace of the Nile.

These four views give an idea of the rest. The darkest (*bluish*) patches have been Herschel Strait, Schmidt Bay, the Meridian Bay, De la Rue Ocean, Christie Bay, Terby Sea, Maraldi Sea, Flammarion Sea and the Hourglass Sea. A light grey tint extended over the flooded lands of Jacob Land. To the north of the grey patch which is made up of Herschel Strait, Maraldi Sea and Flammarion Sea can be noted a white zone. Secchi Continent, yellow-orange; Herschel Continent, yellow; Lockyer Land, yellow; Beer Continent, yellow-orange; Mädler Continent, the same; Browning Lands yellow-orange; Fontana Land and Rosse Land, white. We can recognize the streaks corresponding to the canals Chrysorrhoas, Phasis,

Agathodæmon, Ganges, Indus, Araxes, Læstrigon, Astapus, Alpheus, Nile and several others, often rather as the limits of large patches of different tints rather than as detached lines.

1879.—C.E. Burton: Observations and Drawings[37]

Barton's observations continued those which we have analyzed for 1871 and 1873. They covered a period of 3 months, from 5 October 1879 to 5 January 1880, and were made with a 6-in. (152^{mm}) Grubb equatorial refractor, an 8-in. (203^{mm}) Newtonian reflector by John Brett, and another reflector 12 in. (305^{mm}) in aperture; magnifications, 220×–514×. Burton's memoir is accompanied by 24 drawings and a map. The nomenclature used is Green's.

No comparison with the results of other observers was made before the drawings were completely finished. In my opinion, this is the best method. It prevents illusions caused by preconceived ideas.

The atmospheric conditions were generally good, and the greater altitude of the planet in 1879, compared with 1877, compensated for the greater distance and smaller diameter.

Burton thought that he could attribute almost all the differences in aspect to the varying projection of the planet, and to temporary, partial obscuration by fogs, mists or snows, depending on the seasons.

Burton was assisted in these observations by J. L. E. Dreyer, who made a certain number of drawings.

It does not seem necessary to give the drawings by Burton and Dreyer here, because they are combined into the following map (Fig. 29), contributed by the two observers. One can note some surprising features. Thus, for example, the Hourglass Sea is shown in a form to which we are not accustomed; moreover it is detached like a leg, entirely separated from the Herschel Strait. Schiaparelli Sea is not recognizable. The outline of the Terby Sea is not circular. This is almost a new map, although the background is Martian.

Several canals can be noted: (1) At the Meridian Bay, recalling Schiaparelli's Hiddekel. (2) At Burton Bay, recalling the Indus or the Hydaspes. (3) At Christie Bay, the Channel. (4) Below the Terby Sea, undoubtedly the Araxes, out by a long streak descending obliquely on my map, and which corresponds to the Pyriphlegethon. I believe I can next recognize the Gigas—but shown incomparably too broad—then the Titan and the Tartarus, but very different, even in position, from those on the Milan map; at the extremity of the Maraldi Sea, the Huggins Bay, very much turned back, joining on to the Oudemans Sea; the author identified this, perhaps overconfidently, with the Cyclops canal. Finally, toward Gruithuisen Bay, there is another canal, perhaps corresponding to the Lethes. These comparisons lead

[37] *Physical observations of Mars*, 1879–90 (*Scientific Transactions of the Royal Dublin Society*, 1880. See also *The Astronomical Register*, May 1890, p. 116.

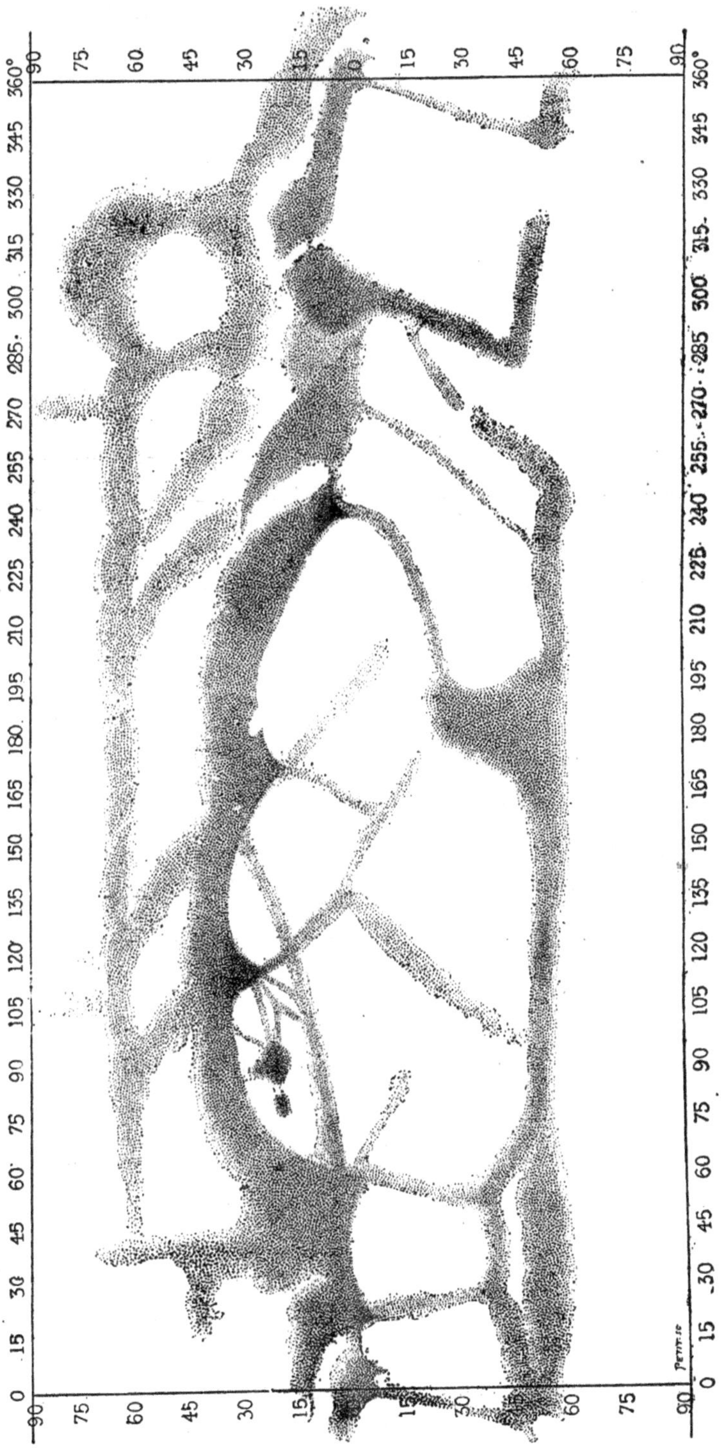

Fig. 29 Map of Mars, by Burton and Dreyer, in 1879

one to think that there has been bad seeing, and that the observer has not drawn the features where they really are, so that everything has changed in a very singular manner as though disturbed by atmospheric formations, condensations glistening on the surface of the soil! However, Burton has *clearly and certainly* seen and drawn the Huggins Bay (meridian 240°). Compare with the map (Fig. 29).

Burton thinks that the canals are identical in nature with the seas:

> I have little doubt that these canals are identical in nature with the seas, though the connection between them is occasionally singularly complex and difficult to define accurately.

The north polar snow is shown in all the observations as much more brilliant and much larger than the southern, although the north pole lay beyond the edge of the disk, on the invisible hemisphere, while on the contrary the south pole was on the near side, on the visible hemisphere. Burton estimated that on 10 December the northern snows extended over a circle 90° in diameter. The south polar cap was seen as clearly eccentric to the pole. The clouds, fogs or mists which from time to time veiled certain regions were not white like the snows, but were of the colour of the continents, i.e. yellowish, and did not stand out as white. On 5 January, the southern polar snows were no longer white or brilliant; they were yellowish, and badly defined. Perhaps this was a fading caused by covering-up by mist, since for several months the region had been exposed to the rays of the Sun.

1879 Various Observations: Dr. O. Lohse, Nicolaus Von Konkoly, E. Hartwig, Etc.

Lohse, whose earlier observations have already been described, made a new series of studies[38] from 17 September to 2 December. These studies were summarized in six drawings and a map. The map is reproduced here (Fig. 30), and bears no resemblance to its predecessors.

Among other discrepancies, the famous circular lake, which has been so familiar for so long, is not round, nor oval, but square (can this form be due to a confused view of the canals which lead there?). The Hourglass Sea is almost cut off in the middle of its length, and is crossed by a white furrow; Herschel Strait and the double Meridian Bay represent nearly all of the rest of the general configuration, with Lockyer Land, which can be made out above the shortened Hourglass Sea.

These observations were made, at the Potsdam Observatory, with the great 298^{mm} equatorial. Magnification from 120× to 350×.

M. de Konkoly made a certain number of observations of Mars during the same opposition, and published three drawings, made on 19 October at $11^h\ 20^m$, 29 October at $9^h\ 40^m$ and 13 November at $8^h\ 56^m$ (Fig. 31). The last is very vague, and features on it are difficult to recognize. The first two—particularly the second—are

[38] *Beobachtungen und untersuchungen über die physische Beschaffenheit der planeten Jupiter und Mars* (*Publ. des astr. Observ. zu Potsdam*, Vol. IX, 1882).

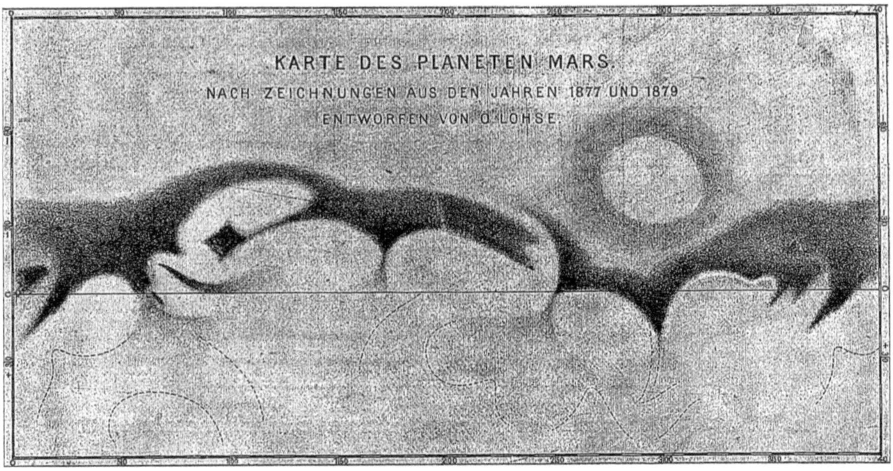

Fig. 30 Map of Mars, by M. Lohse, in 1879

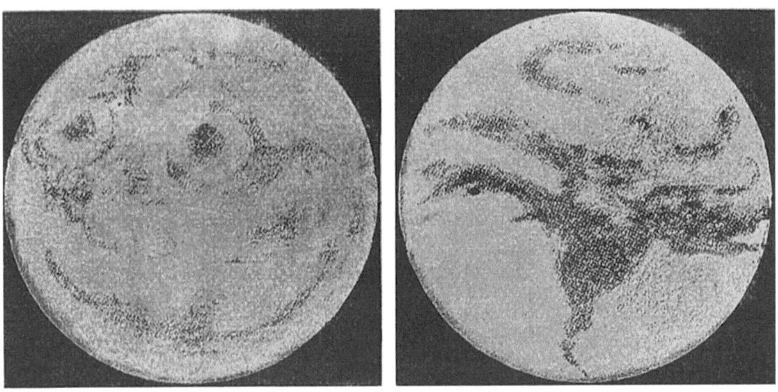

Fig. 31 Sketches of Mars, by M. de Konkoly, on 19 and 29 October 1879

better, and are reproduced here by photogravure. The observations were made with a 6-in. (152^{mm}) refractor, with a magnification of 216×.

On the first, the circular Terby Sea can be recognized, though doubled in some way. Evidently we have here a large personal equation. The second shows the Hourglass Sea and all the adjacent configurations.

M. E. Hartwig made new micrometrical measurements of the diameter of Mars.[39] Combined with those of Arago, Bessel and Main, these measurements give:

Polar diameter = 9″.349.—Ellipticity doubtful.

[39] *Astronomische Gesellschaft*. Lepizig, 1879.

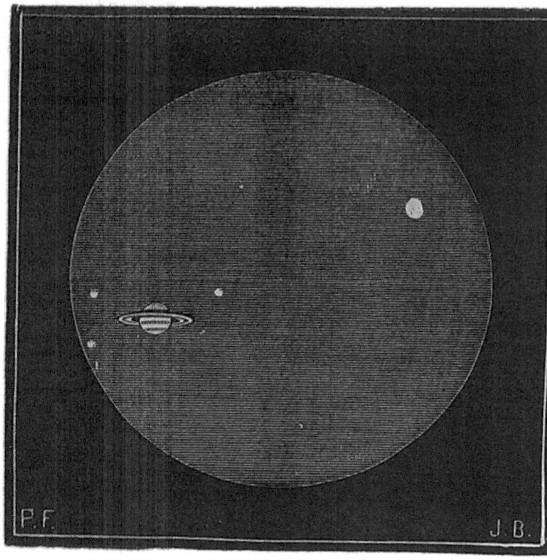

Fig. 32 Conjunction of Mars with Saturn, on 30 June 1879

Hartwig's measurements:

Polar diameter	9″.311
Equatorial diameter	9″.519
Flattening (combined with the results of Encke and Galle):	1/96

On 30 June 1879, Mars and Saturn were close together in the sky (Fig. 32); their minimum distance was reduced to 87″ from centre to centre. The phenomenon was studied by a great number of observers *quorum pars minima fui*. The main comment with regard to the conjunction was the striking contrast between the *red* colour of Mars and the leaden tone of Saturn, which appeared green by contrast. The aspect of these two bodies in the same telescopic field was indeed marvellous.

The south and north polar caps of Mars were of a brilliant whiteness.

There had previously been a conjunction of the two planets on 3 November 1877, but it had not been so close, and the two had never been in the same telescopic field.

The closest approach was at about $7^h\,30^m$ in the evening.

1879.—J.-H. Schmick: Studies of Mars[40]

The author of this little work aimed to summarize the observations made by Schiaparelli in 1877. His work was divided into six chapters, entitled:

[40] *Der Planet Mars, eine zweite Erde, nach Schiaparelli*. Leipzig, 1879.

I. The planet Mars considered as a member of the family of worlds of the Solar System.
II. The observation of Mars in astronomical instruments.
III. What specialized observation of Mars reveals:—(a) from the viewpoint of snowy regions.—(b) dark regions.—(c) the luminous ring which surrounds the disk.
IV. Results of the study of Mars with regard to the solid surface of the planet.
V. What does knowledge of Mars tell us about the evolution of the Earth?
VI. Continuation of the same subject.

Chapter I describes the cosmography of Mars and its movements, real and apparent.

Chapter II is devoted to the patches, the rotation, the snows and the atmosphere. In Chapter III, Schmick reviews the variations in the polar caps, and the effects of the seasons, describing the results of the observations; principally those of Schiaparelli. The dark regions are regarded as seas.

The bright patches or continents form the subject of the following chapter. The author is of the opinion that the planet has lost part of its water, and he comments upon the preponderance of continents in the northern hemisphere. He thinks that the north pole contains a basin, similar to that of the south pole, but smaller.

Finally, in the last two chapters, Schmick considers whether the observations made up to now are favourable to the theory of d'Adhémar and Croll with regard to the Ice Ages, and concludes that they are not.

1879.—C. Flammarion: Studies of Mars[41]

In describing our knowledge of Mars in my *Astronomie populaire*, and in particular returning to the arguments already put forward in *Terres du Ciel*, I insist, as a characteristic of the constitution of this neighbour world, upon the differences in tone between the seas, and upon the variations in these tones, as well as certain changes in form and aspect which seem to me since that time to have been established with certainty, by observation. It has been noted, not without curiosity, that some astronomers who have not spoken of what we know, criticize with complete disregard of the results acquired by our specialized study of the planet. Among the most variable regions of Mars we can list, above all, the Herschel II. Strait and the Terby Sea. Here is an extract from this chapter.

> Another difference from the Earth is shown by the variability of the geographical configurations. The constant study of Herschel II Strait leads us on to very curious results in this respect. In 1830, Mädler often drew it very clearly and very distinctly, as shown previously (Fig. 33). In 1862 Lockyer saw it equally clearly, as in the next drawing, and in 1877 Schiaparelli observed it as shown in the third drawing. This feature, seen round, black and clear in 1830, so clear indeed that Mädler chose it as the zero for Martian longitudes as

[41] *Astronomie populaire*, 1st edition, 1879, p. 484.

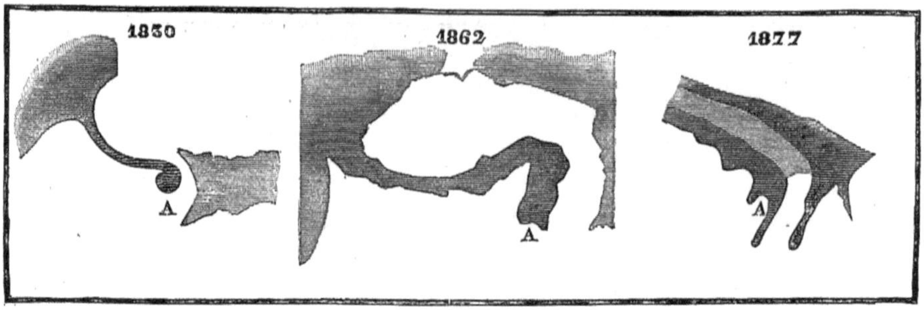

Fig. 33 Variations in the seas of Mars. The Herschel II. Strait in 1830, 1862 and 1877

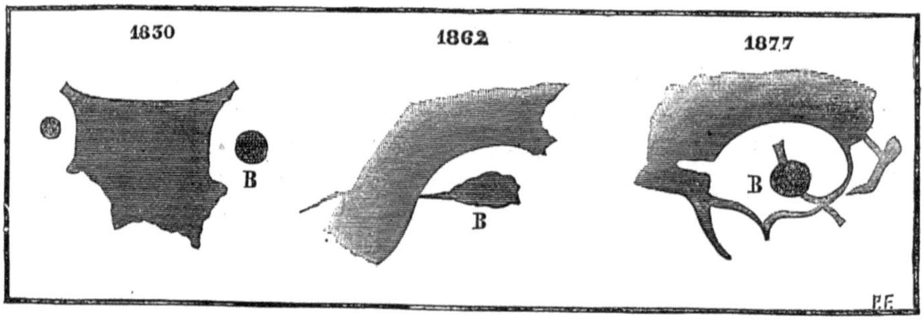

Fig. 34 Variations in the seas of Mars. The circular sea of Terby in 1830, 1862 and 1877

being the blackest region on the planet, and seen earlier in the same form by Kunowsky in 1821 and also indicated as a black globule by Schröter in 1798, could not be distinguished by Secchi in 1858, even though he made a special search for it. The same feature was seen bifurcated by Dawes in 1864, and this is certainly correct; but the region adjoining it to the south appears to be covered with swamp and to be variable in aspect over the years; all the drawings made in 1877 show it as a black disk suspended from a wavy thread, but this thread is broadened to a degree which makes all comparisons difficult; the gulf is as broad at the centre and at its origin as at its eastern extremity.

Actually, the blackest and clearest patch, which I would have chosen in preference as the zero for Martian longitudes, would have been the circular Terby Sea; I would certainly have given it preference over the Meridian Bay. In 1830, preference was however given to the Meridian Bay, and on several drawings of the two may be seen similarly to either side of the ocean (Fig. 34). These drawings do not agree with the aspect now.

This gives us a first variation. The second is demonstrated even by the appearance of the patch; in 1862 different observers saw it elongated from east to west; in 1877, on the other hand, it was perfectly round—allowance being made for perspective—and certainly not elongated in the former sense. A third variation appeared in 1862; it appeared to be joined to the neighbouring ocean by a strait, but in 1877 instruments of the same power and observers of equal skill could see nothing of this strait, though they distinguished another to the north-east. And here is yet another example of variability: in 1862 and 1864, excellent observers saw a luminous point in the De la Rue Ocean which seemed to be due to an island covered with snow, and which I believe ought to have been shown on my first map. Nobody has ever seen it since then.

Undoubtedly we cannot regard all the differences between observers as indicating real change. Thus, for example, in 1877 several observers saw the Hooke and Maraldi Seas joined toward the west, while the separation between the two remained visible to others; eyes are different, and there are bound to be discrepancies. But when attention is specifically focused upon certain special points which should be easily visible in the instruments used—and even then there are differences which appear incompatible with errors of observation—the probability swings in favour of the effective reality of the changes reported.

Of what nature are these variations?—this is what we must try to decide in the future. We can only make vague conjectures in this respect. But whatever they may be, they do not mean that the principal configurations of Martian geography are anything but permanent and consequently real; they appear just as our forebears saw them and drew them more than two centuries ago.

Another comment is no less interesting. This neighbour planet seems to have much less cloud than our own world.

There is in this respect a great contrast with our globe, because there are years when we have true water-layers. In one full year, from August 1878 to August 1879, we had in Paris 167 days when it rained, and only 37 days of clear or only slightly cloudy sky. Only 37 days were made for astronomers! On the southern hemisphere of Mars, the situation was quite the opposite during 1877; the surface features were clearly visible whenever our own weather was fine. We must not indeed forget that for the observation of Martian geography to be possible, two conditions are essential: we must have fine weather, with a clear sky, but it must also be fine on Mars, as otherwise we would be no better able to penetrate its layer of cloud than we could go up in a balloon and then see villages lying beneath a layer of cloud. Well, it is remarkable that on Mars, nine complete months passed by almost without any clouds at all, permitting us to perfect the geographical knowledge which we wanted to obtain of this neighbour world.

We found ourselves in September and October 1877 in the middle of summer in the southern hemisphere of Mars, then sharply inclined toward us, and in the middle of winter in the northern hemisphere, turned in the other direction. All the clouds seemed to be restricted to this latter hemisphere. On this globe, even more than on our own, summer is the season of clear atmosphere, and winter that of bad weather. The permanent patches show up strongly, vivid and clear, during summer in the hemisphere in which they lie; winter comes, they become vague, confused and feeble; this is, without doubt, because the atmosphere of Mars becomes disturbed in winter and stays very transparent in summer. Also, a preference can be noted for the clouds to form over the swamps and shallow seas tinted grey on my map, rather than over the shaded deep seas, and this is what holds back the precise knowledge which we want to gain with regard to the country lying above the Herschel II. Strait, but we cannot regard these areas as constantly cloudy and rainy, analogous to the zones in our temperate regions where it rains throughout the year.

As for the thickness of the atmosphere relative to the disk of the planet, it is inevitably too thin to be visible from here, even if it were much higher than our own. Assuming a height of 80 km, its thickness would amount to only 0″.3 when the planet is at its closest to us; therefore its refraction effects would be too slight to be noticed.

Such was, in 1879, the general trend of ideas with which I conducted my studies of Mars. In the same year, the astronomical review *The Observatory*, edited by Christie, astronomer at the Greenwich Observatory of which he is now Director, took care to temper the view which classed the dark patches of Mars as seas and the bright regions as continents:

> These so-called seas, are they not just as imaginary as in the case of the Moon? The continents appear rather rounded like our seas, while the seas reproduce the pointed forms of our continents. We ignore the real character of these configurations, which would be more scientifically called simply patches, as with Jupiter.[42]

We see that all astronomers have their own ideas. To me, there seems, however, no doubt that the dark patches represent waters on Mars. In the book in which I have discussed this question (*Astronomie populaire*), I have published a second map of Mars.[43]

1879.—Huggins: Photography of the Spectrum of Mars[44]

The learned doctor has succeeded in obtaining direct photographs of the spectra of the principal stars as well as of the planets Mars, Jupiter and Venus. The slit of the apparatus with which these photographs were taken has two shutters, so that an object can be photographed on a plate with one shutter closed; a second spectrum for comparison purposes, can then be put on to the same plate. This second spectrum might be that of the Sun, the Moon, a known star, or a terrestrial element.

In the photography of planetary spectra, Huggins worked before sunset, so that he could obtain first the spectrum of the sky and then that of the planet. By this method, all differences between the light of the planet and that of the sky would be recognizable. He obtained the spectra of Mars, Jupiter and Venus, but found no difference or modification of the solar spectrum in the region between the G and O lines.

The same procedure, applied to small regions of the Moon's surface, has revealed no trace of a lunar atmosphere.

---

[42] *The Observatory*, November 1878, p. 209.

[43] The following details may interest readers of this book who are particularly concerned with the cartography of Mars. From all the observations made after the chart drawn by Proctor from Dawes's observations alone, I constructed my first geographical map of the planet in 1876, I am not here discussing the 1862 sketch, reproduced here, which represents one hemisphere and cannot be regarded as a true map. It was published in the first edition of *Les Terres du Ciel*, 1876, p. 424, and in the *Comptes rendus de l'Académie des Sciences*, 27 August 1877, p. 478. I gave a second map in 1879, *Astronomie populaire*, 1st edition, p. 380; a third in 1882, *Revue d'Astronomie*, 1882 pp. 170–171; a fourth in 1884, *Les Terres du Ciel*, enlarged edition frontispiece. In the same year I published a globe of Mars including the most well-authenticated details. Finally, in 1889 I constructed my last map (*Astronomie populaire*, 10th edition) also published here; it included corrections made according to the most recent observations, and in which I adopted Green's nomenclature. It is not without regret that I have abandoned the nomenclature of the old charts which had been dignified by the history of astronomy and its apostles: Copernicus, Galileo, Kepler, Newton, Huygens, Cassini, Hooke, Maraldi, Lacaille, Lalande, Laplace, Lagrange, Herschel, Schröter, Beer, Mädler, Arago, Secchi, Le Verrier, Faye, etc. But since Green's chart is generally adopted, and Schiaparelli's used with it, it seemed to me important to avoid all complications and all confusion in a subject which is moreover too remote to be completely clear, and so it is preferable to leave only two systems of nomenclature, those of Green and of Schiaparelli.

[44] *The Observatory*, February 1880, p. 295.

1879.—Schiaparelli: New Observations[45]

During the opposition of 1879, the industrious Milan astronomer continued the series of studies undertaken in 1877 and summarized above. The line of work was the same, and the same instrument was used.

These new observations extended from 30 September 1879 to the end of March 1880. The cold weather this winter made the air calm and transparent, yielding excellent images.

To help make conditions as favourable as possible, Schiaparelli came to illuminate the telescope field strongly, thereby suppressing the false effects of contrast between the brilliance of the planet and the adjacent darkness, as well as the change from a dark field to the brightness of the paper upon which the drawings had to be made. Secondly, he did not keep his eye at the eyepiece for longer than was necessary for the best view, and he rested it from time to time, so enabling him to work for several consecutive hours when the atmospheric conditions were really good. Finally, he found it advantageous to place a yellow-red glass in front of the eyepiece. The objective is perfectly achromatic for the red rays, which is an advantage when making observations of Mars.

These new observations completed and modified those of 1877.

The apparent diameter of the planet did not attain 25″, as in 1877, but at maximum 19″3, at the time of opposition (12 November). It decreased to 5″.9 at the end of March. The inclination of the planet was 9°5′ at the beginning of the observations, 18°5′ at its maximum (20 December) and only 3° at the end of March.

Position of the South Polar Cap.—Schiaparelli made 89 observations of the position of this patch, from 30 September to 2 December. The result of these observations, combined with those which he had obtained for 1877, gave, for the projection of the north pole of Mars on to the celestial sphere, the following position (equinox 1880):

RA=318°7′.8; Dec. + 53°37′.1.

The inclination of the Martian equator according to this position is:

To the orbit of Mars	24°52′.0
To the orbit of the Earth	26°20′
To the terrestrial equator	36°22′.0

For the inclination of the axis of Mars to the plane of its orbit, William Herschel had found 24°42′ and Bessel 27°16′.

[45] *Osservazioni astronomiche e fisiche sull'asse di rotazione e sulla topografia del pianete Marte Reale Accademia dei Lincei*. Memoria seconda. 1 vol. large in octavo, 110 pages and plates. Roma, 1881.

According to this new measurement by Schiaparelli, the inclination is only 24°52′, which still produces the same climatological and seasonal conditions for Mars as for the Earth.

The planet passes its perihelion when its heliocentric longitude is 333°49′, and the southern solstice when this longitude is 356°48′. The time-interval from the former to the latter is 36 days.

The dates of the solstices should be put back 8 days from the old determinations, becoming:

Date of southern solstice	
1830	18 September
1862	9 September
1877	26 September

Areographic Triangulation of Fundamental Points

The 1877 micrometrical measurements have been given above. Recommencing these measurements and comparing them with the preceding ones, Schiaparelli obtained the results given in the following tables which were much more accurate than those of 1877.

No.	Name	Longitude°	Distance from So.pole°	No.of obs'ns
1	Vertice d'Aryn	0.92	90.98	8
2	Secondo corno del Golfo Sabeo	4.49	95.77	6
4	Canale di Deucalione punto di mezzo	11.91	86.92	3,4
5	Golfo delle Perle bocca dell'Indo	22.07	95.80	7
5a	Divisione dell'Indo e dell'Oxo	14.06	112.10	3,2
5b	Bocca del Gehon nel Nilo	10.76	120.13	3,2
5c	Bocca del'Indo nel Nilo	27.33	125.63	1
6	Bocca dell'Idaspe nel Golfo delle Perle	24.44	88.26	3
6a	Corno d'Oro	19.31	90.75	1
6b	Bocca del'Indo nel Nilo	34.92	120.86	3
7	Capo degli Aromi	38.66	80.05	7
7a	Bocca della Jamuna nel Golfo dell'Aurora	51.40	82.45	3
7b	Bocca della Jamuna nel Lago Niliaco	41.74	112.77	3
8	Capo delle Ore in Argyre	40.40	50.20	8
8a	Centro d'Argyre	29.12	43.17	2
8b	Canale fra Argyre e Noachide (2°)	15.77	42.73	1
9	Capo delle Grazie in Argyre	47.10	38.78	7
9a	Centro d'Argyre II	60.86	24.08	2,5
10	Golfo dell'Aurora bocca del Gange	56.20	84.58	7
11	Punta dell'Aurea Cherso	60.90	var.	6

(continued)

(continued)

No.	Name	Longitude°	Distance from So.pole°	No.of obs'ns
11a	Bocca dell'Agatodemone	66.12	var.	5
12	Bocca del Nettare nel Mar Eritreo	66.42	67.94	6
12a	Confluente dell'Agatodemone e del Chrysorrhoas	78.57	80.38	5
12b	Confluence de Agatodemone e del Fídella Fortuna	82.38	77.35	1
13	Lago della Luna centro	65.98	117.00	7
13a	Isola Sacra	68.04	118.13	1
13b	Golfo Ceraunio parte australe	97.27	119.45	6
14	Lago del Sole centro	90.87	67.00	12
14a	Bocca dell'Ambrosia nel Mare Australe	89.05	44.74	5
15	Lago dellíFenice centro	107.94	73.76	8
16	Bocca del Fasi	111.70	51.47	6
16a	Divisione del Fasi e dell'Arasse	112.09	65.76	1
17	Colonne d'Ercole bocca esterna	124.33	42.89	8
18	Centro d'Icaria	120.89	52.80	6
19	Bocca de Arasse nel Mare dell Sirene	129.15	61.14	9,8
19a	Bocca australe del canale dell Sirene	131.80	59.98	1
19b	Confluente del canale dell Sirene I con Eosforo II	130.97	77.27	1
19c	Neve Olimpica, 1879	129.41	110.63	6
19d	Canale Flegetonte mezzo	127.26	122.31	1
20	Primo punti di Thyle I	141.81	28.80	3
21	Colonne d'Ercole bocca interna	135.87	52.08	7,5
21a	Bocca del Termodonte nel Mare dell Sirene	139.74	54.04	1
21b	Bocca del Termodonte nel Mare Cronio	137.83	38.15	1
22	Centro di Thyle I	158.57	29.58	2
23	Base australe di Atlantide I	156.00	53.64	4
24	Bocca del Simoenta nel Mare Cimmerio	168.70	52.09	5
24a	Primo punto del Mare Cimmerio	161.45	50.37	4
24b	Bocca del Simoenta del Mare Cronio	172.46	37.28	5
25	Golfo del Titani	170.17	70.67	8
25a	Bocca del Fídell Gorgoni nel Mare Sirene	152.08	59.01	3,4
25b	Punto dell'Erebo	162.96	143.84	2
26	Ultimo punto del Mare delle Sirene	175.80	63.25	8
26a	Base inferiore d'Atlantide I	180.39	60.58	5
27	Stretto d'Ulisse mezzo	189.36	24.27	2
29	Principio della palude Stigia	198.74	108.01	7
29a	Bocca della palude Stigio nel Mare Boreale	306.83	131.65	5
29b	Capo di Buona Speranza	205.73	128.98	1
30	Bocca del canale dei Lestrigoni nel Mare Cimmerio	199.95	68.49	11
30a	Golfo dei Lestrigoni, 1879	183.55	60.09	4
30b	Base d'Atlantide II, 1879	187.92	39.64	4
31	Bocca dello Scamandro sul Mare Cronio	203.43	35.53	5,1

(continued)

(continued)

No.	Name	Longitude°	Distance from So.pole°	No.of obs'ns
31a	Bocca dello Scamandro sul Mare Cimmerio	202.66	51.31	1
31	Base australe d'Esperia	213.68	52.26	3,2
35	Capo boreale di Thyle II	221.51	30.72	1
36	Centro di Thyle II	223.43	23.07	1
36a	Ultimo punto di Thyle II	242.66	12.02	1
37	Golfo e bocca del canale dei Ciclopi, 1877	224.64	80.57	1,2
37a	Golfo di Ciclopi, 1879	229.81	76.47	2,3
37b	Bocca del canale dei Ciclopi, 1879	223.50	73.48	4
38	Primo punto del Mare Tirreno	227.60	53.67	7
40	Bocca del Xanto nel Golfo di Prometeo	234.89	38.04	6,5
40a	Boaca del Xanto nel Mare Tirreno	236.17	50.91	4
41	Ultimo punto del Mare Cimmerio	239.97	79.48	6
41a	Bocca del canale degli Etiopi nell Eunosto	242.49	121.92	2
42	Base settentrionale d'Esperia	249.23	79.48	5,4
43	Piccolo Sirte	257.29	83.29	8
43a	Bocca del Golfo Alcionio nel Mare Boreale	225.09	132.90	1
44	Capo Circeo in Ausonia	266.72	74.90	8
45a	Primo punto dell'Adria	264.49	48.51	4
46	Lago Tritone	265.21	106.32	8,8
46a	Neve Atlantica	269.00	107.06	2
46b	Bocca del Thoth nel Golfo Alcionio	261.23	124.15	1
46c	Bocca dell'Eunosto nel Golfo Alcionio	258.13	–	1
47	Primo punto dell'Ellade, 1877	270.48	43.41	3,2
47a	Bocca del Peneo nel Mare Adriatico, 1879	280.14	46.49	5
48	Lago Meride	275.72	95.57	2,1
49	Biforcazione d' Ausonia	278.90	74.77	4
50	Congiunzione del Nepente col Nilo, 1877	286.15	121.26	1
50a	Bocca del Nepente nella Gran Sirte, 1879	285.03	100.16	3
51	Gran Sirte et bocca del Nilo, 1877	290.31	110.09	6,5
5la	Punta australe di Merso, 1879	290.98	105.16	5,4
51b	Bocca dell'Astabora nella Gran Sirte	299.16	101.91	1
51c	Divisione del Nilo e dell'Astapo	281.72	128.34	3,4
52	Bocca australe dell'Alfeo	–	35.01	0,1
53	Centro o croce dell'Ellade	294.93	46.38	7
54	Bocca settentrionale dell'Alfeo	298.91	61.76	1,3
55	Ultimo punto del Mare Tirreno	297.37	96.70	6,5
56	Ultimo punto dell'Ellade, 1877	314.97	48.92	4,3
56a	Bocca del Peneo del Mare Australe	315.99	44.95	3
57	Corne d'Ammone	317.99	81.21	13,1
57a	Bocca del Tifone nella Gran Sirte	306.26	94.99	1
57b	Palude Sirbonide	327.22	104.01	2
58	Scilla e Cariddi	323.46	69.75	3

(continued)

(continued)

No.	Name	Longitude°	Distance from So.pole°	No.of obs'ns
59a	Novissima Thyle	355.10	19.25	1
60a	Centro dello Noachide	344.93	53.24	3
61	Bocca del Phison nel Golfo Sabeo	336.28	85.24	6,5
61a	Uscita del Phison della palude Coloe	302.52	128.29	3,4
61b	Ingresso del Nilo nella palude Coloe	303.59	134.29	2,1
61c	Bocca dell'Eufrate nel Golfo Sabeo	337.88	82.74	3
61d	Divisione deíEufrate e dell' Oronte	336.93	105.60	2
61e	Bocca dell'Eufrate nel Nilo	334.30	132.38	4
62	Prima corno del Golfo Sabeo	357.17	96.30	8
62a	Golfo di Edoni	345.24	85.98	2
62b	Canale ed istimo di Xisutro (mezzo)	347.90	81.48	4

There are 114 points, determined by 482 observations.

It is seen from the Table that the Meridian Bay, called Aryn by Schiaparelli, is, as we have seen, at longitude 0°.92 instead of 0°.00. If therefore we want to relate all the other positions to the prime meridian, we must subtract 0.92 from their longitudes.

From these positions, and with the help of 30 complete disks and 104 partial sketches obtained during the period, the eminent observer constructed the map reproduced here (Fig. 35), drawn not only for the features but also for tones, and whose object it was to give an improved representation of the Martian surface features. This map is more extended than the first; instead of stopping at latitude 40°N it goes as far as 60°. We will now examine it in detail.

First, with regard to the nomenclature of seas, lands, rivers, canals, gulfs and lakes, the author appears to regret that in his previous memoir he gave the impression that these terms could be taken literally. In fact, they mean no more than in the case of the Moon. They serve to indicate observations: that is all. Searching for what they represent is to abandon science and enter into hypotheses, and this the author declines to do. This declaration must be accepted, not without regret. However, it is fair to add that the role of observer does indeed stop strictly there. That of the researcher goes further, and involves not only making observations but also drawing conclusions from them.

As for the nomenclature, Schiaparelli continues that which he had adopted, extending it for newly-formed features.

The general results obtained, fall into three categories. First, all the aspects observed in 1877 have been seen again, even more minutely, with the exception of the Hiddekel Canal and the Juventæ Fons. One may believe that the Hiddekel has been distinguished, but with confusion and uncertainty. The Juenesse Fons was seen only once in 1877; in spite of all the searches made for it, nothing was seen of it in 1879. These two features are therefore omitted from the new map.

As we must note, a second result which modifies but does not destroy the previous one is that the configurations observed, while remaining basically the same,

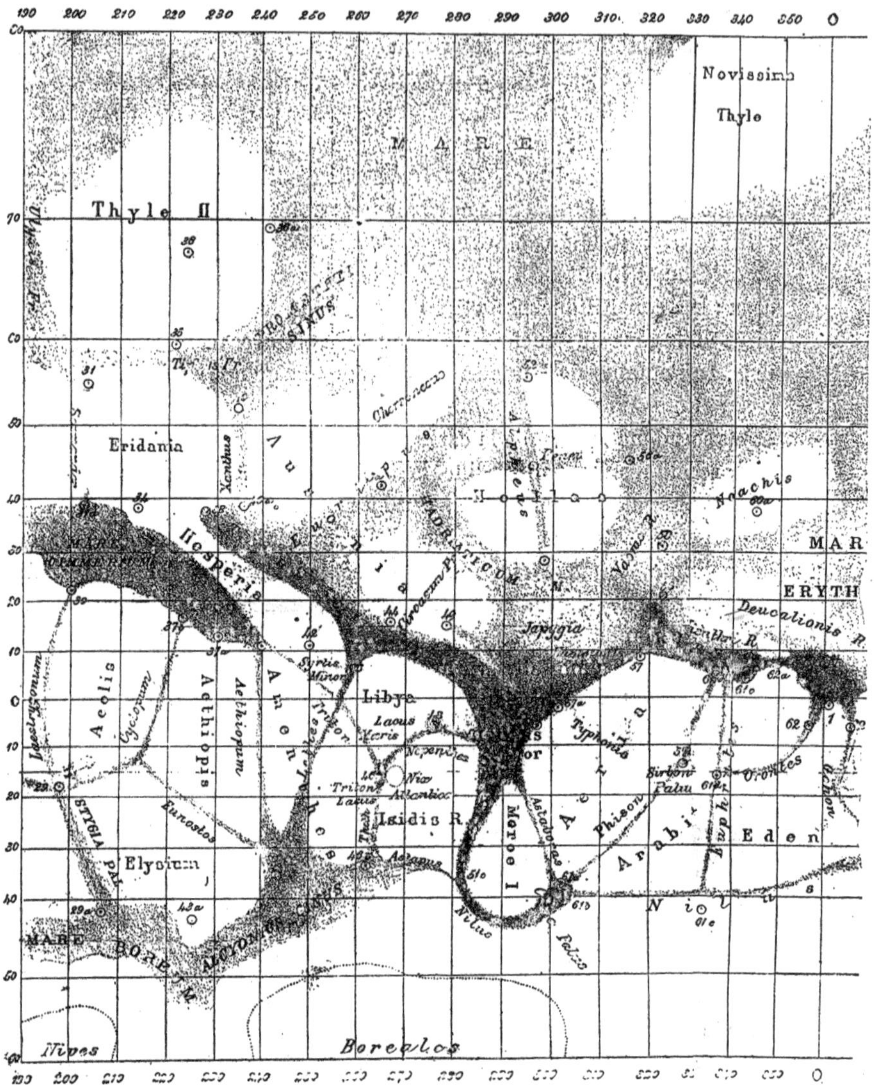

Fig. 35 Schiaparelli's second chart of Mars, based on his observations of 1879

have for the most part changed in aspect, tone, degree of visibility and even—for several canals—breadth. The inclination of the planet to our line of sight, and to the sunlight, can explain the difference of visibility and whiteness of certain regions, such as Argyre and Hellas. Perhaps, also, local causes can result in variations in the degree of whiteness of certain areas. Many of these variations in tone from yellow to either white or gray are real; and Schiaparelli thinks it probable that they are periodical. These variations, he says, are probably the key which will lead to our solving the secrets of the physical constitution of the planet.

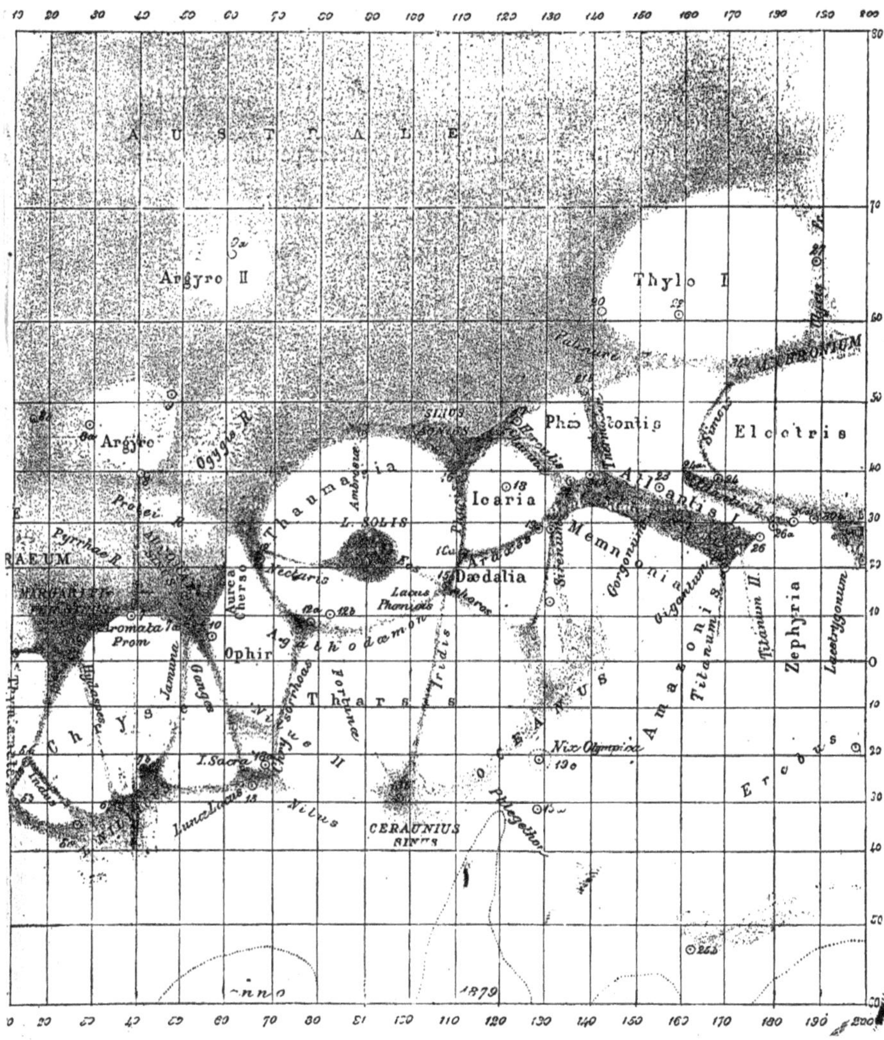

Fig.35 (continued)

The third result has been to confirm observations made by Dawes in 1864, who, among others, recovered the *Bessel Pass* which corresponds to the vertical canal drawn to the right of the Solis Lacus, by Phasis and Iris. This canal, Iris, does not appear on the 1877 map. Change?

Among the details explored we must note the Hydaspes, which Schiaparelli is led to identify with the *Franklin Canal* described by Secchi, at first identified with the Ganges. Such is also the opinion of Dr. Terby.

The Solis Lacus looks a little like the shape of a pear, because of the junction with the Nectar canal, already drawn by Mädler in 1830, Kaiser and Lockyer in

Fig. 36 The Araxes, on 25 September 1877

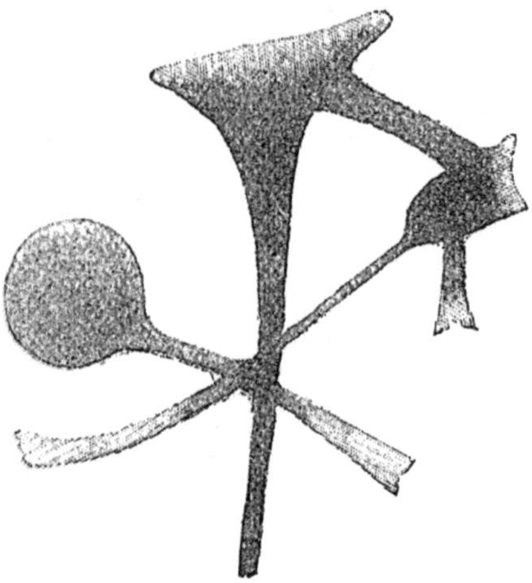

Fig. 37 The Araxes, on 22 December 1879

1862, Dawes in 1864, etc., but absolutely invisible in 1877. This narrow, light canal makes an angle of 15°–20° with the parallel of latitude.

This new display of real variation, absolutely incontestable, confirms what has been said above (Fig. 34). The canal Ambrosia is shown as thin and black during all the observations of 1879, while it was shown as broad and grey during all those of 1877: another example of *certain variation*.

On the 1877 chart, we have seen that the Araxes is separated from the Phasis, to enter the Mare Sirenum by a sinuous course, drawn by Schiaparelli with special care. This course was no longer the same in 1879. Instead of being curved, it was almost straight, and lightly concave toward Icaria. Compare the drawing of 11 November 1879 (Fig. 39) or that of 22 December of the same year (Fig. 37) with that of 25 September 1877 (Fig. 36). Between 1877 and 1879 the course of the river was therefore modified toward the Mare Sirenum. Schiaparelli stressed the precision of his drawings, these evenings having been particularly excellent and the views perfectly clear (Figs. 36, 37, 38 and 39).

Fig. 38 26 December 1879, at 4^h 47^m

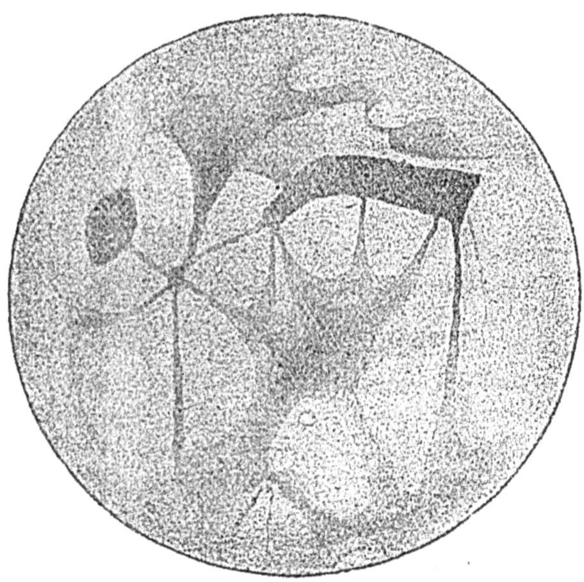

Fig. 39 Mars, on 11 November 1879, at 5^h 35^m

Course modified? How? By what agency? This is what we must try to find out. But the facts are there.

The Lacus Phoenicis shows changes which are equally definite, as can be seen by a comparison of the drawings.

But the most important discovery made in these regions during the opposition of 1879 was perhaps that of the Iris canal (see map, at longitude 105°), which was

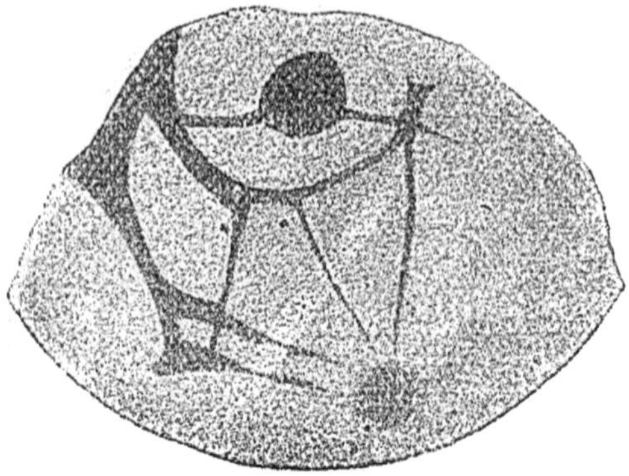

Fig. 40 A white streak crossing the Fortuna canal and the two Niles

absent in 1877. The first observation of it was made on 11 November 1879 (Fig. 39) and is represented on the drawing above. This is the vertical canal which can be seen to the left of the disk, descending from the little lake situated to the right and below the Terby Sea or Solis Lacus. It was clear and black. It was observed until 26 December, when it was less black but broader. As we have seen above, Schiaparelli thinks that it can be identified with the *Bessel Pass* on Proctor's chart.

During the observations of 1879, the region of Tharsis several times showed short-lived white veils which, if not due to some meteorological precipitation or efflorescence, must have been caused by some atmospheric disturbance. Analogous white veils have been seen in a great number of other regions. They are not always as bright as the polar snows, *molti inferiore a quello delle non polaro.* On 26 December, with calm, clear atmosphere, Schiaparelli discovered a white streak from 8° to 10° broad, crossing the region of Tharsis from Lacus Phoenicis to the double Nile and appearing to join on to the remnant of the polar snows (indicated by stippling on the chart, near latitude 65°). As the view was excellent, Schiaparelli tried hard to see whether this white streak passed over the Niles and interrupted them, or whether it was interrupted by them, as if the streak were melted snow. The Niles were not interrupted, but their breadth was noticeably diminished, and was reduced to the state of two almost imperceptible threads, as shown on the drawing (Fig. 40).

Schiaparelli made no attempt to explain this phenomenon.

It is difficult, however, not to ask oneself if there could not have been a streak made up of fairly light snow, inasmuch as it appeared as a kind of extension of the north polar snow. If the streak were snow and the Nile water-filled, the snow would be completely melted, and the streak would be cut by the Nile. Would it not have melted only in the centre of its course? Or would its brightness be diminished along its breath, by irradiation? Or, even, could this water not be of the same kind as ours? But spectral analysis seems to prove that it is identical. Evidently these are some of the questions which we must ask.

During 1879, especially in March, Argyre I and Argyre II were brilliantly white, rivalling the polar snow—just as if they were snow-covered.

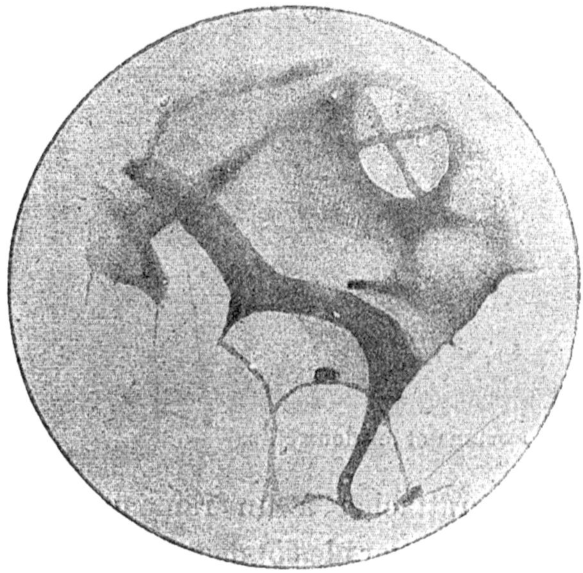

Fig. 41 Mars, on 28 October 1879, at $7^h\ 0^m$

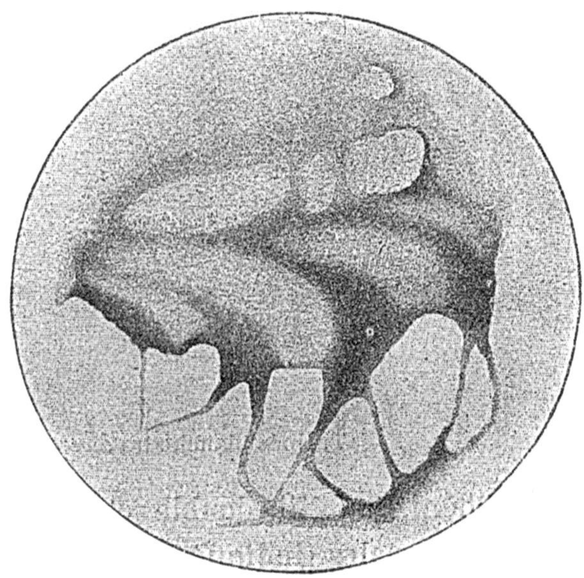

Fig. 42 Mars, on 28 November 1879, at $8^h\ 40^m$

To the left of Deucalionis Regio, Schiaparelli again saw the serpentine form drawn by the old observers, notably by Lockyer in 1862, Kaiser in the same year. This aspect, absent from the 1877 map, is recognizable on that of 1879.

On the same chart, at point 62a, the Herschel Strait is narrowed by a remarkable constriction.

At longitude 129°, below a grey streak which Schiaparelli has named Ocean, may be noted a small circle called by him Nix Olympica—the *Olympic snow*.

This is a white point, which has been observed nine times. It is very small (half an arc sec.), but also white like the polar snow. It is seen, represented rather higher, on the drawing made on 11 November (Figs. 41 and 42).

The Læstrygon canal, at longitude 200°, does not seem to have changed in position, but its estuary, which in 1877 flowed out at the extremity of the sea separated from the Mare Cimmerium by Atlantis II, in 1879 led into the Mare Cimmerium itself. Its banks and beaches seem to have been subject to considerable variation; compare the two charts. Between 21 October 1877 and 10 November 1879, the Læstrygon passed from the left of Atlantis II to the right, following the changes in its shores!

Such changes are of paramount importance in our studies of the true nature of the dark and bright patches on the planet.

The Simoïs, very difficult to see in 1877, was one of the most prominent of the canals in 1879. The curvature remained the same.

The peninsula of Hesperia was less easy to observe in 1879 than it had been 1877. It appeared lightly shaded.

The canals of Æthiopis and Lethe were perfect visible, though in 1877 they had been seen only with difficulty.

As the map indicates, both must surely flow out on to the Gulf of the Alcyons, as does the Eunostos; this streak is the correct one, while that shown on the first chart is not certain.

However, it will be found, on the chart to be described later, that the form of Sinus Alcyonus is by no means stable.

Syrtis Minor, or Gruithuisen Bay, appeared on 1 October 1879 to be separated from the Hourglass Sea or Syrtis Major by a bright region crossing the Flammarion Sea at longitude 280°, as had already been observed by Lockyer, Kaiser, Rosse, Schmidt and Burton.

The Hourglass Sea was distinctly broader in 1879 than in 1877. In 1879 it corresponded better with the classical aspect with which our readers will have become familiar from the very first page of this book.

This is a very important point in our study of the planet, and for the ideas we must form about its physical constitution. The extent of the Hourglass Sea varies. There is no begging the question or evading the facts, though we cannot yet explain it fully. The variation is certain, as are those which have already been described in the course of these comparisons; moreover, we can give degrees of certainty.

If the dark patches on Mars represent seas, these increases in extent correspond to flooding, and lead us on to the conclusion that the banks—especially the left bank of the Hourglass Sea, along the Flammarion Sea—are very flat.

These floodings, lasting for several months, cannot be due to tides.

To avoid these conclusions, we would have to assume that the patches on Mars are not seas. Very difficult!

Because of the broadening of the Hourglass Sea on its banks to the left, the Main Sea, or the Nepenthes Lake, was found to be closer to the Hourglass Sea in 1879 than in 1877, the point of land named by Schiaparelli Osiris Promontory, having almost disappeared. Compare the chart of 1877 with that of 1879.

Ausonia, which corresponds to Cassini Land and Dreyer Island, is shown on 26 and 28 October, crossed by a grey streak—the Euripus—variable in tone and in breadth. This also is a changeable formation which, for most of the time, does not exist. Schiaparelli did not see it even once in 1877. However, Green did see it on one occasion, on 10 September.

Hellas appeared very brilliant, sometimes as brilliant as snow. This is one of the regions destined to give the best information about the changes on Mars due to the seasons (Fig. 43).

South Polar Snow in 1879.—The observations of the south polar snow give 5°.0 as its distance from the geographical pole, and 49°.7 as its longitude. Comparing this with preceding estimates, we have the table below. Despite the difference in longitude, it can be seen that when the southern snow is reduced to its minimum, after the summer, it occupies sensibly the same areographical position. From 30 September 1879 to 9 March 1880, Schiaparelli made no less than 180 observations of the south polar cap. It was reduced to its minimum (4°) during the second half of November, that is to say 3 1/4 months after the summer solstice, which took place on 14 August. Several irregularities were observed in its contour, notably on 24 October under excellent conditions. If we agree that irradiation doubles the apparent diameters of patches as brilliant as the polar snows on Mars, then 4°, reduced to 2°, still represents some 120 km.

Year	Dist. from pole°	Longitude°	Observers
1830	6.6	21.5	Bessel
1862	4.3	15.5	Kaiser, Lockyer, Linsser
1877	5.6	25.5	Hall, Schiaparelli
1879	5.0	49.7	Schiaparelli

North Polar Snow in 1879.—During the observations made at this opposition from 30 September to 24 March, the northern snows seem to have passed through their maximum spread, with certain curious fluctuations. Six branches were seen at the edge of the circle.

The first branch was observed below the Nile, at longitude 0°. Its distance from the pole appeared to be less than 20°.

The second was also below the Nile, at longitude 65°. It seemed to extend to up to 30° from the north pole.

The third was seen at longitude 119°, and at a polar distance of 36°. Perhaps it is not without connections with the Nix Olympica.

The fourth was at longitude 155°, and attained a polar distance of 30°.

The fifth was at longitude 189°, polar distance 29°.

The Sixth was the broadest of all, stretching from longitude 230° to 300°, that is to say over 70° of longitude, and up to about 30° polar distance.

These six branches, therefore, extended the north polar snows up to 30°–36° polar distance in certain places, that is to say, out to latitude 60° and 54°. The terrestrial polar snows can also be considered as being prolonged sometimes to this extent, in winter in Siberia.

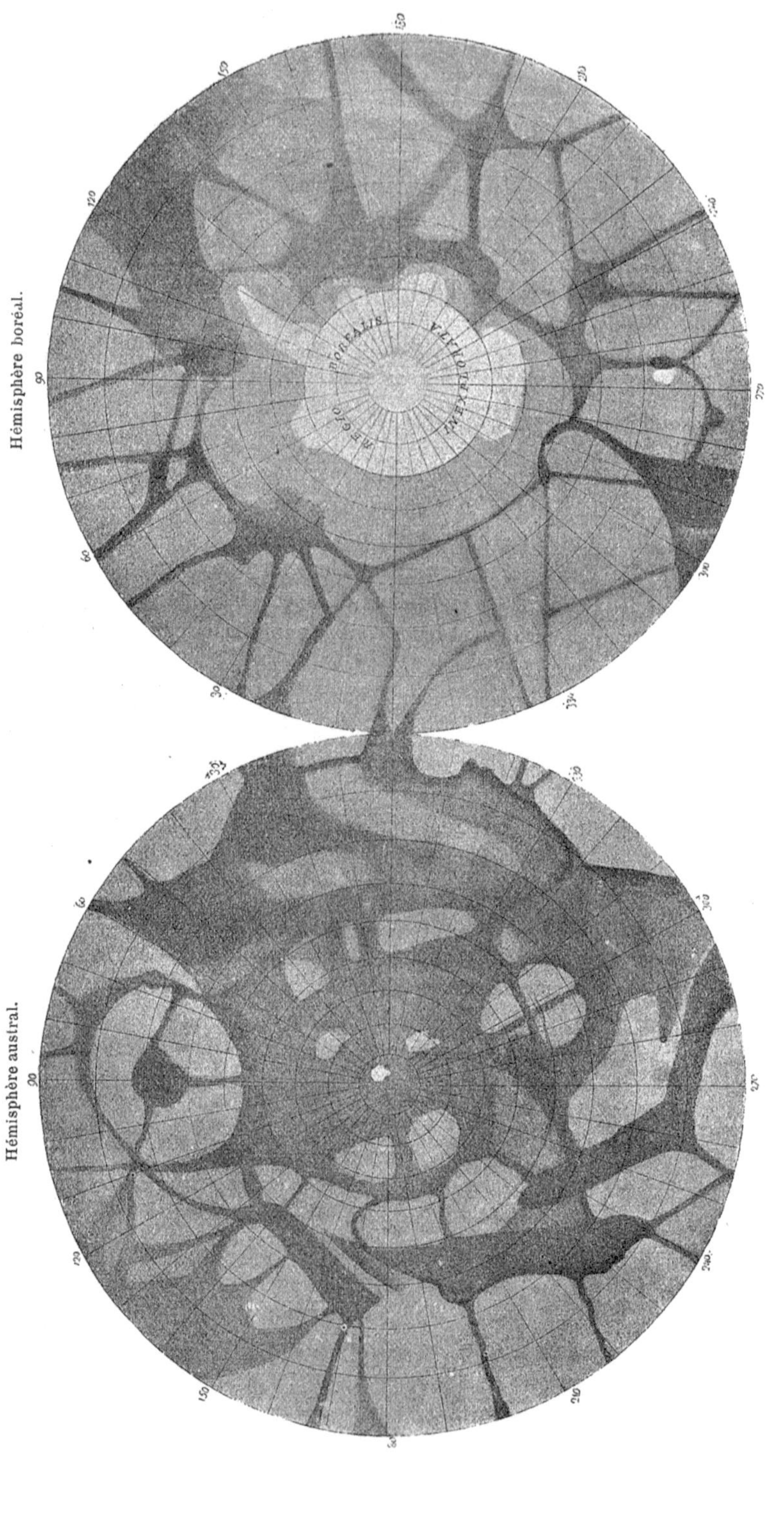

Fig. 43 Polar mappemonde of the planet Mars, in 1879, by M. Schiaparelli

According to the same observations, the northern snows attained their greatest extent in the middle of November, that is to say 3 months after the winter solstice. Subsequently they diminished again. The maximum of the north polar snows and the minimum of the southern took place at almost the same time, as theory would naturally indicate.

Such were the splendid observations of Mars made by the eminent Milan observer during the opposition of 1879, and the conclusions which can be drawn from them. The following opposition, that of 1881–1882, was richer still.

1879.—Satellites of Mars: Observations and Measurements

At the meeting of the Academic Sciences on 1879 November 10,[46] Asaph Hall reported on his new observations of the satellites of Mars. In comparing the measured positions with those which he had calculated from the elements of 1877, he found for Phobos a slight difference of $-1^s.074$ in its period, which thus was revised to

$7^h\ 39^m\ 13^s.996$.

Among the observations made of the satellites apart from those of the discoverer, A. Hall, at Washington,[47] we must first note those of Common at his observatory at Ealing, near London, with a 36-in. reflector, magnifications 220×, 240× and 380×. We will not give the positions determined, which do not concern us here, but we will give the results concerning the brightness and colour of these little globes.[48]

Deimos seemed to him to be as bright as Enceladus. Phobos a little less brilliant than Tethys. But the character of their light is different. While the satellites of Saturn show a steady clarity, those of Mars are sharper, more sparkling, almost stellar. This aspect could be due to the absence of an appreciable disk, or to contrast with the full brightness of Mars. Deimos appeared slightly bluish; Phobos almost white. They did not seem to share the colour of Mars.

Common tried to photograph Mars, but did not succeed in recording any detail.[49]

At the Greenwich Observatory, the second satellite was observed with the great 12¼-in. equatorial, but not the first.[50] When turned toward Mars in 1877, this telescope had been unable to show either satellite. In 1879, it showed only the one that is more distinct.

We know that in March 1877, the director of the observatory of Melbourne, with a telescope whose mirror measures no less than $1^m.20$, failed to make out either satellite. In 1879, however, he made out the outer satellite.[51]

[46] *Comptes rendus*, 1879, vol. II, p. 776.

[47] *Monthly Notices*, March 1880, p. 272.

[48] *Id.*, December 1879, p. 95.

[49] *Monthly Notices*, February 1880, p. 225.

[50] *Id.*, January 1880, p. 161.

[51] *Id.*, February 1880, p. 237.

In November 1879 at the Oxford Observatory, Plummer was able to observe the outer satellite with the aid of the 12¼-in. (311^{mm}) equatorial, magnification 125×.[52]

At Princeton (USA), Young and Brackett made numerous micrometrical measurements of the two satellites.[53] Pickering, with the 9 1/2-in. (241^{mm}) equatorial at Harvard College, made a series of no less than 207 measurements.[54]

According to Pickering's observations,[55] Deimos appeared more brilliant in 1879 than in 1877, and also was rather brighter at the eastern or following side of its orbit than on the preceding or western side. There were more observations on the former side than on the latter. The diameter of the satellite appeared to be 6 miles (9.6 km), that of Deimos 7 miles (11.3 km), according to the photometric measurements. Deimos was not reddish, but bluish.

1879.—G. H. Darwin, D. Kirkwood, E. Ledger: Satellites of Mars, Tides and Rotation Periods[56]

A fluid planet will assume a spherical form under the influence of gravitation.

But if a satellite revolves in a circular orbit around its primary, in the plane of the equator, it will produce tides such that a spheroid will be deformed into an ellipsoid with three unequal axes, the longest (equatorial) axis remaining permanently turned toward the satellite.

Thus the figure of the planet revolves with the satellites, while each molecule of fluid turns with the planet. In consequence, the result is a rise and fall time for each revolution of the planes relative to the satellite.

Now, suppose that the fluid is subjected to friction. Then, the particles of the fluid which at one moment make up the equatorial bulge will not be able to fall back quickly enough to return to their original position before friction takes over. The major axis of the equator will advance together with the satellite.

The Earth was originally fluid, and in the course of its cooling became viscous, giving rise to the friction to which I have referred. Nowadays it is probably doughy, almost solid, and internal tides no longer occur.

If the fluid making up the Earth had been frictionless, the equatorial bulges caused by the pull of the Moon would have remained in a straight line with the Moon. But friction modified the primitive condition, and consequently the Moon exerts a force upon these protuberances tending to draw them backwards, while the axis of the bulge of the equator is always in advance of the Moon. The Moon exerts a breaking action on the diurnal rotation, and reciprocally the Earth tends to Speed

[52] *Id.*, March 1880, p. 292.

[53] *The Observatory*, January 1880, p. 270.

[54] *Astronomische Nachrichten*, no. 2312.

[55] *Annals of Harvard College Observatory*, Vol. XI, part II.

[56] *The Observatory*, July 1879, p. 79.

up the motion of the Moon. The first result is a slowing down of the Earth's rotation. The second is an acceleration of the linear speed of the Moon in its orbit, and an increase in its distance from the Earth.

These two results produce an increase in the length of the day and of the month, especially the former, in the state of the Earth-Moon systems.

Delaunay and Adams had already come to the same conclusion, though they attributed it to the friction of oceanic tides rather than external tides. Ancient internal tides were regarded as the main or even sole cause by G.H. Darwin.

In the actual state of affairs, the rate of diminution in the angular velocity of the Earth's rotation is much greater than that of the angular velocity on the orbital motion of the Moon. If the former were exactly 27 1/3 times greater than the latter, the day and the month, would in the past, have been reduced on the same proportion. We find that when the length of the day 15 1/2 hours, that of the month should have been 19 days.

Darwin calculated that at a certain time in the past, the length of the day was 6^h 50^m, and that of the month 11^d 14^h. This enormous change has been achieved in 56,000,000 years.

Going back still further, Darwin reached an epoch when the day and the month were identical—only 5 1/2 hours long. The Moon was very close to the Earth, with a surface-to-surface distance of only 8,000 km. Earlier still, the two were almost in contact. In fact, the Moon was born from the Earth by the centrifugal force due to rotation, according to Laplace's theory.

If we turn to the future instead of the past, we find that according to the same principles the month and the day will eventually become equal, at a length of 50 times our present day.

Darwin ended his memoir by commenting that the abnormal satellite of Mars, the closer-in, which has a revolution period quicker than the axial rotation of the planet; offered a confirmation of his views; because it seemed probable that its extreme smallness kept it as lasting evidence of the original rotation period of Mars on its axis

To this theory by the English mathematician concerning the evolution of satellites, and above all its application to the inner satellite of Mars, the American astronomer Daniel Kirkwood has replied in the following terms[57]:

> The ratio of the mass of the inner satellite of Mars to that of our Moon is 1 to 30,000,000. Consequently, it is in fact, from this absence of tides that Mr. Darwin infers the constant distance of the satellite. What, then, is the cause of the lengthening of the rotation period of Mars from less that 8 hours to almost 25?

Darwin replied[58]:

> My memoir is only an extract from a long work, now in press. I can however answer the question asked by Mr. Kirkwood, as to how the tidal theory can explain the fact that the Martian day is longer that the month measured by the inner satellite. Such a result is derived

[57] *The Observatory*, September 1879, p. 147.

[58] *Id.*, November 1879, p. 204.

from the effects of friction caused by solar tides. I hope to be able to produce proof of this numerically in the particular case of Mars.

Again, with regard to the satellites Ledger has studied the tides and the eclipses which they can produce at the surface of the planet.[59] He wrote: "The smallness of their masses would, despite their closeness, prevent the production of sensible tides; and they are unable to prevent the Martian seas from being stagnant." With regard to their light, he estimates 1/40 of that of our Moon for the first satellite, and 20 times less, or 1/800, for the second. This is not much. Phobos remains above the horizon for 5 1/2 hours at a time, during which period it is in eclipse for 53 minutes. Deimos remains above the horizon for 60 hours consecutively, during which time there are two eclipses and sometimes three.

1879.—J. C. Adams, F. Tisserand: Inclination of the Satellites[60]

The orbits of the two satellites of Mars are slightly inclined to the plane of the planet's equator. Professor Adams has asked himself whether this state of affairs is permanent. The plane of the orbit of Mars is inclined by 27°–28° to that of its equator. If therefore the planes of the satellite orbits keep a constant inclination to the plane of the orbit of the planet, as will happen if the perturbations by the Sun represent the only force tending to alter their planes, then their inclination to the plane of the Martian equator, and hence their inclinations relative to each other, will alter considerably over the course of time.

Marth has calculated[61] the movements of the nodes of the orbits of the satellites along the orbit of the planet, due to solar action, and has concluded that if there is no other force depending upon the internal structure of Mars, capable of modifying the Sun's action, then the nodes of the satellites should be at opposition to each other in 1000 years, and the inclinations of their orbits one to the other should have risen to 49°.

In this case, the actual near-coincidence between the planes of the Martian equator and the satellites orbits would be fortuitous and transitory but this is highly improbable.

If there has been no external perturbing force, the flattening of the planet can make the nodes of the orbits of its satellites move retrograde along the plane of the planet's equator, while the orbit maintains a constant inclination to the plane. Laplace has shown that if we also take into consideration the action of the Sun and the ellipticity of the planet, we find that the orbit of the satellite moves in such a way as to keep an almost constant inclination upon a plane fixed by the intersection of

[59] *Id.*, November 1879, p. 191.

[60] *On the ellipticity of Mars, and its effect on the motion of the satellites* (*Monthly Notices*, November 1879, p. 10).

[61] Astronomiche Nachrichten, no. 2280.

the equator of the planet with the plane of the planet's orbit, and stays between these planes, so that the nodes of the orbit of the satellite show almost uniform retrograde movement in the fixed plane.

Adams went into detailed calculations applicable to the satellites of Mars, and studied the effect of the flattening of the planet upon the orbits of the two satellites. He found that the movement of the nodes of the orbits of the satellites produced by the ellipticity of the planet greatly exceeded that produced by the action of the Sun, so that the fixed planes for the two satellites are always lightly inclined to that of the Martian equator.

Considering the measurements of the diameter of the planet and the greatest inclination of the satellites, together with the rotation period of Mars and the revolution of the satellites, we find that the ratio of gravity to the centrifugal force at the equator of Mars is about 1/220. It follows that if the planet were homogeneous, its flattening would be about 1/176. If, instead of being homogeneous, the density varied according to the same law as with the Earth, in such a way that the difference between centrifugal force and gravity would be the same as for the Earth, then the flattening would be 1/228. According to all probability, the real value would lie between the limits.

Adams has calculated the following table for the annual movements of the nodes of the two satellites, caused by solar action and by the ellipticity of the planet, for the flattening values previously quoted. (Note that he has even included Kaiser's value of 1/118, which is certainly much too great.)

Annual movement of the node due to solar action:

Satellite I	0°.06	Satellite II	2°.24

For the ellipticity	1/118	1/176	1/228

the annual movement of the satellite's node becomes:

Satellite I	333°	182°	113°
Satellite II	13°.4	7°.3	4°.5

and the inclinations corresponding to the fixed plane to the equator of the planet:

Satellite I	17″	31″	50″
Satellite II	27″	50′	1°19′

From the Table, it is seen that the orbit of the first satellite keeps a constant inclination on a plane inclined to that of the Martian equator by less than 1′, and that the orbit of the second satellite also keeps a constant inclination on a plane with inclination barely exceeding 1° to that of the Martian equator.

The flattening of Mars also produces rapid movements in the line of apsides of the orbits of the satellites—particularly for the first—and one day it may even be possible to determine the flattening by making measurements of this movement.

Tisserand has reached the same conclusions as Adams.[62] The ellipticity of the planet keeps the satellites almost in the plane of the equator. Such is also the opinion of Hall.[63] Tisserand has written the following note:

> The two satellites move almost in the same plane, which differs very little from the plane of the planet's equator. Is the near-coincidence of these three planes fortuitous, or should it always exist? This is an interesting question, which has been treated in part by Mr. Adams at the Royal Astronomical Society of London (14 November 1879). By another analysis I propose to return to the question discussed by the eminent Director of the Cambridge Observatory, and I believe that I will be able to reach more precise conclusion, despite the uncertainty of our knowledge of the true position of the equator of Mars. The analysis has already been used in a comparable study of one of the satellites of Saturn.
>
> Up to now, observations have not shown appreciable flattening for Mars; if this flattening were indeed nil, then because of perturbations produced by the Sun, the orbital planes of Phobos and Deimos, assumed to coincide at a given moment, would end up by being different from each other by a considerable amount. I will show that assuming the density of the interior of Mars to follow the same law as with the Earth, and attributing to the planet a flattening too slight to be measured directly, then the orbital planes of the two satellites will never be far from the plane of the planet's equator. For each satellite, the perturbing force R is caused by the action of the Sun and the bulging out of the equator of Mars. I am here concerned only with secular inequalities. By virtue of these inequalities, we have the integral R = const. Neglecting the orbital eccentricities of the satellites, which are extremely slight, if not zero, we can write
>
> $$K\cos^2\gamma + K'\cos^2\gamma' = \mathrm{C}, \tag{1}$$
>
> where K and K′ have the following values:
>
> $$K = 3/8Ma^2 / a_0\left(1 - e_0^{\,2}\right)^{3/2}, \tag{2}$$
>
> $$K' = ½ma'^2 / a^2\left(\rho - ½\varphi\right),$$
>
> where M is the mass of the Sun, m that of Mars, a is the semi-major axis of the orbit of the satellite, a' the equatorial radius of Mars, a_0 the semi-major axis of the orbit which Mars describes around the Sun, e_0 the eccentricity of this orbit, ρ the flattening of the planet at its surface, and φ the ratio of the centrifugal force to attraction for points on the equator of Mars. Finally, γ is the angle made by the orbit of the satellite with respect to the orbit of Mars, and γ′ is the angle of the satellite orbit with the plane of the planet's equator.
>
> The term $K\cos^2\gamma$ arises from the action of the Sun; $K'\cos^2\gamma'$ is due to the action of the equatorial bulge of Mars. If we disregard the action of the Sun, we could say that γ = const.; the orbit of each satellite would make a constant angle with the orbit of Mars. If, on the other hand, we disregard the planet's flattening, the orbit would make a constant angle with the equator of Mars. If we take both actions into account, the pole of the orbit of each satellite describes a spherical ellipse; this is because of Eq. (1).

[62] *Comptes rendus*, 8 December 1879, p. 961.

[63] Monthly Notices, March 1880, p. 278.

Let us try to find the ratio K'/K; from Eq. (2) we see that

$$K'/K = 4/3\,(n/n_o)^2\,(a'/a)^2\,(1-e_0^{\,2})^{3/2}\,(\rho - ½\varphi), \tag{3}$$

calling n and n_o the mean movements of the satellites and of Mars; n_o and e_0 are well known; n and a have been given by Hall for the two satellites; finally I take, following a memoir by Hartwig, where he takes all former determinations into account, $2a' = 9''.352$, corresponding to a distance between Mars and the Sun equal to 1.

Expression (3) gives

$$K'/K = (3.91061)(\rho - ½\varphi)\,\text{for Deimos}, \tag{4}$$

$$K'/K = (5.99065)(\rho - ½\varphi)\,\text{for Phobos;}$$

φ is easily determined with the data given above, together with the well-known value for the rotation of Mars; we find

$$\varphi = 1/219. \tag{5}$$

Up to now there has been nothing hypothetical. At this point, we consider two hypotheses:

Hypothesis I.—Mars is homogeneous; then $\rho = 5/4\,\varphi$. From expressions (4) and (5) we deduce

$$\log\ K'/K = 1.44567\,\text{for Deimos},$$

$$\log\ K'/K = 3.52570\,\text{for Phobos}.$$

Hypothesis II.—The law of density is the same for Mars as for Earth; then

$$\rho/\varphi = \rho_1/\varphi_1,$$

where ρ_1 and φ_1 indicate the values corresponding to ρ and φ for Earth. The result is that

$$\rho = 1/228$$

and it follows that

$$\log\ K'/K = 1.23650\,\text{for Deimos},$$

$$\log\ K'/K = 3.31654\,\text{for Phobos}.$$

On the sphere, let D be the north pole of Mars, D' that of its equator, M the orbit of one of the satellites; further, $DD' = A$, the angle of the orbit and the equator of Mars, and C a point lying on the arc of the great circle $D'D$ and determined by the equation

$$\tan 2i = K\ \sin 2A / K' + K\ \cos 2A, \quad \text{where } i = CD'. \tag{6}$$

The point C will be the centre of the spherical ellipse which will be described by the pole M; we see immediately that in the two hypotheses considered, K' / K being large, the point C will be close to the point D'.

Let 2 $\rho'$and 2 ρ'' be the major axis and the minor axis of the ellipse; and giving γ_0 and γ'_0 the initial values of γ and γ', B and N being the auxiliary angles defined by the formulas

$$\sin 2B = 2\left\{\left(\sqrt{KK'}\right)/\left(K+K'\right)\right\}\sin A, \tag{7}$$

$$\sin^2 N = K\ \sin^2\gamma_0 + K'\sin^2\gamma'_0 / K + K' \tag{8}$$

we have

$$\cos\ \rho'' = \cos\ N/\cos\ B, \cos 2\rho' = \cos 2N/\cos 2B \tag{9}$$

The value of the ratio K' / K, from angle B, drawn from formula (7), will always be small. Formula (9) shows that $\rho'$and ρ'' will always be much the same. Indeed, if we calculate $\rho'$and ρ'' according to the positions given for the Martian equator by various authorities, we find that the difference ρ'- ρ'' will be only a very few minutes of arc. In short, we can show, with sufficient precision, that the point M describes a small circle having for its centre of the point C defined by Eq. (6), and for the radius of the ratio of ρ' is determined by the following equation:

$$\cos 2\rho' = \left(K\cos 2\gamma_0 + K'\cos 2\gamma'_0\right)/\sqrt{\left(K+K'\right)^2 - 4KK'\sin^2 A}. \tag{10}$$

If ρ' is greater than i, the value of γ' will be between the limits $\rho' - i$ and $\rho' + i$, which differ by 2 i.

If ρ' is less than i, the value of γ' will lie between the limits $i = \rho'$ and $i + \rho'$, which differ by 2 ρ'.

I have made the calculations, taking for the position of Mars's equator, the position given by W. Herschel from the observations by Bessel as calculated by Oudemans, and finally the figures given by Marth (*Monthly Notices*, Vol. XXXIX, p. 473). (Herschel's and Bessel's figures are published here; Marth's are as follows: inclination of the plane of the equator of Mars to that of the Earth = 36°.260, node 47°.945.) The corresponding positions of the equator of Mars differ considerably; however, in all these cases I reached conclusions which are little different. Let γ'_1 and γ'_2 be the upper and lower limits of inclination of the orbit of Deimos on the equator of Mars. I found the following results

		Hypothesis I	
	Herschel	Oudemans	Marth
γ'_1	4°.9	2°.7	0°.1
γ'_2	6 .6	4 .4	1 .4
$\gamma'_2 - \gamma'_1$	1°.7	1°.7	1°.3
		Hypothesis II	
γ'_1	3°.9	1°.9	0°.2
γ'_2	6 .7	4 .5	2 .2
$\gamma'_2 - \gamma'_1$	2°.8	2°.6	2°.0

We see that in each case, the inclination of the orbit of Deimos on the equator of Mars can oscillate only between limiting distances of 3° at most. For Phobos the limits are even more restricted.

We may, then conclude that if Mars is homogeneous, or if the interior follows the same law of densities as for the Earth, *the orbits of the two satellites always coincide with the equator of Mars*, or, at least, never depart from it by more than a small amount.

The same will apply if the flattening of Mars lies between the two limits which apply to Hypotheses I and II.

1879.—Transit of the Earth Across the Sun: As Seen by the Inhabitants of Mars

To all these Martian documents we can add the following, which shows interest of another kind.

On the opposition day of Mars in 1879, the Sun, the Earth and Mars lay in an exactly straight line, so that to the Martian inhabitants the Earth and Moon were projected against the disk of the Sun, just as Venus and Mercury sometimes are for us.

Here, following the calculations by Marth, are the circumstances of the transit:

INGRESS			EGRESS		
12 Nov.			12 Nov.		
GMT.	Degrees		GMT.	Degrees	
$1^h 49^m$	125.7	First contact of Moon	$9^h 40^m$	225.3	Internal contact of Moon
$1^h 55^m$	126.4	Internal contact of Moon	$9^h 46^m$	226.1	Last contact of Moon
$4^h 16^m$	123.3	First contact of Earth	$11^h 39^m$	225.9	Internal contact of Earth
$4^h 37^m$	125.9	Internal contact of Earth	$12^h 00^m$	228.5	Last contact of Earth

The apparent radius of the Sun, seen from Mars, was 650″.5; that of the Earth, 18″.1, and that of the Moon 4″.9. The angles are reckoned from the point of the Sun's disk marking the direction of the north pole of the orbit of Mars.

The transit would not be visible with the naked eye, by eyes analogous to ours. When Venus passes in front of the Sun as seen from Earth, its disk is 60″ in diameter; that of the Earth from Mars is only 36″. But a low-powered instrument would suffice to show it.

An analogous phenomenon occurred on 8 November 1800, and will happen again in May 1905 and May 1984.

From Mars, transits of Mercury and Venus in front of the Sun are much more frequent, but they are not of the same importance; Venus is sensibly smaller than the Earth, as seen from Mars, and Mercury is smaller still.

In April 1886, there was a transit of Mars across the Sun as seen from Jupiter.

Opposition of 1881–1882

During this opposition, Mars remained further from the Earth than in 1877 or 1879; but the greater distance was in part compensated for by the more northerly declination of the planet, so that it rose higher above our horizon. The northern hemisphere of Mars was better presented for observation, the latitude of the centre of the disk being +7° in November 1881 and the north pole being full on the disk, at 7° from the edge. But this inclination diminished quickly, because at the beginning of December it had decreased to 5°, and at the end of that month the north pole again cease to be visible; the equator occupied the centre of the disk at the moment of opposition. In January and February, the planet presented less of its face, and the two poles were similarly placed.

Date of Opposition: 26 December.
Presentation of Mars :

At first the south pole was tilted toward the Earth, but the north pole was Earthward-inclined later. On 6 January, the Earth passed through the plane of the Martian equator.

	Lat. of centre	App. dia.	Phase defect
22 October 1881	+6°.7	10″.7	1°.2
26 December zero (Opp.)	+1°.5	15″.5	0°.0
6 January 1882	0°.0	14″.9	0°.1
1 February 1882	−2°.0	12″.0	0°.6
27 February 1882	0°.0	09″.4	0°.8
21 March 1882	+3°.4	07″.7	1°.2

Martian Calendar		
	Southern hemisphere	*Northern hemisphere*
1 July 1881	Summer solstice	Winter solstice
8 December 1881	Autumn solstice	Spring solstice
25 June 1882	Winter solstice	Summer solstice

1881.—Webb: The Planet Mars[64]

This careful and distinguished writer, of whom we have already heard (1856 and 1873), and whose death a few years ago is deeply regretted, devoted a chapter of his excellent treatise to Mars. This chapter is illustrated by the map by Burton and Dreyer, given here. One of his comments refers to the analogy between Mars and

[64] *Celestial Objects for Common Telescopes.* 4th edition, London,1881.

the Earth, and, without too much temerity, he regards Mars as habitable by the human race. To him, the sombre patches are tinted with a greyish blue, and represent seas; the bright and yellowish regions represent continents. The proportion of land is relatively greater than in our world, so the habitable area must be much more expensive than might be supposed from the planet's diameter. The seas are in communication with each other via narrow canals, the observation of which is, however, so difficult that nothing can be concluded about them with certainty. Perhaps they are only the borders of slightly-tinted regions. Maps of Mars must be regarded as approximate and provisional only. The polar snow was very brilliant and vary with the seasons. Among curious observations, Webb cites one made by Ward, on 22 December 1879, in which the circular lake (the Terby Sea) showed up as black, and as clearly defined as a shadow of a satellite of Jupiter, although at this time the general definition of the planet itself was bad. As for the observed variations of dark and bright tones, Webb thought that clouds seen from above would reflect light more strongly than lands or water. This is a question which is very important, and also crucial in any explanation of the changes observed. It certainly seems that clouds seen from above, lit by the Sun, should always appear white, as we always find when observing from a balloon. However, cannot we imagine mists made up of sombre particles? Cannot the smoke of certain carbons on Earth give rise to flakes which are grey, sometimes nearly black?

1881–1882.—Schiaparelli: Observations and Drawings

During this opposition, the skilful Milan astronomer continued his series of astonishing discoveries and this time he passed from marvel to marvel, as we will see.

All his observations of this period were published in 1886, and a third memoir, following up the two earlier analyses which had been described above. But in the middle of 1882, he made known, via the Roman Academy of Lynxes, the most curious phenomenon of this new series of observations: the doubling of the canals of Mars.

Here is the first summary, which I was myself authorized to publish under Schiaparelli's signature,[65] following an announcement he had been kind enough to send me:

> The last opposition of Mars has been observed from Milan under excellent meteorological conditions. October and November were not favourable, but from 26 December 1881 to 13 February 1882 I had fifty particularly good nights. The high atmospheric pressure prevailing during this period produced a series of beautiful nights, calm and serene, especially favourable for observations. During sixteen nights I could use the highest power on my excellent equatorial, and during forty other nights the atmosphere left nothing to be desired. Thus although the apparent diameter of the planet attained only 16″, whereas it exceeded 19″ in 1879 and 25″ in 1877, it has been possible, in this third period which I have described, to obtain a mass of data with regard to the physical nature of the planet which, by their novelty and interest, exceed anything which I have previously observed.

[65] *L'Astronomie, Revue mensuelle d'Astronomie populaire*, 1st year, 1882, August, p. 126.

The series of interior seas contained between the bright equatorial zone and the southern sea is shown as better drawn than in 1879. In the Mare Cimmerium may be seen an area of island or a luminous streak which divided it along its length, giving it an appearance analogous to that of the Mare Erythraeum. The Mare Chronium has undergone very notable modifications since 1879. More surprising still is the change in appearance of the Syrtis Major, which has invaded Libya and extends, in the form of a black ribbon, down to latitude 60°N. The Nepenthes and the Lacus Moeris have increased in breadth and darkness, while there remain few traces of Coloe Palus, which had been so obvious on the 1879 map. Thus, hundreds of thousands of square kilometres of the surface have become dark, while, on the other hand, there are a number of dark regions which have become bright. Such metamorphoses prove that the cause of the dark patches is a mobile, variable aspect on the surface of the planet, whether it be water or some other liquid, or vegetation which is propagated from one point to another.

But these are not the most interesting of the observations. On the planet, crossing the continents, there are a large number of dark lines to which I have given the name *canals*, though we do not know that this is what they are. Many observers have previously recorded some of them, notably Dawes in 1864. During these last three oppositions, I have made a special study of them, and I have recognized a considerable number—at least sixty. From one to another these lines connect the sombre patches which we regard as seas, forming a well-defined réseau on the bright or continental regions. Their disposition appears invariable and permanent, at least insofar as I can judge after having observed for 4 ½ years; however, their aspect and their conspicuousness are not always the same, and depend upon circumstances which our present knowledge does not enable us to define with certainty. In 1879 I saw many which had not been visible in 1877, and in 1882 I recovered all those which had been previously observed during previous oppositions, as well as discovering several new ones. Sometimes these canals show up as shaded, vague lines, while at other times they are as clear and precise as pen-strokes. In general, they are traced on the globe as lines of grey circles; others show definite lateral curvature. They cross each other, obliquely or at right angles. They are at least a good 2° in breadth, or 120 kilometers, though some are as much as 80° or 4800 kilometres broad. Their tint is almost the same as that of the seas, though normally a little brighter. Each canal terminates, to either end, in a sea or in another canal; there is not a single example of a canal stopping short in the middle of a land-mass.

This is not all. At certain seasons these canals are doubled, or, more accurately, double themselves.

This phenomenon seems to come at a definite time, and to occur almost simultaneously over the whole extent of the continents on the planet. No indication of it was seen in 1877, during the weeks before and after the southern solstice on Mars. A single, isolated case was noted in 1879; on 26 December of that year (a little before the spring equinox, which occurred on Mars on 21 January 1880), I noted the doubling of the Nile, between Lacus Lunae and Ceraunius Lacus. These two regular streaks—equal and parallel—caused me, I admit, profound surprise, becoming even greater because several days earlier, on 23 and 24 December, I had carefully observed the same region without detecting anything comparable. With curiosity I awaited the return of the planet in 1881 to see if an analogous phenomenon would show up in the same place, and I was able to see the same phenomenon on 11 January 1882, a month after the spring equinox on the planet (which had occurred on 8 December 1881); the doubling was still evident at the end of February. At this same date, January 11, another doubling had already occurred; that of the middle section of the Cyclops canal, by the side of Elysium.

To my even greater astonishment, on 19 January I saw the Jamuna canal, which was then at the centre of the disk, made up of two straight parallel lines, crossing the space which separates the Niliacus Lacus from Aurorae Sinus. At first I thought that this must be due to eye-fatigue, or a new kind of squinting, but it was very evident. After 19 January I had surprise after surprise; in succession the Orontes, Euphrates, Phison, Ganges, and the majority of the other canals showed up as clear and incontestable doubles. There were not less than 20 examples of doubling, of which seventeen were observed over a period of a month, from 19 January to 19 February.

In certain cases, it was possible to observe some preliminary symptoms which were not lacking in interest. Thus on 13 January, a light, poorly-defined shading extended alongside the Ganges; on the 18th and 19th I could distinguish these only by a series of white patches; on the 20th, this shading was still imprecise, but on the 21st the doubling was perfectly clear, and I continued to make note of it until 23 February. The doubling of the Euphrates, the Titan and the Pyriphlegethon also began in an indefinite, nebulous way.

These doublings are not an optical effect depending on the increase of visual power, as happens in the observation of double stars, and it is not due to the canal being divided into two along its length. This is what is presented: To the right or left of an existing line, without any change in the course or position of the line, we see the appearance of another line equal and parallel to the first, at a distance generally ranging from 6° to 12°, that is to say 350 to 700 kilometres; it may be even closer, but the telescope is not powerful enough to enable this to be distinguished with certainty. Their tint appears to be reddish-brown, fairly dark. The parallelism is sometimes rigorous and exact. There is nothing analogous in terrestrial geography. Everything indicates that this is an organization peculiar to Mars, probably associated with the march of the seasons (Fig. 44).

These are the observed facts. The distance of the planet and the bad weather hinders my attempts to continue the observations. It is difficult to form a precise opinion as to the intrinsic constitution and the geography of Mars, which are certainly different to a great extent from those of our own world. If the phenomenon is really linked with the Martian seasons, it is possible that it will again occur during the next return of the planet. On 1 January 1884 the position of Mars, with respect to the seasons, will be the same as it was on 13 February 1882, and the diameter will be 13″. Any instrument capable of showing, on a bright background, a black line 0″.2 in breadth, and of separating two such lines at 0″.5 from each other, can be used for these observations.

In the actual state of things, it is premature to make conjectures about the nature of these canals. With regard to their existence, I need hardly say that I have taken all possible precautions to avoid all chance of Illusion; I am absolutely certain of what I have observed.

Such is this skillful astronomer's account in his first article about these strange observations. One need only to look at the map which accompanies the article (Fig. 45) to be absolutely astonished at these discoveries, and even to doubt one's very eyes. It is easy to understand the skepticism which was aroused—these features will be discussed in detail; but we should first describe all the details of Schiaparelli's observations, as given in his third memoir.[66]

These observations extended over a period of six months, from 26 October 1881 to 29 April 1882. I again saw all the canals seen in 1877, among them the Hiddekel, which had been doubtful in 1879, and the Juventae Fons, which had been invisible in 1879. There were many new features, which were probably seen because of the different position of the Sun. The bright red colour mixed with white, which in 1877 occupied the whole of the equatorial zone to the north of the Great Diaphragm, which was considerably extended in 1879 and disappeared almost completely in January and February 1882. In this luminous veil I began to distinguish vague shadings surrounding ill-defined patches which were orange in colour; these shadings became

[66] *Osservazioni astronomiche e fisiche*, etc. Memoria terza (reale Accademia de Lincei. Roma, 1886).

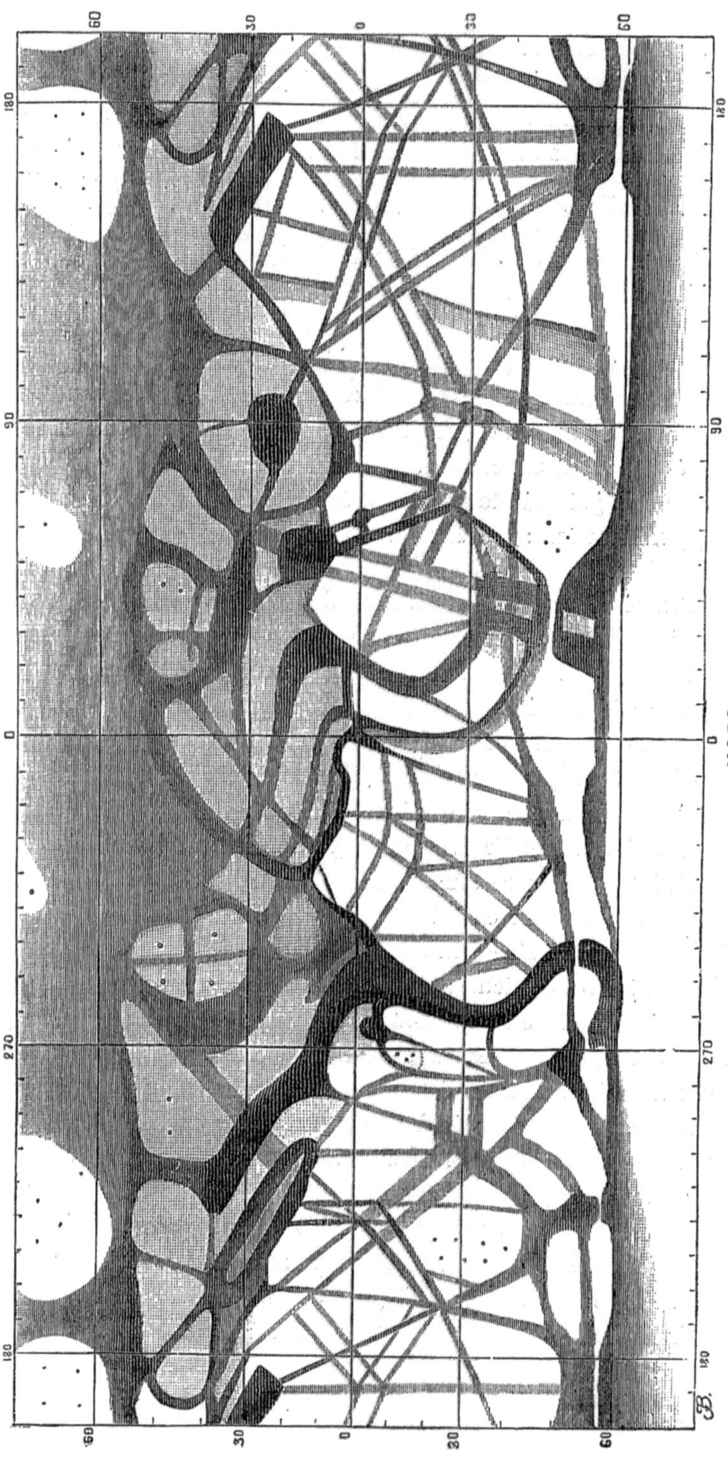

Fig. 44 The enigmatic canals observed on the planet Mars (drawn by M. Schiaparelli) (At the points marked by *small circles* 000, *white spots*, as of snow, were observed)

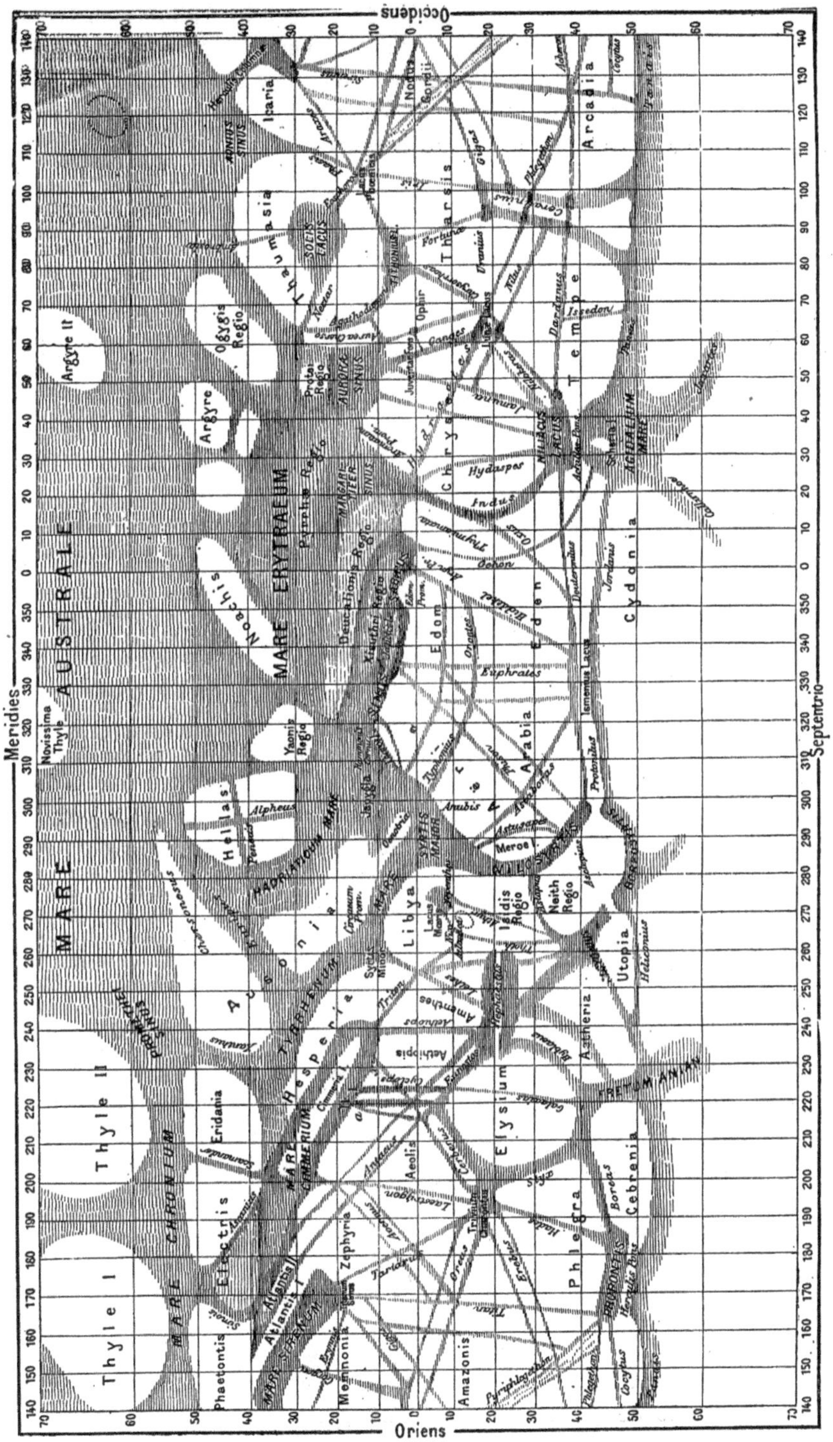

Fig. 45 A new chart of the double canals on the planet Mars, by M. Schiaparelli, from his observations between 1877 and 1886

gradually darker, and, apart from a few exceptions, ended by taking up the whole equatorial zone. The vast extent called the Nodus Alcyonius, which in 1879 appeared grey and indeterminate and seemed rather to be of maritime character, was resolved into very complicated clumps of small lines. Then came the disclosure of the curious and unexpected gemination of the canals, which probably led to a considerable modification about current opinions of the physical constitution of the planet.

Schiaparelli repeated the determination of the direction of the axis of rotation, and found results which absolutely confirmed those which have been described above, according to the measurements made in 1877 and 1879.

During the opposition of 1881–1882, 162 partial sketches were made, and 15 drawings of the full disk. It is decidedly preferable, when saying is excellent, not to waste time in making full-disk drawings.

The map given here (Fig. 45) was published in Schiaparelli's memoir, which is analyzed here, and is constructed for the southern part and up to latitude 20°N. from the observations of 1877 and 1879, and for the northern part from those of 1881, 1884 and 1886, thus completing a full survey of the Martian globe.

When trying to make a map of Mars during this opposition, a great difficulty was encountered due to the singular changes in the aspect of the planet which began in the middle of January, notably because of the doublings of the canals. To avoid total confusion and a single representation of the phenomena, which probably relate to different physical conditions, it has been necessary to divide the whole series of observations into two periods and to draw a chart for each. For the first period, however, the observations are insufficient. The geminations refer only to the second period, though perhaps certain aspects go back to the first. Schiaparelli has made a map from the total of all the observations of the opposition, without regard to time, and has combined everything into one chart (given as Fig. 6 in Part II, Chapter 9 of the present work) representing the northern hemisphere. The chart published above (Fig. 44) was provisional only.

Readers of this book will already have commented on the first of these maps, and on the second will have noted the lower extremity of the Hourglass Sea, which is outlined in the form of a serpent. The Italian astronomer gave the name of Syrtis Major to this sea, that of the Nile to the river which is attached to it; he gave the name of Nilosyrtis to this extremity, which is so singularly broadened and darkened. The continuation of the serpent was called Boreosyrtis. The Nilosyrtis resembles the Scorpion's Tail on the drawings made by Secchi in 1858. Compare also a drawing made by Dawes in 1864, my own in 1873, and those by Green also in 1873. But to me the Boreosyrtis appears decidedly indefinite, and indicates variations more considerable than the preceding ones.

Schiaparelli has similarly added new names for the newly-drawn configurations.

Our readers will know that one of the brightest parts of the planet is the Beer Continent shown on my map, which extends to the right of the Hourglass Sea. In general, this is a brilliant uniform region. During the opposition of 1881 Schiaparelli made some very curious observations of it.

Early on (9–14 November), he found here no notable differences from what he had seen 1879. The same canals were seen, not all equally distinct, and the only important difference was the aspect of the Ismenius Lacus, which was first seen on 12 November in the form of a patch, at the point where the Euphrates cuts the course of the Protonilus. In the second period of the observations (14–29 December) a veil of unknown nature seemed to be drawn back from the region; the Protonilus, which had at first had the aspect of a single line, showed up as being separated into two parallel courses each coming from the Ismenius Lacus. During the third period of the observations (17 January–4 February) the Orontes, Euphrates, Phison and Typhenius appeared double; the Hiddekel, invisible in 1879, reappeared; and the Oxus was extended beyond the Gehon as far as the Deuteronilus (see map). Thus we have a patch, Ismenius Lacus, which showed up as very sharp and single on 12, 13 and 14 November, without anyone having suspected a trace of gemination, and on 23 December two equal lakes could be seen, which extended latitudinally until by 28 and 29 December they corresponded to the aspects drawn on the map. The appearance on 22 January was the same.

The Orontes was one of the most evident canals. It was doubled on 18 January. The Euphrates and the Phison remained equally clear, single and obvious until 18 January. On the 19th, they appeared broadened and indefinite. On the 20th, observation was prevented by clouds. On the 21st, both were double, and splendidly clear. Their colour was not the same as that of the seas, but a kind of reddish brown, *una specie di bruno rossegiante*. These canals had not changed in position, but a second, absolutely parallel line, had been formed not far from each.

The Indus appeared very broad, all through the opposition, about half as broad as the Nilosyrtis. (This is the Burton Bay on my map: same breadth.)

Niliacus Lacus was shown as separated from the Mare Adidalium by a yellow isthmus which Schiaparelli named Achilles Pons. The contours of these two patches were not completely defined, except at Achilles Pons; Niliacus Lacus was not black, but yellowish brown.

This patch (Niliacus Lacus and Mare Acidalium) can be recognized on the drawings made by Knobel in 1873 and Bœddicker in 1881–1882.

In observing the Ganges, the presence of a black point to the right was noted many times; this was the Juventae Fons. It was joined to the Ganges by a thread. Then the doubling of the Ganges passed across it, from Aurorae Sinus to Lunae Lacus.

At the Lunae Lacus, which appeared to be formed simply by the intersection of thin lines which crossed each other, the Nile showed up as double after 12 January, and very clearly after 19 January, as grey threads crossing the snowfield. It was the same on 18 February. This gemination of the Nile had already been observed, as a unique case, in 1879 on 26 December; this was a month before the equinox, which occurred on the following 21 January. In 1881, the phenomenon began in an indecisive and confused manner, on 11 January—a month after the equinox, which had occurred on 8 December. If, therefore, the phenomenon is associated with the orbital revolution of Mars, the association is not narrow and rigorous, but rather a relationship analogous to our terrestrial seasons, in which we observe irregularities of greater or lesser extent.

Above the Solis Lacus, Thaumasia was yellowish-brown in colour; the colour resembled—among others—that of Libya, a tone very different from the bright yellow or near-white of Ophir and Tharsis.

The Solis Lacus was not round, as in 1877; neither was it pointed, as in 1879; but oval, as is shown on my map. These variations in form are irrefutable. Schiaparelli tried without success to re-observe the quadrilateral or rhomboidal form drawn by Lohse and Burton. It was very dark, and darker when near the edge of the disk than when near the centre.

The Araxes showed the rectilinear from of 1879, not the sinuous curve of 1877.

The Ceraunius, with the Isis and the Phasis, occupied the position of the Bessel Pass on Proctor's map.

Dawes' Snow Island, or Argyre, always appeared very white, as in 1877. The Herschel Strait was again seen in the serpentine form, drawn by Kaiser in 1862, on 31 October and 10 December.

The land of Ogygia, or which only light traces had been seen in 1879, but much whiter and more brilliant near the edge of the disk than in the interior:—Clouds?

In the Mare Erythraeum, there were again certain dark, but not black, regions such as the Deucalionis Regio, Pyrrhae Regio and Protea, showing that on Mars there are regions of transition between the shaded and the bright.

The Titan canal was the subject of observations which were very perplexing, and more extraordinary than the preceding ones. It was seen along the 170° meridian; this line was visible until 9 January. From 10 January to 10 February, a second canal was seen beside it, also coming from above, from the Titanum Sinus, but directed toward the right-hand extremity of the Propontis. In a third period, from the 12 and 13th February, this second canal had disappeared, but another line could be seen, this time parallel to the first. What part can illusion play in this?

The Olympic Snow of 1879 was not recovered.

On his 1879 chart, Schiaparelli gave the name of the North Polar Sea (also shown on my map) to a long grey patch which seemed, in effect, to surround the North Pole. During his observations made in 1881–1882, he became convinced that this was not a vast stretch comparable with our South Polar Sea, but rather several seas or lakes, such as the Mare Acidalium, Propontis, Sinus Anian, Tanais, Alcyonius—not forming a continuous sheet, and probably leaving a sealess land at the North Pole.

The Maraldi Sea or Mare Cimmerium was seen in its usual form, and was very dark at its edges. But in its middle region, it was so bright that Schiaparelli regarded this region as a long island, resembling the tail of a comet; narrower and brilliant to the right, broad and less bright in its extension to the left.

The two islands of Thyle were shown as bright white patches as brilliant as the polar snows, and which had changed in position.

To the right of Elysium could be seen a curved double canal, the Hyblaeus. This is an instance of curvilinear gemination, on the planet.

As has already been noted above, the lower part of the Hourglass Sea—which we have designated Nilosyrtis—was seen during this opposition as broadened and darkened, attaining almost the breadth of the Mare Tyrrhenum, which was not the case in 1879. This broadening had been observed earlier by Secchi in 1858, Burton in 1871, Burton again in 1873, and Green also in 1873. We have here an instance of certain variation.

The same conclusion can be reached with regard to the neighbouring region named Boreosyrtis.

Libya presented a dark red hue, and its surface gave the impression of a plush cloth, or, if you like, vegetation being strewn with little pores.

The Atlantic Snow was visible throughout the opposition. Moreover, the region of Isis showed other white patches; in particular, a promontory which formed an angle between the Hourglass Sea and Nepenthes. The Coloe Palus was not recovered.

Summing up, the most curious of the discoveries made during this opposition, apart from the variations of tone and extent noted above, was that of the doubling of the canals, which should take up our main attention. There were more than 30 well-attested cases. Several had taken place before the very eyes of the observer, and the phenomenon had been completed in 24 hours. If we reflect that we are dealing with lines about 100 km broad and over 1,000 km long, the rapidity with which the phenomenon appears merits our most serious attention.

We are not dealing with the analogous effect of the doubling of a star when seen through a more powerful eyepiece, nor with the division of a simple line into two others, but with the addition of a new line to the side of the original one—and parallel to it, at a distance of 4–12° (that is to say, 240 to 700 km).

At the intersections of the double lines there is a darkening of the colour of these lines. We seem to see here a geometrical réseau of perfectly regular lines, made with a ruler, compass and Chinese ink.

This regularity, together with the transitory and probably periodical character of these strange formations, does not allow us to class them as geographical formations in the same way as with the patches which have been called seas, lakes, continents or islands. It also seems that the geninations regulate the original lines and make them uniform. Thus the Euphrates, seen as single in 1879, had several irregularities or undulations; double in 1882, it was perfectly clear-cut and regular. In 1879 the Jamuna was not of uniform breadth, but it was uniform in 1882, after its gemination. Before its doubling, the Haephestus was an elongated irregular patch, but after gemination it was seen as two perfectly regular streaks.

Gemination is usually preceded by a nebulous appearance of the canal. It seems that though this cloudiness is a preliminary, giving rise the phenomenon, it ends in the separation into two. It is rather like a group of soldiers who align themselves into two columns.

We have, therefore, variable formations, due to local causes and which are periodically reproduced in the same manner. By combining the dates of observation, we find that the phenomena correspond to certain seasons on Mars, beginning to manifest themselves toward the spring equinox in the northern hemisphere, which occurred on 8 December in 1881, and is particularly noticeable in the second month after the equinox; after having lasted for several weeks or even several months it disappears, so that nothing of it remains by the time of the northern solstice. These geminations therefore occupy all of the season which we call spring in the northern hemisphere. Is there any analogous phenomenon during the autumn? This is something which we cannot decide from the observation given above.

We must note that over the planet as a whole, there is a great tendency toward dualism and symmetry. Lakes are divided into two by isthmuses; the Herschel Strait has been seen longitudinally whitened in its mid-region, as has the Maraldi Sea; the Hourglass Sea ends at Burton Bay; the Meridian Bay is double, etc., etc.

As for the explanation...**There is nothing analogous on Earth**.

After the publication of Schiaparelli's three charts (Figs. 24, 35 and 45), the English astronomical review *The Observatory*, edited by Christie and Maunder, published a special article about them (1 May 1882), in which it was concluded that of these maps the second agreed better than the two others with the well-known streaks on the planet, so that it should be preferred to those of 1877 and 1881; also that certain canals were evidently the boundaries of areas which were slightly shaded, while others were illusions perhaps due to the use of so high a magnification. They added: "Nor would it be the first time that a distinguished astronomer has fallen into that error."

In general, English astronomers shared this feeling of skepticism with regard to the réseau of lines drawn by the Italian astronomer on his charts, as can be seen by reference to other specialized periodicals such as the *English Mechanic* and *Nature*.

At the meeting of the Royal Astronomical Society of London on 14 April 1882, reported in the same issue of *The Observatory*, there was a very interesting discussion about Schiaparelli's observations, between Green, Maunder and Rand Capron. In the *Times*, Proctor published an article about the canals and their doubling, and suggested that the inhabitants of Mars could be engaged in engineering works on a vast scale, so that the canals ran in all directions and kept an astonishingly regular distance from each other. Green commented:

> I have no intention of introducing a pleasantry in so serious a subject, but I believe that we cannot accept this singular appearance of Mars as being real until other observers have recorded it with certainty. The canals which have been seen during the last few years show constant changes, both according to the drawings of the same observer and according to several observers. The lines are shown in the drawings by Dawes, but the lines drawn by Dawes do not appear in those by Schiaparelli. On the other hand, Dawes' lines are again found in the drawings by Burton; I do not think that these lines are imaginary, but it seems that we are not dealing with permanent features on the planet.

Maunder, for his part, expressed the view that the canals drawn by the Milan observer were not real lines. Several could be due to optical illusions; others could be due œto the borders of shaded districts. Schiaparelli seemed to prolong the lines beyond their real length; for example, when two dark lines ran toward one another, he would prolong them until they met, which did not appear to be correct.

Green agreed with Maunder that the canals were simply the borders of lightly-shaded patches.

The English astronomers insisted that new observations must be made before they could accept the real existence of this strange réseau of straight lines which interrupted each other in various ways.

Same Opposition, 1881–1882.—Otto Bœddicker: Observations and Drawings[67]

At the 17 April 1882 meeting of the Royal Society of Dublin, Lord Rosse communicated the results of the observations made at his observatory at Birr Castle by Otto Bœddicker. These observations extended from 19 November 1881 to 23 January 1882, and were accompanied by 18 drawings. They were made with the great 3-ft reflector, magnification 216×.

In general, the bright or continental patches appeared orange, and the dark or maritime regions appeared blue.

The Hourglass Sea appeared to be bounded, all along its preceding border, by a zone which was distinctly bright.

Among the drawings are two which are reproduced here (Fig. 46). They are notable for the details shown, but at the same time they appear to be transitory and indecisive images. In the first, made on 20 December, the Meridian Bay lies in the centre; it resembles the leaf at the end of a ribbon, and recalls the old drawings of 1830. To its right may be seen the Arago Strait and Burton Bay running down to the Knobel Sea; this is Schiaparelli's Indus. On the second drawing, made on 26 December at $11^h 36^m$, the Hourglass Sea is presented; at the lower extremity the narrowing and the curvature are greatly exaggerated.

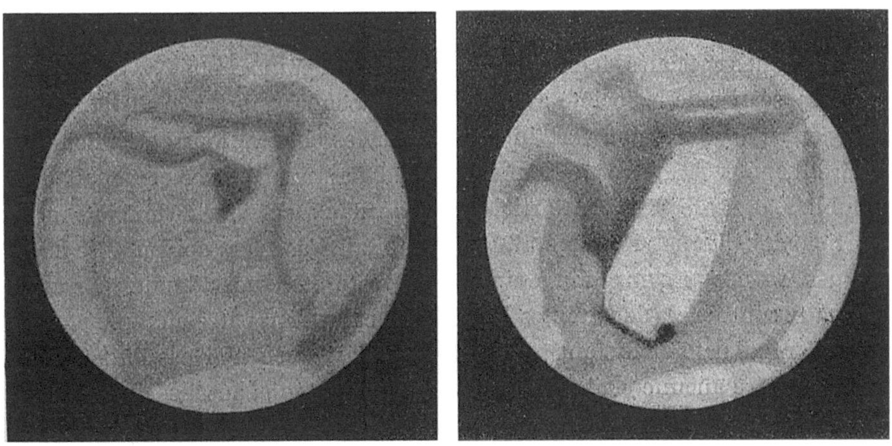

Fig. 46 Drawings of Mars, by O. Bœddicker, on 20 and 26 December 1881

[67] *Notes on the physical appearance of the planet Mars, Birr Castle Observatory. Scientific Transactions of the Royal Dublin Society*, 1882.

Same Opposition, 1881–1882.—C. E. Burton: Observations and Drawings[68]

These observations, made in February, March and April 1882 with a 9-in. (229^{mm}) reflector, magnifications 270× and 600×, were presented to the same Society on 17 April 1882. The author first commented that the north polar snow had been seen constantly, above all in two evenings of excellent definition, in a complicated and lobate form, a crevasse being well seen in the elliptical contour toward longitude 300°, as if the white material had melted more quickly than in the other areas under the influence of the Sun, then almost at the solstice.

The drawing made on 13 March is reproduced here (Fig. 47); it shows the snow of the lower pole as double-lobed. Two other white regions are visible, one in the region of the north end of Burckhardt Land and the other corresponding to the Atlantic Snow. The Hourglass Sea appears to be bordered by a white zone to the left or following side, as we have already noted in other circumstances

In this drawing, the Hourglass Sea is seen to stop at the Flammarion Sea, as if Libya, instead of being invaded by the grey tint, advanced into the sea. On 11 March, the whiteness was even more marked.

Burton believed that he had identified several of Schiaparelli's canals, but he could not confirm the doubling.

Fig. 47 Mars, on 13 March 1882, drawing by M. Burton

[68] *Notes on the aspect of Mars in 1882. Scientific Transactions of the Royal Dublin Society*, 1882.

Same Opposition, 1881–1882.—Niesten: Observations and Drawings[69]

The skilful astronomer of the Brussels Observatory made his observations between 12 December and 16 March, with the same instrument and under the same conditions as at the preceding opposition; they yielded 20 drawings with their descriptive summary, showing a great number of details. It was quite certain in this case also that the observer's eye played a major role in the results obtained. For instance, consider, among his drawings, those which are reproduced here, and which show exactly the same hemisphere of the planet, with the Hourglass Sea at the central meridian (longitude of the meridian 303° in the left-hand drawing, 304° in the right-hand drawing). Here is an extract from Niesten's description. Not without good reason, he used the old nomenclature for the large patches which were definite, and the new for the canals, which appeared to be variable.[70]

31 January.—A very curious drawing (Fig. 48A). The shadings appear as thin grayish lines which are doubled. The Hourglass Sea is darkest toward the east.

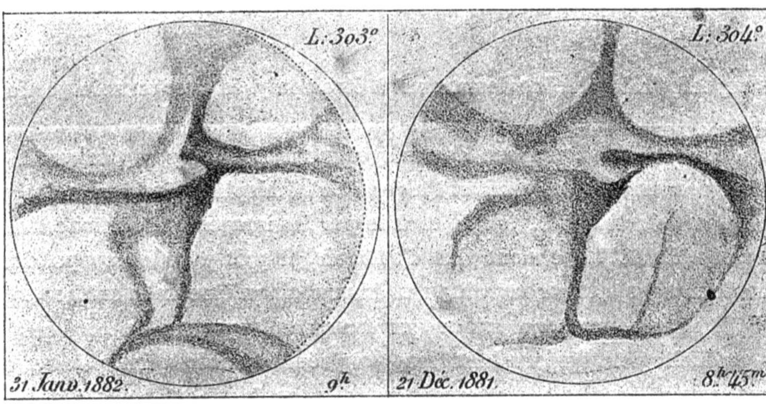

Fig. 48 Drawings of Mars, by M. Niesten, 31 January and 21 December 1881

[69] *Observations sur l'aspect physique de la planète Mars en 1881–82. Annales de l'Observatoires royal de Bruxelles*, Vol. VII, 1890.

[70] With regard to these nomenclatures, here is a report from the *Royal Astronomical Society of London*, dated February 1884: "It is most desirable that there should be some agreement established among astronomers on the question. The principle adopted by Proctor, of designating the 'lands' and 'seas' by the names of astronomers, was provisionally a convenient one, and this was continued by Green and Flammarion in their maps, but with modifications. Prof. Schiaparelli has adopted the divisions of land and water, but selected his names from ancient geography and history, and the confusion in the nomenclature thus introduced renders the discussion of any particular region of the planet rather difficult. It is desirable that these different systems should not continue, and that some agreed nomenclature should be generally adopted."

Libya is tinted with grey, that is to say that the Main Sea extends as far as the Flammarion Sea. The Thoth is very evident, as are the Protonilus and the Arethusa.

Thus in this drawing, the Thoth, which is seen to the left of the Hourglass Sea, is of equal size with the sea. This is not probable, and the effect must have been produced by an imperfect view of the region, with the details confused in a sort of grey shading. We are here at the limit of visibility.

This drawing is very curious because of the spacing of the longitudinal doubling of the Herschel Strait, which is rare but real.

The adjacent figure (Fig. 48B), of 21 December, shows a definite difference from the preceding one. The Hourglass Sea has turned back to its classical form: it is sombre to the east, grey to the west. The Main Sea is prolonged, turning toward the north-west and making up the beginning of the Thoth. The Herschel Strait appears to finish in a gulf the Hourglass Sea. We can recognize the Protonilus, from which rises the Euphrates, while the country immediately to the east is tinted with grey. In another drawing, made on 21 December, we find an analogous aspect; the canals seem to be the boundaries of tinted or veiled regions, as several English observers believe.

As we have seen, these observations are very valuable and they also extend the limits of features observable on Mars as far as possible. They confirm the Milan observations, without increasing the precision attained and giving a greater impression of vagueness.

M. N. Konkoly has published[71] the observations made at his observatory by N. A. de Gothard. There were 36 of Jupiter and 9 of Mars, from which three are reproduced in his article; those of 10 November and 22 and 25 December. They point out nothing in particular, except that the Hourglass Sea is shown as broad, particularly on the latter date. The instrument was a good Merz refractor. The observer remarked in particular that the colours of the patches were more evident in the central region of the disk than near the edge. The north polar cap constantly appeared as a beautiful white colour.

1882.—Trouvelot: Comments About the Planet Mars

We have already noted the observations made by Trouvelot in 1873, and we have begun to give an account of his deductions drawn in 1882 in his excellent handbook.[72] Let us continue with his exposition before turning to his observations of 1884.

There are aspects of the planet that necessitate its being examined with particular care.

We can easily mistake clouds for the polar snows. The observer noted that during winter in the Southern Hemisphere, the polar snow was invisible for most of the

[71] *Beobachtungen angestell am astrophysikalischen Oservatorium in O Gyalla*, Vol. IV. Halle, 1882.

[72] See p. 224.

time, hidden by the clouds which piled up in these regions. In 1877, for more than a month, he saw the polar cap hidden by clouds, which covered at least 1.5 of the total surface of the disk; he recognized his error only when the clouds gradually dispersed with the approach of summer, first very vaporous and then perfectly clear-cut, in the form of a cap much smaller than the earlier cloud-cover. These clouds resembled cloths of cumulus, forming in autumn and winter and dissolving in spring.

The observer believed that the polar ice disappeared completely in summer, and that the disappearance was total in 1877. This observation is not in accord with the others. Actually, a little snow always remained, a patch about 120 km in breadth, eccentric to the pole.

The author also concluded that the snows and ices persist on Mars at a temperature above that which would cause melting on the Earth, because here the Arctic and Antarctic snows never melt completely. "If the polar caps are composed of a white substance melting under the rays of the Sun, as seems altogether probable, the melting point must be above that of terrestrial snow," the author wrote. (In 1865 I outlined a view decidedly different. The conditions of atmospheric pressure, gravity, etc., being different from ours, the seas also ought to be of a different physical nature and should melt at a lower temperature. This means that for the mean temperature of Mars the zero for the planet—the freezing point of water—would be, as indicated elsewhere, above ours, because the effects produced by this temperature are analogous to those produced on Earth by a higher temperature.)

Most of the sombre patches on Mars, Trouvelot continued, in particular those whose north banks form an irregular band along the equatorial regions, are shown broadened to the side by a white border which follows all the sinuosities of the shore. This white border is variable. Sometimes it is excessively brilliant, particularly at certain points, where they are almost equal to the brightness of the pole; sometimes it is so feeble that it can scarcely be recognized, in spite of the transparency of the Martian atmosphere and the visibility of the patches. Adopting the view outlined by Green after his observations of 1877, Trouvelot attributes these white fringes of the coasts of Martian seas to condensation of vapour over the summits of *high mountain chains* bordering them, making them analogous to the Andes and the Rocky Mountains which border the Pacific Ocean. These elevated plateaux sometimes even produce projections along the terminator; the mountainous district of greatest altitude lies between latitude 60° and 70° south, near the western end of Gill Land, between longitudes 180° and 190°.

Hall Island, sometimes snow-covered and sometimes invisible, is without doubt also very high; it seems to be joined onto the coast.

In general, there are few clouds on Mars. But from time to time, mists veil the transparency to a greater or lesser extent. Once, for 8 consecutive weeks, from 12 December 1877 to 6 February 1878, one hemisphere remained completely foggy, while the other remained clear.

In short, Mars shows a very close resemblance to the planet upon which we live.

We shall continue later with Trouvelot's observations.

1882.—Downing and Pritchett: Diameter of Mars

From 537 measurements of the vertical diameter of Mars made with the meridian circle at Greenwich Observatory between 1851 and 1880, Downing gave a value of 9″.697 for the diameter.[73]

During the opposition of 1879–1880 and also 1881–1882, Pritchett made a good series of measurements with the Morrison equatorial (USA). His results are as follows[74]:

	Equatorial diameter	Polar diameter	
1879–1880	9″.638	9″422	
			{at unit distance
1881–1882	9″.635	9″.394	

If we neglect the flattening, we have: 9″.486±0″.033 for 1879, and 9″.484+0″.036 for 1882.

This value agrees with that found by Hartwig (9″.352) and that derived by Downing from the Greenwich observations given above.

Satellites.

During the opposition of 1881, Professor Pickering found for the brilliancy of Deimos, reduced to mean opposition distance, a magnitude of 13.13, that of the planet being taken as 1.29. The magnitudes found in 1877 and 1879 were 13.57 and 13.66.[75]

The brightness of the satellites of Mars varies in the following proportion, taking as unit the brightness on 1 October 1877 according to this author's observations:

1877	October 1	1.000
1879	September 21	0.490
1879	December 18	0.372
1881	November 16	0.303
1881	December 14	0.399
1882	January 13	0.330

They were observed at Ealing, near London, by A. A. Common on 21 September 1877, and by Hall at Washington until the following December.

Deimos, the outer satellite, was observed during the 1884 opposition by Asaph Hall. This means that the satellite has been seen at every opposition.

According to observations made by Pickering in 1881, the red colour of Mars is not shared by the satellites, notably with regard to Deimos.

73 *Monthly Notices*, Vol. XLI, p. 42.

74 *Astronomische Nachrichten*, no. 2652.—*The Observatory*, 1885, p. 135.—*Publications of the Morrison Observatory*, Glasgow, Missouri, Vol. I, p. 74.

75 *Astronomische Nachrichten*, no. 2437.

1883.—Marth. Rotation of Mars

The skilful and zealous calculator whose observations gave the ephemeris and precise positions for Mars for each opposition noted that the figure of the rate of the diurnal rotation of Mars, 350°.8922, which had been used for his calculations ever since 1864, and which was deduced from Kaiser's period of 24^h 37^m $22^s.62$, was remarkably precise.

This can be confirmed by the drawings made by Maraldi in 1704, which, in spite of their rudimentary aspect, do show that the patch which came to the centre of the disk in October 1704, slightly to the north of the middle, is Schiaparelli's Titanum Sinus, which came back to the same apparent position in 1877, 1894 and 1909. Comparison of these observations by Maraldi with those of Schiaparelli in November 1879, when the patch crossed the central meridian to the south of the centre, shows that the rate of rotation adopted is almost correct, because from 1704 to 1879 the difference did not exceed 6°.3948. The most accurate possible calculations for the rate of rotation give 350°.89217 for the tropical zone, leading to a sidereal rotation of

24^h 37^m $22^s.626$.

Opposition of 1884

Date of Opposition: 31 January
Presentation: the North Pole is tilted toward the Earth.

Positions of Mars				
Dates	Lat. of centre°	Diameter″	Phase defect″	Angle Sun/Earth°
1883 October 31	+16.3	7.6	0.9	39
1884 January 31	+14.8	13.9	0.0	3
1884 April 30	+17.6	7.4	0.8	37

Calendar of Mars		
	Southern hemisphere	Northern hemisphere
1884 May 13	Winter solstice	Summer solstice

This opposition, coinciding with the aphelion of Mars, was the complement of that of 1877, as can be judged from Fig. 49. The Sun is at one of the foci of the orbit of Mars, C is the centre of the orbit, M*s* the perihelion of Mars, M*n* its aphelion. The northern hemisphere of the planet is presented to the Earth.

We will begin our description of the observations made during this opposition with the work of Trouvelot, carried out at the Meudon Observatory.

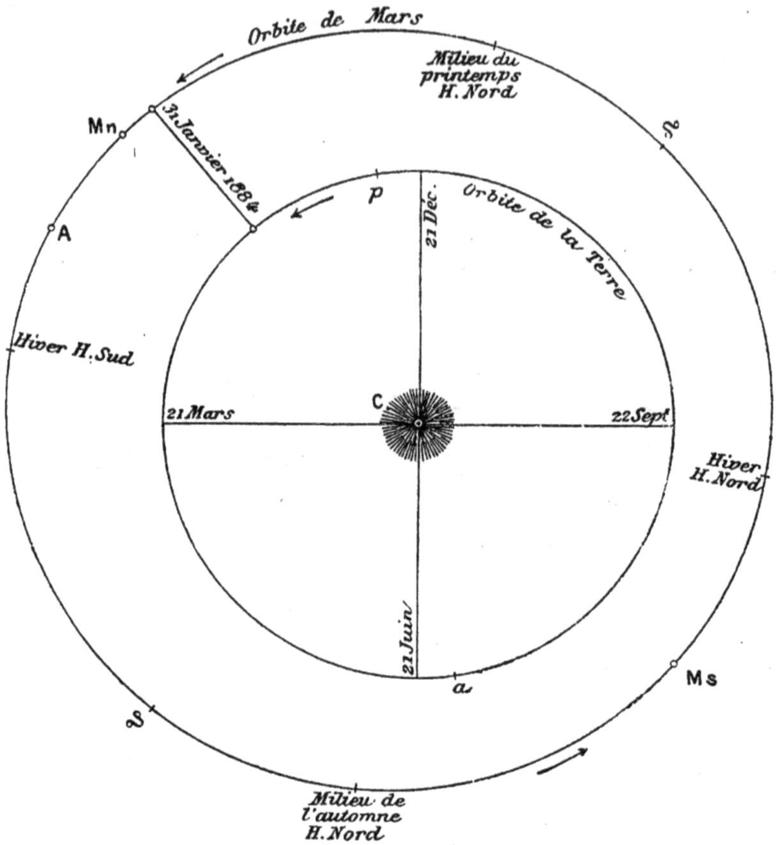

Fig. 49 Orbit of Mars for an aphelic opposition

1884.—Trouvelot: Observations and Drawings[76]

Here is the article published by Trouvelot, and the four views accompanying it.

> The Southern Hemisphere of Mars is sufficiently well known to astronomers today; there remains only the need to study some of the surface details, and the numerous variations caused by the seasons and by Martian meteorological phenomena. But it is not the same with the Northern Hemisphere, which, because of the greater distance of the planet at the times when this hemisphere is tilted toward us, is much more difficult to observe and is consequently less well-known. The observations of Mars made during the present year are of particular interest, mainly because the planet presented its little-known Northern Hemisphere and observers were concerned with studying its configurations. As the observers have carried out the work, we may perhaps hope that the results obtained will suffice for the completion of a general map of this interesting planet.

[76] *L'Astronomie, Revue mensuelle d'Astronomie populaire*, September 1884.

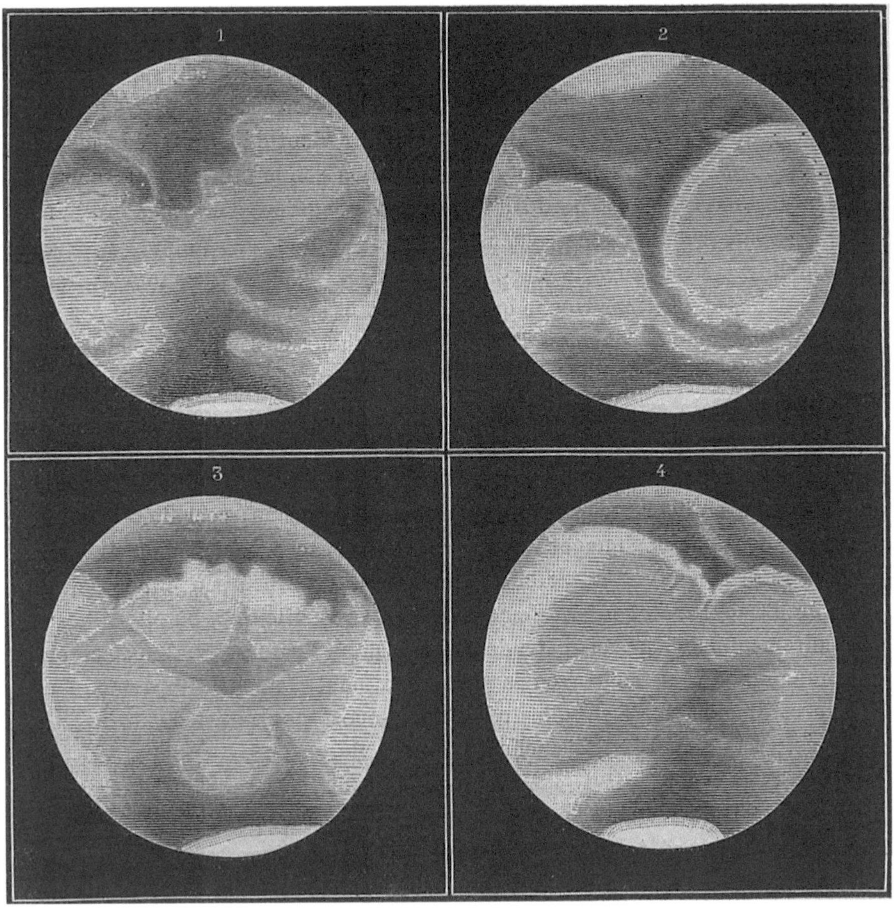

Fig. 50 Telescopic aspects of the planet Mars in 1884 (Observations and drawings by M. Trouvelot)

Since the end of last year, I have set to work, and although the atmospheric conditions have not been as favourable as I would have liked, my series of observations covers a reasonably extended period of time, so that I have seen the different points on the surface at several returns and am almost certain that I have recognized all the important features.

Among the fairly numerous drawings that I have obtained during this opposition, I have chosen four, which are reproduced here (Fig. 50, no. 1, 2, 3 and 4), because they show the whole circumference of the northern hemisphere of Mars and allow for the recognition of the principal features visible in this hemisphere.

To make these drawings understandable, I give here the text of the original observations; this will allow the reader to identify the features which are already known.

[The figure numbers below refer to those in our Fig. 50.-WS.]

Fig.1, 16 March, $7^h\,20^m$.—At the south-west is seen the eastern extremity of the Herschel II. Strait, which ends at the Meridian Bay. To the south-east, the De la Rue Ocean may be distinguished, extending up to the terminator. Burton Bay forms the extreme northern point, which lies a little to the west of the central meridian. Between the massif which abuts on Burton Bay, may be seen a narrow whitish band which joins Beer Continent to Phillips Island. To the south-west of these large dark patches and near the edge, is seen a white patch, undoubtedly due to vapours. The north polar cap has diminished; above it to the south are the Campani Sea and the Knobel Sea, which appear very sombre, and are clearly detached from Rosse Land, which is however less brilliant this evening than usual. The Knobel Sea comes back a little to the east, toward the Tycho Sea, from which it is separated by a whitish band, which is fairly broad but also very vague. The Tycho Sea appears as a sombre quadrilateral which, at the top, is surmounted by a paler, angular patch which is separated from the quadrilateral by a whitish band. To the east, this quadrilateral is mainly separated, by a whitish band, from a grey patch, which extends as far as the terminator and is associated with the Airy Sea. To the north-west, on the edge, may be seen the end of Lassell Sea and Le Verrier Land.

Fig.2, 15 February, $6^h\,45^m$.—The Hourglass Sea lies on the central meridian. As always, it is much darker—almost black—on its eastern edge, which is the border of an irregular, very brilliant fringe. Above, the brilliant fringe penetrates the Tycho Sea, and forms Banks Cape, which goes some way into the interior. The Flammarion Sea, to the west, is similarly fringed with white, as is the Hooke Sea, which lies above it. The Flammarion Sea is seen to be separated from the Hourglass Sea, to the east, by a narrow isthmus which, to the south, broadens out and forms a whitish triangle, at the middle of the latter sea. The bay which forms the Main Sea is visible, but decidedly vague. Toward the lower or north end of the Hourglass Sea, where the sea is very narrow, a widening to the east produces the Nasmyth Strait; it seems that this narrow sea is separated from the rest by a small white band. This must be caused by mists or clouds crossing the strait, because I have never seen it before. The north polar cap is bordered by the Delambre Sea, which, toward the west, is strongly accentuated, and rises toward the south, where it has an angular end in the neighbourhood of the Main Sea. Laplace Land seems to be connected with the great Herschel I. Continent by a narrow, whitish tongue. Between the south-west of the Main Sea and the Huggins Bay there is a fairly conspicuous white patch.

Fig.3, 27 February, $7^h\,45^m$. —To the south, not far from the edge, may be seen part of the Maraldi Sea, which extends from Burckhardt Land as far as Trouvelot Bay. The northern border of this long sea is fringed by a luminous band which has numerous waves along its length. A little to the west of the enormous area formed by this patch, Noble Cape may be clearly distinguished, forming a brilliant white serration on the Maraldi Sea. Not far from the centre of the disk may be seen a very singular grey oval patch, very diffuse at the edges which, at the east and at the west, is joined on to the bays of Huggins and Trouvelot by a narrow, vague, greyish band which curves back toward them. This remarkable oval patch was certainly not visible in 1877, 1878 and 1879, *although Mars was then closer to us.* [77]This oval patch is again attached to the Maraldi Sea by another narrow greyish band,

[77] This article has been reproduced in *The Observatory*, December 1884, and the editor comments that this oval patch is the Schiaparelli Ocean, while the three canals attached to the Maraldi Sea are the Titan, the Læstrygon and the Cyclops, all seen in 1877 by Schiaparelli. The Ocean was noted in the same year at Greenwich and the Cylcops canal in 1879 and 1882. The same aspect had been seen in 1877 at Potsdam and in 1879 by Burton. Trouvelot replied to this comment (*The Observatory*, 1885, p. 26) by saying that the oval patch agreed well in position with the river Ocean, but did not resemble it, in its ordinary appearance. The patch was darker and almost isolated.

which goes from north to south, and which I have often observed previously. At the edges of the north polar cap are seen two angular patches which go up toward the south. The more easterly of the two runs in the direction of the oval patch and curves back to the west, fading out a little before reaching the patch. The more westerly of the two, forms a very pronounced curve, and coming back toward the east, fades out gradually; it is joined to the oval patch by a band which is hard to see. To the west of this curved patch, coming back and running as far as the limb, is a very brilliant white patch.

Fig.4, 2 March, $6^h\ 40^m$.—To the south may be seen the western part of the Maraldi Sea, the Trouvelot Bay forming an angle slightly to the east of the central meridian. To the southwest, near the limb, is the large, narrow patch which descends from the south and goes to the terminator below the Terby Sea. From the Trouvelot Bay extends a vague greyish patch, already identified, which broadens and curves back toward the west. This vague patch is joined to the Maraldi Sea by a narrow feeble grey band, which lies a little to the east of Trouvelot Bay. The north polar cap is bounded to the south by a large sombre patch (undoubtedly the Oudemans Sea), which goes up toward the south, where it is seen to be separated by a narrow whitish band. Then, continuing on its way even though more indefinite, it forms an angular patch, whose outlines are very diffuse and difficult to recognize. To the east of the Oudemans Sea, near the limb, is seen Fontana Land, which is not very luminous. To the west of this same sea, a little above the polar cap, lies an elongated white patch, very easily visible, which is brilliant near the Oudemans Sea, and loses its brilliancy as it approaches, so that it is confused with the edge of the disk. The place where the terminator meets the south limb of the planet is *manifestly deformed*, because its curve, instead of being elliptical, as it should be if the surface were perfectly spherical in this region, makes a very pronounced obtuse angle, which indicates for this point a considerable elevation of the surface. This part of the limb seems, also, to be more luminous than the other regions.

Such is the summary of these observations. A casual glance is enough to show that the Northern Hemisphere of Mars differs notably from the southern, from the geographical point of view. On the latter hemisphere, the sombre patches are much larger, more numerous and more vigorous and better defined than those of the Northern Hemisphere. Here, it is only the Knobel and Delambre Seas which show up with reasonable sharpness, while to the south nearly all the patches are of remarkable sharpness, particularly along their northern borders. In general, the sombre patches of the Northern Hemisphere are vague and diffuse along their borders, so that it is difficult to decide upon their form.

According to my observations, it seems certain that the patches vary in their form and in their colour. Up to now we do not have sufficient data to decide with certainty on the causes of these changes; whether they are due to effects of illumination, or whether they are associated with the seasons, rains, or mists or clouds. Future observations will no doubt enable us to resolve these various problems.[78]

---

[78] In a note published in the *Comptes rendus*, 31 March 1884, Trouvelot was inclined to the view that certain patches on Mars could be due to vegetation, undergoing seasonal changes. The great continents of the Northern Hemisphere are occupied by grayish, more or less feeble patches which are spread over them. Judging from the changes which I have seen in these patches from year to year, one could believe that these variable grayish patches are due to Martian vegetation which is affected by the cycle of the seasons.

As for the disappearance of the polar snow, it is scarcely 3 months after the summer solstice in the Southern Hemisphere that I have on several occasions seen the solar patch disappear completely.

Same Opposition, 1884.—Knobel: Observations and Drawings[79]

The northern hemisphere of Mars is the less well-known, because the inclination of the axis—much the same as that of the Earth—means that the north pole is presented to us when the planet is furthest away. It is therefore doubly important to study these regions carefully, bearing in mind the disadvantageous conditions.

The observations by Knobel, our hard-working colleague of the Royal Astronomical Society of London, were made in January, February and March 1884, near the opposition date, when the planet—at its maximum opposition distance—passed within 100 million km of us, showing a disk of only 13″–14″. Among the numerous drawings made by Knobel with an 8 ½-in. (216^{mm}) reflector, using a silvered mirror made by Browning, and equipped with eyepieces giving magnifications of from 250× to 450×, I have chosen the four most interesting with regard to the northern regions, about which we are so anxious to find out more. These drawings (Fig. 51) and especially the chart constructed by the author (Fig. 52) fill in part of the gap on the maps of Mars, leaving out only the circumpolar regions.

First, let us note that the southern hemisphere differs geographically and perhaps meteorologically, from the northern not only because it is richer in sombre patches or seas, but also because these observations do not, even once, show a perfectly sharp geographical contour, if we sometimes except the northern extension of the Hourglass Sea, longitude 200°, latitude 30°–40°. All the contours are shown as vague and badly defined. This effect could perhaps be due to lesser atmospheric transparency, or perhaps to coasts which are genuinely less sharp, less abrupt, and less rugged. Knobel supports the idea that, without doubt, the cliffs in the southern hemisphere are steeper, and loftier and that the waters are held firmly between the coastlines, whereas in the northern hemisphere the borders and beaches are gentler and flatter and the rivers alter in level more gradually. The observations were made during summer in the Southern Hemisphere. As we have seen, we have here a vital point of special interest in our knowledge of the planet.

Though Knobel did not succeed in recognizing the canals noted by Schiaparelli, the following observations are worthy of note.

The canal designated under the names of Huggins Sea and Cyclopum Mare (longitude 200°–223°, crossing the equator) was observed very clearly on several occasions. It is absent from Green's chart. It emerges from the Maraldi Sea and runs in the direction of the Oudemans Sea. Drawing A (Fig. 51) was made under excellent conditions.

On this drawing, as on the chart, may be more noted a second canal, which corresponds to the Læstrygon. Knobel wrote:

> The space lying to the east of these canals is shown to be covered with a sort of very delicate réseau; not only does it seem mottled and marbled, but the borders of this mottling seem to be made up of faint lines. I cannot distinguish the straight parallel canals, but when I made

[79] *L'Astronomie*, June 1886, p. 201.—*Memoirs of the Royal Astronomical Society*, 1885, Vol. XLVIII, p. 2.

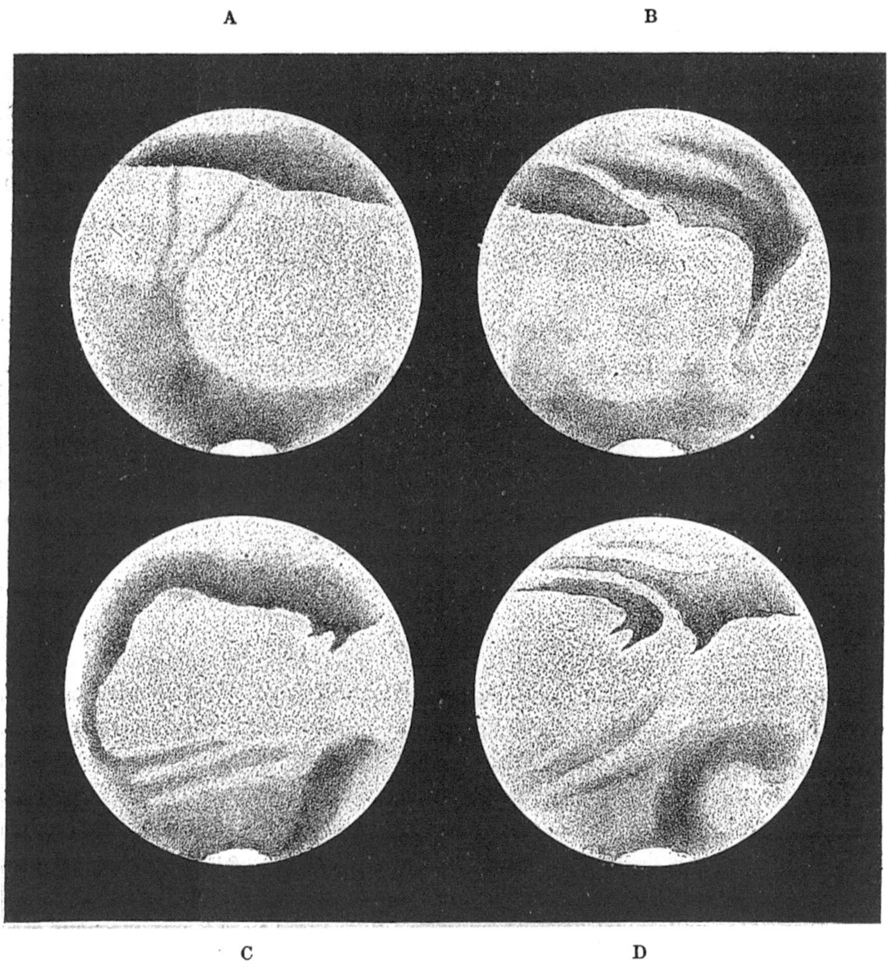

Fig. 51 Aspects of the planets Mars, after observations made in 1884, by M. Knobel A. 29 February 10^h, Long. 215°. B. 26 February 11^h, Long. 256°. C. 11 February 7^h 15^m, Long. 334°. D. 11 February 9^h 30^m, Long. 7°

the drawings the aspect was not so very different from the general aspect of the Milan drawings, which resemble a spider's web.

However, it is right to note that on this day 29 February, there was nothing to be seen of Fontana Land (200° to 238° and 13° to 46° B.) and perhaps clouds, which undoubtedly hid this region, have produced the aspect to which I refer.

On 26 February, Burckhardt Land—Hesperia— (220° to 255°, 40° to 10° A.) was perfectly visible. On the same date, the tone of the sombre regions west of the Hourglass Sea, known as the Flammarion Sea was not uniform. The lower part was certainly less dark than the upper part.

The Meridian Bay lay on the central meridian of the disk on 17 February, at 7^h 50^m. The English astronomer proposed to adopt the Terby Sea as the zero for Martian

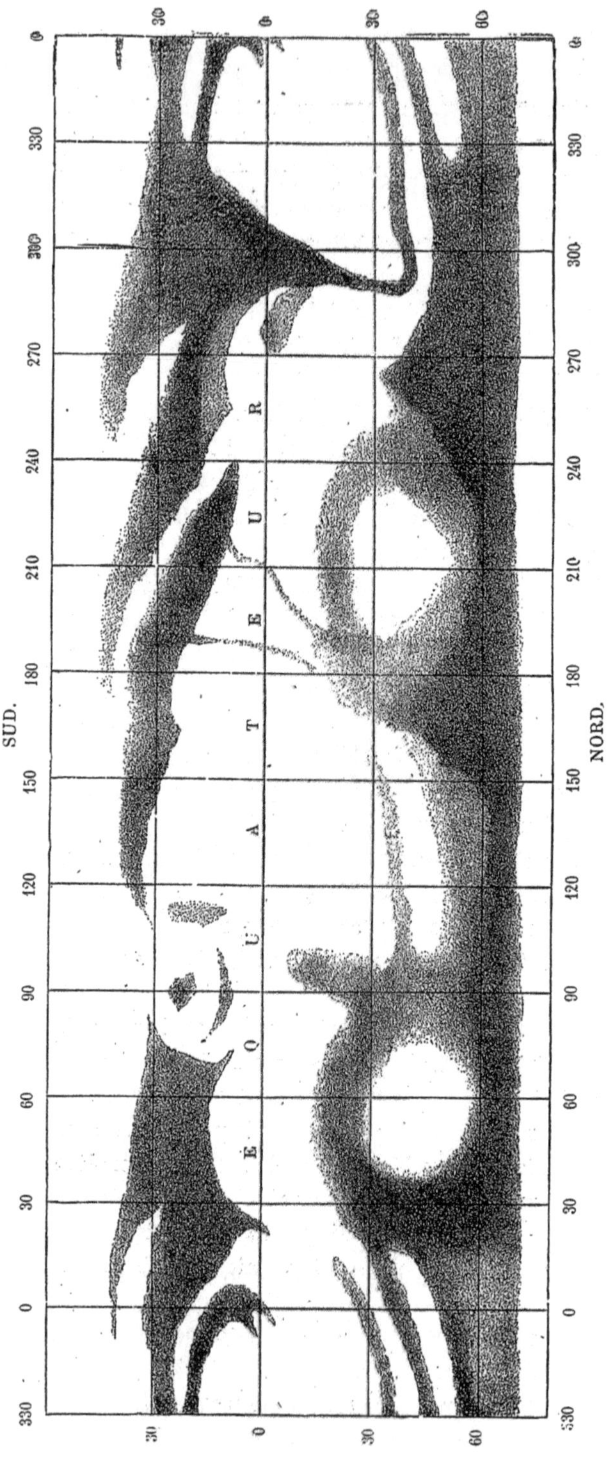

Fig. 52 Geographical chart of Mars, especially for the Northern Hemisphere, constructed by M. Knobel, from his observations between 1873 and 1884

longitudes, instead of the point adopted by Beer and Mädler, Proctor, Schiaparelli, etc. because the Terby Sea was better separated and could be more certainly identified. I believe that it is useless to make a change. In the time of Beer and Mädler, the Meridian Bay was the most characteristic feature upon the whole planet; it has not altered sensibly in this respect and in the future it could again become very dark.

Knobel again examined the curious prolongation of the Hourglass Sea known as the Nasmyth Canal. On 11 February only, this extremity was well seen; it was curved back on itself, not abruptly but almost imperceptibly, in the direction of Burton Bay, though without extending far enough to reach it.

Phillips Island (Deucalionis Regio), above the Meridian Bay, was shown as attached to the continent on 10 and 11 February; this was so clearly seen that there could be no doubt about it. However, the author on 21 October and Green in 1877, saw this region well separated from the continent by a grey tint. Variations!

The Knobel Sea (longitude 20° latitude 30°–65° B.) appeared to extend as far as the north polar snows. Its configuration differed from that shown in the old drawings by the absence of the white streak seen in 1873 and the white spot seen to the east of its centre. The most scrupulous attention was paid to the region, in order to try to verify the observations made in 1873 of the narrow, sombre bands crossing Le Verrier Land, to the west of the Knobel Sea. In no case was there a view of these bands, which had been so clearly traced in 1873; however, the drawings made on 11 February (Fig. 51) undoubtedly confirmed the old observations. It may be noted that the drawings made by Mädler in 1839, Jacob in 1854 and Schmidt in 1873 all showed narrow bands in this region, which led me to modify Green's chart in this respect. In 1884, each time the Knobel Sea was observed, a sombre space was seen touching the western coast—either homogeneous, or else divided into bands. The white space called Le Verrier Land was not seen.

Under excellent conditions of observation, the Terby Sea was seen very definitely on 5 and 6 February; so was the little dark patch to the north, named Agathodaemon by Schiaparelli. On the latter date, at $11^h 45^m$ GMT, the centre of the Terby Sea passed across the central meridian; this gives it a longitude of 83° instead of 90°.

On the same evening, 6 February, the Airy Sea was very distinct; it extended well to the south. According to the observations made on 24 and 29 January and 8 March, the western limit of the Oudemans Sea extended more than 10° to the west of the limit shown by Green.

The Acheron was seen as a broad grey streak, from 100° to 160° longitude, by 35°N. latitude, but it was shown as much broader than on Schiaparelli's chart; the two representations scarcely resembled each other, and Schiaparelli wrote:

> L'Acheronte è uno dei canali di Marte che ebbero la sorte di esser veduti distintemente da più di un osservatore; trovasi, infatti, disegnato con tutta la possibile chiarezza del signor Knobel sulla carta che accompagna le sue osservazione areografiche del 1848. (III, 46.)

Such were Knobel's observations. They confirmed the preceding conclusions: *certain changes in detail do occur on Mars.*

1884.—Terby: Comments About Mars[80]

"The fact which strikes me and astonishes me most," wrote Terby in discussing the drawings of Mars made by Schröter, "is the presence, in the drawings in the *Areographische Fragmente, of several patches which could be mistaken for the Hourglass Sea.* Schröter made 73 drawings of the planet in 1800 and 1801, and among these I have found 35 which seem, *at first sight, to show evident representations of the Hourglass Sea and Dawes Ocean.* Yet close examination shows that these 35 drawings *do not show the same region and therefore indicate the presence of several patches, all having the same appearance*… How can we explain the presence of these numerous patches shown by Schröter as ending in points toward the north?—*Patches which strongly resemble each other, but which correspond to different positions on the surface*? As we know, they greatly perplexed this skilful observer himself. I must comment that on Proctor's chart we find, beyond the Hourglass Sea, several other bays and straits directed toward the north: such are the Huggins and Bessel Channels, the Beer and Dawes Bays and Dawes Strait. But none of these is anything like so large as the Hourglass Sea.

The same singular features are indicated in the drawings by W. Herschel.

As an admittedly imperfect explanation, the thought has come to me that these bays may have *diminished in size* since the observations made by these two illustrious astronomers, but this opinion seems to me to be untenable. It is impossible to regard the drawings by Herschel and Schröter as being as reliable as those made in modern times with better instruments. Therefore, the problem remains unsolved.

In his first discoveries made in 1877, Schiaparelli gave a valuable clue: the bays concerned are prolonged to the north by canals—very delicate, it is true, but rather more like the objects seen by Schröter. The marvellous observations made from Milan in 1877 and 1879 combined show the broadening of the Hourglass Sea and the changes in detail of the features supposedly fixed to the planet.

The work carried out by Schiaparelli in 1881–1882 confirms all these marvels and in his map made at this time we find the Indus so developed, so broadened and so shaded that it is suddenly almost identical with the Hourglass Sea. This canal has undergone a great broadening and increase in prominence since 1877. With this astonishing modification we have, simultaneously, the mysterious phenomenon of the *gemination* or doubling of almost all the other canals.

1884.—Otto Bœddicker: Observations and Drawings

From Lord Rosse's observatory at Birr Castle, Bœddicker made a series of sketches from 24 February to 2 April. These sketches, 13 in number, were presented to the Royal Society of Dublin on 16 June, and published in its *Transactions*.[81]

The 3-ft reflector was used, magnifications 144× and 216×.

---

[80] *L'Astronomie*, June 1886, p. 207.

[81] Vol. III, Series II, 1885. *Notes on the aspect of the planet Mars*, etc.

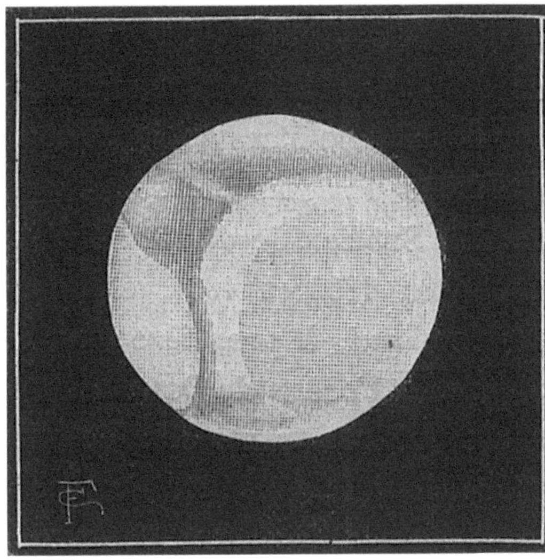

Fig. 53 24 February 1884

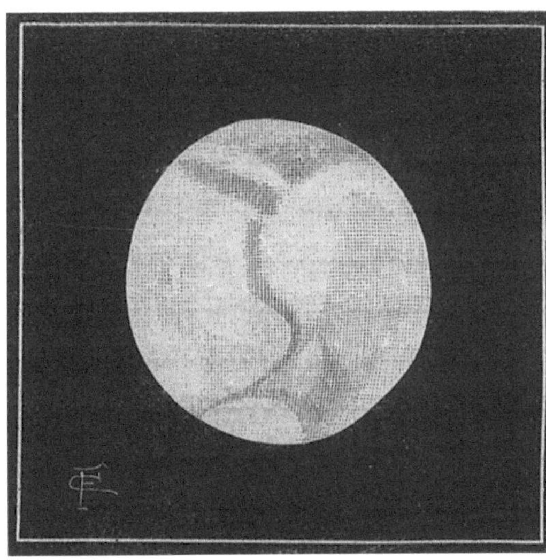

Fig. 54 22 March 1884

It is rather remarkable that the great reflector at Lord Rosse's observatory showed so little detail. This can be judged from the two best drawings, reproduced here. The first (Fig. 53) was made on 24 February, 21 days after opposition, and the second (Fig. 54) on 22 March. On the first the Hourglass Sea can be recognized, with the whole coast of the Herschel Strait and on the second (central meridian longitude, 25°) the Arago Strait and Burton Bay of my map and Schiaparelli's Indus descending to the Niliacus Lacus, as well as the Deuteronilus. This is all that we can say with certainty about these little drawings, made by a skilful observer with the aid of one of the world's largest telescopes. Are powerful instruments less valuable than smaller ones for studying Mars? Does atmospheric turbulence and warmth affect the most delicate images?

1884.—Denning: Length of the Rotation Period of Mars[82]

In spite of its relatively small diameter, and the slowness of rotation, Mars does offer remarkable opportunities for measuring the length of the rotation period. There is certainly no other planet which presents circumstances so favourable in this respect; the principal patches on Mars are definite and keep the same characteristic forms, whereas the details discernible on the other planets are due to temporary atmospheric

phenomena—and these are unfavourable circumstances which will render them less distinct, thereby hindering the observer in his attempts to measure the rotation period. Moreover, we can be certain that the details seen on Mars are permanent, genuine surface features, while the patches seen telescopically on other planets appear to be nothing more than effects produced by changes in their atmospheres.

The length of the rotation period of Mars has already been given with great precision, and it may seem superfluous to return to the subject; but it is always interesting to check whether recent observations are in agreement with the old results. The Hourglass Sea, which is generally regarded as the most easily seen feature on the planet, serves admirably for studies of the length of the rotation period. In 1869, Denning observed its transits of the central meridian with a 4¼-in. refractor; on 2 February it was central at 10^h, on 4 February at 11^h and on 5 February at $11^h\ 30^m$.

Denning observed the same feature in February 1884 with a magnification of 252× on a 10-in. reflector and found that transits occurred at the following times:

14 February 1884	$5^h\ 55^m$
15 February "	$6^h\ 35^m$
19 February "	$9^h\ 5^m$
22 February "	$11^h\ 4^m$

Denning compared his observation of 4 February 1869 with that of 14 February 1884. The interval corresponds to 5,487 days 18 hours 55 minutes = 474,144,900 seconds. A correction must be made for the difference in longitude between Mars and the Earth at the epoch and also for the phase.

It is useless to apply a correction for the speed of light, because at the two dates chosen for the comparison the apparent diameter of the planet was 16 seconds, so that its distance from the Earth was practically the same. After having made all these corrections, Denning found the length of the rotation period to be;

$24^h\ 37^m\ 22^s.34$.

This value, derived from an interval of 5,349 rotations, is in satisfactory agreement with the periods calculated by Kaiser, Schmidt and Proctor, from a much longer series of observations. The principal earlier determinations are as follows:

[82] *L'Astronomie*, Vol. III, August 1884, p. 296.

1837	J. H. Mädler	$24^h\ 37^m$	$23^s.8$	*Astron. Nachrichten*, no. 349
1864	F. Kaiser	"	$22^s.62$	*Astron. Nachrichten*, no. 1468
1866	R. Wolf	"	$22^s.90$	*Astron. Nachrichten*, no. 1623
1869	R. A. Proctor	"	$22^s.735$	*Monthly Notices*, Vol. XXIX, p. 232
1873	J. F. J. Schmidt	"	$22^s.57$	*Astron. Nachrichten*, no. 1965
1873	F. Kaiser	"	$22^s.591$	*Annalen der Leidenen Sternwarte*, Vol. III
1884	W. F. Denning	"	$22^s.34$	

Obviously Mädler's period, $24^h\ 37^m\ 23^s.8$, is about a second too long. If we take the mean of the six other values, which differ from each other by $0^s.6$ or less, we find a period of

$24^h\ 37^m\ 22^s.626$

which differs very little from that given earlier and is absolutely identical with the corrected period given by Marth.

1885.—Van De Sande Backhuyzen. Rotation Period of Mars[83]

Van de Sande Backhuyzen, Director of the Leyden Observatory, has produced a careful memoir about the rotation period of Mars in which, after an exhaustive discussion of a great number of observations made between those of Huygens in 1659 and Schiaparelli in 1879, he gives a still more accurate value than those listed above. His method consists in accepting the value of the period given by Proctor and the position of the north pole of Mars given by Schiaparelli and next working out the areographic longitudes of the principal features on Mars by using these elements. Comparing this rotation value with the observations, he obtained the corrections indicated by the latter. The value he obtained is

$24^h\ 37^m\ 22^s.626 \pm 0^s.0132$,

which is in exact agreement with that of Kaiser,

$24^h\ 37^m\ 22^s.626$.

The value obtained by Proctor seems, however, to be a little too great, as can be seen in the following table of the mean longitude of the north point of the Hourglass Sea, calculated for different oppositions:

Dates	Observers	Long°	Weight°
1661	Huygens	315.7	1.5
1666	Hooke	296.4	0.3

(continued)

[83] *Untersuchungen über die Rotationszeit des Planeten Mars und uber Aenderungen seiner Flecke.* Leyden, 1885.

(continued)

Dates	Observers	Long°	Weight°
1782	Herschel	305.8	3.0
1799	Schröter	303.3	16.3
1830	Beer and Mädler	299.5	3.0
1862	Kaiser, Lockyer, Rosse	294.9	7.5
1864	Kaiser, Dawes	294.3	3.0
1877	Schiaparelli, Lohse, Green, Nies-ten, Dreyer	289.6	15.0
1879	Schiaparelli	288.4	8.0

There is, with the exception of Hooke's value, a slow decrease of longitude with time; and as it was on Hooke's two drawings that Proctor's work was based, the excess in his determinations is easily explained.

Backhuyzen ended his memoir with an appreciation of the changes observed on the surface of Mars. He established an important fact with regard to the variations which have so often been noted—that the elongated sea called Huggins Bay on my map and which Schiaparelli called Cyclopum, was much broader in the time of William Herschel and Schröter than it is now; and it was then comparable in form and extent with the Hourglass Sea. Schröter seems to have observed the Læstrygon canal, which he could hardly have done if the region had not been better marked in his day than it is in ours.

Opposition of 1886

Date of Opposition: 6 March
Presentation: The North Pole is inclined toward the Earth.

Calendar of Mars				
	Southern hemisphere		Northern hemisphere	
31 March 1886	Winter solstice		Summer solstice	
	Lat. of centre°	Diameter″	Phase defect°	Angle Sun/Earth°
6 January	+23.3	9.20	0.75	33
6 February	+22.5	12.18	0.40	21
6 March (opposition)	+21.9	13.95	0.01	2
6 April	+21.9	12.43	0.50	23
6 May	+23.4	9.82	0.88	35
6 June	+25.3	7.84	0.89	39

1886.—Denning: Observations and Drawings[84]

During March and April 1886 Denning made a series of observations of Mars from his observatory at Bristol, using a reflector with a silvered 10-in. (254^{mm}) mirror by With of Hereford. The magnifications used were from 252× to 475×; but there was

[84] *L'Astronomie*, September 1886, p. 321.

little advantage in using the latter, which seemed to be too high. In general, the 252× eyepiece was sufficient, though under certain circumstances a power of 350× was advantageous.

The planet came to opposition on 6 March, but during the previous 3 weeks of this month conditions were so icy that it was impossible for Denning to begin his observations before the end of March. The position of Mars was favourable, though the apparent diameter was small. But the interest of the actual observations lay in the fact that it was the northern hemisphere which was presented for examination and this hemisphere had not been studied so completely as the southern; neither did it offer such clear features. The latitude of the centre of the disk was about 22°N during March and April.

The patches observed were numerous and varied; there was evidently *a quantity of details* on the planet, but it is extremely difficult to combine them into a satisfactory representation. A large number of very feeble patches strike the eye strongly enough for one to be certain of their existence; but they cannot be distinguished with enough clarity and precision for their outlines to be recognized, or the relative positions to be assigned. Only the best-marked features could be drawn in a satisfactory manner. The small diameter of Mars during these observations certainly contributed in large measure to the uncertainty of the physical aspect of the disk. Another cause of uncertainty lay in the rarity of really good telescopic images. Not only must the atmospheric conditions be particularly favourable if the images are to be perfectly clear, but there must also be a complete absence of wind. The slightest vibrations hinder the following and studying of a complicated system of patches and details. Finally, as a telescopic subject Mars is much less satisfying than Jupiter or Saturn. All these circumstances explain the uncertainty of the observations and the discrepancies which are found between drawings of the details visible on the surface.

Here are Denning's observations:

From 23 March to 30 April the planet was examined on 22 evenings and a considerable number of drawings made. During this period, there was an exceptional series of good nights and every time the image was sufficiently good, the observer's details were carefully noted; then the results were compared with each other, as with analogous work carried out earlier.

My drawings agree exactly with each other and show good agreement with the charts by Green, Schiaparelli, Flammarion, Knobel, etc. I have also compared them with the views given in Terby's *Areography*, and with the drawings obtained by Bœddicker in 1881 and 1884 with Lord Rosse's 3-ft. ($0^m.915$) reflector. This comparison has given me a new confirmation of my work. Some discrepancies are more marked than might seem probable; but experience has shown that it is useless to hope for uniformity in the representation of planetary details.

During the five weeks of my observations, I found no certain proof of change in any of the patches; but the period was too limited, and the circumstances under which the observations were made were too unfavourable, for me to make any certain pronouncement on this point. The slight differences shown on my drawings are simply of the same order as those which are due to changes in the local atmospheric condition. During a poor night, very feeble markings, distinguished earlier, were effaced, while during the best nights I saw delicate details which it had been impossible to suspect under less favourable conditions. I am convinced that such changes in the conditions of seeing exercise a considerable influence upon the apparent configuration of the planet, more considerable even than observers generally admit. It is sometimes concluded too hastily that real changes have occurred; true modifications can be established only after very careful examination and according to indisputable proof.

Most of the better-defined seas, show very brilliant outer borders, with very sharp limits. These brilliant borders recall the luminous regions on Jupiter which often confine sombre patches, only on Mars, they are more extended and more permanent and also of dissimilar forms. As a special case of one of these brilliant borders, I can cite the region which extends along the eastern bank of the Hourglass Sea. I have sometimes seen it as so luminous, that in brilliancy it has rivalled the white cap at the North Pole. It extends several degrees to the east of the shaded contour of the sea, and is bounded by a feeble, irregularly condensed patch which is prolonged to the north and is inclined to the east, starting from a point at longitude 290°, immediately to the east of the northernmost extremity of the Hourglass Sea (see Fig. 55, II and III). This streak is very long; it extends almost to just below the Meridian Bay and Burton Bay, to both of which it is connected by faint ligaments which recall Schiaparelli's canals (see Fig. 55, IV). This particular patch which is not shown on Green's

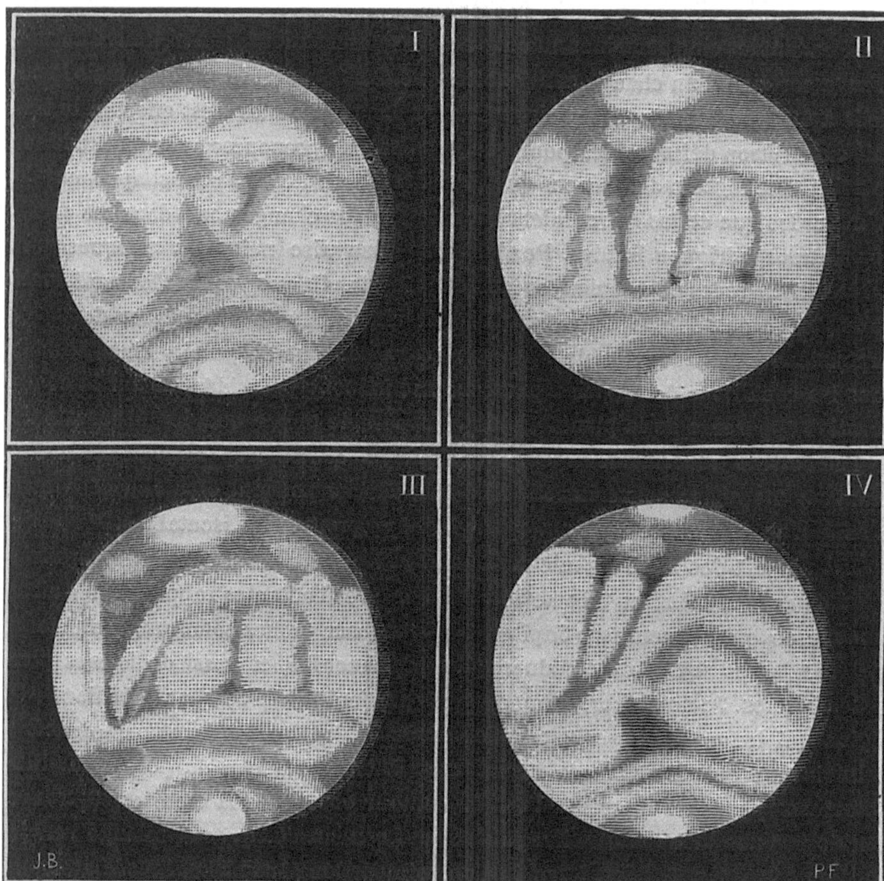

Fig. 55 Aspects of the planet Mars, from observations in 1886, by M. Denning. I. 27 April, $8^h\ 40^m$, Long. 187°. II. 13 April, $8^h\ 0^m$, Long. 305°. III. 13 April, $9^h\ 50^m$, Long. 332°, IV. 3 April, $7^h\ 30^m$, Long. 28°

map, is perhaps identical with the réseau of narrow dark bands drawn in this region by Schiaparelli on his map. It is also shown more or less clearly defined on some other drawings, notably in one by Schmidt.

As for the Hourglass Sea, it is shown as very feeble and very narrow, if not suddenly broken off in the region at 10° to 15° south of its northern extremity (see Fig. 55, II and III). This particular fact was represented well in the drawings by Bœddicker. On the other drawings, I could not find the appearance shown adequately. It is evident, at least, when the region concerned is near the centre of the disk, as during the last opposition.

Knobel's drawings of 1873 agree better in general with mine than do those of the same author made in 1884. On Green's map, the Knobel Sea is separated at its southern extremity from the weak curved band which extends to the east, as in drawings Nos. 6, 7, 8 and 9 of 1873. This break does not appear in the later drawings of 1884, so that this region seems to have undergone some change in aspect, at least if the difference in inclination is not the case of the lack of agreement between the observations. It is probable, however, that this is the true cause, because the inclination of the planet in April and May 1873 was almost exactly the same as in March and April 1886, and it is precisely in these two periods that the drawings most closely resemble each other in the most noticeable forms. I see the northern bank of the Knobel Sea distinctly separated from the shaded longitudinal band immediately adjacent to the north polar cap (Fig. 55, IV). Drawing No.12 of 19 May 1873, by Knobel, draw in the whole mass of shading which surrounds the north pole as being shaded without interruption; but these differences in aspect are undoubtedly due to changes in inclination.

With regard to the canals observed by Schiaparelli, I distinguished a large number of features which strongly indicate the existence of a similar configuration; but the drawings made in Italy during the three months from October 1881, to February 1882 give them a definite character, and quite apart from their doubling, a straightness of form and a general uniformity in tone which my observations do not confirm. To me, the more complex and more delicate details of the planet appear, under the most favourable conditions, as linear shadings, which are extremely feeble, with evident gradations in tone and with irregularities which produce breaks or condensations. If they exist in the same form and with the same certainty of direction as represented by Schiaparelli, they should have been easily seen here every time the definition was sufficiently good; because the features are indicated as being easily observable in the be 8.6-in. (218^{mm}) refractor of the Milan Observatory in February 1882, when the diameter of the planet was only 13″. The doubling of these lines was also seen under the same unfavourable conditions. What is more astonishing is that the eminent Italian astronomer should have discovered such marvellous details on the surface of the planet—because these details undoubtedly exist—and to have observed such a complex and difficult configuration at a time when Mars was particularly badly placed for observations of so delicate a nature.

During the last months, the north polar cap of Mars has been very brilliant; it often presents a striking contrast with the less reflective regions of the surface. There were also other noticeably brilliant parts of the disk. These luminous regions of Mars deserve at least as much attention as the shaded parts, because it is probably in their appearance that changes can be most clearly observed in so far as these are real modifications on the surface. We do not attach sufficient importance to these white patches.

Most of our principal treatises on astronomy say that Mars has a dense atmosphere; during my observations I saw nothing of a nature to confirm this theory. It seems to me much more reasonable to admit that the atmosphere is extremely rarefied. The principal patches are invariably visible, and the differences observed seem to be due rather to the influence of our atmosphere than to that of Mars. Jupiter and Saturn are undoubtedly enveloped in thick vapours which hide their surfaces from our eyes. The patches which we observe there are atmospheric, though in some cases very persistent; they are constantly subject to modifications in aspect and changes in position due to longitudinal currents. On Mars, the nature of things is quite otherwise. Here, the features observed and incontestably geographical configurations and they show none of the variations which are so remarkable with the details on Jupiter. It is probable that most, if not all, of the changes which we believe we have seen in the appearance of these patches on Mars are due simply to the diversity of the conditions under which the planet has necessarily to be studied. If the conditions of observation were always the same, then there would be much greater uniformity in the results obtained. The very clear-cut character of the patches and their great permanence, are entirely opposed to the idea that, the planet can be surrounded by a thick, cloud-laden atmosphere.

Such are the interesting observations made by Denning in 1886. Despite the excellent reasoning—reasoning which may be accepted without reserve—there can be no doubt that the surface of Mars is subject to real variations, considerable and frequent. At the meeting of the Royal Astronomical Society of London on 14 May 1886, Green made important comments about Knobel's observations of 1884. He stated that in 1886 he had confirmed most of the observations, but had nevertheless found certain curious differences. Knobel had stated that Green's map had shown a sign of strengthening near the Knobel Sea, where he had found dark bands instead of the bright area named Le Verrier Land. Green had recognized the bright area very clearly during the opposition of 1886. He wrote: "The comparison of the two series of observations shows that changes occur from time to time in several regions of Mars. The Lassell Sea, which during the last opposition was almost as distinct as the Eye, and Huggins Bay may be cited as examples; the latter is shown as broad and well marked."

One of the most remarkable facts observed during the opposition of 1886 has been the frequent appearance of luminous mists near the limb, which never reach the meridian; and also, orange patches seen at the meridian which become white when reaching the edge. Can we not conclude that this is due to cloudy condensations first seen on the right hand side of the planet and which are dispersed when they reach the meridian beneath the Sun?

1886.—Perrotin: Observations of the Canals

This skillful French observer reported on the observations at the Nice Observatory as follows[85]:

During the last opposition of Mars, Thollon and I have devoted several evenings to studying the configurations on the planet, with the aid of the $0^{m}.38$ equatorial at the Observatory of Nice.

Beginning only at the end of March, because of bad weather, the observations were followed through until the middle of June on every occasion when circumstances permitted. Our aims were to recognize the single or double canals discovered by Schiaparelli but scarcely seen by any observers other than him.

Conditions for observation were very unfavourable, because of the planet's small apparent diameter, which even at opposition, on 6 March, never exceeded 14″, while it attained 25″ when the eminent Italian astronomer made his observations in 1877.

[85] *Bulletin astronomique*, July 1886, p. 324.

Our first attempts to see the canals were not encouraging and after several days of fruitless searching, explicable partly by the bad quality of the images and partly because of the actual difficulty of an investigation of this kind and after once having given up and subsequently returned to the investigation, we were about to abandon the attempt definitely when, on 15 April, I managed to distinguish one of the canals situated to the west of the Hourglass Sea, Schiaparelli's Syrtis Major, connecting the sea with the Herschel Strait (Sinus Sabæus).

Thollon saw it similarly soon afterwards.

By the end of that night, under good conditions, we had been able to recognize successively, several canals presenting, in nearly all respects, almost the character attributed to them by the Director of the Milan Observatory.

These canals are described by Schiaparelli, and as we have seen them, make up in the equatorial region of the planet a réseau of lines which seem to follow the arcs of great circles. They cross the zone of the continents in all directions and form connections with the seas of the two hemispheres or, alternatively, the canals between them. They cut each other at all angles, and show up against the bright background of the disk as greyish tinted lines of different degrees of darkness.

On the chart published in *L'Astronomie*, reproduced here, Perrotin has lettered the configurations recorded (p. 355, original). Details are as follows. Magnifications used, 450× and 560×. Observations made generally between 8^h and 10^h.

First Region, lying between 290° and 350° areocentric longitude.

On 15 April, we distinctly saw canal AB (*Phison*) [Fig. 56] and at moments we believe we suspected a finer line CD, parallel to the first. We similarly saw FEA (*Astaboras*) and HG and DK (*Euphrates*), the two latter being parallel and not diverging as in the drawing

Fig. 56 Caals observed by MM. Perrotin and Thollon, in 1886

On 19 and 21 May, when the region again passed across the centre of the disk at a convenient hour, we saw the same objects and also the canal FG, which cut the Phison canal at a right angle. FG did not seem to have its origin at F, as shown on the map, but at a point closer to the equator, almost at Lacus Moeris.

Second Region, between longitude 180° and 260°. On 23, 24 and 25 April we distinguished LM (*Stygia Palus*) and LN, LO and OP (*Cyclopum*) as single canals. At moments we believe that we saw LO as double, but this was a fugitive impression. We again saw the same canals on 25, 26 and 31 May and 1 June; on the first two days we saw, beyond them, RQ (*Æthiopum*) and R′Q′, which, contrary to the drawing, is a straight continuous line parallel to RQ. On the 26th, I succeeded in seeing a section of the double canal QO (*Eunostos*), which comes from the northern extremity of the single canal QR. On 1 June, M. Gautier saw LO at the same time as us. Since our first observations, the canal LN has undergone a considerable change; it can be distinguished only as a feeble stretch and to the northern side only. This canal was shown on Schiaparelli's map of 1882, but not on that of 1879. Our observations therefore confirm that there are almost constant changes and also that these changes can occur over a short period of time.

Third Region, between longitudes 30° and 100°. On 11 May, the double canal R6 (*Nilus II*) and TU (*Iridis*) appeared clearly. M. Trépied, who was visiting the Observatory, saw them without too much difficulty and although he had no knowledge of the chart, he is the first to have noted the two shaded parallel lines which make up the double canal TU. Thollon only suspected the doubling. In the canal R″S, the two lines making up the portion R″Z seem to us to be finer than as indicated on the drawing; the two lines of the portion ZS seem, on the other hand, to be darker. We similarly saw the line VZ. On the 16th, I saw, with certainty, the double rectilinear canal XY (*Jamuna*). On the other hand, on neither the 11th or the 16th could we see the canal XZ (*Ganges*), shown as double on the map. On 12 June we easily distinguished the canal TT′ (*Fortunae*), which could well be double. During these observations, the Nile appeared clearly throughout its length, much more marked than as shown on the map. The canals which we have enumerated, most of which were seen twice or more by several observers, are in the positions drawn by Schiaparelli in 1882. In general, their appearance differs little from that shown on the map; but several shown as double appeared to be single, no doubt because of the greater distance of Mars at this opposition. It seems, therefore, that at the equatorial region of the planet there is a state of things which, if not absolutely permanent, is not modified in any essential manner.

Changes observed on Mars.—During our studies of the canals, there was a notable but short-lived change in the region of the Hourglass Sea, which is worthy of being placed on record. When we made our first observations, this part of the surface was sombre as the seas are, and sensibly in agreement with the map; but when we saw the area again, on 21 May, the aspect was very different. The part of the Syrtis Major which extends between N latitude 10° and 55° was hidden by a luminous veil, of the colour of the continents, but softer and less vivid. It could be said that the clouds or mists were arranged in parallel regular bands, running NW - SE. At moments these clouds became transparent and allowed us to glimpse the outlines of the prolongation of the Syrtis Major. On 22 May the clouds were arranged more uniformly than before; they were again seen on 23, 24 and 25 May, but with diminishing intensity. They probably extended a long way back on the continents, to the east and west of the sea, because from one day to another, or sometimes even during the course of the same evening, the neighbouring sombre regions, among others the Lacus Moeris to the east and the Nile to the west, were sometimes visible and sometimes not. On 25 May we were again able to see the isthmus drawn in the prolongation of the Syrtis Major, beyond its junction with the Nile, at 300° longitude and 52°N. latitude and which had remained hidden up to this time. On the same date, we detected an increased darkening of the continents in the immediate neighbourhood of the sea. During these singular appearances, the southern part of the Syrtis Major, which had not been covered by clouds, became darker and showed a characteristic bluish-green tint. Are phenomena of this kind really produced by clouds or mists, circulating in the atmosphere of Mars? It seems probable. In any case, they are due to an element appertaining to the atmosphere on the surface of the planet, capable of movement and modification over a relatively short period. While making the observations described above, we noted two or three brilliant points round the white patch at the North Pole, a short distance from the cap, between longitudes 200° and 280°. They were similar to those noted by Green in 1877, from Madeira, around the southern cap at the time of the winter solstice on Mars. Our observations, made about fifty days after the summer solstice, supports that of the English astronomer, apparently indicating that during the decrease of each polar cap, at the time of the corresponding solstice and particularly after the summer solstice, we see a phenomenon not peculiar to this apparition.

Such are all the facts regarding the observations by Perrotin and Thollon at Nice. In every respect, their study confirms the beautiful discoveries of Schiaparelli on the singular physical constitution of Mars. Remarkable also are the clouds observed over the Hourglass Sea, which are truly rare.

1886.—Walter Wislicenus: Studies of the Rotation of Mars[86]

Wislecenus, astronomer at the Strasbourg Observatory, first reviews all the observations made of the planet; he then examines the published maps and discusses the various systems of nomenclature, giving a synoptic table of them. He next calculates the projections of the globe of Mars as seen from the Earth.

According to the observations made by Winnecke at Strasbourg in 1877, the position of the south polar cap was:

Distance from areographic pole	4°.43 ± 0°.591
Areographic longitude	20°.67 ± 5°.711

The direction of the axis of Mars on the celestial sphere is, for the North Pole:

Ascension	317°55′.1	Declination	+50°15′.7

Comparing the principal observations of the positions of the patches on Mars, from Huygens in 1659 to Bœddicker in 1881, Wislicenus gives for the precise value of the rotation period:

$24^h\ 37^m\ 22^s.655 \pm 0^s.00861$,

with a degree of accuracy which does indeed seem excessive.

To the preceding studies, we can also add those of less specialized observers, such as Guiot and Soissons (*L'Astronomie*, October 1886, p.393), and also Lihou, at the Flammarion Scientific Society at Marseilles, etc., who show great skill even though using only modest instruments.

---

[86] *Beitrag zur Bestimmung der Rotationszeit des Planeten Mars*. Leipzig, 1886.

Note also that on 13 April 1886, Mars passed in transit across the face of the Sun as seen from Jupiter, and the Earth was in transit on 12 November 1879.

Opposition of 1888

Date of Opposition: 11 April[87]
Presentation: the North Pole was turned toward the Earth.

	Lat. of centre°	Diameter″	Phase defect°	Angle Sun/Earth°
11 Feburary	+19.4	9.6	0.77	33
11 March	+18.6	12.7	0.47	22
11 April (opposition)	+21.1	15.4	0.00	1
11 May	+24.0	14.2	0.59	23
11 June	+24.8	11.4	1.16	37
11 July	+23.4	9.2	1.20	42

Martian calendar		
	Southern hemisphere	Northern hemisphere
16 February 1888	Winter solstice	Summer solstice
15 August 1888	Spring equinox	Autumnal equinox
8 January 1889	Summer solstice	Winter solstice

1883–1888.—O. Lohse: Observations and Drawings

Lohse continued his studies of Mars during the oppositions of 1883, 1884, 1886 and 1888 and published his results in 1891. He was principally concerned with measurements of the position angle of the north polar cap and also made a large number of drawings of the planet, of which 36 were published in his memoir, followed by a map drawn from them.

According to his observations the position angle of the axis of Mars was as follows:

1884	8 February	$0^h\ 0^m$	(GMT)	357°.226	±0°.185
1886	22 February	$0^h\ 0^m$	Id.	21°.84	0°.310
1888	24 May	$0^h\ 0^m$	Id.	30°.66	0°.491

[87] *Conjunction of Mars and Uranus, 5 May.*—Several observers, notably Bruguière at Marseilles, Guiot and Soissons at Valderrama, (island of Tenerife), observed the conjunction of Mars and Uranus on 5 May 1888. Mars passed 35′ to the north of Uranus, so that the two planets were visible in the same field, Mars reddish and of the first magnitude, Uranus bluish by contrast, of the sixth magnitude and overpowered by the brilliance of Mars. This curious conjunction took place near the fifth magnitude star Theta Virginis.

Fig. 57 Chart of Mars, by M. Lohse, from his observations of 1883–1884

The drawings were made from 15 September 1883 to 17 March 1884, and from 30 January to 7 April 1886. In 1888, he found a value of 289°.56 for the longitude of the Hourglass Sea.

The chart given here (Fig. 57) is from Lohse's observations made in 1883–1884 with the 11-in. equatorial of the Potsdam Observatory. It inspires the same reflections as Lohse's first and, to be honest, bears no resemblance to Mars. The Meridian Bay, the Herschel Strait and the Hourglass Sea are the only features which can be recognized. But what has become of the Terby Sea, which on his other chart Lohse had shown as quadrangular? What can be recognized in the region to the left of the Hourglass Sea? Lohse listed 18 landmarks identifiable on his chart and that of Schiaparelli. They are as follows:

		Long.°	Lat.°
1	East coast of the Margaritifer Sinus	27	−7
2	Region of Proteus. Middle	38	−27
3	Tempe. Middle	76	+47
4	Solis Lacus. Middle	81	−31
5	Arcadia. Middle	125	+47
6	Elysium. Middle	218	+34
7	Sinus Alcyonius. S. point	265	+35
8	Lacus Mœris	265	+7
9	Circe Promontorium	275	−10
10	Nilosyrtis. East coast	293	+30
11	Hourglass Sea. East coast	294	+7
12	Sinus Sabæus	326	−14
13	Forked Bay. West point	346	0
14	Fastigium Aryn	357	−10
15	Forked Bay. East point	357	+2
16	Forked Bay. East coast	7	−12
17	Margaritifer Sinus. West coast	7	−23
18	Margaritifer Sinus. Point	9	+9

For all the undoubted skill of the Potsdam astronomer, the difference between his chart and the general aspect of the planet is truly enormous. This only goes to show that observations of Mars are very difficult.

1888.—Proctor: The Canals of Mars and a New Map of the Planet (His last work)[88]

At the meeting of the Astronomical Society of London on 13 April 1888, Proctor contributed the following note about the canals of Mars[89]:

> Mars was carefully observed in June and July for the doubling of the canals, since the Martian autumn was approaching. Regarding these curious double dark lines (or, rather, bright lines in between darker, feebler lines to either side) as diffraction images of rivers when mists are held in suspension over them, as I have interpreted them for the last four years, I could hardly wait to recover the phenomenon at the approach of autumn and after the beginning of spring, for the northern hemisphere, at which times these double canals have been best seen.
>
> I suppose that nobody regards these double canals as real: but, on the other hand, they can no longer be regarded as optical illusions. If we regard them as phenomena of diffraction, that is to say optical products, we can explain their variations in appearance (when the rivers appear dark, as is normally the case, over a clear bottom, the duplication is not seen); their synchronization with the seasons and the fact that they are visible only with instruments of adequate aperture. This last consideration suggests an effective method of verifying the diffraction theory.
>
> It would be useful if the appearances shown by Schiaparelli could be seen and drawn by observers with real artistic skill. Nobody observing Mars with a good instrument will accept the unnatural features shown by Schiaparelli. The drawings by Dawes, Burton, Knobel, Denning and Green are much more satisfactory.

Proctor was concerned with Mars in the greater part of his researches, and had just begun to publish his latest results when he died. To the ingenious deductions we have already described, we can add the following comments from his last book.

Everything led him to believe that the dark patches represent seas and the bright yellow regions continents.

Proctor proposed to assume that the quantity of water and air ought to be proportional to the masses of the planets and that Mars, with nine times less mass than the Earth, should have nine times less water and air. As the surface of the Earth exceeds that of Mars in the ratio of 7 to 2, the total quantity of air and water over each hectare of the planet Earth exceeds that over each hectare of Mars in the proportion of 18 to 7.

(But nothing authorizes us to believe rigidly that the original conditions of formation of the two planets were the same.)

---

[88] R.A. Proctor, born 31 March 1837, died 12 September 1888.

[89] *Monthly Notices*, Vol. XLVIII, p. 307.

When we consider the density of the atmosphere at sea level, we must take the surface gravity of Mars into account. The proportion is again 18 to 7.

Thus, while we can expect 7/18 of the water and air which exists on Earth, per square metre, the ratio should be 7^2 to 18^2—that is to say 49 to 324, or 5 to 33.

If we admit that the atmosphere of Mars is as tenuous as this, while the quantity of water per square kilometre is only 7/18 of that which exists on Earth, and remembering that the radiation from the sun is twice as weak as with us, it would be difficult to see how there could be enough water vapour in the Martian atmosphere to be detectable with a spectroscope. But if we double the quantity of air and water, the difficulty is at least reduced.

Although the atmosphere of Mars is probably much thinner than ours, it should extend to a greater height, because it is held down by a force which is weaker than that of terrestrial gravity. On our planet, a height of 4000 m is enough to reduce the atmospheric pressure by half; on Mars, we would have to go up to 10,400 m to achieve the same result. Here, at a height of 21,000 m above sea level, the atmospheric pressure is reduced to 1/32; at the same altitude over Mars, the reduction would be only 1/4. If we assume that at sea-level on Mars the pressure is only 1/7 of what it is on Earth, then above a height of 29,000 m the Martian atmosphere should be the denser of the two. At greater heights, the difference in favour of Mars would be even more pronounced.

It is not easy to decide what happened on Mars, since we believe we have recognized meteorological signs, such as clouds which form and dissolve and morning and evening mists, as well as the phenomena which seem incompatible with the idea of extreme cold; even the presence of ice and snow implies the action of warmth. Cold alone, as Tyndall has shown, cannot produce glaciers; the most rigorous north-east winds can blow throughout winter without bringing a single flake of snow. For cold to produce snow, there must be water vapour in the air and water vapour is produced by the action of warmth. Therefore, on Mars, the Sun must provide enough heat to raise a certain amount of water vapour into the atmosphere; and this water vapour is transported in some way or other toward the polar regions, where it is precipitated in the form of snow.

But on the other hand, the entire surface of Mars should apparently be above what we call the snow line for a planet analogous to the Earth, because the whole region of Earth where the cold is as great as we believe it to be on Mars and where the atmosphere is rarefied, is certainly above the line of permanent snow. Why then does the snowfall on Mars make itself evident, giving us variable snowy areas and reddish regions? To this, Proctor replies as follows:

> The amount of snow on the surface of Mars could be small, since the Sun's heat is less effective in producing water vapour. There would be no accumulation of snow similar to that which exists above the permanent snow line on Earth, but on the Martian surface, except near the poles, there could be a layer of thin snow, or rather a layer of white frost. Now, the Sun's radiation on Mars, though incapable of causing large quantities of water vapour to rise into the atmosphere, could nevertheless melt and vaporize this thin layer of snow or white frost. When the Sun is high enough, its heat coming through so rarefied, an atmosphere should be sufficient, and the atmospheric pressure is so slight that vaporization should be easy, due to the low boiling point. Consequently, during most of the Martian day,

> the layer of frost or light snow which had fallen during the preceding night could be completely melted, and the reddish soil or greenish seas of ice will again be visible from Earth. The limb regions of Mars' disk appear whitish, since they are regions where the Sun is low over the horizon.
>
> If we adopt this view of the Martian climatology, the most characteristic fact will be the daily melting of the layer of white frost or light snow before midday and the precipitation of a new white layer when evening approaches. During most of the day the atmosphere will remain clear, at least so far as we can judge according to the telescopic appearance of the planet, while nothing can prevent the appearance of light cirrus or clouds of snow, mainly during the afternoon. In fact, the phenomena which we generally regard as being due to the precipitation of rain over the oceans and continents of Mars can be much more probably due to the evaporation of cirrus by the Sun's heat. The polar regions could be permanently covered with snow, the limits of the polar caps varying with the seasons and undoubtedly presenting accumulations of snow much less to those which exist on Earth.

Such are the considerations of the English astronomer about this interesting subject. I will return to them later, to discuss them in more detail. As will be seen, Proctor has also examined the curious observations by Schiaparelli of the canals and their doubling. He does not believe that the canals are real. He writes: "We cannot regard them as objective realities; this is manifestly incredible."

Proctor believes them to be optical images; not illusions, but images explicable by the known laws of optics. He regards them as diffraction images, produced in the observer's eyes to each side of the rivers of Mars when these rivers become whitened by frost or clouds along their courses. The existence of the canals has led Proctor to recast his first chart and to replace it by that reproduced here (Fig. 58), in which a large number of rivers may be seen, with Dawes' observations (1864) that this eminent observer regards the Meridian Bay as being formed by two points, giving the impression of two very broad river mouths. Thirteen years later, in 1877, Schiaparelli believed that he had seen these rivers, so vainly sought by Dawes and

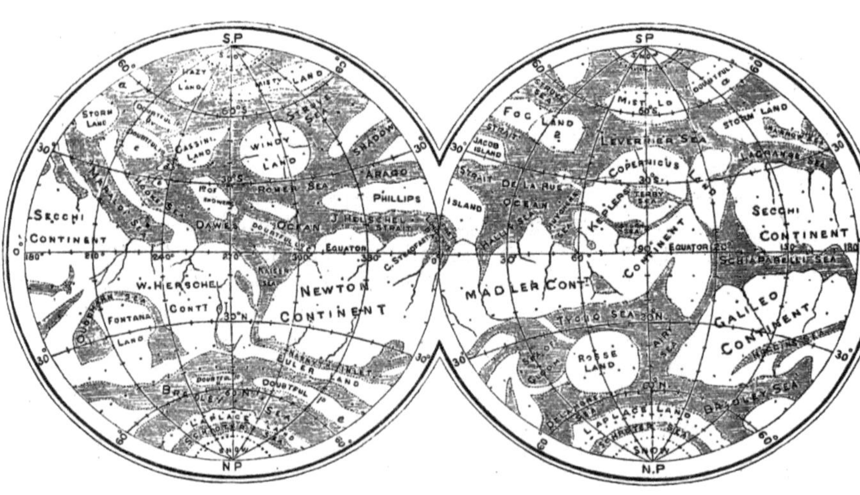

Fig. 58 New chart of Mars, by Proctor, in 1888

regarded them as canals, to which he gave the names of Gehon and Hiddekel. The idea of rivers is indeed straightforward and natural: Proctor logically returned to it and his map, thus modified, presents an appearance which is not without analogy to the principal features of our terrestrial geography. But one great objection always remains: this is that the canals begin nowhere, go from one sea to another and cross each other... These rivers, if rivers they are, do not resemble ours. Proctor, profiting from the criticisms which had been levelled at him for using the same name several times and names which were largely English, modified the nomenclature of his first map. It would have been better if he had followed Green's chart.

Proctor's conclusions in 1888 were that Mars is more advanced than the Earth in its evolution; that for a long time it has not had any heat of its own; that the heat it receives from the Sun is only half that received by the Earth; that this heat produces a special climate, decidedly cold, because on the Martian temperate regions there is white frost and perhaps snow every night, melting every morning; that the atmosphere is rarefied; that the oceans are without doubt frozen, as are the rivers for most of the time. Proctor does not seem to think that the atmosphere of Mars could be constituted differently from ours; it possesses the gases and vapours capable of conserving the heat received from the Sun, acting as a thermal greenhouse in much the same way as the terrestrial atmosphere, shutting in the long-wavelength rays and giving the planet a temperature little different from that of the Earth.

1888.—Perrotin: The Canals of Mars with New Changes (Inundation of Libya)

In 1888, Perrotin wrote[90]:

> With good images, it is possible, with our great refractor (the $0^{m}.76$ equatorial) to see parts of the canals of Mars which I had observed in 1886.
>
> They are in the positions where I saw them in that year and present the same characteristics; they are seen against the reddish background of the continents of the planet, following dark straight lines (probably arcs of great circles), some single, others double—the two components, in the latter case, being most often parallel—cutting each other at various angles and appearing to establish communications between the seas of the two hemispheres or between the different parts of the same sea or even between the canals themselves.
>
> In general, their aspect is the same as it was in 1886. However, some of the canals appear to be feebler, while others have partially disappeared.
>
> From then to now, I must note three important modifications which have occurred since 1886 in the appearance of the surface of the planet—modifications which are quite certain, as they apply to regions to which I paid particular attention in 1886.
>
> First, the disappearance of a continent which then extended along the equator, at longitude 270° (Libya on Schiaparelli's map). This continent was somewhat triangular in form, it was bounded to the south and west by a sea and to the north and east by canals.
>
> Clearly visible two years ago, it does not exist today. The neighbouring sea (if sea it is) has totally invaded it. The reddish white continental tint has been replaced by a black or,

[90] *Comptes rendus*, 16 July 1888, p. 161.

rather, dark blue colour characteristic of the Martian seas. A lake, the Lacus Mœris, situated on one of the canals, has similarly disappeared.

The area of the region whose aspect has thus completely changed is about 600,000 square kilometres in area, a little more than the surface of France. In flooding the continent, the sea has drawn back in the south from regions which it formerly occupied and which now show a tint intermediate between those of the continents and those of the seas, with a bright blue colour, similar to that of a slightly misty winter sky.

The inundation (or other phenomenon) of the continent of Libya, which I believe I show on earlier drawings (1882), may well be a periodic phenomenon. If so, observations will show its frequency.

Next, to the north of the vanished continent, at latitude 25° N., we come to a single canal which is not shown on Schiaparelli's chart, even though that eminent astronomer noted canals which were much feebler and which I have not seen during this last opposition. This canal, 20° long and from 1° to 1.5° broad, is undoubtedly of recent formation. It is parallel to the equator and directly continues a branch of an already-existing double canal which is in communication with the sea.

The third modification consists of the unexpected presence, on the white cap at the north pole, of a sort of canal which seems to connect, in a straight line across the polar ices, two seas adjacent to the pole.

This canal, which shows up with great clarity on the surface of Mars, cuts the spherical white cap, following a chord which corresponds to an arc of about 30°.

A new communication from Perrotin, addressed to the Acadèmie was accompanied by the following drawings:

The difference between drawings 1 and 2 of this year (Figs. 59 and 60) and the corresponding drawing 3 of 1886 (Fig. 61), he wrote, is striking with respect to the region of Schiaparelli's Libya. Over an interval of a month, drawings 1 and 2 also indicate notable modifications in the same region.

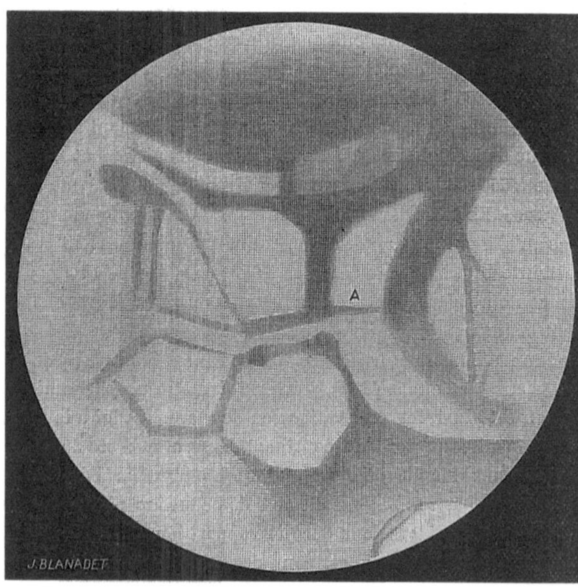

Fig. 59 Mars in the great equatorial of Nice, by M. Perrotin (Drawing no. 1. 8 May 1888)

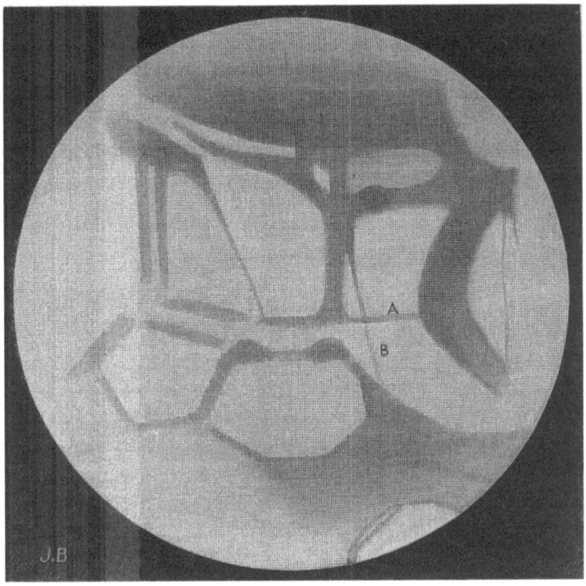

Fig. 60 Mars in the great equatorial of Nice, by M. Perrotin (Drawing no. 2. 12 June 1888)

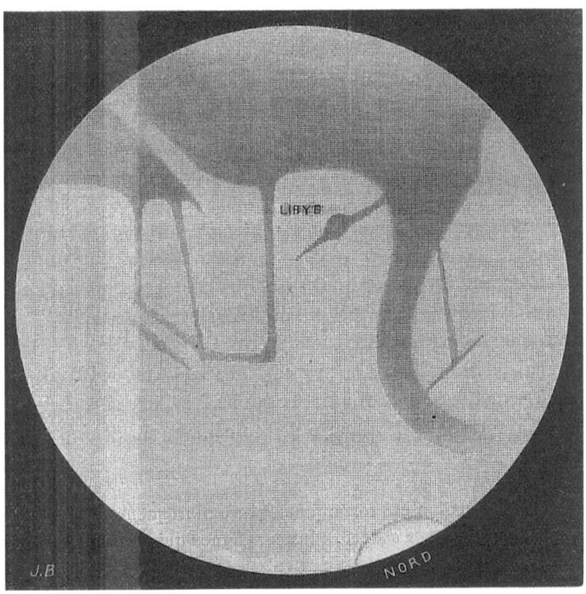

Fig. 61 Mars in the great equatorial of Nice, by M. Perrotin (Drawing no. 3. 21–22 May 1886)

The first two drawings show the new canal A and the canal on the white polar cap.

In drawing 2 we find, beyond, a single canal B, seen for the first time on 12 June.

Drawing No. 4 (Fig. 62) shows four single canals and three doubles, of which one is double over only part of its length, though all are quite characteristic.

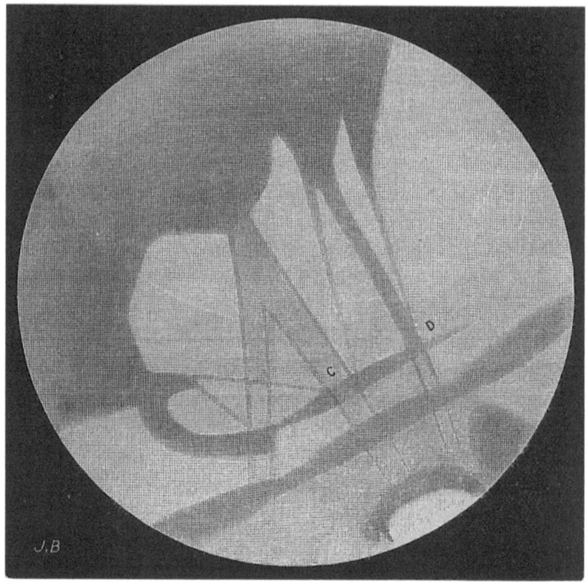

Fig. 62 Mars in the great equatorial of Nice, by M. Perrotin (Drawing 4. 4 June 1888)

In the latter drawings, C and D, parts of the regions near the equator and almost following the meridian (longitude 338° for one, 5° for the other) are lost in the environs of the white cap of the North Pole.

Are these canals, in the generally accepted sense of the term? It seems to me that the two remarkable double canals which I have noted can give us useful information on this score. If they are true canals, they effectively prove that profound modifications can take place according to the season, particularly at the time when, under the influence of the Sun's rays, the white patch at the north pole tends to disappear—to melt, as certain astronomers think.

Thus considered, the canals in question, and two others of the same kind, are particularly recommended to the attention of observers.

So wrote Perrotin. The main point brought out in his 1888 observations was the claim of the existence of new canals and also an analogous sombre streak crossing the north polar cap along a chord. As Faye commented, it is though it had been put there to form a link between the two sides of the pole.

On 12 July the eminent discover of the canals, Schiaparelli, wrote to me from Milan:

> The opposition has been remarkable for the frequency of double lines; many more than in 1884 and 1886. Several lines which have remained single at all previous oppositions (Læstrygon, Nepenthes, Astaboras, Heliconius, Callirrhoe) have been double this time.

These appearances therefore clearly depend upon certain critical epochs. Four new drawings are given to follow the preceding ones.[91]

Here are the co-ordinates of the centre of the planet at the times when the drawings were made:

[91] *Comptes rendus*, 10 September 1888.

Numbers	Long.°	N. Lat.°
5	195	24
6	140	24
7	120	24
8	90	24

Perrotin continued:

Drawing No. 5 shows a very uneven part of the planet's surface, particularly in the region of the north polar ice cap and at the same time a region R′ made up of a sort of pentagon formed by canals and which, by its white colour and its brilliancy, contrasts in striking manner with the reddish colour of the adjacent regions. (The brightness is almost as vivid as that of the polar cap. This was not the case, or was not noted, when I made drawings 1 and 2; but it is probable that we are dealing with a notable change which occurred during the observations made this year.)

Drawing No. 8 (Fig. 66) shows two canals, one single and the other double, KL and MN, analogous to those referred to above. These canals come from the equatorial regions and run, almost following the meridian, toward the North Pole.

This drawing seems to agree with No. 4 in the first series (Fig. 62). Moreover, it shows regions adjacent to No. 4, lying rather further to the east on the planet.

Drawings 6 and 7 (Figs. 64 and 65) are unfortunately incomplete. I give them here because they show evidence of the existence of a new canal which, like that noted above (Figs. 59 and 60), cuts in a straight dark line across the white cap of the polar ices.

This new canal is perhaps a little less sharp than the first; but its existence and its character are not in doubt. (In bizarre fashion, the new canal begins on the periphery of the cap at the same point where the previously recognized canal ends.)

Drawing No. 5 (Fig. 63) shows the two canals; that to the right is the old one, while the new canal is to the left.

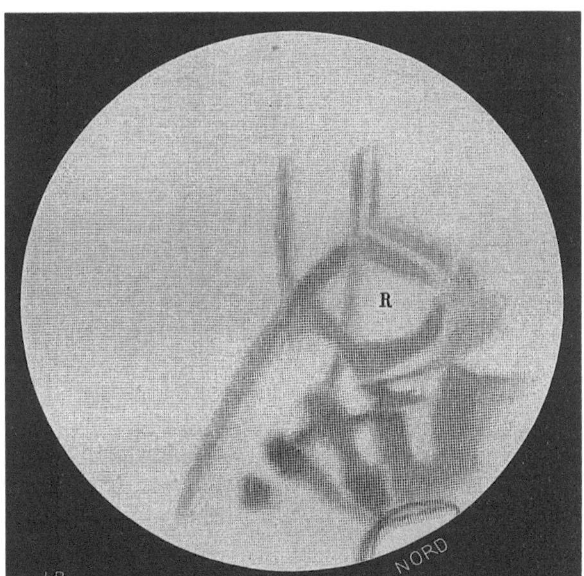

Fig. 63 Mars in the great equatorial of Nice, by M. Perrotin (Drawing no. 5. 12, 13 and 14 May, 18 and 19 June 1888)

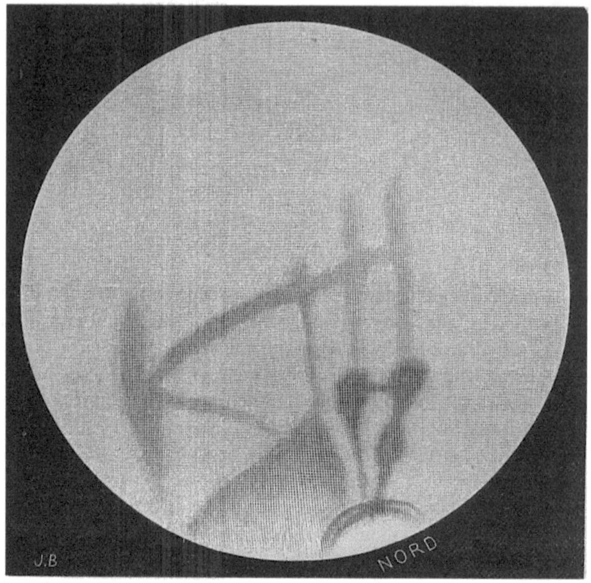

Fig. 64 Mars in the great equatorial of Nice, by M. Perrotin (Drawing no. 6. 17 May, 23 June 1888)

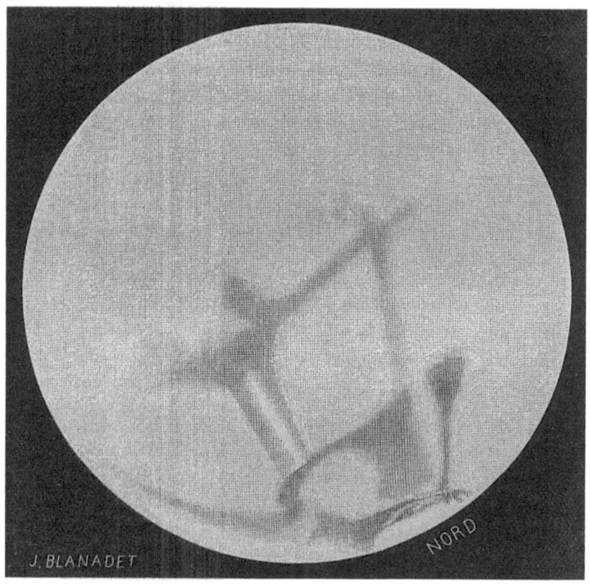

Fig. 65 Mars in the great equatorial of Nice, by M. Perrotin (Drawing no. 7. 18, 20 May 1888)

The latter is better shown in Drawings 6 and 7 (Figs. 64 and 65). In Drawing 7 it is seen in its full development.

I greatly regret that the atmospheric conditions did not allow me to re-observe the area of Libya, in July, with good images. What I glimpsed led me to believe that there had been new modifications since June in this part of the planet and I very much hope that it will not be too late to recognize their nature. This is the continuation of changes to which I drew

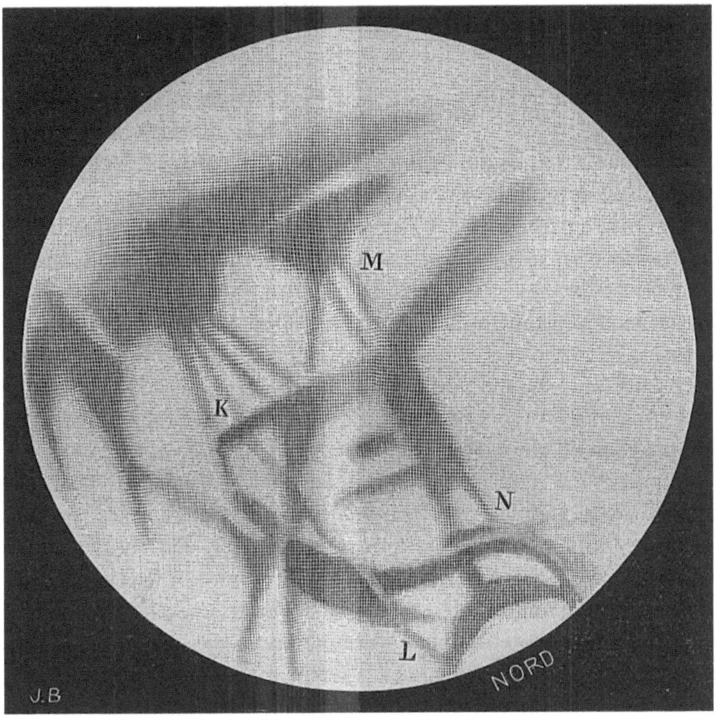

Fig. 66 Mars in the great equatorial of Nice, by M. Perrotin (Drawing no. 8. 25, 26, 27 May; 2 July 1888)

attention last May, and which are, without doubt, parts of the changes which periodically and frequently take place on Mars. These have occupied my attention during my long observational sessions and I have seen several, particularly in the neighbourhood of the ice cap. These changes, which sometimes occur from one day to the next, do not modify the general aspect, but only the details; in the main they affect the sombre parts of the surface (Fig. 66).

I have also found changes of a different nature: thus on 18 and 19 June I saw from time to time, during the course of my observations, the region R in drawing No. 5 covered and uncovered by a kind of reddish mist which, extended as far as the adjacent canals, stood out; the rest of the planet's surface continued to show up with great clarity and sharpness of detail

I can best compare this phenomenon with that which we often find here during the summer, with sea mists after a hot day; these mists cover the coasts for some minutes, but subsequently soon disappear.

I need not stress that even in our great refractor these phenomena are not obvious and that to see them one needs close attention, a good instrument and above all images which are not merely good, but excellent.

1888.—Niesten: Observations and Drawings

With regard to the observations published by Perrotin, Niesten presented to the Belgian Academy some comments drawn from observations which he had made at the same time at the Brussels Observatory. He gave two drawings relating to regions of the planet where certain modifications had occurred.

These two drawings show the Hourglass Sea; to the left on 29 April (Fig. 67), to the right on 5 May (Fig. 68). Libya has not disappeared. On 5 May it was well seen, yellow-orange in colour.

Niesten thinks that the differences in aspect often observed are due mainly to the variations in angle of the regions seen under different conditions of obliquity and solar illumination. In 1888 the northern regions were presented very obliquely.

Note, also, that in Niesten's two drawings—particularly the second—there is not the usual impression of Mars. The Hourglass Sea can be recognized, to the left in the first drawing and to the right in the second, but this is all and the features of lesser importance—even short lived and indefinite ones, of which many are uncertain—show a geography which is almost imaginary, particularly with the drawing of 5 May. The pencil cannot fix features which are scarcely glimpsed. But what else can one do? One can at least distinguish certain slight shadings, even if it is impossible to be sure of their outlines; and extra detail appears only in rare and fugitive moments of perfect transparency. Illusion or reality? It seems that such telescopic views depend only upon thought. We indicate the features with the pencil and those which are fugitive, uncertain and perhaps atmospheric take on the same significance as those which

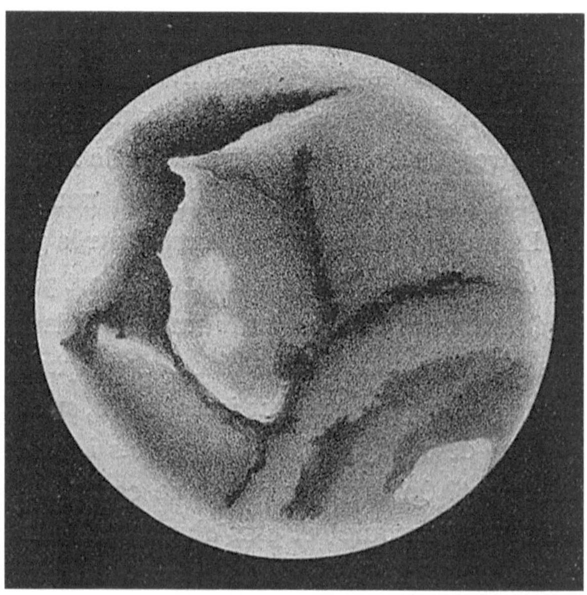

Fig. 67 Mars, on 29 April 1888, at 9^h 15^m (316°). Drawing by M. Niesten

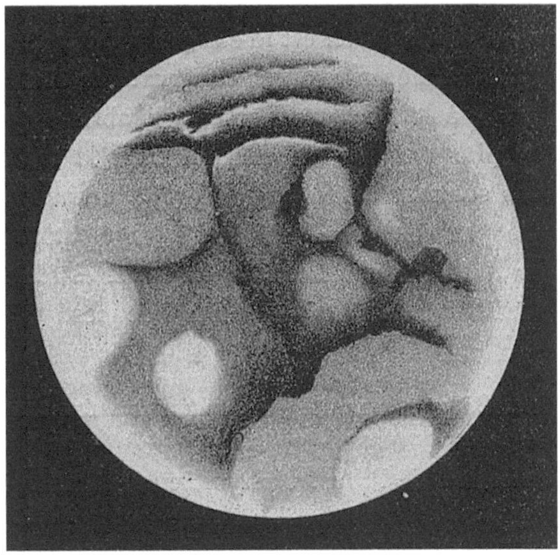

Fig. 68 Mars, on 5 May 1888 (275°). Drawing by M. Niesten

are incontestable and permanent. The same comment applies to the drawings made at Nice; here also we see uncertain features and such is general with drawings of Mars.

In spite of the difficulties inherent in observations so delicate as this, we can nevertheless see, that thanks to the perseverance and the skill of the observers, we can learn more and more about this neighbour world. In this respect we have reached out to a planet which is extremely interesting and we can do our best to interpret the numerous facts gained from the observations. The essential need is to avoid drawing back; and this is the point I made to the Académie des Sciences.

1888.—Fizeau: An Explanation of the Canals

Here is the communication from the distinguished physicist of the Académie[92]:

> The singular appearances observed on the surface on Mars by Schiaparelli and by several others, notably by Perrotin at the Nice Observatory, have recently added new data without any plausible explanation. We agree in calling them canals, from their distant resemblance to irrigation canals, but without prejudging their nature.
>
> It seems, however, that the most recent observations now enable us to try to resolve this enigma, as follows.
>
> First, it is generally accepted that there is water on the Martian surface and we agree that this water plays an important role in the changes observed there. We know that the polar caps are snowy in appearance, growing and decreasing, following the course of the seasons;

[92] Meeting of 25 June 1888 (*L'Astronomie*, August 1888, p. 287).

we know, moreover, that spectral analysis of the light of Mars has enabled Janssen to show that the presence of water is very probable.[93]

The canals of Mars appear as lines, more shaded than the rest of the surface; straight, often parallel or else cutting each other at various angles. The réseau of lines is not fixed and even over short intervals, is shown very differently on different drawings; the changes recall those of the more extended patches (called continents or seas) which appear to be modified or may even vanish over intervals of several months. Quite recently, a very clear line has been noted, crossing the polar ices in a chord with regard to the polar cap turned toward the Earth.

It seems natural to compare these singular appearances with the varied phenomena which we note on our own globe at the surfaces of large glaciers, such as the sea of ice on Mont Blanc, the Rhone Glacier and above all on the vast icy region of Greenland, to cite only the best known. We know that among the unceasing changes produced on the ice surfaces by the march of the seasons, we also find—very significantly, in our present context—parallel ridges, crevasses, straight cracks extending for considerable lengths and cutting each other at various angles. In Greenland, Nordenskiöld has recently noted phenomena of this type which are very remarkable for their size and for their very precise characters, which allow us to assign them to regions subjected to a glacial regime.

In this, approaching the principal circumstances which link the canals of Mars with what is observed on our glaciers, we may note that the analogies and the resemblance between the two orders of phenomena are really so marked that we may believe, with great confidence, that the two are due to the same cause: a glacial state.

We are therefore led to the hypothesis of the existence of immense glaciers on the surface of Mars, analogous to those of our globe, but of considerably greater extent and whose movements and ruptures are also more pronounced. We may also note that the long duration of the seasons on the planet (double those on Earth) clearly favour the development and the periodical upsetting of icy masses, under the influence of expansion and contraction due to changes in temperature—effects which augment those due to the planet's low surface gravity (1/10 of that of the Earth).

But, on the other hand, does this hypothesis agree with the other well-known facts about the constitution of the planet?

---

[93] Janssen is not alone in the spectroscopic observations cited above and when I read this comment by Fizeau I wrote to ask him for information. Here is the reply from the eminent Director of the Meudon Observatory:

"My dear Colleague: You ask me where I have published my observations of Mars with regard to the presence of water vapour in the atmosphere. I announced the presence of water vapour in *Comptes rendus,* Vol. LXIV, 1867, p. 1304.

"The studies which I carried out were made at the Paris Observatory with the Foucault telescope which Le Verrier had put at my disposal in 1863, on Etna, where I stayed for three days (to eliminate any possible effect due to the Earth's atmosphere) and at Palermo, where I used the equatorial of the Marseilles Observatory with the Foucault 0m.80 telescope. It was only after having observed under these varied conditions and above all after having observed the clear spectrum of water vapour at the La Villette Works in 1866 that I regarded myself as able to announce its presence. Other observers have subsequently been able to do no more than show the presence of a telluric spectrum en bloc."

Here is the text of the letter in *Comptes rendus* to which Janssen refers. It is from Vol. 1, 1867, p. 1034; the following paragraph being at the end of a letter written by Janssen to Charles Saint-Claire Deville on the island of Santorin:

"I cannot end this letter without telling you that I climbed Etna to make spectroscopic observations from a great altitude, to nullify the major part of the influence of the Earth's atmosphere. From these observations and those which I have made at the Observatories of Paris, Marseilles and Palermo, I believe I can notify you of the presence of water vapour in the atmospheres of Mars and Saturn."

First, the distances from the Sun of Mars and the Earth are in the ratio of three to two; the intensity of radiation is as four to nine; the solar radiation on Mars is therefore 4/9 of that on Earth. Without trying to decide what our climates would become if the Sun sent us only 4/9 of its rays, we can be sure that all the mean temperatures would be lowered and that a large part of our globe would enter a glacial period. The temperatures on Mars should therefore be much lower than those of the Earth, even if we assume that the planet has an atmosphere similar to our.

Moreover, there are sound reasons for thinking that the atmosphere of Mars is less developed than that of the Earth.

First, the absence of equatorial bands shows that regular atmospheric movements are not produced there, as on our globe; this seems to indicate an atmosphere of limited extent and consequently less able to absorb and conserve the warmth than in the case of our own atmosphere.

Next, we note that the light of Mars has a red tint, recognizable at all times and by all observers. Now, this red colour furnishes a new proof that the atmosphere of Mars does not have a constitution similar to that of our atmosphere; this is what we must conclude after considering the colour of *the lumière cendrée* which the Moon receives from the Earth at certain days of the first and last quarters. This light is in effect borrowed by the Earth, directly illuminated by the Sun and can give us a reasonably exact idea of the colour of the Earth, surrounded by its atmosphere and seen from space. Now, according to Arago the *lumière cendrée* is of a bluish green and is certainly not red, as it would be if our atmosphere were similar to that of Mars. Therefore, the red tint indicates, with a high degree of probability, the relative predominance of water vapour in the gases in the atmosphere of Mars. We see that the hypothesis of the glacial state of Mars seems to agree well with the principal physical charts that we possess at the moment concerning the planet.

At the following meeting, that of 2 July, I put forward some counter arguments to this conclusion.

1888 C. Flammarion: Comments and Observations

Here is the response[94]:

I ask the Academy's permission to submit the following facts, in response to the considerations which were presented at the last meeting by one of our most illustrious members.

The polar ices melt on Mars to a greater extent than on the Earth. This is a fact shown by constant observation. While on Earth the hardiest and most adventurous expeditions have never approached to within less than 7° of the North Pole and stay even further away from the South Pole and while our two poles are constantly surrounded by ice, on Mars the melting of the ices with the increasing height of the Sun above the horizon seems to be complete during the summer at the two poles of the planet, particularly at the South Pole, where summer occurs at perihelion.

In this year 1888, the planet still had its Northern Hemisphere turned toward us, because of the axial inclination. The limit of the north polar ices was sharply determined; it gradually approached the pole during February, March, April and then May. I estimate that at the end of May, at its minimum, the diameter of the polar cap was about 300 kilometres. (For the Northern Hemisphere, the summer solstice occurred on the previous 16 February and the autumnal equinox fell on the following 15 August.)

The snows at the two poles have long been the object of scrupulous attention and very precise measurement. It has always been known that they melt considerably, much more

[94] *Comptes rendus*, 2 July 1888.

than on our planet. The total observations show, however, that the minimum is reached from two and a half to three months after the solstice. (We know that the year on Mars lasts for 687 days.) The phenomenon is therefore exactly of the same order as that which occurs at the terrestrial poles, but is more marked.

The micrometrical measurements of the south polar cap made by Schiaparelli in 1879 showed that the cap was reduced to 4° in diameter by the end of November (the summer solstice occurred on 14 August). Admitting that because of irradiation the 4° represents double the real dimensions of the polar cap, we see that in 1879 the cap was reduced to 2° or 120 kilometres in diameter. It varies at least in the proportion of from 900 to 120 kilometres in diameter.

As with Earth, the pole of cold does not coincide with the geographical pole, but is eccentric to it; it lies about 6∞ from the geographical pole, near the intersection of the 84th degree of latitude with the 35th degree of longitude.

As with the south cap, the north cap is subject to variations corresponding to the seasons and the temperature.

This melting of the polar caps during summer is in flat contradiction to the hypotheses that the Martian continents are ice fields, and that the temperature of the planet is lower than that of the Earth. It proves the contrary, if we admit that these snows and these waters are of the same nature as ours—which is not absolutely certain, in spite of spectral analysis investigations, because the melting and saturation points and the chemical composition of the atmosphere and of liquids could show original and permanent differences from those which exist on our planet.

This is perhaps the place to comment that the temperature of a place is not regulated solely by the distance from the Sun, but also and above all, by the physical properties of the atmosphere which covers it. There is a great deal of water vapour in the atmosphere of Mars, as is shown by the absorption lines in its spectrum (though the colour of the planet is not due to this cause, because it is strongest at the centre of the disk, where the light passes through a thinner layer of the planet's atmosphere than near the limb). Now, it is water vapour which plays the greatest role in the consequence of the warmth received from the Sun. We know that the absorbing power of a molecule of aqueous vapour is 16,000 times greater than that of a molecule of dry air. Without water vapour or some analogous protection, our own planet would remain permanently frozen. The vapours of sulfuric acid, formic acid, acetylene, elephant gas, ethyl iodide and carbon dissuasively have the same properties, according to experiments by Dentally.

Note also that the aspect of the continents of Mars differs considerably from that of the polar ices and snows which sometimes whiten certain regions. The snows and ices glitter with a shining whiteness, while the continents are of a very warm yellow colour, recalling the tone of cornfields seen from above in a balloon.

All in all, the observations of Mars and the application of the resulting information to the study of the physical constitution of the planet leads us therefore to conclude that the polar ices do not invade the entire surface of the globe, but that, on the contrary, they are themselves highly subject to the influence of temperature; that relative to the physical constitution of the shows and the waters, the temperature there produces effects at least as evident as on our planet; that the world of Mars is not in a glacial state, and that the canals are not crevasses in glaciers.

Among the observations which I made during this opposition from my observatory at Juvisy, I give the four sketches reproduced here (Fig. 69).

I could distinguish no canals with my 245^{mm} equatorial. As with my earlier sketches, my drawings of this year simply show the general outlines of areography. I have represented only that which I am absolutely sure I saw and nothing which to me seemed doubtful. The North Pole of the planet, of which our knowledge is less than might be desired, was well placed for observation. The summer solstice in this hemisphere having taken place on 16 February, the melting of the polar snows ought

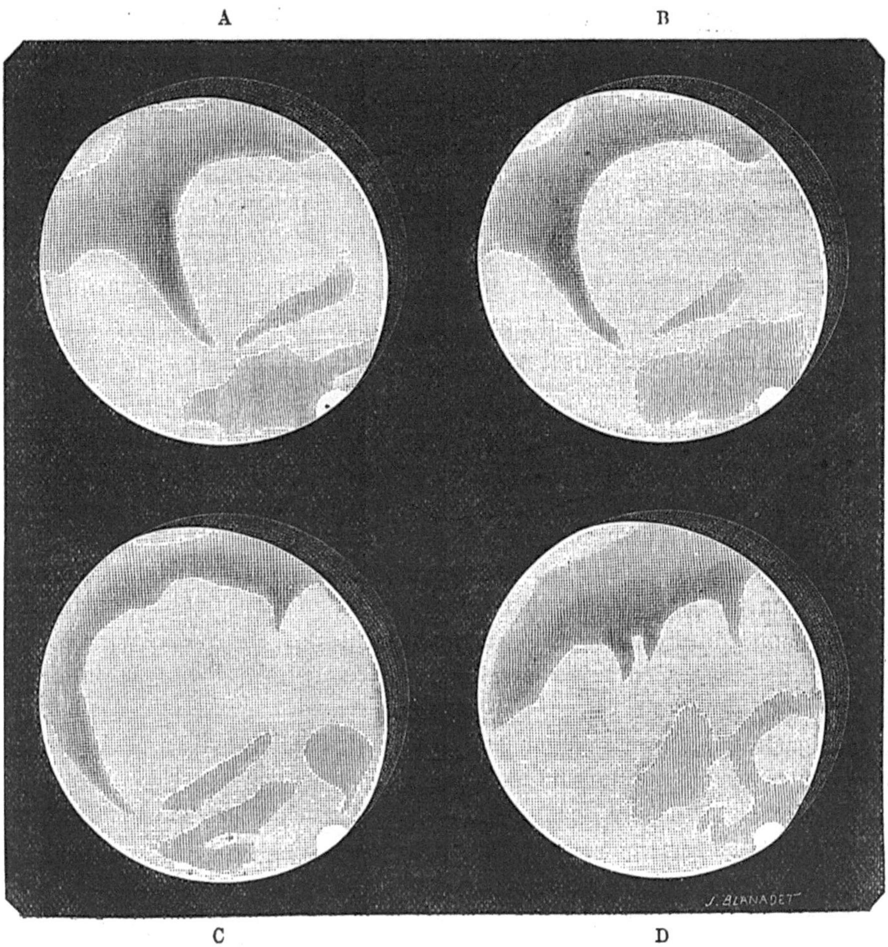

Fig. 69 The planet Mars, in the month of June 1888 (Sketches by M. Flammarion)

to have reduced the ice to its minimum at the end of May. Nevertheless, there remained a definite quantity of it; as I have mentioned earlier and as can be judged from the four drawings reproduced here (Fig. 69), made on 2 June, from six to nine o'clock in the evening, some snow remains.

With regard to these times, it is not inappropriate to comment here that when the air is calm and transparent, views taken during the daytime, in full sunlight, are as good and as clear as those made at night.

In the first of these drawings (A = $6^h\ 0^m$), the Hourglass Sea had just crossed the central meridian; below it we can see a trace of its prolongation, toward the east (Nasmyth Channel), and lower still, the north polar sea; the Hourglass Sea was the most sombre of the features. The lower pole was of a sparkling whiteness; the Herschel and Beer Continents were very evident and were of a very bright yellow ochre colour; the seas were grey and very varied in tone and at the top of the disk, near Cassini and Webb Lands, the regions were whitish (clouds of snows in the southern hemisphere?) The rest of the disk appeared very pure and cloudless. I could see no trace of the inundations reported to the west or left of the Hourglass Sea, and it was the same during all my June observations.

In the second of these drawings (B = $7^h\ 0^m$) the Hourglass Sea had advanced toward the west; we can see Laplace Land.

In the third ($8^h\ 0^m$) the sea had almost reached the west limb of the disk, and the Meridian Bay had come into view in the east; the Knobel Sea shows above the pole, and the whole upper region of the disk is whitish.

In the fourth drawing (D), made at 9^h in the evening, the Meridian Bay was crossing the central meridian; above the pole and to the east of the Knobel Sea, may be seen a sea which appears to envelope the pole.

2 and 3 June—when the thermometer showed 33 °C (91 °F) in the shade—were remarkable for the transparency of the atmosphere and the steadiness of the images. These observations were made with a magnification of 400x.

1888.—Terby: Observations and Drawings[95]

The eminent Louvain astronomer has sent me the following résumé with regard to the drawings which he had been able to make with the 200^{mm} equatorial.

Mainly, the canals have been seen with such difficulty and under such unfavourable circumstances, that one has to search through the details to find them; you will recall that M. Perrotin had found the same in 1886. However, I can guarantee that my drawings show only the lines really observed, nothing being due to prior knowledge of the region under observation. I can cite two proofs: first, the imperfections of my drawings, which will not escape the notice of a practised eye when two of the drawings are rigorously compared with each other; in these cases I have not attempted to make the details agree and have only tried to represent exclusively what I have seen; for instance, the invisibility of the Euphrates and its gemination. I had perfect knowledge of the existence of this canal, of its gemination and the exact form, from the drawings which Schiaparelli had sent me, so that I could provide confirmation; he believed that I would see the Euphrates easily. Yet in spite of all my efforts, which I obstinately continued, not only could I see no trace of the gemination, but I could not even see the Euphrates itself. It was not the same with the Phison, as I have already said.

Each drawing was the result of several days' observation; therefore, not all the features shown were seen at the same time.

I rarely used a magnification of more than 200x or 300x; if I did so it was without making things better, because of the abominable conditions of observation this year. You must know that I made a very large number of drawings; I have chosen those which best show the greatest part of the surface. Here are the regions observed:

Figure 70A (12 May, $9^h\ 18^m$).—At the centre, the Trivium Charontis is in the form of an almost black spot; it is prolonged at one end by the Erebus, at the other by the Cerberus; the latter is extended, in a perpendicular direction, by the Styx. The rest of the outline of Elysium is marked by the Eunostos and the Hyblæus. In the upper figure, the Maraldi Sea is seen, with the Sinus Titanum near the left edge; the Bay of Læstrygonum lies in the vertical diameter. A filament joins this bay to the Cerebus; this is the Antæus. The two grey

[95] *L'Astronomie*, September 1888, p. 324. See also *Ensemble des observations faites à Louvain en 1888*, by Dr. Terby. Bruxelles, 1889.

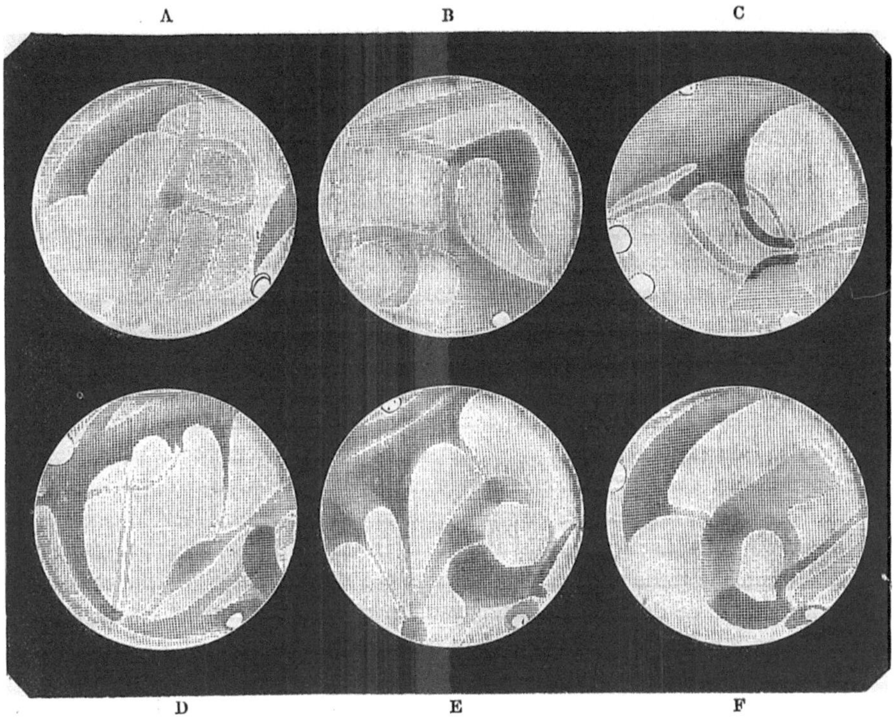

Fig. 70 Geographical aspects of the planet Mars (Drawing by M. Terby at Louvain)

rectangles in the lower right quarter of the disk make up the Propontis. The black thread in the polar cap can be seen in the position which had occupied on 12 May, at $9^h\ 15^m$. Figure 72 shows the displacement, by the planet's rotation, at $10^h\ 48^m$. Figure 73 gives the appearance of the polar cap on 13 May.

I similarly observed white points at the lower limb, at the prolongation of the Erebus; when they reached the limb, they shone and protruded in the manner of the polar cap. A third white point had already vanished over the limb by the time that this drawing was made. The white regions and the reddish areas can be distinguished. To me the Cerebus often appeared rosy, as did the perimeter of Elysium.

Figure 70B (9 May, midnight), shows, a little above the centre, the confluence of the Triton, the Thoth and the Nepenthes; below, Alcyonius; to the left, Elysium. On 12 June, the Eunostus was prolonged by a smoky line as far as the Thoth.

Figure 70C (29 April, $8^h\ 16^m$). Triple confluence of the Triton, Thoth and Nepenthes to the left of the centre; Libya well seen; Mare Tyrrhenum very pale; Astusapes visible; Protonilus, with Ismenius Lacus, at the right limb; a little lower, Callirrhoe. On the central meridian: Nilosyrtis and Boreosyrtis. The break in the Nilosyrtis, on the central meridian, seems to be temporary effect produced by a cloud in the atmosphere of Mars. Schiaparelli has paid great attention to this feature with his 18-in., and when the region was again available for observation, he did not confirm the break. Above, there shines a small snowy patch.

Figure 70D (29 April, $11^h 38^m$).— Hellas well visible at the left border; the bay of the Phison can be seen; the Forked Bay well doubled momentarily during April. Next, the Margaritifer Sinus, with Edom Promontorium very white. The Phison appeared double on 1 and 3 June; the Oxus seen up to Ismenius Lacus. This latter lake is reddish. Tempe at the right limb.

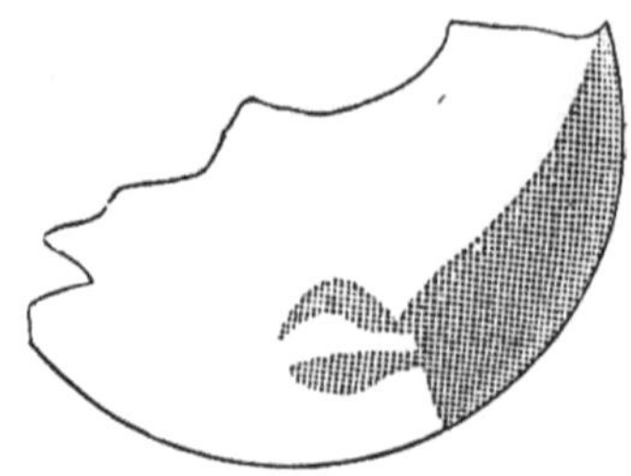

Fig. 71 The Gulf of Aurora

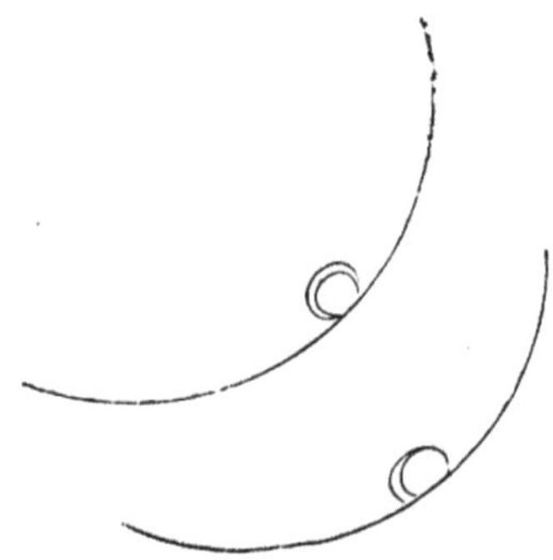

Fig. 72 Aspect of the polar spot on 12 May, from $9^h\,43^m$ to $10^h\,48^m$

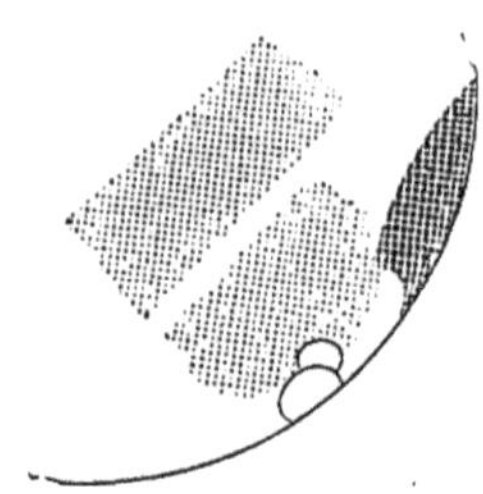

Fig. 73 Aspect of the polar spot on 13 May, at $9^h\,3^m$

Figure 70E (27 May, $8^h\,4^m$).— Argyre brilliant in the upper part of the disk, like a polar cap. Deucalionis Regio well seen, between Sinus Sabæus and Margaritifer Sinus; a brighter region in the large upper patch corresponds to Pyrrhæ and Protei Regio. From left to right and going upward, we find the Gehon, Indus-Oxus and the Ganges. At the lower limb, to the left of the vertical diameter, the little black patch is Ismenius Lacus; the Deuteronilus is rosy, joining it to Niliacus Lacus; the latter is separated from the Mare Acidalium by Achillis Pons; below the latter is Lacus Hyperboreus; again the polar patch; the Ganges flows out at a black patch, the Lunæ Lacus. The rest of the outline of Tempe is marked by the Nilokeras, the Nile (reddish) and Ceraunius. The shading which comes down from Mare Acidalium toward the lower limb is Callirrhoe.

Figure 70F (23 May, $8^h 28^m$) shows the partly polygonal form of Tempe; we still see Argyre and the Ganges coming up to the Lunæ Lacus; the principal regions of the preceding drawing are shown and there is an improved view of Tanaïs, which comes from Ceraunius to the right limb. Under the Tanaïs we see the Iaxartes.

Figure 71 represents the Aurorae Sinus, with the Solis Lacus; also the Tithonius Lacus, which I saw only once, on 16 April. Nectar and Agathodæmon are well seen.

Figures 72 and 73 show the variations in perspective of the polar cap, due to the planet's rotation.

1888.—Schiaparelli: New Observations[96]

These new observations by Schiaparelli were reported in the form of letters to our eminent colleague Terby, of Louvain, whose own researches have just been described. Schiaparelli's observations were made with the new 19-in. ($0^{m}.49$) Milan equatorial, whose power of definition had been shown by its performance with double stars: Epsilon Hydrae had been found to be a triple, with the separation between the components 0″.25 and 0″.20 (less than a quarter of a second of arc!).

Our Fig. 74 represents the appearance of Mars on 8, 9 and 10 May 1888. Here is what Schiaparelli wrote:

> Two great changes have occured in the region of Propontis. With low magnifications only confused shady streaks could be seen; using 500x and 650x, these were resolved into a curious triangular space, of which one side was double; this triangulation even continued to the left, where there were at least two more triangles. The sides were shaded, the apices being formed by visible black spots, sometimes elongated in form; the background was yellow.

Apart from the existence and the gemination of the canals, comments Terby, this triangulation ranks as one of the greatest mysteries of Mars. There is an obvious

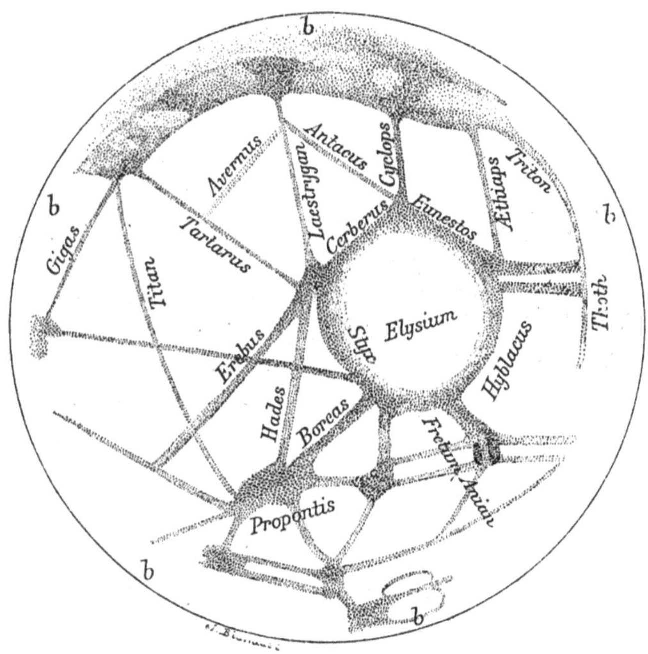

Fig. 74 Telescopic view of Mars, after M. Schiaparelli, on 8, 9 and 10 May 1888. (b = white spaces)

[96] *Ciel et Terre*, August 1888.

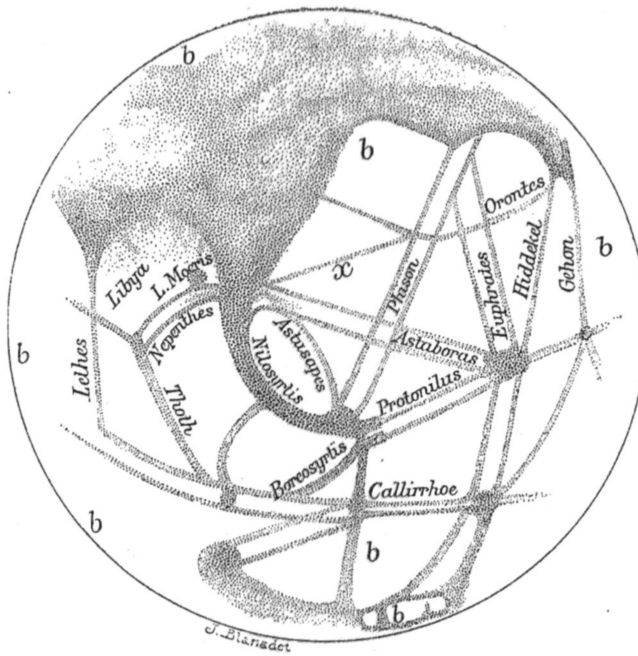

Fig. 75 Telescopic view of Mars, after M. Schiaparelli, 2, 4 and 6 June 1888

analogy between this figure and the trigonometrical canvas of our geodetic surveys. And the double side! And the darker summits! (Figs. 75 and 76)

Schiaparelli judged that the Triton had changed into a very broad gulf of the Mare Cimmerium; this was a striking and extremely interesting development. In a letter of 21 May, the Italian astronomer wrote:

> On the 20th the Solis Lacus, near the central meridian, was very pale and barely visible; Tithonius Lacus was better seen. Iris, Fortuna, Chrysorrhoas, Ganges, Jamuna and Hydaspes were all visible, Ganges and Chrysorrhoas being particularly easy; single and straight, but with very marked undulations on their two edges, both of which could be distinguished. However, the colour was so light (I believe that there must be a trace of red) that there was some difficulty in confirming all these lines and I doubt whether they would have been visible in an 8-in. telescope; with 200x and 350x I could not see them, but with 500x and 650x they were distinct. The same light tone made it difficult to see the two lines of the Nile, but it was definitely double; the two bands were fairly broad and their positions were exactly as shown on my map; they were well drawn. Also double, but badly defined and shaded, was the Nilokeras. The line to the right of Nilokeras cut the Achillis Pons, which was here interrupted at its right extremity; the rest of the bridge still exists, but has become smoky and badly defined. The Mare Acidalium and Lacus Hyperboreas have several complexities and extensions which I saw for the first time.

Thus Schiaparelli now distinguished the two borders of certain canals, the borders being a little darker than the interior. This new discovery must be regarded as another mystery. On 28 May, the Director of the Milan Observatory wrote again:

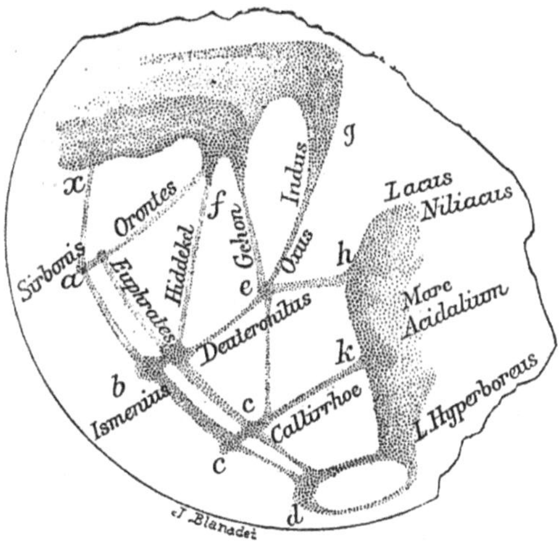

Fig. 76 Fragment of the geography of Mars, after M. Schiaparelli (Drawn on 27 May 1888)

> Yesterday the long streak of the Euphrates looked rather as shown on the attached drawing (Fig. 76). The base *ax* appears to be single and lies at a direction slightly different from the normal; *ab* is a very regular gemination, not very dark, but distinct; *bc*, a very strong gemination although ill-defined, white on both sides and in the middle; *cd*, a strongly elongated triangle, with lines which were strong but ill defined; *d*, another Hyperborean Lake; *b*, Lacus Ismenius, double; *c*, another large lake, double; *a*, the Lacus Sirbonis, seen in 1879, double; *af*, Orontes, well seen; *fb*, Hiddekel, less certain; *ec*, a delicate but well marked line; *beg*, Oxus and Indus, easy to see; lake *e*, a beautiful black point; *beh*, Deuteronilus, well seen; *ch*, Callirrhoe, well seen. As is evident, all these lakes are doubles!

Figure 75 was accompanied by a note dated 7 June, in which the Italian astronomer commented as follows:

> I believed that I saw the planet well enough at $9^h\ 25^m$ on 27 May and began to be almost satisfied, having confirmed at least three or four geminations. But I had a happy surprise on 2 and 4 June, and it was only then that I had any idea of the power of an 19-in. aperture for Mars! I then saw that the memorable days of 1879-1880 and 1882 had come back for the first time and that I could again see those prodigious images presented in the telescope field as an engraving on steel; again there was all the magic of the details, and my only regret was to have the disk reduced to 12 inches in diameter. Not only could I confirm the gemination of the Nepenthes (*quantum mutatus ab illo*!) and the reappearance of the Triton of 1877, but I could again see Lacus Mœris, reduced to a very small point, but sometimes perfectly visible and scarcely separated from the Syrtis Major. Hephæstus had suddenly disappeared. The Nilosyrtis had no break in it; it is however true that in the Syrtis Major, in the last days of May, I could see a small, fairly brilliant island (Œnotria), which, with low magnifications, gave the impression of a bridge, though this impression disappeared with 500x. However, the phenomenon ceased suddenly and Œnotria is not now always visible. The Euphrates is again double throughout its length, but was no more conspicuous than on 27 and 30 May. Yesterday, however, it was again fairly well presented, the two streaks lightly shaded and the part of it below Ismenius Lacus better seen than the part above... Callirrhoe and Protonilus are two very narrow geminations, though geometrically perfect and very black, particularly Callirrhoe; with 650x the latter was seen without the slightest doubt. The two streaks of Callirrhoe were equal; in the Protonilus, the upper streak is a little thinner than the lower, although perfectly traced. The Phison is double, rather close, as with the

Euphrates; Astaboras also double, but more easily visible to the left; Tiphonius and Orontes are single; also single is a new line marked *x* in the sketch.

The Oxus has faded considerably and latterly I saw it no more, while the Indus reappeared. The Hiddekel is almost invisible; the Gehon is slightly smoky, going into a little lake, from which two lines emerge to the right toward Lacus Niliacus. But—and this particularly extraordinary and particularly neglected—there are the changes which have taken place over the month in the Boreosyrtis and the neighbouring regions; the sketch which I have given is not very definitive, because there are some small details which I would need to check with further examination; however, their geminations and arrangement are beyond doubt. What strange confusion! What does it mean? Evidently the planet has fixed geographical details similar to those of the Earth, with gulfs, canals, etc. Then comes a certain moment when all these disappear, to be replaced by the grotesque polygonal forms which seem to refer to an earlier state; but this is a gross masque and something which I ought to ridicule. From this point of view, a study of the Nepenthes is very informative and with the Boreosyrtis we have the same kind of thing; however, in this case the great obliquity of view makes studies more difficult. In 1884 and 1886 I tried in vain, with an 8-in., to examine the area. An aperture of 18 in. is needed for this.

Here are some observations which appear to have been made with the most rigorous precision. How can one believe that an author capable of this analysis could be deceived by optical illusion? The application of his equatorial to the measurement of double stars, as with the examination of the details of Martian geography, proved that the results obtained corresponded to the magnification used, according to the position of the object and to the transparency of the atmosphere. In the one case, as in the other, it is very difficult to question the value of these observations.

And yet, since the gigantic equatorial of almost one metre in aperture ($0^{m}.91$) has been installed at Mount Hamilton, the astronomers have there directed their attention to Mars in order to verify these beautiful discoveries—and they have not succeeded. They have certainly seen the canals, but as broad and vague streaks, scarcely distingishuable and never doubled.

We must therefore admit that Schiaparelli has extraordinary eyesight, perseverance according to which he will wait for a long time to utilize the fugitive moments of perfect seeing and, in the third place, an excellent telescope, I do in fact admit this.

But what can one conclude from these marvels?

This is what will be examined in Part II of this work, where I will also return to the general dissertation relating to the preceding observations.[97]

1888.—Holden, Schæberle, Keelr: Observations Made at the Lick Observatory with the Most Powerful Refractor in the World[98]

These observations almost make one despair. The more one gives time, study and labor to analyzing the numerous and varied observations made of this mysterious planet, the more difficult it becomes to give a definite opinion. However, there is no

[97] See *L'Astronomie*, November 1888, p. 416.

[98] *L'Astronomie*, May 1889, p. 180.

other way to get this information. Examine, compare, discuss; but nothing much can be deduced from such investigations. Well! It would be easy to avoid all bother. This would involve disregarding as irrelevant all the observations which do not agree with the most reliable drawings and with the certainly known configurations on the planet and not to inquire into the discrepancies, attributing them simply to errors. This would be an easy and a rapid method. But it would be dangerous, because it may be precisely from these divergencies and difficulties that we will be shown the way toward understanding the special physical characteristics of this singular and enigmatical planet.

The most powerful refractor in the world has been used to study Mars by a distinguished astronomer who is very experienced in observation: Holden, Director of the Mount Hamilton Observatory. The immense $0^{m}.91$ objective has given the best-ever results with the measurements of double stars and has remarkable powers of definition. Yet we must admit that the drawings obtained by Holden, Schæberle and Keeler with the aid of this colossal equatorial, using magnifications of 350× and 700×, correspond neither with those of Schiaparelli at Milan, nor Perrotin at Nice. Can it therefore, be that each astronomer has his own particular way of seeing? However, there are limits to the personal equation. Astronomy is the most exact of the sciences. It must not lose its reputation. Observers can differ in certain appreciations of hues, extent, forms and even positions when they are dealing with objects which are only just visible, but we certainly cannot admit that they will see features which do not exist at all.

Observations of Mars were begun at the new Mount Hamilton Observatory only on 16 July 1888, that is to say more than 3 months after opposition, which had taken place on 11 April. The planet had already receded from the Earth; its disk had diminished to 9″ and the zenith distance was 60°. Certainly these conditions were not good. However, the satellites were well seen.

From 16 July to 10 August the Californian astronomers made 21 drawings of Mars between them. From these, I have chosen a series of eight for representation here. The Director of the Lick Observatory has himself recognized that they are not in agreement with other published work. Holden wrote:

> For Libya, in particular, the observations of 25, 26, 27, 29 and 31 July, which agree well with each other, differ materially from those made at Nice in April and May, which were published in *L'Astronomie* for June 1888, p. 214. Our drawings are closer to those made by Schiaparelli in 1877 and 1878.

In particular, consider drawings 1, 2, 3 and 4 (Fig. 77). The longitudes of the central meridian are respectively 305°, 310°, 318° and 278°; that is to say, the Hourglass Sea is to the right of the central meridian on the first of the drawings, has just reached the centre on the second, has slightly passed the centre on the third, and has passed the centre more obviously on the fourth.

[Note: the figure designations below refer to those in our Figs. 77 and 78.—WS.]

It is impossible to consider these four views of Mars—made respectively on 27 July at $8^{h}\ 0^{m}$, the same evening at $8^{h}\ 15^{m}$, the 26th at $8^{h}\ 18^{m}$ and the 29th at $7^{h}\ 28^{m}$—without being struck by their dissimilarities. Thus for example, *figs.* 1 and 2 were made at almost the same time and with the same instrument, the first at $8^{h}\ 0^{m}$ by

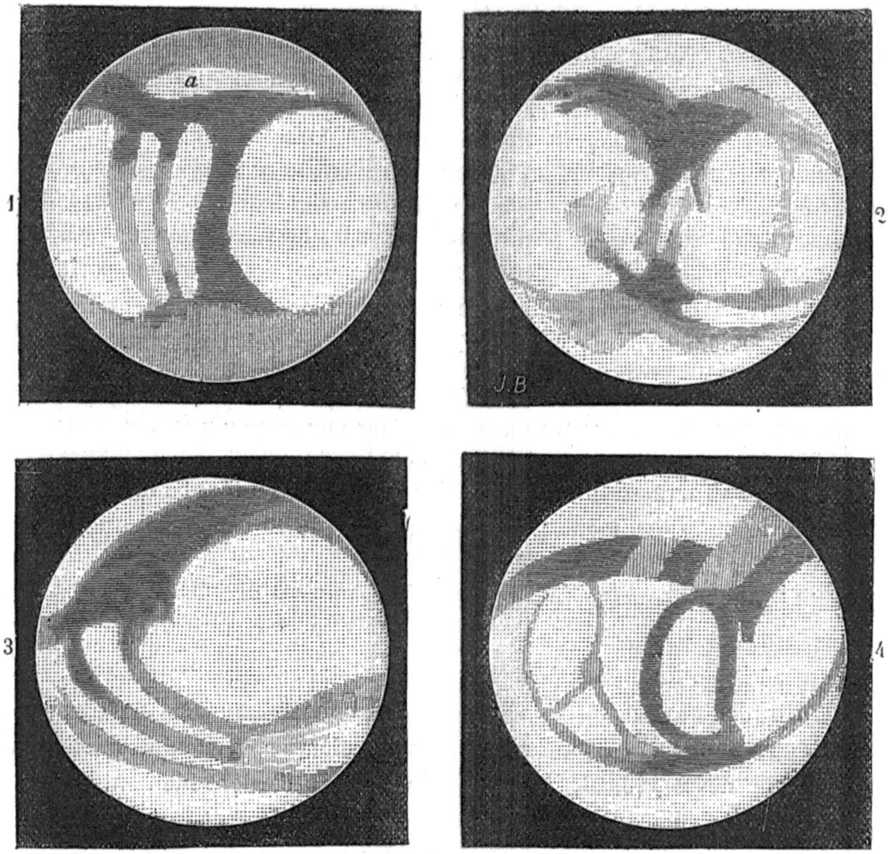

Fig. 77 Drawings of the planet Mars, made at the Lick Observatory, in 1888. 1.—27 July, at $8^h\ 0^m$ (M. Holden); *a*, whitish island. 2.—Same date, at $8^h\ 15^m$ (M. Keeler). 3.—26 July, at $8^h\ 10^m$ (M. Holden). 4.—29 July, at $7^h\ 28^m$ (M. Holden)

Holden, the second at $8^h\ 15^m$ by Keeler. We must certainly think that the difference between the two drawings proves that the observers have different ways of seeing, because it is inconceivable that the planet could have changed so much in a quarter of an hour, even taking atmospheric clouds into account.

The two drawings are completely different: the breadth of the sea, the canals to the left, etc. These discrepancies show that we must be wary of drawing conclusions about changes on the surface of the planet. In particular, it seems that the two broad vertical streaks drawn by Holden in *fig*. 1 also represented by him in *fig*. 3 and one of which is shown in his drawing on *fig*. 4, are exaggerated.

If we choose another series of four drawings, we receive the same impression. For instance, compare *fig*. 5 (Fig. 78), taken on 5 August at $7^h\ 28^m$ (longitude 211°) with *fig*. 6, taken on the same evening at $7^h\ 40^m$ (longitude = 213°), both made by

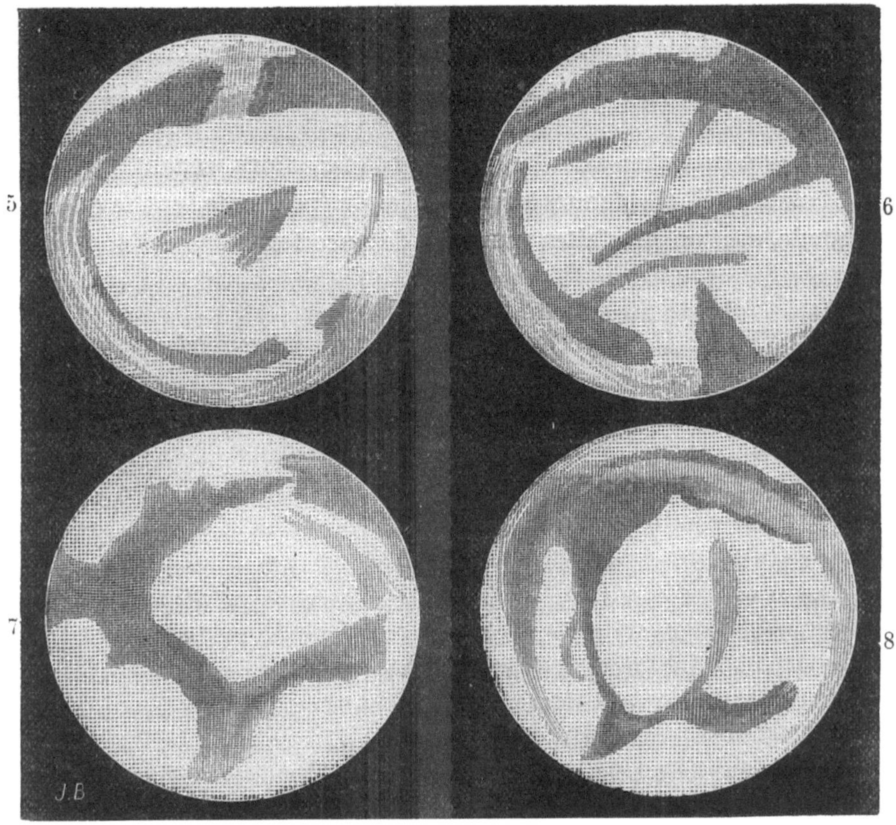

Fig. 78 Drawings of the planet Mars, made at the Lick Observatory, in 1888. 5.—5 August, at 7^h 28^m. Long. = 211° (M. Holden). 6.—5 August, at 7^h 40^m. Long. = 213° (M. Holden). 7.—25 July, at 8^h. Long. = 325° (M. Keeler). 8.—26 July, at 8^h 40^m. Long. = 325° (M. Keeler)

Holden. Can we believe that we are dealing with the same face of the planet? Equally, compare the drawings in *figs.* 7 and 8, the first made on 25 July at 8^h 0^m and the second on the following day at 8^h 49^m; both were made by Keeler, and show the same Martian hemisphere, with the central meridian at longitude 325°. What a difference in aspect! The second sketch does give a reasonable impression of the geography of Mars, but that of the previous day differs essentially. We cannot admit that there has been a change of this kind on the planet in 24 hours; on the same day, 25 July, at 7^h 45^m and 8^h 20^m, two other sketches differed almost as much as *fig.* 7 differs from *fig.* 8; instead of a rectangular appearance, most of the patches seem to be of a perfectly-pronounced oval form.

The authors of these drawings of Mars state elsewhere that they have made only simple sketches and not complete drawings. It is regrettable that the great equatorial was not turned toward Mars until three months after opposition, when conditions would have been more favourable.

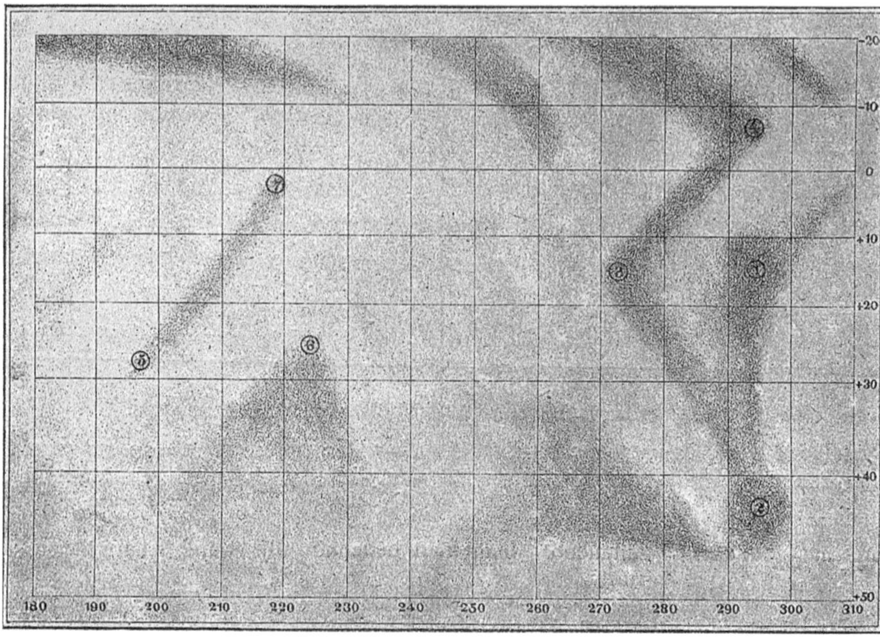

Fig. 79 Micrometric measures by M. Wislicenus, in 1888

We must nevertheless state, without any pretence, that these observations, coming after those made by Schiaparelli and made with the world's most powerful refractor, are a little disconcerting (Fig. 79).

1888.—Wislicenus: Micrometrical Measurements.[99]

Using a new method, Wislicenus, astronomer at the Strasbourg Observatory, made measurements of the positions of seven points on the Martian globe between 2 and 13 May 1888 and drew up the little map reproduced above. He was using a 6-in. (152^{mm}) refractor, with a magnification of 256×. The points were as follows:

	Longitude°	Latitude°
1. Hourglass Sea	294.5	+14.6
2. North point of Coloe Swamp	295.1	+42.4
3. To the left of the Hourglass Sea	271.6	+15.4
4. Œnotria/Iapygia	293.6	−7.9
5. Stygia Palus	198.0	+28.1

(continued)

[99] *Astronomsiche Nachrichten*, no. 2872.

(continued)

	Longitude°	Latitude°
6. Eunostos	223.7	+25.6
7. Cyclopium	219.5	+2.0
Position angle of the axis of Mars		31°.5
Polar distance of the northern snow		3°.5
Longitude of the north polar snow		281°.1

This new chart bears little resemblance to the preceding ones. It is true that it is represented only by a sketch.

1888 Ph. Gérigny: The Tides of Mars[100]

This question has already been discussed above. I asked M. Gérigny, the eminent secretary of *L'Astronomie*, to carry out a new set of calculations. Here is the result of this investigation.

> The very remarkable modifications which have been observed in recent years on the surface of Mars seem to be due to large-scale movements of the liquid mass which covers part of the planet's globe; they seem to be veritable inundations, followed by retreats of the sea from vast areas. Such phenomena naturally suggest the idea of tides, which must be produced in the oceans and to which Mars is no stranger. It is therefore of great interest to calculate the importance of the tides produced by the two satellites and by the Sun. At M. Flammarion's instigation, I have made the calculations. Unfortunately the elements of the satellites are deficient, since we can ignore their masses. We know only that they are very small and very close to the planet. We can believe *a priori* that their closeness compensates for their smallness and enables them to produce appreciable effects on the levels of the Martian oceans, but to calculate these effects we are reduced to hypotheses.
>
> The reason why these satellites are faint is that they are extremely small. I have assumed a diameter of 12 km. for the first satellite and 10 km. for the second, which is a little less bright. I have also assumed that the densities of the satellites are the same as that of the planet. For the distance of the first satellite I have taken 2.771 radii of Mars and for the distance of the second 6.291 radii. These are the values given by Hall. To simplify the calculation, I have assumed a spherical globe having the radius of the planet covered with a layer of liquid of negligible thickness; and I have determined the free surface equation of this liquid, assuming a relative equilibrium under the attractive forces of Mars and of each satellite separately, which enables me to calculate the heights of the two bulges produced on the surface of Mars at the two points where the satellites are respectively at the zenith and at the nadir. We have here a remarkable result which has no analogy with the terrestrial tides; thanks to the closeness-in of the satellites, the two opposite bulges are far from being equal. Thus for the first satellite, the bulge at the nadir is only a little more than half or 5/9 of the bulge from the zenith point. For the second satellite, the nadiral bulge is 41/44 of the zenithal. These results are independent of the masses of the satellites and depend only upon their distances from the planet.
>
> The actual heights of these protuberances must not be confused with the height of the tide; the latter is generally much the greater and depends upon several unknown factors, of

[100] *L'Astronomie*, 1889, p. 384.—*Bulletin de la Société Astronomique de France*, September 1889, where all the details of the calculations are given.

which one essential is the configuration of the coasts. It is that on the Earth we see in the Channel, at a place well away from the point where the Sun or Moon is at the zenith tides amounting to 13 metres, while the protuberance due to the Moon is only $0^m.50$ and that of the Sun $0^m.25$, giving a total of $0^m.75$. At the latitude of Granville, this protuberance will be reduced to about half, at the surface level of the seas. The height of the tide is thus equal to that of the protuberance multiplied by 13/0.75, or about 35.

To interpret the figures to be given here, it is therefore important to keep them as giving the differences of level between the high tide and the low tide; these must be compared with the corresponding values for terrestrial tides, as follows

$0^m.50$ for the lunar tide
$0^m.25$ for the solar tide
$0^m.75$ for the high tide at syzygies.

With the dimensions assumed for the satellites, the heights of the tides are extremely feeble. I have found

1. For the first satellite:
   Zenithal protuberance $1^{mm}.79$,
   nadiral $1^{mm}.05$
2. For the second satellite:
   Zenithal protuberance $0^m.088$,
   nadiral $0^m.082$.

These values are indeed insignificant; they correspond to tides much weaker than those of the Mediterranean and represent only an infinitesimal fraction of the solar tide on Mars, for which I have calculated a bulge of 52^{mm}, almost 1/15 of the solar tide on Earth.

If therefore the satellites are as small as I have supposed, we must give up any idea of tides as an appreciable influence upon the flooding phenomena of which we have evidence; but the dimensions assumed here may be much too small and the height of the protuberances is proportional to the masses of the satellites concerned, that is to say to the cube of the diameters. If we double the assumed sizes and make the diameters 24 and 20 km., the protuberances become eight times greater, that is to say:

$14^{mm}.35$ and $8^{mm}.40$
$0^m.704$ and $0^m.666$.

Tripling the diameters and making them 36 and 30 km., the original figures must be multiplied by 27, becoming:

$48^{mm}.43$ and $28^{mm}.46$
$2^m.38$ and $2^m.21$.

The tides due to the second satellite remain negligible, but those due to the first satellite become comparable with the solar tide. Finally, if we multiply the original dimensions by 10, giving 120 and 100 km., diameters still very small for bodies visible at the distance of Mars and in its immediate proximity, the original heights must be multiplied by 1000, giving:

$1^m.79$ and $1^m.05$
$0^m.088$ and $0^m.082$.

The tides due to the second satellite still remain very weak, but those due to the first become double those of our own oceanic tides.

The preceding calculations show that under any hypothesis the second satellite cannot exert any appreciable influence; its action can be scarcely 1/20 that of the first. But if the mass of the first satellite is great enough, it can produce tides at least as important as those which we observe in our seas.

As we have noted above, the height of the protuberance of the surface level is only one of the factors concerned in the phenomenon of the tides. The sea is never in equilibrium; but under the attraction of the satellite, the liquid mass executes a series of oscillations whose period is equal to the time which elapses between two successive returns of the satellite to the same meridian, and naturally, in this oscillation, it passes across the equilibrium position, so that the height of the tide is necessarily greater than that of the protuberance. It is natural to assume that the more rapid and more furious oscillation will be the movement of the sea and the higher will be the tide. Now, for the first satellite, the only one which concerns us, the length of the revolution period round Mars is $7^h\ 39^m\ 15^s$, while the planet rotates in $24^h\ 37^m\ 23^s$; this means that the satellite returns to the same meridian after $11^h\ 6^m\ 24^s$. The double oscillation of the tides must therefore be completed in this short interval of less than twelve hours; during this time there are two high tides and two low tides, so that only six hours elapse between two high tides, and less than three hours between high tide and low tide. This rapid movement of ebb and flow certainly leads to a great increase in the heights of the tides.

The configuration of the seas on Mars, opening out from a coast and ending in narrow canals, leads itself admirably to the increase in level at full tide. In these elongated seas, there is a phenomenon analogous to that produced in our Channel. The rising wave, produced in the heart of the ocean, is propagated in a basin which becomes narrower along its length; the water-mass is therefore more and more broken up and the wave must necessarily become higher as it moves along, finally attaining a considerable altitude. The satellite turns more quickly than the planet and so the Martian tides are propagated in the opposite sense to ours, that is to say from west to east. When we turn to the map and examine the outline of the Hourglass Sea and recognize that the flow coming in from the Dawes Ocean engulfs the Hourglass Sea from south to north, we see that the level becomes higher and higher as the flow passes down the narrowing coasts. An analogous phenomenon can be seen on many regions on Earth. It is true enough that in long straits, it can produce in each sea a tidal current with bars similar to those of the Seine or the Amazon. Moreover, since the oscillations of a liquid mass are always propagated in the same manner and as the arms of the Martian seas link the different oceans, the seas should receive at least some of the flows produced in the oceans, propagated in opposite directions. We can judge what will happen at the meeting of waves moving in opposite directions!

Finally, the solar tide, though weak, also exists on Mars and is propagated in a direction opposite to that of the tides due to the satellites. Twice a day, in the same place, the lunar and solar tides, in opposition to each other, will meet and will add to the grandeur of the phenomenon.

There is moreover another important factor with regard to the tides of Mars. We have already noted that in these oscillations the level of the sea passes through the equilibrium position, but in reality the problem of the movements of the ocean is much more complicated than the simple determination of the form of equilibrium. The greatest analysis of the last century, Lagrange, Laplace and Legendre, considered that important question, and Lagrange was able to show that the relative densities of the sea and of the solid core of the planet had great influence upon the result. He proved that the amplitude of the oscillations of the ocean remained between certain limits, provided that the density of the seawater was less than the mean density of the Earth. In the contrary case, if for example the ocean were made up of Mercury instead of salt water, the whole mass of the seas would leave the ocean beds and would spread in a formidable flood over the continents. Undoubtedly this extreme case is not found on Mars; the density of the seas must be less than that of the planet, but the density of Mars is only three fourths of that of the Earth, while the water is presumably of the same density as ours. The ratio of the density of the sea to that of the inner core is therefore about four thirds of that on Earth. According to Lagrange's analysis, the increase should produce a corresponding increase on the amplitude of the oscillations of the sea level.

For all these reasons—if, even on the Earth, the height of the tides can reach up to 35 times the height of the bulge at the equilibrium state—there is nothing astonishing in finding that on Mars, the difference in level between high and low tide can reach up to 50 and even 100 times the height of the equilibrium bulge.

In short: the solar tides on Mars are certainly much inferior to those on the Earth; the tides due to the second satellite attain only one twentieth of those due to the first. As for these satellite-produced tides, they depend upon the masses of the satellites. It may be that they are insignificant, but it may also be that they are considerable. In either case, if the mechanical influence of the satellite is comparable with that of the Moon on the Earth, the movements of the seas of Mars are certainly more tumultuous and more important that those of our oceans and if moreover we agree that the relief of Mars is slight—as is generally supposed—and that the coasts are low, then the tides ought to produce, four times a day, floods covering vast stretches of the coastal areas. Astronomical science cannot yet give a definite solution to the problem, because we are ignorant of the masses of the satellites of Mars, but we may be assured that the solution will be found before long since, very fortunately, there are two satellites which will exert an influence upon each other. After a sufficient number of years of observation, we will certainly be able to determine the mutual perturbations of these two bodies, and so deduce their masses. We can then find a reliable basis for the calculations which need to be made and decide upon the importance of the Martian tides and the role they play in the surface modifications which we observe.

1888.—Schiaparelli: The Physical Constitution of Mars[101]

In *Himmel und Erde* and *L'Astronomie*, the eminent Italian astronomer has published a general résumé of his researches. It is my duty to reproduce here the most important extracts from it and at the same time I am happy to offer my readers the last two charts drawn by Schiaparelli himself from the total of his observations.

A.—REGIONS OF INTERMEDIATE TONE AND THEIR VARIATIONS.

The regions which we call seas or continents, according to whether they are dark or bright, occupy the greater part of the surface of the planet; but there are other regions whose appearance is variable and have sometimes the apparent character of seas, sometimes of continents, sometimes both at the same time.

Among others, such are the two zones in the Mare Erythræum known as Deucalionis Regio and Pyrrhæ Regio; also the two islands named Hellas and Noachis. Of the same nature are the two islands Iaygia and Œnotria in the Syrtis Major, and in general all the parts of the sea which are shown on the map as being brighter than the rest. Mare Cimmerium and Mare Acidalium each contain a country of this type. According to the differences in the time and in visual angle, these regions can present either completely or in large part the diverse hues which we see on the continents and on the seas of Mars and this make up a series of transitions. They do not all appear to be of the same character, so far as I can judge from my observations up to the present time. It seems that some are predominantly of a maritime nature, while others are continental. These regions are not always clearly separated from the neighbouring continents or seas, but they often connect the one to the other by a series of insensible gradations of luminosity and tone, as can be seen from the various examples on the page.

One of the most remarkable of these intermediate regions is the Deucalionis Regio, which lies in the Mare Erythræum, where it forms a peninsula at a right angle. It is clearly

[101] *Himmel und Erde*, 1888.—*L'Astronomie*, January, February, March and April 1889.

delimited on the side touching the continent, while on all the other sides it loses itself in degraded tones. Its colour is midway between those of the continents and the seas; it is sometimes yellow, sometimes grey; near the border, we can sometimes see a greyish-white colour. In any case, it seems to me to be bright enough to be distinguished from the dark background which surrounds it. Reference must also be made to the Pyrrhæ Regio, which can become so dark—particularly in the part adjoining the continent—that it cannot be distinguished from the rest of the Mare Erythræum.

The island of Cimmeria, a long band which, on Plate II, occupies a considerable part of the Mare Cimmerium, is more remarkable in this respect than any of the other mixed regions.

In 1877, the whole of the Mare Cimmerium appeared very dark; it was then classed as one of the darkest areas on the whole surface of Mars. In 1879 it showed little change and all that was noticed was that although the colour remained very dark, it was less sombre than in 1877. Toward the end of 1881, the area contrasted sharply with the yellowness which surrounded it, but on 3 February 1882, when this part of the planet became visible, I was, for the first time, a long band in the form of a comet, which extended over more than 30°, from longitude 205° to 235°. This observation was confirmed on 4, 5, 6 and 7 February; later there was no chance to re-observe the area. In my notebook I find no mention of the island of Cimmeria during the opposition of 1884. In 1886 and 1888 the region was seen at a very oblique angle; also, the observations were not very precise. My impression was that the island of Cimmeria was visible.

The metamorphoses of the large island of Hellas are more complex, but no less remarkable. In 1877, near the end of the southern solstice of Mars, the island formed a very regular round or slightly elongated form, whose diameter was not less than 30°; generally yellow, it appeared whiter when near the limb than when near the central meridian. Once (16 December 1877) I saw it almost as white and brilliant as the polar region; on 21 December, however, the former colour had already been restored. During the opposition of 1879-1880 it was approximately round, but instead of a brilliant surface it showed an uneven and unequal brightness, becoming duller toward its upper left part (in a reversed telescopic image). It was crossed by several clearly visible canals, one of which was almost parallel to the meridian and another to the latitude circle. (In 1877 I saw only the first of these canals and even this only with difficulty.) The island was therefore divided into four parts. In January 1880, only the two lower quarters were yellow; the others were darker, the left-hand lower quarter more so than the right. Also during this opposition (1879-1880) Hellas seemed brighter when near the limb than when near the centre of the disk; several times it appeared white. In 1879 and 1880 it seemed, to the eye, a little smaller than in 1877. During the opposition of 1881-1882, I found that its brightness had sensibly diminished; its colour was bright ashy grey. The outlines sometimes lost their sharpness, appearing only as a cloud which dissipated. Only on a few occasions, when near the central meridian, did it take on a brownish-yellow colour, like that of the Deucalionis Regio. It was throughout, divided by the two canals which made up a cross, but the size had noticeably diminished and in several places the former limits were more or less occupied by sea, so that it assumed the form of a trapezium with rounded angles, as shown in Plates I and II. During the following oppositions, Hellas was seen at a more oblique angle; it had the aspect of a nebulous whitish patch of imprecise form. Its diameter certainly did not exceed 12° to 15°. Yet it was sometimes whiter and more brilliant than usual and could have been confused with the north polar cap.

In certain respects, the region called Libya can be included in the category we have been discussing. It lies near the equator and can consequently be well observed at all oppositions, whatever may be the axial inclination of the planet. In 1877, Libya was bounded, on the Mare Tyrrhenum side, by an elegant, regular arc ending toward the north in a thin, elongated point (Osiridis Promontorium). The surface of this point was covered with a shade which seemed to darken further along toward the end. Toward the north, Libya was bounded by a canal which took almost the form of a semicircle, at the middle or summit of which was a feature resembling a sombre point; I have given this feature the name of Mœris Lacus. In 1879, I found that a part of Libya had been invaded by the Syrtis Major, which extended as

Fig. 80 Variations observed on the planet Mars on the shore of the Hourglass Sea (Syrtis Major)

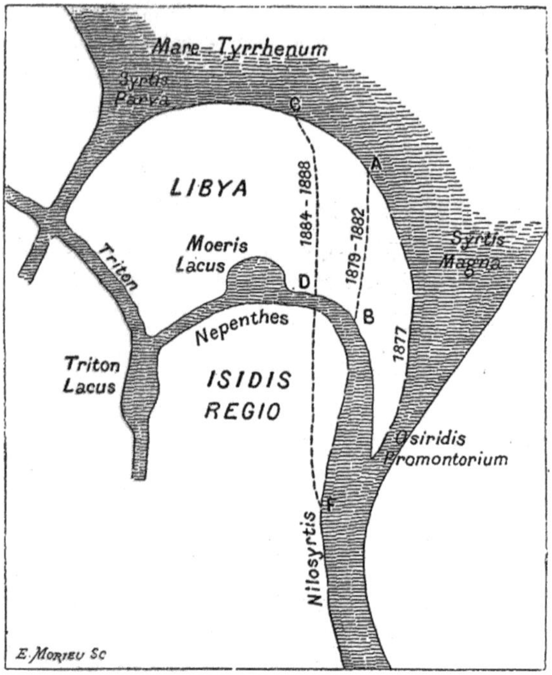

far as the line AB (Fig. 80); the region of Libya to the right of AB had become quite dark instead of the yellow colour which it had formerly shown and it had taken on the tint of the neighbouring sea which had flooded it. The Osiris Promontorium had been covered by the invasion of the sea; the course of the Nepenthes had been shortened and its mouth transferred to B; the coast of the Syrtis Major had assumed a different curve and was notably closest to the Mœris Lacus. Finally, the indefinite shade which in 1877 had covered the Osiris promontory had advanced as far as the middle of Libya; it also enveloped Mœris Lacus, which had previously lain beyond it. The other part of Libya (that is to say, the left-hand half) had assumed a red colour, considerably darker than at the previous opposition. During 1881 and 1882 I saw no change. I noted only that the surface of Libya, always red in tint, resembled a cloth filled with small patches which could not easily be separately distinguished. Since the opposition of 1884, the invasion of the sea has progressed as far as the line CDF, as can be seen from inspection of the drawing (Fig. 80), so that a great part of Libya and a small part of Isidis Regio have disappeared. Mœris Lacus, which in 1877 lay in the middle of the Nepenthes, is now almost at its mouth. Instead of presenting a beautifully curved arc, Libya forms a promontory between the Syrtis Major and the Mare Tyrrhenum, resembling an angle with a blunt point. In 1884 it also kept, independently of the dark colour which distinguished it from its immediate surroundings, the appearance of a speckled cloth, as if the region were covered with innumerable small patches which could be confused with each other. During the opposition of 1886, the state of affairs was in general different from that of 1884; I should however add that for part of the time the weather conditions were unfavourable. Finally, in May 1888, Libya appeared very much shortened in the neighbourhood of the meridian, as was also recorded in Nice by Perrotin. However, the observations of 6, 7 and 8 May showed a dirty white colour when the area lay near the limb, which completes the analogy with the other areas we have been discussing. Mœris Lacus remained visible, but was very difficult; it made almost less than a right angle with Libya, near the mouth of the Nepenthes in the Syrtis Major. On various occasions the Isidis

Regio (below the Nepenthes) appeared very bright and the contrast with the brownish colour of Libya became very marked. During the same opposition, the colour of the Syrtis Major was not as black as it had been during the preceding oppositions from 1877 to 1884; instead it was a brighter grey, except in some small streaks which will be described shortly. There was therefore not a great difference in tone between Libya and Syrtis Major, though the colours were not the same and the boundary between the two remained obvious.[102]

From my notebook I can give several other analogous cases; but the two which I have cited, Hellas and Libya, suffice to give a general idea of the variations observed. For these two cases, the series of events which I have described occurred during a period including six oppositions from 1879 to 1888. From this we cannot perhaps conclude that the variations are slow, taking place over periods of secular duration. All the same, in certain cases it is true that the phenomena recur at each revolution of Mars. But at each new opposition, the planet lies 48∞ in longitude ahead of the orbital position at the previous opposition; consequently the Martian season is advanced by 1/8 of the whole period between one opposition and the next; and this circumstance allows us to follow the series of phenomena taking place on Mars, although one part of the observational sequence belongs to one revolution and another part of the sequence to the following revolution. In the same way, a meteorologist could study the shift of climate of a region if he could put together the observations made in different months over several years, as for example January 1888, February 1989, March 1890... and so on to December 1899.

B.—CONTINENTAL REGIONS WHICH WHITEN WHEN SEEN OBLIQUELY.

We have seen that the regions of doubtful character are often brighter when seen obliquely, than when at the central meridian; this also applies to regions which are purely continental in character. In this respect I may particularly cite the polygonal or almost round region which on the map are given the names of Elysium and Tempe. These regions are of variable whiteness, but always less brilliant than the poles. This whiteness is more generally seen when the regions are the edge of the disk of Mars and I have often observed them even when, several hours before or afterwards, during their transit of the central meridian, the regions showed nothing remarkable.

The analogous transformations of the island of Argyre are particularly interesting. Under certain circumstances the island becomes so brilliant at its border that some observers have mistaken it for a polar cap. This intense brilliance of the island was noted by Dawes as long ago as 1852; the English observers of Mars called it Dawes' Snowy Island. On the other hand, I have often seen it as yellow or even dark red when near the central meridian. I consider it as being similar in nature to the neighbouring island. The latter is called Argyre II; it is smaller and lies further south; its existence became known to me only on 8 November 1879. It was then near the left-hand edge of Mars and its brilliancy was less than that of the polar cap. When it crossed the central meridian, it was of an uneven red colour and of low brightness.

Independent of these changes in colour subordinate to the diurnal rotation, we have analogous changes in the continental regions; but these latter are much slower and often extend over larger areas. Such was that which took place in the years 1877 to 1878-1879 in the large region which extends between longitudes 220 and 170 degrees as far as latitude 40°N and which is known as the Mare Sirenum. Over all its surface—notably in the upper part, adjoining the sea previously mentioned—it is much brighter than some of the other continental regions.

To this class also belong observations which I made from 1877 to 1882 of a small, bright, white spot at the north end of the Nepenthes, at longitude 269°, latitude 17°N. I saw this spot for the first time on 14 September 1877; it had a diameter of about 8° and it seemed

[102] What I have said about the grey tone and changes of Libya can also be recognized in the drawings by Lockyer (1862), Kaiser (1864), Dawes (1864) and Green (1873)—as I have already commented.

almost square; it shone much more brightly than any other part of the planet and also showed very distinct outlines. This spot, whose whiteness I unhesitatingly compared with that of the south polar patches, was still visible on 14 October. I observed the same phenomenon, in the same place, during the following opposition from November 1879 to January 1880; the size was unaltered, but the shape seemed to have become a little rounder. Surprised at the persistence of this bright patch, I gave it the name of Nix Atlantica. I saw it again, during the opposition of 1881-1882, from November to March, but not always with the same ease; it showed differences of appearance and variations in brilliancy which could not always be attributed to differences in the clarity of the telescopic image. But I sought it vainly during the following oppositions and it was still invisible this year. If its visibility depends upon the seasonal period of Mars, we ought to see it again during the oppositions of 1892 to 1897, and it is easy to appreciate the importance of the reappearance for the study of the physical constitution of the planet.

An analogous patch, though much smaller and more difficult (Nix Olympica) was shown with great persistence in 1879, at longitude 129°, latitude 21°N; the diameter was 4° or perhaps more. I have never seen it again. Other patches appear, sometimes here and sometimes there, in various parts of the continental regions; they are of various degrees of white and sometimes clear, generally lasting for several days though apparently following no law. This has often happened during the last two oppositions, along the right bank of the Syrtis Major and along the coast which runs from here to the sinus Sabæus, as well as in other places. Often a definite region seems to be sprinkled with white patches, as, for example, in the land between the Ganges and the Iris on 18 and 19 January 1882 and between the Nilosyrtis and the Indus on 31 January. It is also found that the white bands in the form of regular belts, of uniform breadth, spread out rather obliquely from Northeast to Southwest, are of low inclination to the meridians.

C.—THE VARIATIONS IN TONE OF THE SEAS.

Between them, the seas present equally remarkable changes in colour, but more slowly and more regularly. At this point at which we have arrived and where my own studies have commenced, I am prepared to state that when the seas cross the central meridian in oblique positions, because of the diurnal rotation, they do not change in colour. I have at least followed the colour change of the island of Argyre, which goes from dark red to the most brilliant white according to the way in which the Sun's rays strike it, without any change of colour or brightness being seen in the neighbouring regions. I have observed the same phenomenon in the island of Œnotria, in the Syrtis Major. This proves that the sea-surfaces are, in a way, different from those of the other regions I have discussed; in any case, they are fundamental in the study of the physical nature of Mars.

It is no less certain that from one opposition to another, very remarkable changes of tone are seen in the seas. Thus there are the regions named Mare Cimmerium, Mare Sirenum and Solis Lacus, which, during the years from 1877 to 1879, were among the darkest areas on the planet, but which became less and less black during succeeding oppositions, until in 1888 they had become a bright grey which scarcely sufficed to render them visible when obliquely placed. From 1877 to 1879 the Syrtis Major and the Nilosyrtis appeared very black, but in 1888 the Nilosyrtis had not varied, while the Syrtis Major (in a small streak below the north of the Nepenthes and in some other very narrow zones) had become so bright that it was hardly to be separated from the adjacent regions, notably Libya. The Mare Erythraeum had also become very bright, with the exception of the three gulfs, the Sinus Sabæus, Margaritifer Sinus and Auroræ Sinus, which in consequence must now be classed not as gulfs, but rather as three large isolated islands. On the other hand, at the same time, the Mare Acidalium and the Lacus Hyperboreus have become very dark; the latter appears indeed very black, although it was seen more obliquely than the Syrtis Major and the seas mentioned above. The state of the regions which we call seas is not constant. This is undeniable. Perhaps the modifications which occur there are associated with the Martian seasons.

D. – THE CANALS.

When, on Map I, we consider the great gulf lying below the equator of the planet, by longitude 290°, we see that it is prolonged toward the north by a long appendage called the Nilosyrtis. This is a band normally very sombre, which may even appear virtually black by contrast with the bright regions which surround it. Its breadth is from about 4° to 5° and seems exactly uniform in the northern part as far as latitude 20°N. Its borders are well defined, and their general course is curved in a regular manner; there seem to be small serrations along the whole of the length, but it is never possible to see these indentations individually. If the sombre patches on the planet are seas, such a formation as the Nilosyrtis should be regarded as a canal. I use the term without pronouncing upon the nature of the feature.

The Nilosyrtis is not the only canal on Mars, but it is much the broadest and the most easily visible; it was shown on the drawings by Schröter and, during the past thirty years, by a large number of observers. Secchi in 1858 and Dawes in 1864, recognized, more or less definitely, the existence of several other analogous formations; their number has been multiplied several times in an unexpected manner, and there is now no longer any doubt that these canals from a very complicated réseau covering all the continental regions of the planet.

Planisphere I gives a schematic representation of this réseau, containing almost all the canals whose existence I can confidently affirm from my observations from 1877 to 1888. By the term schematic, I mean to infer that the lines or bands of the réseau are drawn in a manner giving approximately the length and the direction of each canal, their relative positions and the polygonal formations which result, without taking into account their colour, darkness or breadth (with the exception of the Nilosyrtis, whose breadth is altogether exceptional), or the sharpness of their outlines, or the doubling which affects some of them at certain epochs. In fact, these conditions of visibility, breadth and form are variable from one opposition to another and even from one week to another during the same opposition; and the variations are not simultaneous for all the canals, since even in the same region at the same time there may be great differences between adjacent canals. It follows that it is possible to give a representation of these canals for a certain epoch; but it is impossible to draw up a permanent chart. Therefore, we cannot expect perfect or even near-agreement between Planisphere I and the Martian canals; such absolute agreement is impossible over a long period. Each canal on the chart is drawn simply as a line, or rather as a narrow band, upon which different features can develop at different oppositions. We can therefore see that with regard to the canals, the chart is only a sort of topographical index, necessary for the understanding and co-ordination of the very numerous and very variable details which are observed at each moment in different regions. Such a representation does not serve as a description of the physical appearance of the canals; but it suffices to show the geometrical and topographical characteristics of the réseau.

It is clear that for the most part, the canals follow courses which are not very different from great circles on Mars. There are, however, some exceptions. Of these, the Phasis, the Simoïs, the Gehon, the Indus, the Boreosyrtis and above all the Nilosyrtis are the most remarkable.

There is also another more or less general property: each canal leads, at each of its ends, into a sea or a lake, another canal, or an intersection of several other canals. I do not recall ever having seen one of these lines stopping short in the middle of a continental expanse, in the form of an isolated trunk and without any further connections. This fact is of the greatest importance in the study of the nature of these formations.

The canals can cut each other at all possible angles. On the planet there are several regions where three, four, or even six or seven canals meet in one small area; this area is then usually marked by a very sombre patch, a lake, whose size and appearance can vary within certain limits. A very important 'knot' in such an area is the Phœnicis Lacus (longitude 108°; latitude 16°S), formed by the meeting of seven canals; Agathodæmon, Eosphoros,

Phasis, Araxes, Eumenides, Pyriphlegethon and Iris, which diverge from it in the manner of a more or less regular star. Another less regular 'knot', the Trivium Charontis (longitude 195∞, latitude 17°N) is formed by the more or less eccentric meeting of the Cerberus, Læstrygon, Tartatus, Orcus, Erebus, Hades and Styx. In the Ismenius Lacus (longitude 335°, latitude 40°N) converge the Euphrates and its northern prolongation, Protonilus, Deuteronilus, Astaboras, Hiddekel and Jordanus. It is easy to recognize several other examples on the map, such as Propontis, Lacus Niliacus, Lacus Tithonius, Lacus Lunæ and Nodus Gordii, the most extensive and the most imperfectly defined of all.

On examination of the chart, it can also be seen that the lengths of the canals show a side range; some scarcely exceed 10° or 15° (Xanthus, Scamander, Eosphoros, Nectar, Ambrosia, Issedon). Others, however, are very long and without sensible irregularities; they may extend for a quarter of the circumference of the planet. Such are the Euphrates, which, with its northern prolongation, extends from the equator almost up to the north pole and the Erebus-Acheron, which occupies at least 90°; if we regard the Dardanus to one side and the Cerberus to the other as prolongations of the Erebus-Acheron, since they appear to join it without any appreciable break in continuity, we have a line stretching for about 160°, from Niliacus Lacus as far as Mare Cimmerium.

The great uniformity and the composition of the whole system is strange and difficult to explain and we must look for a law about the distribution of the lines which will be as simple as that which …like de Beaumont thought he had found for the direction of the great mountain ranges on Earth, with his famous pentagonal réseau. I believe that there is probably a law of the same kind, but we must not forget that my sketch is neither exact nor complete.

E.—VARIATIONS IN THE ASPECTS OF THE CANALS.

(a) A canal may become invisible for a period which may be either short or long. We are concerned here not with invisibility due to poor conditions of observation, but genuine invisibility, which persists even when the atmospheric conditions are so good that the canal would show up at other times. Moreover, the idea of invisibility should be related to the optical power used, since during my researches I have used various telescopes, that is to say, we must not exclude the possibility of seeing the same object with a more powerful instrument. Here is a very striking example of such invisibility. During the evenings of 2 and 4 October 1877, with excellent seeing and with Mars showing a diameter of 21″, the continental region between the Margaritifer Sinus and the Auroræ Sinus was very bright, and devoid of canals, without the slightest traces of any. Indus, Hydaspes, Jamuna and Hydraotes were completely invisible. This state of affairs persisted until 7 November, when the apparent diameter of the planet was 15″. Several months later (21-26 February 1878) the Indus was perfectly visible, even though the apparent diameter of Mars had been reduced to 5‴.7. During the opposition of 1879, the Indus was always very evident; on 21 October (diameter of Mars 19″), the Hydaspes appeared for the first time; and on 27 November (diameter 17.5 inches) I had my first view of the Jamuna. On 28 November all three, the Indus, Hydaspes and Jamuna, were broad, black and visible at first glance. The Hydraotes was discovered in 1882, when the apparent diameter of Mars was 14″. All these canals remained more or less visible during the following oppositions, but latterly (1888) the Indus and the Hydaspes have again become very difficult. Without wearying the reader by citing other similar cases I regarded it as established that the canals of Mars can become invisible at certain times.

(b) In most cases the presence of a canal is first detected in a very vague and indeterminate manner, as a light shading which extends over the surface. This state of affairs is hard to describe exactly, because we are concerned with the limit between visibility and invisibility. Sometimes it seems that the shadings are mere reinforcements of the reddish colour which dominates the continents—reinforcements which are at first of low intensity and which become visible only because of their breadth, which is considerable even though we cannot measure it exactly. At other times, the appearance may be more that of a grey, shaded band made up of a lightly shaded oblong cloud. It was in one or other of these indeterminate forms that, in 1877, I began to recognize the existence of the Phison (4 October), Ambrosia (22 September), Cyclops (15 September), Eunostos (20 October) and many more. Analogous examples were not lacking during the following oppositions.

(c) Very often a canal had the appearance of a dark band shaded on its two sides, having a maximum intensity in the middle, which seemed to be sombre enough to give the impression of a reasonably well-marked line. This state of affairs varied according to the preponderance of this central line compared with the nebulous lateral parts, under the double relationship of their breadth and their intensity. The bands thus formed were normally regular enough, though there were sometimes some possible anomalies in their breadth and intensity of shading—anomalies which could ordinarily be suspected with the power of the telescope used, though full evidence could rarely be obtained. Cases of a structure differing on its two sides are very rare; but it was shown indisputably on 30 January 1882 for the Gehon, only one side of which was shaded, the other being ill-defined; and for the Euphrates, on the 19th of the same month, which was nebulous to the right and well-defined to the left. In 1879 several canals were shown with unequal structure along their courses, which changed little by little from one extremity to the other. Læstrygon, Tartarus, Titan, Ganges, Gorgon and Sirenius were thin, black and well-defined at their southern ends, which debouched into the Mare Sirenum or the Mare Cimmerium in penetrating to the north; in the continental region, they broadened into the form of a comet's tail, ending as a broad, ill-defined shading at the north end. In the same year, the Astapus came out from the Nilosyrtis as a very thin, badly-defined canal; it broadened considerably and went on to lose itself in the Alcyonius as a very broad, very light shading. There were similar lacks of uniformity in the neighbouring canals; the bright region of Elysium often took on a circular form, although it was set in a pentagonal space bounded by five canals.

(d) The most perfect type of canal, which I regard as the normal or ideal state, is a well-defined, sombre line (sometimes virtually black), which looks like a pen-stroke on the yellow surface of the planet. The appearance of the canals in this phase of their existence is very uniform throughout their length, with few exceptions; the general course is regular; in the very rare cases when it is possible to make out the two boundaries individually, I have been able to note very small sinuosities and indentations. This was shown in 1879 by the Euphrates and the Titan and in 1888 by the Ganges. Each border was perfectly well-marked, indeed as perfectly as the borders of the continents or the seas. The breadth is very different from one canal to another, ranging from the Nilosyrtis, which can attain or even surpass 5° or 300 km., down to simple lines without appreciable breadth, as with the Galaxias, Issedon, Anubis and Erinnys in 1882, and Æthiopis in 1888, when the probable breadth did not exceed 1° (60 km.). This breadth is uniform, with few exceptions; however, the Jamuna and the Iris in 1879, Hades and Athys in 1882 and Nilokeras in 1886 have provided definite cases of canals which are broader at one end than at the other.

As a very remarkable case, let us site the Simois, which was invisible in September 1877 and was seen in October as an extremely fine line. In 1879 it was broad enough and black enough to count among the most prominent of the Martian canals. At the beginning of January 1879, the Simoïs was as broad and black as the Nilosyrtis; the breadth was estimated as 4°. At the same time the canal Ascanius appeared to the right of the Simoïs and the part of the continent between the Ascanius and the Simoïs took on a tint much more sombre than those of the neighbouring regions (see map). Unfortunately this part of the planet could not be well observed in the following years, because it was too far south and too near the edge of the disk.

An almost identical case is that of the Triton. In 1877 I could see only the right half, between the Lethe and the Nepenthes. At the following oppositions, it was possible to follow the whole of the Triton from the Nepenthes as far as the Mare Cimmerium, though not all parts of it were equally easy. But lately (May 1888) it has become extraordinarily broad and forms a vast strait. And—which is even more remarkable—the Syrtis Minor has broadened considerably at the expense of Libya, while Libya has become very dark, as I have noted above. This coincidence of a broadening of the Simoïs and the Triton and the darkening of a vast adjacent region, is probably not due to sheer chance. For the rest, all the canals on the planet seem to be subject to similar variations in a greater or lesser degree. The Nilosyrtis itself was at maximum breadth in 1882 and minimum in 1886, but here the difference between maximum and minimum was much less marked. From the observations by Dawes and Secchi, we also know that in 1864 and 1858 the Hydaspes was one of the most easily seen of the canals, but this was no longer true in my period of observing (1877ñ1888). And M. Van de Sande Backhuyzen has recognized in Schröter's drawings some considerable sombre patches which were no longer observed in our own time, and were no doubt bound up with phenomena of the same nature.

A similar feature was also shown on a vast scale in the neighbourhood of the North Pole, during the opposition of 1884-1886. Around the snowy polar cap, several canals had become very black and very broad, and at the same time the intervening spaces had become very dark. When the definition of the telescope is inadequate, the confusion of all these details produces a grey zone around the polar cap and it is probably a similar observation which indicates traces of a non-existent polar sea.

The variations in intensity of a well-marked canal also affect it throughout its length. But when, by intersection with other canals, it is split up into several parts, it can happen that the intensity, uniform in each part, may differ from one section to another. We have already noted that in 1877 the Triton was visible only to the right of the Lethe, and the section between the Lethe and the Mare Cimmerium was invisible. In 1879, the Phison was very black in the northern section between the Nilosyrtis and the Astaboras, while it was much less evident in the southern part, between the Astaboras and the Sinus Sabæus. In 1882, the Hydraotes was very delicate in the section to the left of the Jamuna, but broad and easy (and even double) in the section to the right of the Jamuna. In this case, the change of intensity in passing from one section to another is quite abrupt, without any appreciable transition; each section is ordinarily quite uniform throughout its length.

F.—THE DOUBLINGS OR GEMINATIONS OF THE CANALS.

We will now consider the last and the most remarkable of the transformations of the Martian canals: that which causes the geminations. These phenomena strain the imagination when we try to draw an analogy between the facts we know on Earth and the physical study of Mars. A canal which has been recognized in one of the forms described above, or even in several successive forms, can in a few days (or even hours), by a process of transformation about which we do not yet know the details, become double, made up of two bands close together, ordinarily equal and parallel; cases of divergence or of differences in breadth are rare. In several cases it is possible to state, from a close comparison with the neighbouring details, that one of the bands keeps to the original canal (or very nearly so). But latterly, in 1888, I have become convinced that this role is not general, and that it can be that neither one nor the other of the new formations coincides with the original canal. The identity of general direction and location is, then, only approximate; all trace of the original canal disappears, to be replaced by the two new lines.

The distance between the two parallel lines is very different from one gemination to another; the upper limit has been estimated at 10 or 12 degrees, even 15° for some very long and imperfectly-marked geminations such as those of the Titan in 1882 and the Gigas in 1884. As for the lower limit, it can be determined only by the power of the telescope used and to the circumstances of observation; in 1888, Protonilus and Callirrhoe were resolved into two components separated by 3° at most. It sometimes happens that one can assume that a line is double, according to its particular appearance, without being able to separate the two lines completely. The doubling of a line can therefore escape even an attentive observer.

The breadth, ordinarily equal and uniform for the two bands, is very different from one gemination to another, since a line is imperceptible if its thickness is below 3°. The radio of the breadth of the bands to the bright region which separates them is very variable. The

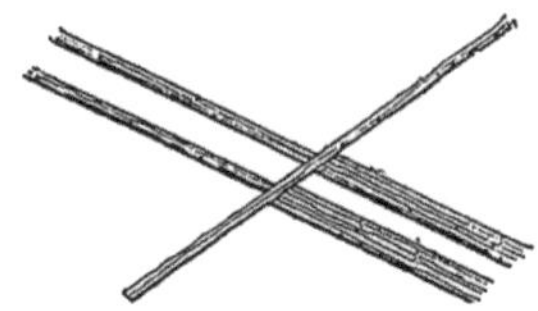

Fig. 81 Enlargement of the two bands of a canal, near an intersection

usual separation is broader than either of the bands, but often it is equal, or even a little narrower.

The colour is almost always the same for the two bands, with respect to both quality and intensity; but there are considerable variations between one gemination and another. The colour is generally black, or almost so, with geminations made up of very thin lines; broader bands are rarely black or brown (a remarkable case was the gemination of the Cyclops in 1882, which was so strong and so well-marked that no other feature on the disk could be compared with it); they often show a brick-red tint, some darker than others. Some bands have been so pale that they are hard to observe against the yellow background of the planet, in spite of their considerable width, which amounts to several degrees. On various occasions I have been able to show that an intersection of these very pale bands with another canal produces a noticeable reinforcement of colour at the place of intersection. I am led to believe that with all the double canals, the colour is always the same in quality, and that the differences relate only to intensity.

If a double canal is cut into two sections by another canal, and if one of the bands is broader and more intense to one side of the intersection, then the other band will also be so, as shown in Fig. 81. Such have been the Antæus-Eunostos in 1882, and the Euphrates in 1888. If one of the bands is very thin, or scarcely visible to one side of the intersection, the other will also be so-and in this case it may be that one of the two will be lost completely, becoming invisible. We have then the example of a canal which is double along one section of its course and single in another section. Cerberus, Hydraotes and Acheron were examples of this in 1882.

Sometimes the two lines are regular and their axes perfectly parallel; but both are surrounded by an area of penumbra, as with Cerberus in 1882, Hebrus in 1888. But in the great majority of cases, the two lines are traceable with absolute geometrical regularity; the uniformity of breadth, colour and separation is complete. My examination of them, made under excellent conditions and with magnifications ranging from 322x up to 650x, has not shown me the slightest irregularity or even suspicion of irregularity; each might have been traced with a ruler and compass. In 1882 this was the case with, among others, Cyclops, Euphrates, Phison, Jamuna and Hephaestus; in 1886, Hydraotes; in 1888 Euphrates, Phison, Astaboras, Protonilus and Callirrhoe. If there is any trace of anomaly in the original single canal, it disappears completely after gemination. Canals which are sensibly curved have given rise to perfectly straight geminations, as with the Jamuna in 1882 and the Boreosyrtis in 1888. In a word, there is a pronounced tendency toward more absolute uniformity and suppression of all irregular elements.

The aspect of a gemination can often change over a period of time. In 1882, the two bands of the Euphrates showed sensible convergence in the side toward the north; one of them was directed so as to follow a meridian on the planet. In 1888, the two bands were absolutely equidistant along their whole length between the Sinus Sabæus and Ismenius Lacus; their angle with the meridian was about 8° to 10° on average. They were thin and well defined in 1882; in 1888 the two borders of each band were shaded, the bands themselves being brighter and their separation sensibly less than in 1882. Moreover, with the Hephæstus, the two broad, reddish bands of 1882 were reduced in 1888 to two narrow, more sombre lines, with the separation between them reduced by half. A similar reduction in separation was seen with the Protonilus.

The geminations of the canals are accomplished in a relatively short time and by a rapid metamorphosis. As often as can be ascertained by reliable observations, the change always takes place in a few days or less. Sometimes the metamorphosis is complete within the twenty-four hour interval between two consecutive observations. So far as I can judge, the phenomenon takes place simultaneously along the whole breadth of the double canal.

In a few cases, it has been possible to follow several phases in the process of gemination. During January 1882, the Euphrates had been visible up to the 18th without showing anything remarkable. On the 19th, it seemed considerably broader and somewhat nebulous on its left side. On the 20th, dense fog prevented me from observing. On the 21st, the gemination was complete and very evident. Also in January 1882, the Ganges was single up to the 12th. On the 13th it was accompanied, to the right, by a light, luminous band, which flanked it at a distance of about 5°; all along its length between Lacus Lunæ and Juventæ Fons this was seen. The band was no longer visible on the 18th and 19th; the entire region was sprinkled with white patches. These patches no longer existed on the 20th; but the new band had reappeared, darker, narrower and better defined this time; it resembled the Ganges, though it was a little weaker. The Ganges was double and its appearance did not change further until the end of observations in 1882. The appearance of a white or whitish blanket over one point or another of a canal at the time of doubling has been seen several times; in 1882 for the Thoth, in 1888 for the Protonilus and the Nepenthes; this white blanket shows up very distinctly between the two lines of a gemination.

I have also frequently seen the two lines separate simultaneously from a greyish nebulosity of varying intensity, elongated in the direction of the canal; I am even inclined to think that this nebulosity is an essential phenomenon in the prowess of gemination. But we need not conclude that we are dealing with objects hidden beneath the fog and which become visible when the fog disperses. So far as I can judge, that which appears as cloudiness is not an obstacle to our seeing features which already exist, but rather material which produces features which did not previously exist. To explain what I mean, I may say that the procedure is not comparable with objects which separate from a thinning fog, but rather to a multitude of irregularly dispersed soldiers who, little by little, form ranks and columns. I should add that this must be regarded as an impression, not the result of observation in the proper sense of the term.

If there is a time for the appearance of a gemination, there must be also a time of disappearance, or effacement in some manner. Unfortunately, I have never been able to observe anything with respect to this phase of the phenomenon. I could only say that several of the 1882 geminations were no longer visible during the following oppositions; the canal again appeared single, or even disappeared entirely. In most cases, the remoteness of Mars or the unsatisfactory state of the Earth's atmosphere gives a more or less reasonable explanation of the disappearance of the geminations. I believe that the character of the phenomenon is periodical. Really, such a periodicity could be established with certainty only after several appearances and disappearances have been seen; however, the observations secured up to now do seem to indicate the possibility. In 1877, no trace of gemination could be seen during the weeks which preceded or followed the southern solstice. One isolated case was noted in 1879; on 26 December I saw the doubling of the Nilus between the Lacus Lunæ and the broad streak called Ceraunius. This was in the month before the vernal equinox, corresponding to the passage of the Sun from the southern to the Northern Hemisphere of the planet. This phenomenon surprised me somewhat, but I considered it an exception. During the opposition of 1881–1882, I awaited a repetition of the same thing; in fact, it came a month after the vernal equinox, on 12 January 1882. By this time several other geminations had already appeared and soon the planet was full of them; in two months, from 19 December to 22 February, I was able to find thirty geminations. During the opposition of 1884, I could distinctly see some of them; others appeared probable, but were not so distinct. This was from two to four months before the northern solstice. In 1886 (at the epoch of the northern solstice, a month before and a month afterwards) most of the geminations no longer existed; many canals had become single, others had disappeared; however, some were still clearly double, among them the Hydraotes, very distinctly. Some of these doublings were confirmed at the same time at the Nice Observatory by Perrotin and his colleagues. Finally, in May and June 1888 (two and three months after the northern solstice) there began a new cycle of geminations, during which many canals became double while others remained single; several of the canals which stayed single had been double in 1882. The combined observations give weight to the idea that the phenomenon is regulated by the cycle of the Martian seasons; that it is produced a little after the spring equinox and a little before the autumnal equinox; that after having lasted for several months, most of the geminations disappear, at the time of the northern solstice, and all of them disappear at the time of the southern solstice. The verification of these conjectures need not be long delayed and a chance to check them will occur in 1892. The opposition of that year will be under almost the same conditions as that of 1877 and we may expect a complete absence of geminations.

Plate II gives an idea of the general arrangement of the geminations observed in 1882 and 1888. There should be no need to remind the reader that this chart does not represent the state of the planet at any particular epoch, because the geminations are not all produced together. It is therefore a graphical index of these formations, which does not correspond to all those which can be confirmed at the present time.

I have already commented above that on the planet there are various knots or points of intersection, of convergence, where several canals meet in a more or less regular pattern. The appearance of these knots changes in a manner analogous to that of the canals. When all the canals which flow into a knot are invisible, the knot is invisible also, or is traceable only as a very slight, diffuse shading. The appearance of the canals as single or double lines produces, at the knot, a rèseau of lines whose structure is usually impossible to unravel, because of the great quantity of detail which is crowded into a comparatively small area. The confusion is increased in a great number of cases by an ill-defined area which surrounds the main knot and renders it visible as a patch of greater or lesser intensity, which is sometimes transformed into a true lake, black in colour and with well-defined outlines (Lacus Niliacus 1879-1886, Trivium Charontis 1882, etc.). At certain times, a double, elongated patch may issue from this shading, forming a sort of gemination composed of two short, broad bands, lying near the surface of the shading or the lake in question. In Plate II, the Trivium Charontis and the Lacus Lunæ are shown as transformed in this way. So far as I am able to judge up to the present time (the observations are extremely difficult), the direction of this gemination changes considerably from one epoch to another and coincides sometimes with one, sometimes with the other of the components of the double canal. As this is of the greatest importance in the history of the geminations, I will describe, in detail, some of the examples which I have observed.

The Ismenius Lacus, in its ordinary state, is formed by a sombre patch of oval form, elongated in the direction of the parallels of latitude. On 23 December 1881 I saw it divided into two bands, which formed a short gemination extending in the direction of the Protonilus, which also was double. Protonilus and Ismenius were much the broader, as seen from the sketches above (Fig. 82). On 27 May 1888, a similar phenomenon was seen in the same place; but this time the division into two bands followed the direction of the Euphrates, which was double (see Fig. 83). The dimensions of the Ismenius in the direction of the Euphrates were less, the bands being neither so long nor so broad; in short, the gemination took the form of two small, almost round patches, juxtaposed and aligned in the direction of the Protonilus. Later, with the Protonilus as clearly double as the Euphrates, I wanted to see the Ismenius divided into four; but this did not happen. On 4 June the lake resumed its former oval shape, with shaded, ill-defined outlines.

In 1879 the Trivium Charontis existed only as the meeting-point of the Læstrygon, Styx, Cerberus and Tartarus, the only canals then visible in this region. In 1881–1882 the intersections of the canals were more numerous in this area, all of them being contained in a shading which was confused and fairly extensive, though not sharply bounded. In 1884 the shading was divided into two very strong bands, elongated exactly in the direction of the Orcus. In 1888 (13-15 June) the division into two bands existed, but their orientation fol-

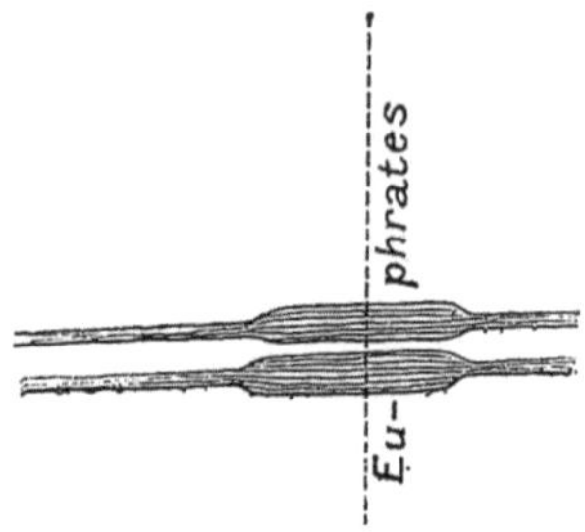

Fig. 82 Duplication of Lacus Ismenius, on 23 Dec. 18881, in the east-west direction

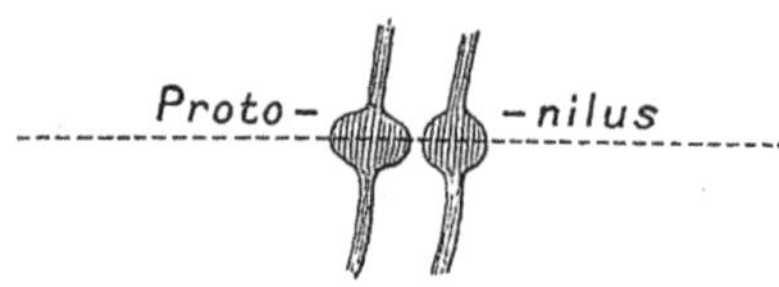

Fig. 83 Duplication of the same lake, on 27 May 1888, in the form of two circular lakes

lowed the direction of the Erebus. These systems of bands are represented in Plate II, superimposed on each other. But such a superimposition was not actually observed.

An identical phenomenon was observed on the Lunæ Lacus, which, in 1879 and 1882, was divided into two strong bands following the double Nile, while in 1884 the orientation was in the direction of the Uranius; these forms are shown superimposed in Plate II. The Nodus Gordii also presented phenomena which were analogous, though much more difficult to observe.

From all this, it seems that the cause of the geminations does not operate only along the canals of Mars, but also on the sombre surfaces of any form whatsoever, provided that they are not too extended; in the latter case, the direction of the same gemination can be very different from one epoch to another, while in the case of the canals it can oscillate only within certain limits. The cause seems also to affect the permanent seas; because the appearance of the island of Cimmeria in the middle of the Mare Cimmerium is essentially a transformation of this sea into a great gemination made up of two dark bands to either side of the island. A similar phenomenon seems to be found in the Mare Acidalium, though with less certainty or regularity.

The tendency to divide a sombre area by a yellow band seems to be shown also by the production of certain diaphragms of luminous isthmuses, of astonishing regularity, which are formed in various places in the northern hemisphere of the planet. Such is the Achillis Pons, which in 1882-84-86 separated the Niliacus Lacus from the Mare Acidalium and which partially disappeared in 1888; and also the break which separates the Nilosyrtis from the Boreosyrtis, a break which shows up when the Protonilus is double and which is really a continuation of the bright band which separates the lines making up the Protonilus. Another similar break in the course of the Boreosyrtis was seen in 1882, but has not been seen since. Finally, the doubling of the Sinus Sabæus and the peninsula of Atlantis, which separate the Mare Cimmerium from the Mare Sirenum, seems to be due to phenomena of the same nature.

G.—PHENOMENA OBSERVED ON THE CANALS.

Such are the various appearances which are presented by the canals of Mars and the analogous formations. Each of them has its own metamorphoses and its own particular history; and this history is undoubtedly linked with that of the neighbouring canals, even though the connection is not always obvious.

To give an idea of the manner in which the phenomena of the canals can develop, I have chosen one example out of fifty: it concerns the canal Hydraotes (Fig. 84) and its

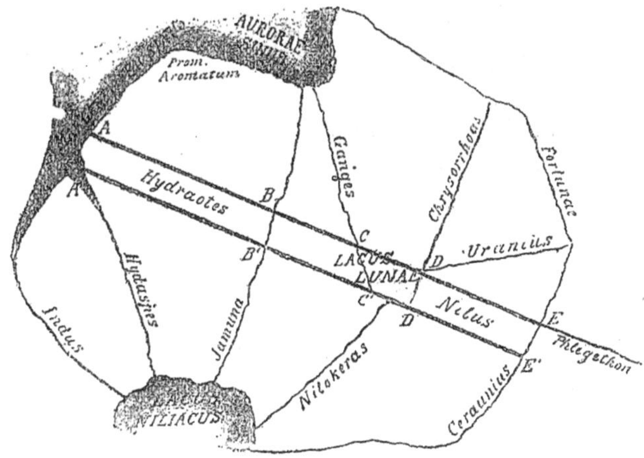

Fig. 84 Phenomena observed on Mars. The canal Hydraotes-Nilus

prolongation, the Nilus. To demonstrate the link between their phenomena and the Martian seasons, I have inserted, among the equinoxes, taking the vernal equinox as the moment when the Sun passes from the south to the north of the planet's equator. For the sake of clarity, I have lettered the different sections of the Hydraotes-Nilus. The canal issues from one side of the Ceraunius, which sometimes presents the aspect of a large nebulous band and sometimes shows an imperfect gemination which is bordered in a trumpet-form to the north; to the other side it stretches to the beautiful gulf of the Margaritifer Sinus. The three canals Jamuna, Ganges and Chrysorrhoas divide it into four sections, AB, BC, CD, DF. It seems that the canal is prolonged still further to the right, as far as the Ceraunius, by the Pyriphlegethon; but I will confine my examination to the part AE. In the environs of this section four other canals, the Ganges, Chrysorrhoas, Nilokeras and Uranius, converge in an eccentric and imperfect manner, producing one of the knots to which I have referred above—the Lacus Lunæ, a shaded patch of variable size and intensity.

Here is an extract from my observations:

OPPOSITION OF 1877.

September 27.—*Southern solstice.*

September 28, October 4.—All the canals invisible, with the exception of the Ganges.

November 4.—First appearance of the Chrysorrhoas, broad and nebulous. Its convergence with the Ganges forms an ill-defined but reasonably prominent patch; this is the first identification of the Lacus Lunæ.

February 21.—First view of the Nilokeras and of the Nilus in the form of sombre bands near the lower limb. Observations difficult; diameter of Mars reduced to 5″.7.

February 24,25.—First view of the Indus. The Ganges still visible throughout its extent; at its meeting with the Nilokeras and the Nilus it forms a triangular patch, the Lacus Lunæ.

March 6.—*Vernal Equinox.*

OPPOSITION OF 1879.

August 14.—*Southern solstice.*

October 13,14 and 18.—Ganges broad; Chrysorrhoas and Nilus well marked. Lacus Lunæ is an ill-formed, very sombre patch.

October 21.—First view of the Hydaspes.

November 27,28.—First view of the Jamuna. Nilokeras very strong. Nilus visible.

December 21.—Lacus Lunæ very large and very black.

December 23.—The Lacus Lunæ takes on the form of a trapezium CC′DD′, formed by four black bands; the bands CD, C′D′ are much broader than the others, but CD is itself broader than C′D′. The luminous island in the middle is well defined and the usual yellow in colour. The Nilus extends in the direction D′E′ in the form of a grey, poorly defined band. Ceraunius has the same appearance. At their point of intersection E there is a large, darker nebulous patch.

December 26.—The Nilus is double; the two tracks perfectly equal and well-defined, following the directions DE, D′E′ of the parallel sides of the trapezium formed by the Lacus Lunæ, but they are less broad and less dark than their two sides.

January 1.—Nilokeras black and well visible.

January 22.—*Vernal equinox.*

OPPOSITION OF 1881-1882.

December 9.—*Vernal equinox.* Ganges and Lunæ Lacus well marked. Nilus C′D′ barely visible; CD does not exist.

December 14.—Hydaspes, Jamuna, Ganges, Nilokeras poorly visible; broad and dark, apparently not double.

January 10.—Nilus and Lacus Lunæ marked by light shadings; fair view of the Uranius.

January 11,12.—Nilus certainly double; the tracks are slightly nebulous. Nilokeras imperfectly double.

January 13,20.—Doubling of the Ganges.

January 13.—First view of the Hydraotes, AB, in the form of a nebulous thread.

January 19.—The Lunæ Lacus has again taken on the form of a trapezium, with its luminous island in the centre. The Nilus forms two lines, DE, D′E′, which are easily recognizable and stand out against a whitish background. The arrangement seems identical with that of last year; Hydraotes AB well seen; Jamuna double.

February 18.—Nilus again double; the upper track seems to be prolonged by the Phlegethon.

February 22.—Hydraotes divided by the Jamuna into two sections AB, BC, of which BC is the broader and more prominent.

February 23,24.—Hydraotes double in the section BCB′C′, but always single in the section AB. The two tracks of BC are the prolongation of the two sides CD, C′D′, of the trapezium formed by the Lacus Lunæ, but are somewhat weaker. The single line AB is the prolongation of BC, but is weaker than BC. Jamuna and Ganges always double.

June 26.—*Northern solstice.*

OPPOSITION OF 1883-1884

October 26.——*Vernal equinox.*

December 31.—Jamuna, Ganges, Nilokeras well seen; nothing seen of the Hydraotes; Lacus Lunæ, a patch which is not very obvious; Nilus very confused, perhaps double.

January 2.—I believe I have seen the whole of the Hydraotes, AC, in confused form without being able to say whether it is single or double; the part BC is the most conspicuous. Ganges beautiful; Chrysorrhoas feeble; Jamuna broad, probably double.

January 29,30.—Uranius double, Nilus single. I can see only D′E′. The Lacus Lunæ forms a confused shading, in which I can see two blacker patches, elongated and following the direction of the Uranius and which form the prolongation of the two hands. Nilokeras sombre, but single. Ganges feeble.

February 3,4.—Hydraotes, as on January 2.

February 5.—Uranius has disappeared, but the Lacus Lunæ is again divided into two bands which follow its direction. Nilus double, but very weak. Hydraotes double at BCB′C′, single at AB.

March 9.—Lunæ Lacus always double in the direction of the Uranius; the latter is single, the upper track being the visible one; Nilus double. Only the track AC can be seen of the Hydraotes. Chrysorrhoas very broad, very probably double; Jamuna and Ganges rather weak (see Fig. 85).

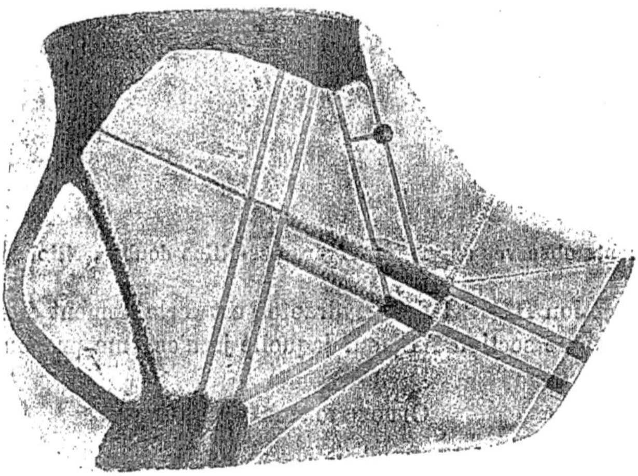

Fig. 85 Changes observed on Mars. Double canals and thickening of the Hydraotes-Nilus, the Nilokeras and the Jamuna. Indus and Hydaspes enlarged

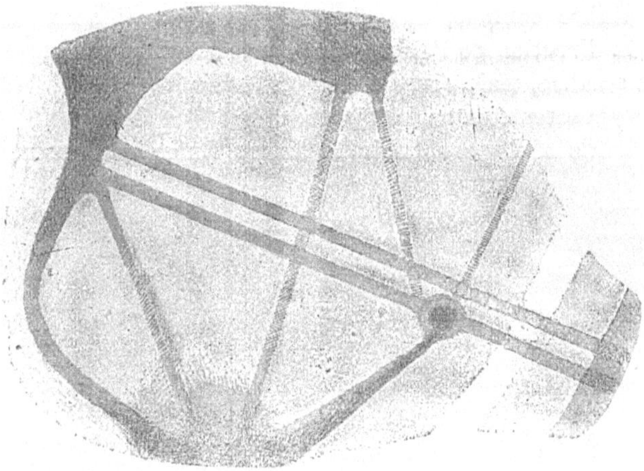

Fig. 86 Changes observed on Mars. Hydraotes-Nilus double. Nilokeras enlarged

April 5.—Despite the reduced diameter of the planet, the Nilus still appears double.

May 13.—*Northern solstice.*

OPPOSITION OF 1886.

March 27.—Hydraotes and Nilus clearly double; they form a single gigantic gemination which can be seen from the Margaritifer Sinus as far as the Ceraunius, as indicated in Fig. 86. The two bands are very broad (perhaps 4°) and of a reddish colour, darker than the surrounding yellow neighbourhood. Their separation (between the midpoint of the bands) is 9° to 10°. Nilokeras black and very strong, debouching at a black point at C′. The other canals Hydaspes, Jamuna, Ganges, Chrysorrhoas and Fortuna are visible, but none appears double.

March 31.—*Northern solstice.*

April 2.—The two lines of the Hydraotes are still visible, though very pale; they are a little darker in the section BCB′C′. Jamuna appears single.

May 7.—The Hydraotes-Nilus band still appears double; at least, it is very broad, though not striking; observing conditions bad.

May 8, 9.—The section BCB′C′ of the Hydraotes is certainly double; it is easier to see than the other section AB, which I cannot record with any certainty.

OPPOSITION OF 1888.

February 16.—*Northern solstice.*

May 23.—I believe I can recognise the part BC of the Hydraotes, which appears reasonably dark and perhaps double; but the atmosphere is bad.

May 24.—Hydraotes fully visible; the part BC is darker. I cannot say whether it is single or double. But the Nilus is certainly double.

June 27.—Nilus always double; the two streaks appear a little weaker near the middle of their course.

July 2.—The pair BC or the Hydraotes is well seen and dark; the other part AB is dubious. Atmosphere bad.

August 15.—*Autumnal equinox.*

These observed variations in the Nilus Hydraotes from 1877 to 1888 show a certain regularity and it is possible to give a history of periodicity and renewal at each revolution, from the southern solstice through to the autumnal equinox.

H.—The Polar Snows.

It has long been known that because of the effect of the seasons, the polar caps of Mars show periodical variations in size a little similar to those which we find in our own polar ice-caps during the analogous seasons. Since 1877 I have observed these patches with particular care and I can confirm that the periodical changes are real. There are however certain phenomena which are different from those which we see on Earth, and these must not be neglected. Here, in summary, are the results of my observations, beginning with the southern cap.

	Date	Days before (-) after (+) the southern solstice	Apparent diameter of the N. polar cap°
1877	23 August	–35	29
	22 September	–5	15
	4 November	+38	7
1879	21 October	+59	8
	28 November	+106	3
	27 December	+135	11

In the first days of January 1880, the polar cap began to disappear in the unlit hemisphere of the planet; during the following years it has always been invisible, because it has been situated on the hemisphere turned away from the Earth. At the top of the disk, white or whitish patches can often be seen; these are the known islands, shining with short-lived brilliance.

The diminution of the southern cap proceeds in a fairly regular fashion.

It would be very interesting to fix the time of the minimum size of the southern cap. In my earlier publications, I have always said that the minimum should come about four months after the southern solstice; but this conclusion does not now seem to me to have a firm basis. In fact, from more plausible considerations, I believe I can say that during the period from 17 to 22 January 1882 (that is to say, 200 days after the southern solstice), the cap cannot have been more than 10° in diameter at the most. It is therefore possible that the minimum can be delayed for more than four months after the southern solstice; this I can say with certainty—in 1879 it was delayed for at least four months.

Next, we must consider the northern cap. As with the southern cap, its decrease is fairly regular. It would be natural to assume a similar regularity in the rate of its increase, which we cannot observe. This cannot be verified; the northern cap, very small at the beginning of January 1882, had already attained its maximum diameter of about 45° by the end of the month and then immediately began a gradual diminution. This important fact merits a more detailed description.

During the oppositions of 1877 and 1879, the North Pole was constantly out of view, on the hidden hemisphere of Mars. No observations relating to it were possible in 1877. But throughout the duration of observations in 1879, I often saw, near the lower limits of the disk, one and sometimes two whitish patches which, from their lack of movement, could be regarded as extensions of the north polar cap, extending out to more than 30° from the north pole. Yet they were neither so bright or so well defined, or so constant in position and outline, as the true visible polar caps ordinarily are. The patches were five in number, spread out in a ring between 30° and 40° from the pole; their connection with a central polar patch and the existence of this central patch itself was not the main object of my observations, because of the unfavourable tilt of the planet's axis. This occurred between October 1879 and February 1880, four months before and one month after the Martian vernal equinox.

During the following opposition, that of 1881-1882, the North Pole was always almost exactly on the limit of the visible hemisphere; if the north polar cap had been even 10° or 15° in diameter it would undoubtedly have been seen in the calculated position. In fact, however, between 26 October 1881 and 25 January 1882, no permanent polar patch was observed in the neighbourhood of the pole. It follows that during this interval the north cap, even if it existed, could not have exceeded 10° to 15° in diameter. It is true that certain whitish features could be seen almost daily at the northern part of the limb. But this time, as in 1879, it was easy to see that these features were not produced by a fixed polar patch. Not only were they ordinarily pale, poorly-defined and variable in brilliancy and size, but, as in 1879, perceptible changes in their direction, due to the effect of the planet's rotation, indicated a greater distance from the pole, and even gave a method for determining their mean distance from it. The irregularity of their appearances, and the simultaneous visibility of two similar patches, fairly close together, clearly showed that we were dealing with not one object only, but with several whitish irregular features similar to those which had been seen in 1879. Close examination showed that the various branches had almost the same longitudes as the patches of 1879; but in 1881-1882 the polar distance was perhaps a little less.

Toward the beginning of January 1882, I began to recognize, in the whole of this system of white patches, the symptoms of a progressive concentration toward the pole. The branches at first shortened and then lengthened; then ended by joining up, forming a compact cap concentric with the pole. On 26 January, after several days of bad weather, the true polar cap appeared for the first time and it remained on view until the end of the opposition. It was made up of a single brilliant mass almost circular in shape and about 45° in diameter, with well-defined regular outlines. This phase of the rapid coagulation of the cap therefore took place a month and more after the northern solstice. An analogy, even if a somewhat imperfect one, can therefore be drawn with the terrestrial polar ices. The progressive diminution after this date is shown by the following table of apparent diameters. Each diameter is the mean of several days of observation.

The patch diminished rapidly in brilliance in July 1888, because of the very great obliquity of the solar illumination, soon followed by the oncoming of the polar night. The North Pole entered the shadow on 15 August, the day of the autumnal equinox.

		Days before (-) after (+)	App. diameter
	Date	the northern solstice	of the N. polar cap°

1882	30 January	–146	42
	10 February	–135	37
	12 March	–106	33
	10 April	–77	26
1883	26 December	–138	38
1884	20 January	–114	36
	15 February	–88	31
	23 March	–51	23
	2 May	–11	15
1886	18 January	–62	25
	26 February	–33	10
	14 March	–17	6
	28 March	–3	6
21 May	+51	5	
	1 June	+62	9
1888	7 May	+81	12
	2 June	+107	11
	July	–	Scarcely Visible

Exact measurements showed that in 1882 the northern cap was exactly centred on the pole; the same—or very nearly so—was found at the following oppositions. In 1888, Perrotin and Terby noted a division into two very irregular parts, which I could confirm by my own observations. (This division is shown on my two planispheres.) The patch has been almost constantly surrounded by a narrow, rather sombre zone, which is due partly to the effect of contrast. But this border has not always been uniform throughout its extent and has sometimes been black or nearly black; this indicates a real coloration of the surface in the immediate area of the boundary of the polar cap. The zone does not seem to me to accompany the cap in its shrinking; if this observation is later confirmed, it will be very important. Later oppositions have at least shown that the environs of the North Pole are not occupied by a large sea, but rather by a réseau of canals and small lakes. It is therefore possible that the conditions in the two hemispheres of Mars are very unequal from a meteorological point of view.

We may therefore ask whether the whitish features which we observe in various longitudes and latitudes, even under the equator—as described in detail in articles B and C above—are phenomena of the same nature as the polar patches. My opinion is that they are not. These features are never sparkling white; they often vary from ashy white to grey and yellowish. When these colorations are found in the continental regions they have usually been poorly defined. Their existence is irregular and transitory. Moreover, their brightness is always greater when the features are near the limb than when they are near the central meridian—exactly the opposite is true of the polar patches. This is particularly evident with the south polar patch, which, being perceptibly eccentric with respect to the pole, can change its distance from the limb during a single rotation of the planet; it always shows its maximum brilliancy when it reaches its minimum distance from the centre of the disk. At its maximum, the southern cap occupies a large area of the sea; on the other hand, the whitish features are produced on the continents and the islands, but never over the sea, as I have noted above.

I can say nothing about the white patches which have been described as being ramifications of the north polar cap and which in 1881-1882 preceded the formation of this cap; but it seems that their greatest visibility is at the time of transit across the central meridian and that when near the limb they become invisible. This leads me to believe that they are of the same nature as the polar cap; in this case they are made up of scattered material which, when joined up in a mass, forms the polar cap properly called. The patches of Nix Atlantica and Nix Olympica are of this type.

It is not difficult to put together several hypothesis capable of explaining, in a plausible manner, the phenomena of the white polar and non-polar patches, bringing in the evaporation from the assumed seas on Mars and also the atmosphere, whose existence cannot be doubted. I believe, however, that it is more useful to say that the white patches of all kinds are the easiest of all the diverse phenomena of Mars to observe really well; they can be studied with an instrument of modest power, used with great care. The various characteristics which I have described with respect to these patches show that we have here a very interesting field of research it is of great importance in the physical study of Mars, and can occupy the attention even of observers who cannot make out the much more difficult details of the canals and their geminations.

1888.—C. Flammarion: The Rivers of Mars and Changes Observed on the Surface

The first of these studies has established that if on Mars there are rains, snowstorms and aqueous condensation and if the water runs on the surface along brooks and rivers to flow into the sea, then these rivers should have broad mouths and these mouths could be the bays which we constantly observe, notably (1) the two points of the forked Meridian Bay, from which flow the Orontes, Hiddekel and Gehon, (2) Burton Bay, from which flows the Indus, (3) Christie Bay, where the Hydaspes flows.[103] We have as evidence the drawings made by Dawes in 1864. This question will be examined in detail in Part II: results Obtained from the General Study of the Planet and the research involved will be treated in a special chapter.

The second study[104] has had as its objective a description of all the examples of changes observed on the surface of Mars, and a discussion and careful analysis of them. An important chapter in the second part of this book will be devoted to the unquestioned changes which take place on the planet. It is therefore unnecessary to summaries this study here.

In 1879 a conjunction of Mars and Saturn was seen. The planets passed within 55″ of each other. Mars was strongly red, Saturn livid yellow. On 25 December 1889 Mars passed not far (55′) from Uranus; by contrast Uranus appeared blue.

[103] *Bulletin de la Societé Astronomique de France*, 1888, p. 111–115; *L'Astronomie*, December 1888, p. 457.

[104] *Bulletin de la Societé Astronomique de France*, 1888, p. 125–159.

The Societé, mentioned here for the first time, was founded on 28 January 1887. It has had as successive Presidents: 1887–1888, M. Flammarion; 1889–1890, M. Fay; 1891–1892, M. Bouquet de la Grye. It has its headquarters in Paris, Hotel des Societeé Savantes, Rue Serpente, and holds its meetings on the first Wednesday of each month. It has an observatory and a library. Already it has over five hundred members.

Opposition of 1890

Date of Opposition: 27 May
Presentation of the planet : The North Pole was inclined toward the Earth until 23 September; then the South Pole.

	Lat. of centre°	Diameter″	Phase defect°	Angle Sun/Earth°
27 February	+ 9.85	8.38	0.86	37
27 May	+9.48	19.02	0.00	1
7 July	+14.30	16.79	0.19	31
27 August	+7.32	11.50	1.70	45
23 September	0.00	9.66	1.49	46
31 October	−11.41	7.81	1.12	45

Calendar of Mars		
	S. hemisphere	N. hemisphere
2 January	Winter solstice	Summer solstice
3 July	Spring equinox	Autumnal equinox
26 November	Summer solstice	Winter solstice

1890.—William H. Pickering: Photography of Mars. Snowfall Photographed. Observations

Several attempts to photograph Mars have already been described. Here are the first satisfactory results. Mr. Pickering has been good enough to send me the photographs obtained at Mount Wilson (California). Seven of these photographs were taken on 9 April 1890, between $22^h\ 56^m$ and $23^h\ 41^m$ GMT; seven others on the following night between $23^h\ 20^m$ and $23^h\ 32^m$. Therefore, the same face of the planet has been photographed in each series. On all the exposures, the geographical configurations can be recognized distinctly, but on those of the second day the white polar cap which marks the South Pole is much larger than as shown on the photographs taken on the first day. It has long been known that the sizes of the polar caps vary according to the Martian seasons, shrinking in summer and growing during winter. But this is the first time that the precise date of a considerable extension of the snows has been recorded. The south limb of the planet was at latitude −85°. The snow extended from the terminator, which had a longitude of 70°, along the −30° parallel as far as longitude 110°; then from longitude 145° and latitude −45° up to the planet's limb. Presumably it extended equally on to the hemisphere which was turned away from the Earth and was therefore unobservable. The visible extent of the snows, wrote Pickering, is truly immense and covers 2500 square miles, or almost the area of the United States.

Fig. 87 Photographs of Mars in 1890, by M. W. H. Pickering

On the morning of 9 April, the polar snows were badly defined, as though they had been veiled by mist or by small separate bodies too tiny to be shown individually; but on 10 April the entire region was bright, equal in brilliancy to the snow of the North Pole.

The date of this event corresponded to the end of winter in the Southern Hemisphere of Mars, which is equivalent to the middle of our February.

The explanation of these observations is naturally linked with terrestrial analogies. We have here witnessed an immense fall of snow in the Southern Hemisphere of Mars.

These appearances are so evident on each of the fifteen photographs that it is sufficient merely to give the date upon which each was taken. I have reproduced two of them (Fig. 87) as well as possible by photogravure, but it is impossible to reproduce the delicate appearance of the originals. These photographs were taken with a 13-in. (330^{mm}) equatorial.

William H. Pickering has already succeeded in obtaining satisfactory photographs of Jupiter and Saturn. On Jupiter the details of the belts can be excellently distinguished; on Saturn, the dusky ring, the Cassini Division at the ansæ and the belts across the planet can be recognized.

This skilful astronomer is also concerned with direct observation of the surface of Mars, using a 12-in. (304^{mm}) refractor.

Pickering has recognized some of the configurations noted by Schiaparelli, but he has objected to the name of canals given to the straight streaks, because, he says, there is not the slightest reason to suppose that there is water there. However, Pickering does not express an opinion as to what the features may be.

Of all the canals, he says, the easiest to see is the Nasmyth Channel, which those readers who know my charts will recognize as prolonging the lower part of the Hourglass Sea, by an almost right-angled turn; the Milan astronomer has given it the names of Protonilus, Ismenius Lacus, Deuteronilus and Jordanis. The Boreosyrtis and Astapus are equally easy to see. With these three exceptions, the other canals which have been discovered are very difficult objects and the author blames this difficulty on the use of too powerful magnifications, plus the way in which these specialized observations are made. When the observer becomes accustomed to the examination of Mars, the canals Styx, Fretum Anian, Hyblæus, Cerberus, Eunostos, Hephæstus, Alcyonus, Cyclops and Læstrygon can be recognized without difficulty.

The Cambridge astronomer does not doubt the doubling, or the discovery of the very feeble canals and he expresses the highest admiration for Schiaparelli, who made the discoveries with a telescope of only 8 in. aperture. He believes that all experienced observers should be able to find the principal canals when using a 10–12-in. refractor and that, except under exceptional circumstances, the magnification used should not exceed 100× or 200×.

In short, Pickering's observations confirm those of Schiaparelli with regard to the existence of these enigmatical lines.

1890.—Asaph Hall: Observations of Mars From Washington

Asaph Hall, the eminent astronomer to whom we owe the discovery of the satellites of Mars, made a new series of observations of these satellites from 28 May to 25 June 1890. He confirmed the orbits.

On several nights he tried to recognize the double canals, but without success. The image of the planet was diffuse and unsteady. We must bear in mind that Mars was always low in the sky.

The great equatorial of the Naval Observatory at Washington has an aperture of 26 English inches, or $0^{m}.66$.

It is strange.

1890 Keeler: White Patches on the Terminator of Mars

An appearance analogous to that which we observe along the terminator of the Moon, when the summits of the lunar mountains and the craters show up outside the region which is fully illuminated, was seen on Mars with the aid of the great 36-in. ($0^{m}.91$) equatorial of the Lick Observatory during the evenings of 5 and 6 July. A sketch by J. F. Keeler, made on 5 July at 10^{h}, shows a very narrow, elliptical white patch measuring 1 1/2″–2″ in length, projecting to the north and forming a slight angle with the line of the terminator. The evening was very fine and the atmosphere excellent. At $10^{h}\ 39^{m}$ this small white patch was fully on the disk and remained visible against a darker background. On the next day, 6 July, the same appearance was observed, with the greatest care. A similar white patch was followed for more than an hour; then the two joined up. Of these two patches, the lower was situated at the end of a long, brilliant band on the planet's surface, stretching to the north of Deuteronilus. The most simple interpretation of these phenomena is naturally to regard this band as elevated above the general surface of the planet. At $10^{h}\ 25^{m}$, the appearance was the same as before and without doubt the cause of it was also the same.

The observers made several sketches. Schiaparelli's principal canals were seen in the form of broad, diffuse bands, weak apart from the Gehon, which was very striking. Double canals were seen (*Astronomical Society of the Pacific*, Vol. II, p.299; *L'Astronomie*, 1890, p.465).

The two satellites of Mars were seen by a woman visitor who had not known of their existence. The planet was in the centre of the field and was not masked by an occulting bar.

1890 C. Flammarion: Observations and Sketches

The planet remained very low in our latitudes and observations were very difficult. Moreover, fine nights were very rare during the summer of 1890; almost constantly the sky remained rainy or cloud-covered. The period near opposition, which should have been very favourable because of the planet's closeness, was largely lost. The eminent and industrious Mr. Huggins, who had promised to make a new spectroscopic study of Mars, wrote to me from London that the low altitude of the planet, combined with the bad atmospheric conditions, had made it impossible for him to carry out his intention.

Among the observations which I was able to make at the Juvisy Observatory, I have selected those of 27, 30 and 31 July, which I offer to my readers as being less bad than the others. Opposition was already past (it had occurred on 27 May) and the phase was very marked. The distance from Earth was 0.666, or 98 million km.

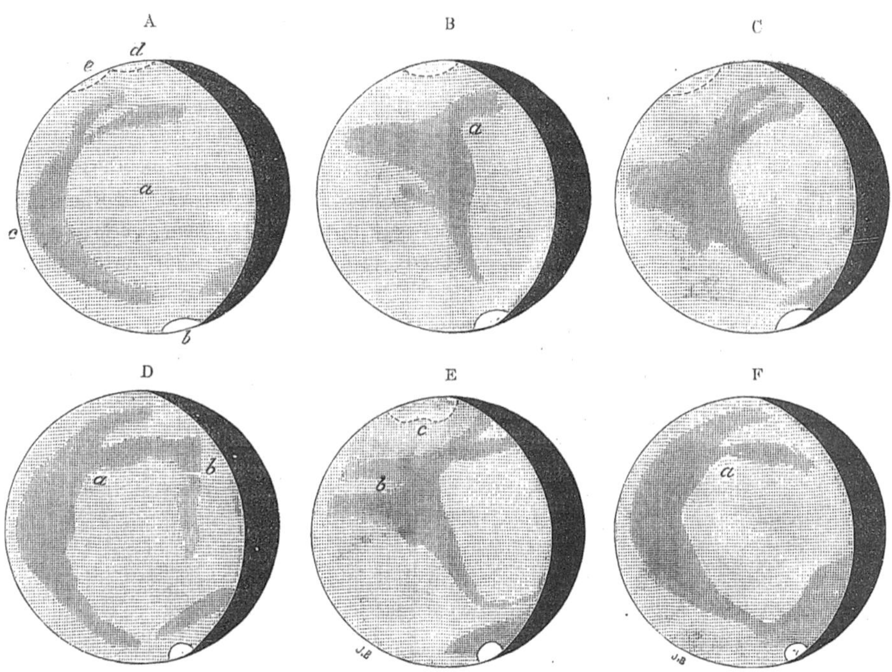

Fig. 88 Some aspects of the planet Mars in 1890 (Sketches by M. Flammarion)

The observations were made with the 240^{mm}. equatorial, magnification 140×, 220× and 300×, near meridian passage and generally before nightfall. Here is an extract relating to the drawings reproduced here (Fig. 88):

> 27 July, $7^h\ 0^m$, 140× (Fig. 88A). —Sky perfectly clear; view quite good. Full sunlight (sunset = $7^h\ 44^m$). The continent *a* is very yellow. The lower (north) pole is very white. The upper pole is whitish at *d* and *e*. The Hourglass Sea, which was well visible on the central meridian or the disk at $5^h\ 20^m$ and had shifted to the west as seen in a drawing made at 6^h, was approaching the western limb. It was darkest in the mid-region, toward the part marked *c*. It had been the same at $6^h\ 0^m$.

Diameter 14″.4.—Culmination $7^h\ 33^m$.

30 July, $6^h\ 45^m$, 140× (Fig. 88B). Sky perfectly clear. Day warm, with a hot sun. Atmosphere calm. Both poles can be seen. The lower is the better marked and whiter. The point of the Hourglass Sea is slanted to the right, or toward the east of the polar cap. It is darkest toward its eastern shore. At *a*, the cap is definite.

Same Evening, $7^h\ 20^m$, 220× (Fig. 88C).—Mars at the meridian. The lower pole is very white. The Hourglass Sea has passed the central meridian. It is darkest in its central region. Herschel II. Strait is detached from it at the bottom and is darkest at the point marked in the drawing. The point of the Hourglass Sea is slanted toward the east of the polar cap.

Same Evening, $8^h\ 45^m$. Night (sunset = $7^h\ 40^m$).—Observed without illuminating the field. Image good. Power 300× (Fig. 88D). The lower pole is very marked. Herschel II. Strait is clearly detached. At *a*, a cape; at *b*, a gulf—*b* is the Meridian Bay; I believe that below it I can see a grey streak. The Hourglass Sea is very dark in the position indicated. The continent is a beautiful yellow, like a wheat-field. The tone of the disk to the left (west) is very bright and almost white.

Diameter: 14″.1.—Culmination: $7^h\ 25^m$.

31 July, $7^h\ 20^m$, 300× (Fig. 88E).—Magnificent day, hot sun, but with a calm atmosphere and a very clear sky. The Hourglass Sea lies on the central meridian of the Earth-turned hemisphere. All the eastern part of it is sombre, almost black. Its lower point is directed not toward the polar cap, but sensibly toward the right. Its extension (the Nasmyth Strait) can be well seen, as can the north polar sea above the pole. The polar cap is very white, but does not project beyond the disk by irradiation. Above, the region is whitish and vague. Above the Flammarion Sea, the region *b* is very pale (Dreyer Island). Lockyer Land *e* is very pale, below the South Pole.

Same Evening, $8^h\ 45^m$, 300× (Fig. 88F).—The image is less steady than it was before sunset. (The best time to draw Mars is certainly during the half-hour before sunset.) The Herschel Strait and the Meridian Bay can be well seen. At *a,* Banks Cape is very evident from time to time. The Hourglass Sea is sombre. Above the polar cap, the sea is grey. Image good. A power of 400× shows nothing extra.

Diameter: 14″.0. Culmination: $7^h\ 22^m$.

These drawings do not give much new information, except perhaps that the weather on Mars was good, with few clouds; and without going too far, we may say that the wind over the Hourglass Sea was not strong during the period when my observations were being made. Indeed, the sea constantly appeared very dark. Undoubtedly the agitation of a sea-surface by wind has the effect of making the surface look less uniform, as the Sun's rays are less absorbed and are broken up by millions of little wavelets reflecting the incident light and in consequence giving the surface, as seen from above, a tone which is less bright than when the sea is calm and uniform. We can therefore judge whether a Martian sea is calm or agitated. But there are also other causes of variations in tone.

The lower or North Pole was shown as snow-covered. However, the winter solstice did not occur until 26 November. It is true that the Northern Hemisphere of Mars had its autumnal equinox on 3 July; summer was at an end and the winter season began.

I have estimated the diameter of the north polar cap as 1/14 the diameter of the disk, which corresponds to about 480 km and this means that the true value is more like 240 km, assuming that irradiation accounts for half the apparent diameter of the cap. (I have often found these snows much more brilliant and more extended—notably during June 1873, when, with an 108^{mm} refractor, they seemed to leave the disk by irradiation.)

During August, my observations showed that the snows at the lower pole grew slowly from week to week. The snow at the upper pole remained almost imperceptible. The latitude of the centre of the disk was + 6°, so that it was natural to see the northern pole better than the southern; but as the latitude of the centre of the disk decreased and the planet appeared more and more face-on, the snow at the north pole should have been seen less and less. Instead, the snows became more conspicuous. Therefore, they must have shown a genuine increase.

At the end of September, and in October, the southern snow could be distinguished; it covered from 25° to 30°. Subsequently, it diminished noticeably. From an observation which I made at the Nice Observatory on 13 December, it was reduced to about 10° and it later decreased even further.

1890.—Terby: New Studies with Observations by Schiaparelli and Stanley Williams

In *L'Astronomie* for November 1890, Terby writes[105]:

> Has the appearance of single or twin canals on the surface of Mars, first noted by Schiaparelli, been truly confirmed by other observers? In spite of the positive results obtained during the opposition of 1888, certain doubts are still expressed by some astronomers. In particular, they cite the partly negative results obtained at the Lick Observatory. They forget that in fact Holden and Keeler had observed several canals, even though they began their observations as late as three months after opposition, at a time when the planet was already a long way away and most areographers had given up observing.
>
> The first news of the results of the 1890 opposition was eagerly awaited. A well-known English astronomer, Stanley Williams, has been able to render full justice to Schiaparelli.
>
> This year, Mars was seen under deplorable conditions; its southern declination (−23°) meant that it never rose higher than about 16° above the horizon, even at the meridian passages;

[105] Académie de Belgique.—*L'Astronomie*, November 1890.

thus until 23 June the continuous unsteadiness of the image meant that I could clearly make out only the main outlines of the features, with no fine detail, instead of the usual views; this in spite of repeated attempts carried on for several hours at least. I am bound to say, with great regret, that my results up to that date were absolutely nil.

On 23 June, for the first time, between 9^h and 10^h, I could use 450x to advantage on my 8-in. I then saw, with great clarity and also for the first time, the bay which Schiaparelli shows to the side of the Syrtis Major and from which come the canals Astusapes and Astaboras; at moments and with complete certainty, I could see the Syrtis divided at that point, one side being continued by the Nilosyrtis, which was easily visible and the other by the Astusapes canal, which came out from the bay and followed around Meroe Island. The Protonilus, with the Ismenius Lacus and the Callirrhoe, was even more prominent.

On 24 June, from 10^h to 10^h 30^h, the image was good enough to bear 250x, 280x and 450x; I again saw the same details as before; moreover, though only at moments, I saw the Astaboras canal running in a straight line from the bay to the Ismenius Lacus. I observed this canal for the first time. The Nepenthes was excellently seen on this occasion, and I believe I could even see its origin in the Lacus Mœris.

On 25 June, from 9^h to 10^h, the quality of the image was mediocre; continuous agitation made the Astusapes and Astaboras canals almost invisible, particularly the latter, though without hiding the bay where these two lines originate; on the other hand, I could see the Boreosyrtis well and I saw—perfectly—the Nilosyrtis and the Nepenthes, as well as the Protonilus and Ismenius Lacus. Callirrhoe was more difficult. The agreement with the chart is remarkable. Only a power of 250x gave the required definition, but it was inconveniently low; 280x, 420x, 450x and 560x gave less sharp images, but were nevertheless more useful. The white region Hellas, well defined, was brilliant at the upper limb, and to the southern limb, under the Callirrhoe, showed up with vivid whiteness.

Such are the only useful observations which I was able to make.

One circumstance to be noted; the view of the observer seems to have an enormous influence upon these delicate researches. It is certain that an essential condition for the visibility of the canals is the absolute clarity of the outlines of the patches; do not forget that under such circumstances the image has been compared with that of a steel engraving. The view of all observers does not indicate such perfect results, and even the first drawings made at Milan are open to objection because of their extraordinary clarity.

Recently, Stanley Williams has published his 1887 observations of Jupiter. Instead of showing the usual vague, cloudy appearance so often depicted in drawings, the features shown by the English astronomer seem a little strange, because of the unusual precision of the outlines.[106] By a happy coincidence, having observed Jupiter independently at the same time, I have been able to identify nearly all these features.

Now, it happens that Stanley Williams has achieved magnificent success in studying Mars this year; he observed from South England with a 6½-in. (165^{mm}) Calver reflector, powers 320x and 430x. On 31 May he was able to identify thirty three canals: Cyclops, Eunostos, Hyblæus, Hades, Styx, Cerberus, Gigas, Chrysorrhoas, Ganges, Nilokeras, Jamuna, Nilus, Indus, Protonilus, Hiddekel, Deuteronilus, Gehon, Lethe, Æthiops, Titan, Erebus, Sirenius, Orcus, Pyriphlegethon, Euphrates, Nepenthes, Phison, Asclepius and Triton.

He had noted the gemination of five canals (Nilokeras, Cerberus, Erebus or Hades, Titan, Euphrates) and suspected that of the Phison.

Finally, the English astronomer also referred to Libya; he observed it on 18, 20, 21 and 24 May and also on 24 June; the area was very dim on 21 May, but on the 24th it was brighter; however, it seemed shaded in comparison with Isidis Regio, which was whiter and more brilliant.

I myself noted this greyish tint of Libya on 23, 24 and 25 June, when I was making the observations of the Nepenthes described above.

---

[106] See *L'Astronomie*, October 1889, p. 361–371 (reference to *fig*. 239–243).

Fig. 89 29 April, from 13^h 38^m to 14^h 15^m

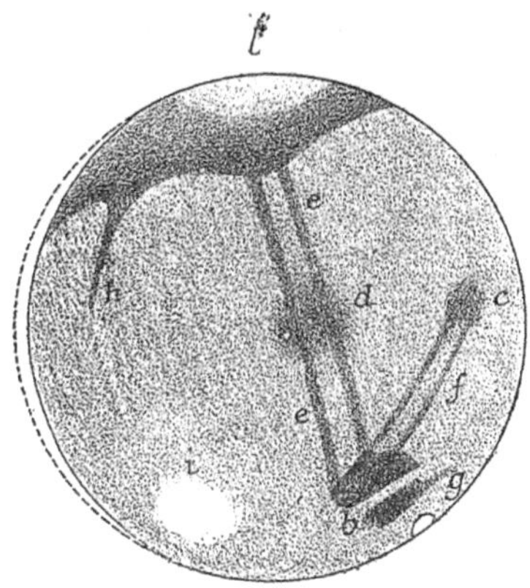

Fig. 90 25 April, from 14^h 30^m to 15^h 15^m

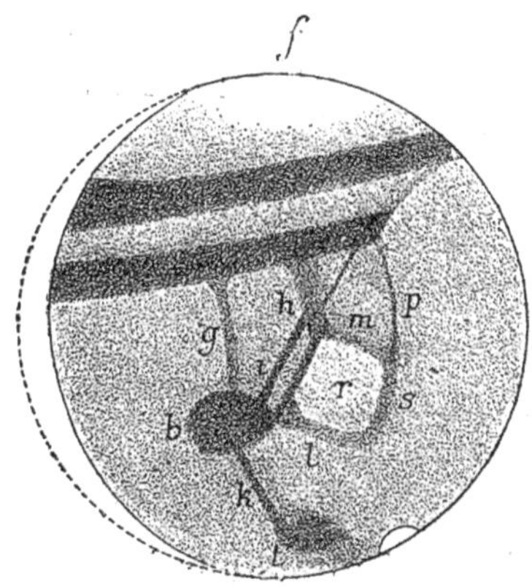

Williams noted that the more delicate canals became visible only at their transits across the centre of the disk; therefore, they were not often seen many times; they were usually observed only with great difficulty.

To this note I have added the five beautiful unpublished drawings made by the English observer (Figs. 89, 90, 91, 92 and 93); they will show, better than any description could do, the extraordinary results which this astronomer had the good fortune to obtain during this unfavourable period for observation.

In addition, Stanley Williams has succeeded in making out ten other lines: Boreas, Agathodæmon (partially), Fortunæ, Nectar, Eumenides, Oxus, Hydaspes, Thoth, Callirrhoe and Astusapes; altogether he verified forty three canals. On 9 June he distinctly saw the gemination of the Gigas.[107] On 31 May, at 11^h 05^m, during a few moments of excellent seeing, he saw the Cerberus and the Erebus crossing the disk and forming prolongations of

[107] 1.) *b*, Propontis; *c*, Trivium Charontis; *d*, very feeble grey patch; *e*, Titan; *f*, Erebus or Hades; *g*, Boreas; *h*, Sirenius; *i*, whitish oval; *l*, very brilliant region.

2.) *b*, Trivium Charontis; *g*, Læstrgyon; *h*, Cyclops; *i*, Cerberus; *k*, Hades; *l*, Styx; *m*, Eunostos; *p*, Æthiops; *r*, Elysium; *s*, Hyblæus; *t*, Propontis.

3.) *h*, Nilosytris; *i*, Protonilus; *k*, Hiddekel; *s*, Libya; *t*, Euphrates; *e*, Ismenius Lacus; *m*, small grayish patch.

4.) *g*, Libya; *h*, Nilosytris; *i*, Thoth; *m*, Astusapes; *p*, Asclepius; *r*, Protonilus; *s*, Phison; *f*, Hellas; *u*, small blackish patch.

5.) *h*, Triton; *l*, Libya; *h*, Isidis Regio; *c*, Nilosyrtis; *d*, Protonilus; *e*, Boreosytris; *f*, Asclepius; *g*, Neptenthes.

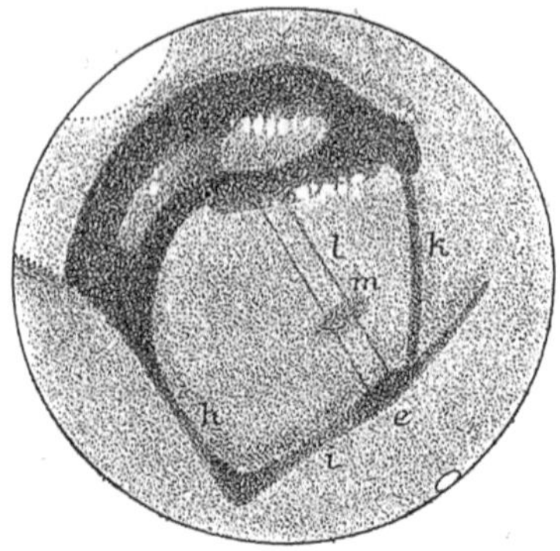

Fig. 91 18 May, from 12^h 15^m to 13^h 0^m

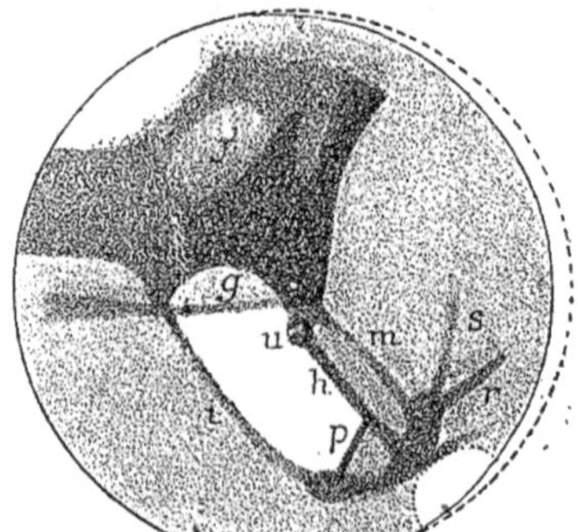

Fig. 92 24 May, from 12^h 0^m to 12^h 50^m

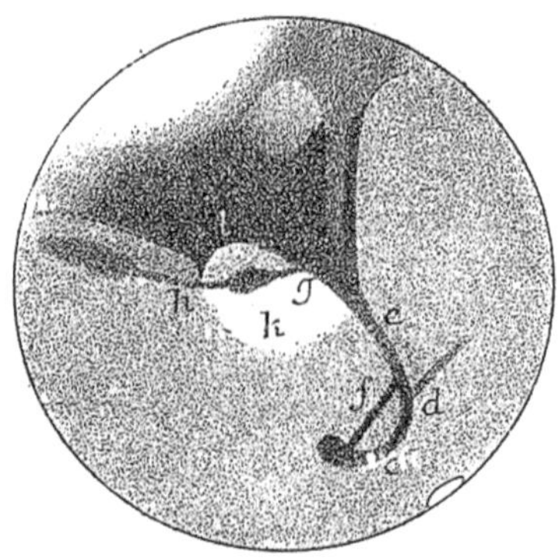

Fig. 93 27 June, at 10^h 0^m

each other, thus making up a single canal which was distinctly double; the two streaks making up the Cerberus being such broader and blacker than those making up the Erebus and giving instead the broadening of the canal breaking up at the point where the Cerberus began.

The north polar cap remained very dark up to the beginning of June; toward the middle of that month it became completely invisible, or else represented by a feeble grace. Thus, on 24 and 25 June (9^h 30^m) it was scarcely visible; on the 27th, on the other hand, it appeared

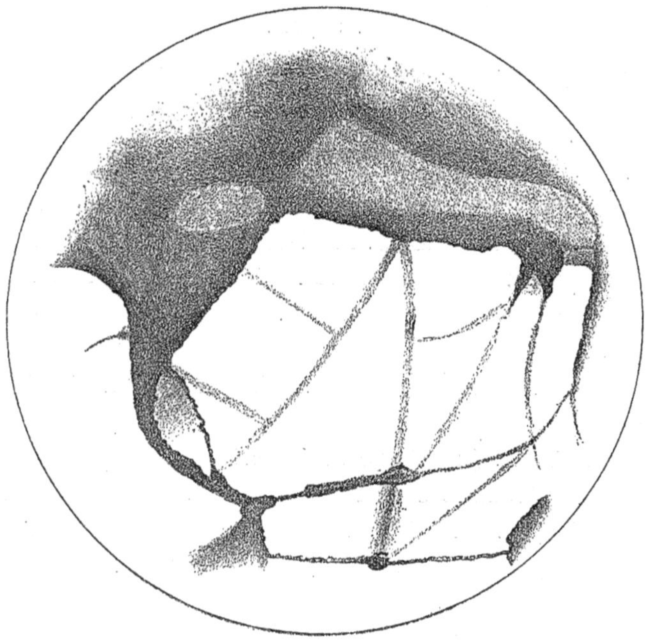

Fig. 94 Mars in 1890.—Canals transformed. Drawing by M. Schiaparelli (16 May)

as shown in Fig. 93. The same drawing shows a very black point in the Nilosyrtis; unfortunately the image was a little unsteady, so that Williams could not study the detail with the necessary precision. I ask myself whether this patch could be due to the presence of the Lacus Mœris.

Schiaparelli, like Williams, let me know about the unpublished new facts and I will end this section by giving extracts from the letters he has sent me. They are accompanied by three superb drawings (Figs. 94, 95 and 96).

Schiaparelli made useful observations only after 16 May. He wrote to me on 12 June:

All that I have seen up to the present time is summarized in the drawings which I have sent you. With regard to the third (16 May) (Fig. 94) I should comment that the canals to the lower part, Protonilus and Deuteronilus, Callirrhoe, Boreosyrtis, Astusapes and Pyramus and the lakes of Ismenius and Arethusa, with the portion of the Euphrates which joins them, were very easily visible, particularly the Callirrhoe and the Protonilus. (The Callirrhoe has been seen from Florence with a 4-in. (102^{mm})) The narrowing of the Protonilus was seen very conclusively. The canals near the right limb (Hiddekel, Gehon, Oxus...) were very delicate, and I could not judge their form or their colour. On the contrary, Euphrates, Phison, Typhon and Orontes had disappeared as canals and in their places could be seen only bands of a slightly darker red than the background—bands which were so ill-defined that I could confirm only their existence and their colour. On the same day the land of Deucalion was very beautiful and—remarkably—much broader at the left end than at the right; this was something I saw for the first time. On the other side everything was confused... Hellas, Ausonia, Libya, etc. But Iapygia was very evident.

On 4 and 6 June I was able to examine, with reasonable clarity, the whole region contained between Iris and Titan (long. 110°, 170°); the Mare Sirenum and the Eurotas (latitude 30°S and 50°N); it was again almost devoid of remarkable objects, as in 1877 and

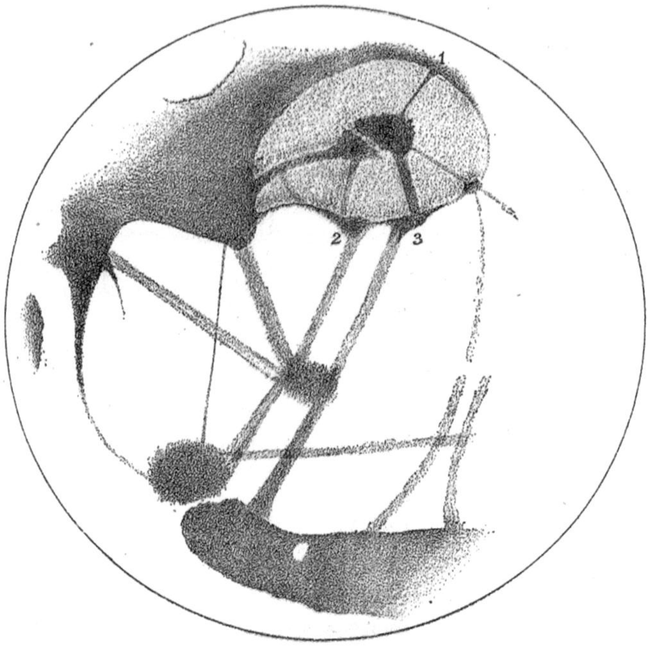

Fig. 95 Mars in 1890.—Lake split in two. Drawing by M. Schiaparelli (9 June)

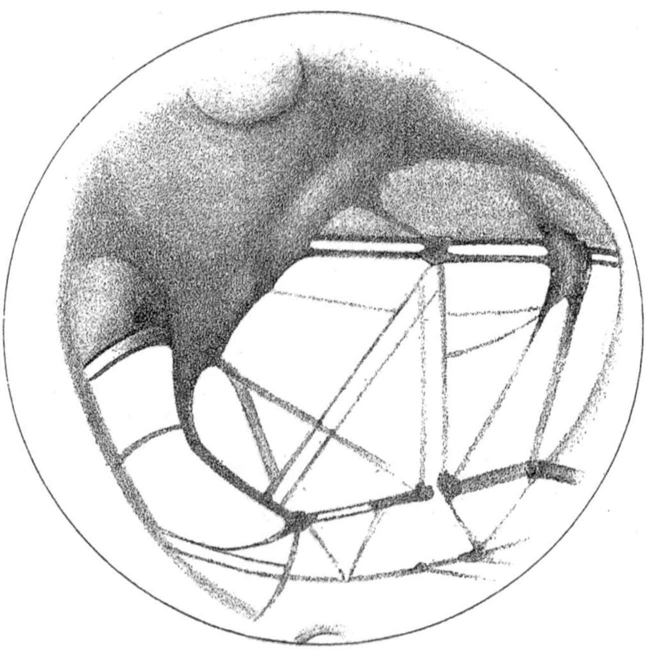

Fig. 96 Mars in 1890.—Strait split in two. Drawing by M. Schiaparelli (20 June)

1879; the canals existed only as doubtful streaks near the edges of the region; the remainder was a medley of red and yellow of different intensities, eventually with a little whiteness here and there without precise boundaries; this was the most difficult and least interesting part of the planet. The Araxes and the Phison existed, though they were very difficult to see; the Iris could scarcely be made out; the double Ceraunius was visible because of its size; but its colour was a very ill marked reddish. The two Nilus were not certainly seen. At the bottom of the disk I saw the Eurotas, which was imperfectly double and the Hebrus, which although double in 1888, is now single.

The evening of 9 June yielded new results, represented satisfactorily enough by another drawing (Fig. 95). You will see the great gemination of the Chrysorrhoas and the Nilokeras, the latter the darker and the more evident, though the Chrysorrhoas also was easily visible; the two lines were not well defined, but were rather shaded, either on the interior or on the outside. The Mare Acidalium showed nothing new, but I noted the total absence of the Lacus Hyperboreus; the regions of Baltia and Nerigos were ill defined and nebulous. By the Hydaspes, the Jamuna appeared as a narrow thread; Ganges and Hydraotes broader; I could not see them double, but their appearance was resolvable. Above, Argyre, very brilliant. But it was Thaumasia and the Solis Lacus which offered the greatest interest. The Lake of the Sun—this beautiful feature, so black and so regular—did not follow the law of gemination which rules the whole planet; it is cut in two by a yellow band which divided it into two unequal parts. The Tithonius Lacus is also divided, into two very dark, shaded nuclei, from which issue the two lines which make up the double Chrysorrhoas.

The old characteristics of the Lake of the Sun have disappeared; I believe that I have seen only a feeble trace of the Eosphoros; but four new outlets have suddenly opened up, of which the furthest left passes onto Aurea Chersonesus. This peninsula, so beautiful at other times, has almost vanished, or has at least been transformed. The region of Thaumasia is sombre yellow, which contrasts sharply with the brilliant surface nearby—particularly the upper limb at Ophir, which is white… see on this drawing the canal marked 1; on 4 June it was much stronger than 6; on the same day, 4 June, there was no visible trace of the canals marked 2 and 3. Forty-eight hours later, they were very much in evidence.

In another letter, dated 21 June, the Director of the Milan Observatory wrote:

Mr. Stanley Williams is quite correct in seeing the Euphrates double; it is effectively so (see Fig. 96) and better than in 1888; the two bands are perfect and the colour is of the characteristic red that I have already noted several times in similar formations, only it is not so intense. With it, and of the same type, are the double Phison, Crontes, Protonilus and Boreosyrtis, perhaps also the Astusapes, Astaboras, Oxus and Deuteronilus. But there are four geminations made up of strong lines; and I am convinced that there are others elsewhere. One is the Nepenthes, which appears as it did in 1888; but the Lacus Mœris is much brighter and more evident than it was then. Two other geminations have made the Sinus Sabæus almost unrecognizable, from the Hammonis Cornu to Dawes' double bay. Finally, the fourth gemination is in the isthmus of the Deucalionis Regio (which this year is brighter and better defined than formerly). The lines of these four geminations are perhaps of the same colour as the others, but this colour is so strong that I would call it almost black. It is like Chinese ink.

This is Schiaparelli's account. The two most surprising facts in his communication seem to be the doubling of the Lake of the Sun and that of Herschel II Strait, which showed itself to be made up of two straight bands, broad and parallel, but very close together and very difficult to separate. (I have already seen an analogous appearance.) Ismenius Lacus and Tithonius Lacus are also doubled in the same way. It therefore seems that there is no formation on the surface of the planet which is not subject to this curious phenomenon of gemination!

1890—J. Guillaume, Giovannozzi, Wislicenus: Observations and Drawings

M. Guillaume, observer at Péronnas,[108] announced that he had managed to distinguish a certain number of canals and even to see some of them as double, with the aid of an excellent With reflector of 216^{mm} aperture and 1^{m}.95 focal length, using a magnification of 195×. He saw them as brick red. He has often observed variations in tint on some of the seas and attributes them to the presence of clouds in the Martian atmosphere. As evidence in favour of this explanation, Guillaume cited his drawing of 19 May, at 10^{h} 05^{m}, compared with that of 23 May at 11^{h}. It is evident, he remarked, that on the 19th the atmosphere was less clear over this region than on the 23rd.

Some of Guillaume's drawings are reproduced in Fig. 97.

Guillaume did not mention the very curious division of the Lake of the Sun observed at Milan, nor that of the Ismenius Lacus or Herschel Strait.

M. Giovannozzi, Director of the Ximénien Observatory at Florence, observed Mars with an 108^{mm} Fraunhofer refractor, armed with magnifications of 105× and even 240×. The little chart which he drew up from the total of his observations is reproduced here (Fig. 98). As with preceding observations, note the brightness seen above the Hourglass Sea and to the right and the lower prolongation of this sea. As for discrepancies attributable to the indistinctness of the details, consider the two drawings of 16 and 24 June compared with those of Guillaume. The first was made at virtually the same time as the sixth drawing on the preceding pages at 11^{h} Rome time (corresponding to 10^{h} 20^{m} Paris time), while the Péronnas drawing was made at 10^{h} 15^{m}. The point *p* indicates the Meridian Bay, and *q* Burton Bay; these two bays appear very short here, but very elongated on Guillaume's drawing. This difference cannot be attributed to the actual views in the instruments, since Giovannozzi saw the Solis Lacus while Guillaume did not. Also note at point *n* a definite difference between the two drawings (Figs. 99 and 100).

The comparison between the two drawings made on 23 June is no less instructive.

Let us conclude, in short, that when dealing with details which are slight, scarcely perceptible, and at the limit of visibility, we should attribute the inevitable discrepancies to the difficulty of observations and our obligation, when drawing a planet, to mark without hesitation everything which has been glimpsed, however vague and indefinite. One day, photography will save us from these uncertainties. But when?

Despite the strong southerly declination of the planet, Wislicenus, astronomer at Strasbourg, succeeded in obtaining some interesting results.[109] The weather was not favourable and the telescope was often being used for other research; Wislicenus was therefore able to make only 20 observations, from 12 April to 1 August. At maximum altitude, the planet was less than 20° above the horizon. The first attempt

[108] Bulletin de l'Académie de Belgique, 1890.—*L'Astronomie*, May 1891.

[109] *Astronomische Nachrichten*, no. 3034.—*L'Astronomie*, July 1891.

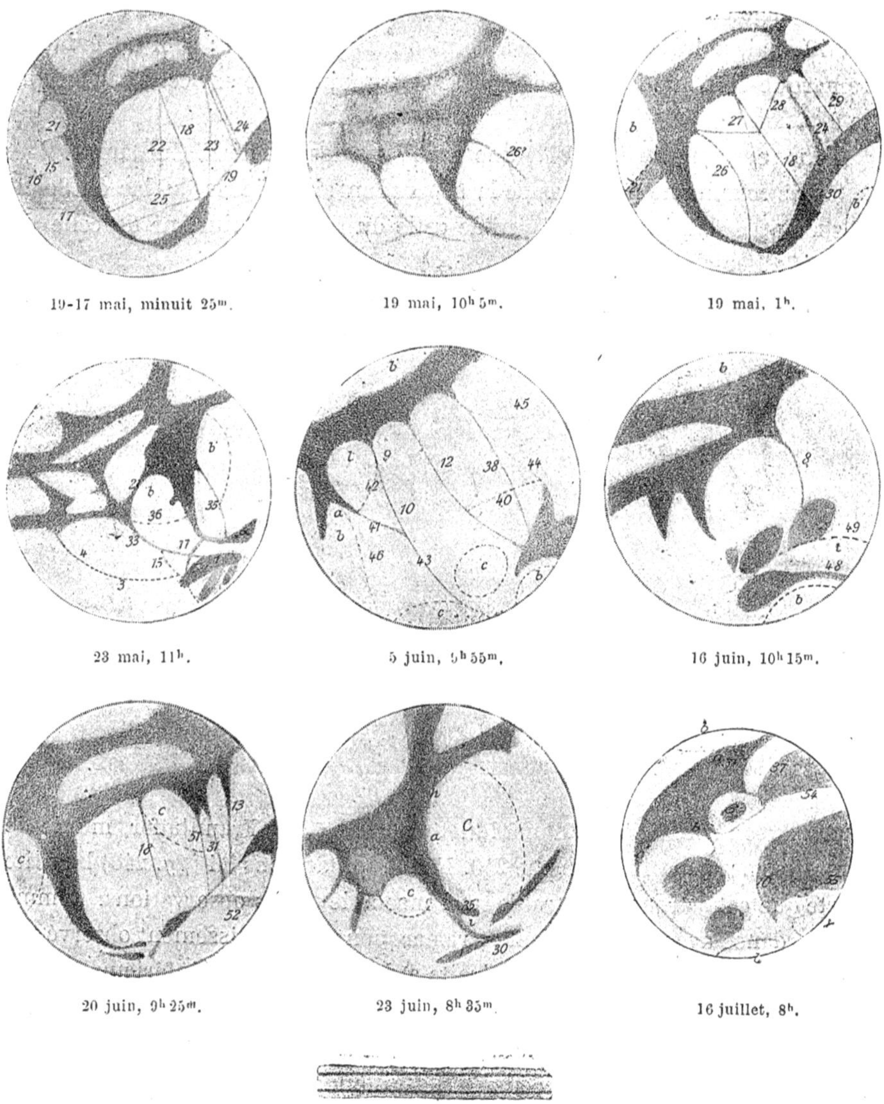

Fig. 97 Drawings of Mars in 1890, by M. J. Guillaume, at Péronnas

was made with an 18-in. refractor; but because of the low altitude and the unsteadiness of the images, there was no chance of using high magnifications, and Wislicenus preferred instead to use a refractor with an aperture of only 6 in. (152^{mm}) with a power of 182×. It was possible to make 21 drawings, to measure the position angle of the axis of rotation on 16 different days, and on 6 days, to make measurements of the patches.

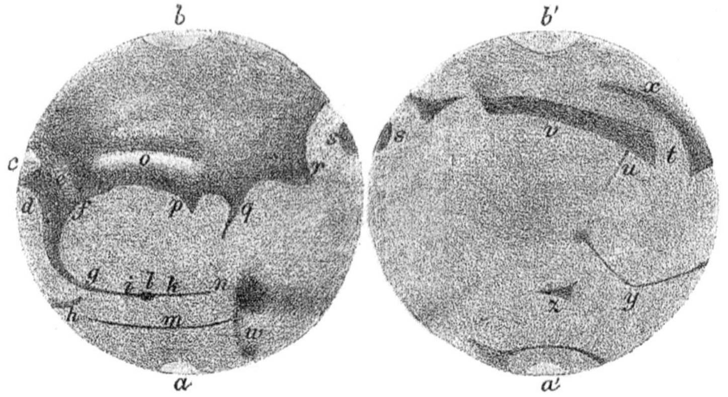

Fig. 98 General aspects of Mars, from the observations of M. Giovannozzi, in 1890

Fig. 99 Drawing of Mars by M. Giovannozzi, in Florence: 16 June at $11^h\ 0^m$

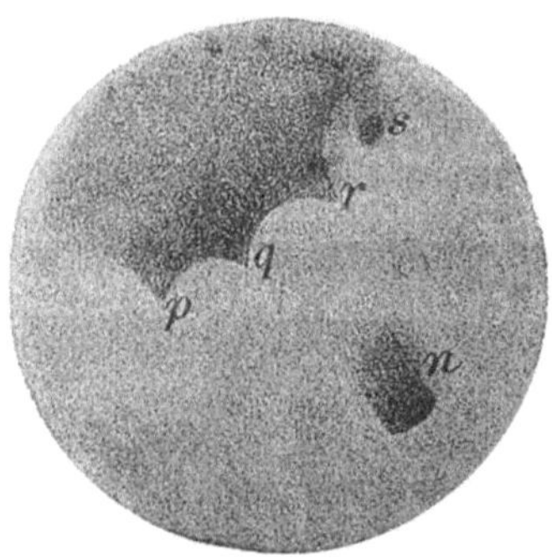

Fig. 100 Drawing of Mars by M. Giovannozzi in Florence: 23 June, at $10^h\ 45^m$

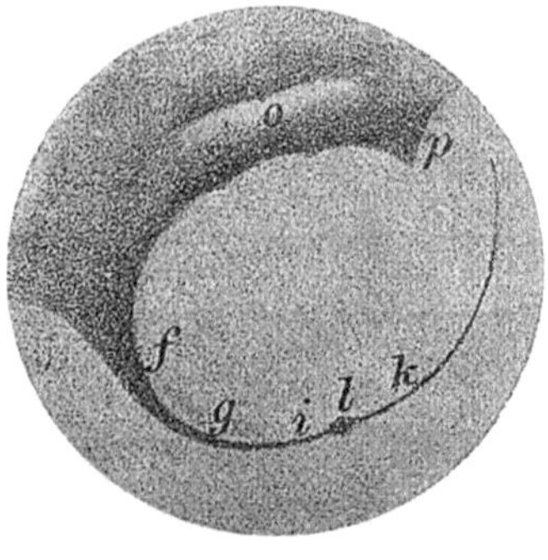

The planet's equator was then so slightly inclined to our line of sight that the snowy patch of the North Pole was scarcely visible. But the patch surrounding the South Pole was visible five times, and Wislicenus tried to measure the position of its centre. On these 5 days, the northern patch had been simultaneously visible on three occasions, so that the locations of both poles could be seen. The snowy northern cap was small and quite well defined, while the southern polar aspects more closely resembled extended snowfields, making it difficult to determine the central point.

In a table, Wislicenus gave the positions of the central points of the northern and southern polar caps and also the position angle of the axis of the planet, from 12 April to 13 July. By reducing the data, he found:

Position of the axis of Mars, angle	28°.09
Polar distance of the northern snow	7.19
Longitude of the north polar cap	199.85

On his first chart, Wislicenus was only able to measure the position of patch No 1 (the Hourglass Sea), but later he was able to add the 13 other points indicated on his chart (Nos. 8–20), and to measure them micrometrically. They are:

Point	Longitude°	Latitude°
1	295.82	+15.07
8	300.63	–16.85
9	321.65	–29.55
10	323.91	–8.12
11	8.52	+7.09
12	11.74	–24.11
13	28.71	+7.28
14	45.77	+19.32
15	48.05	+36.89
16	62.34	–14.40
17	66.46	–7.12
18	94.02	–23.70
19	96.91	–2.61
20	123.28	–9.13

To me, these positions seem a little too far to the right. Longitude 0° is close to point 11 on the chart (Fig. 101). Points 1 (the Hourglass Sea) and 18 (Solis Lacus) should be at 283° and 89° respectively; Point 14 (Niliacus Lacus) at 33°, point 13 (Gulf of Pearls) at 18°. The differences range between +5° and +12°.

Generally speaking, the configurations correspond to those on Schiaparelli's maps. Several points are worthy of comment. On 3 May, the Tithonius Lacus seemed to be attached to the Solis Lacus, but on 9 June this was no longer so. On 3 May a very brilliant white patch was seen on the terminator, at position angle 112°.55 (at $14^h\ 11^m$).

As for the canals—so numerous in the northern hemisphere—the sketches made on 27 and 28 April, 3 and 29 May, 2, 3, 9, 16 and 29 June and 13 July show more or less definite traces of them. But the state of the atmosphere did not allow Wislicenus to make out the details; in particular, he was unable to look for or recognise the doublings.

The question of drawings of Mars was the subject of an important discussion at the British Astronomical Association meeting of 31 December 1890, when comments were made by Green about the very marked differences between drawings.

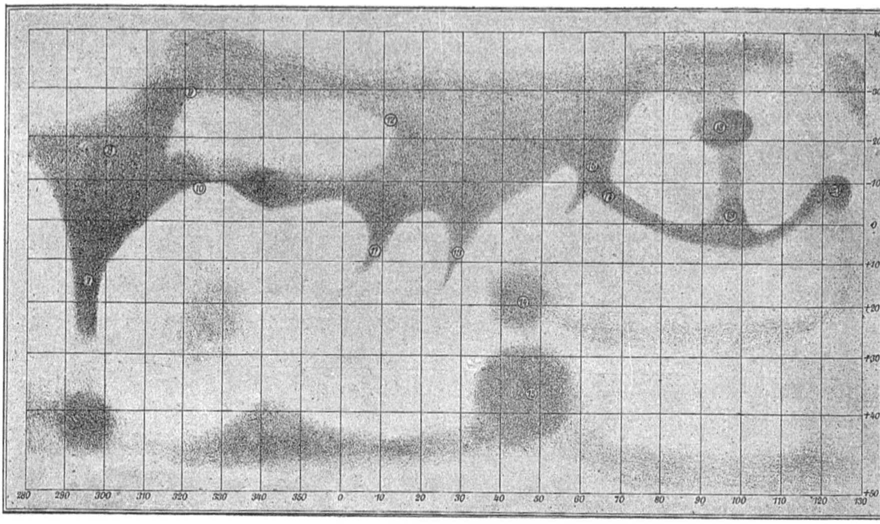

Fig. 101 New triangulation of Mars, by M. Wislicenus, in 1890

To Green, Schiaparelli was not a completely reliable observer, so that only limited confidence could be placed in his representations of Mars. Green commented that ever since childhood he had been used to painting and drawing and he did not hesitate to say that the drawings made by the Milan astronomer differed markedly—even with regard to the general appearance of the planet—from those of all other observers.

This is the comment which W. H. Pickering had already made (*Sidereal Messenger*, 6 October 1890), stating that Green's chart was that which gave the most reliable picture of the planet.

The general outlines drawn by Schiaparelli in his maps, said Green, are not accurate. He went on to say that not only did Schiaparelli's drawings differ from those of others, but also differed between each other. If the broad outlines were not accurately shown, it was impossible to place confidence in the representations of finer details. Green commented that Schiaparelli's chart distorts the form of the Hourglass Sea, particularly in its eastern prolongation, as well as that of the Main Sea. The Knobel Sea, one of the most distinct and most easily recognizable features on the planet, is not shown at all. Instead of this, the planet is covered with a réseau of fine, straight lines.

Green went on to say that the three drawings made by Schiaparelli in 1877, 1879 and 1882 (top, Fig. 102) should agree better with each other than three made by different observers (bottom line, Fig. 102), but this is not the case. It would be more logical to think that it was the lower three which were the work of the same observer. He continued:

> Now, a very interesting question comes before us. What were the forms that the observer at Milan did see—for he must have seen something—that formed the basis of these canals?

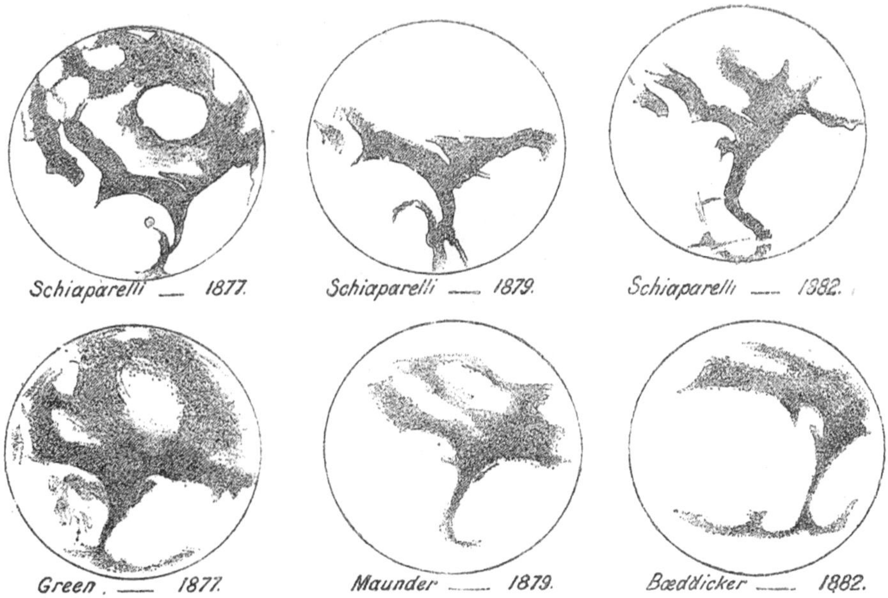

Fig. 102 Comparison of six drawings of Mars

> A careful examination will partly answer this question. I have in my hand a drawing by Schiaparelli of the 1877 series, in which the Oculus (the Terby Sea) is on the meridian. From this, passing directly south, is a faint piece of soft shading. In the 1879 map this is drawn with two hard lines and in 1882 with one sharp line, so that we have three methods of representing a single form. Now, when I was in Madeira in 1877, I had some especially fine views of this portion of the planet and had there been anything resembling these lines, I must have seen it. It may be objected that others have seen them, so that they must be there. That other observers have seen whatever forms the basis of these lines I do not for a moment doubt, but I feel thoroughly convinced that they have not drawn what they have seen, or, in other words, have turned soft and indefinite pieces of shading into clear, sharp lines.

Moreover, in Green's view, if we compare the drawings made in 1877 with each other, we see a vague region representing the Main Sea to the left of the Hourglass Sea in Green's drawing, but Schiaparelli shows it as a pencil-line ending in a point. In 1879 the Milan observer drew it quite differently, and much as shown by Maunder in the same year; but in 1882 we are back to the first aspect. While Green, Maunder and Bœddicker saw this region as a diffuse shading, Schiaparelli showed only its north border, sharply defined.

Green added that his drawings of the South Pole of Mars established that his view and his telescope were not inferior to those at Milan.

Captain Noble (President) confirmed these opinions and said that his own observations of Mars, made over a long period, did not agree with those of Schiaparelli.

The discussion was lengthy, giving the impression that the Milan observer was better at observing double stars than planetary configurations. I have published this discussion without reservation, as I feel that I should put all the various comments into my reader's hands.

1890.—Don Lamey: Variations in the Colour of the Planet Mars

At his observatory at Grignon (Côte d'Or), Dom Lamey made a series of observations which led him to the following results:

> Among the numerous drawings of Mars made at the Grignon Observatory, I have taken care to make some with red, yellow and blue crayons, to represent the dominant tints on the planet. Now, in summarizing all these drawings, the conclusions which I have gradually reached have been drawn from evidence so reliable that I do not believe they will be altered by subsequent observations.
>
> 1. Near opposition, the lower part of the planet is a fairly uniform reddish yellow; the arcs and curves which have been called canals are of a greater or lesser-accentuated blue, turning sometimes to bluish grey. The snowy patches at the poles and the other less brilliant patches of the central regions do not appear to vary much in brilliancy according to their eastern or western position on the disk; however, it is at the centre of the planet that the white equatorial patches shine with the greatest brilliancy.
> 2. Before and after opposition, particularly when the phase is very marked, the white equatorial patches are brilliant and clearly defined at sunrise, become pale and sometimes disappear near midday and again become very brilliant, though vague, near sunset. Their forms also vary; to the east they are rounded, often bordered by a bluish arc. At the extreme west, they are broad and bounded by less brilliant arcs, whose centres are often situated toward the east; the more western arcs show white borders to the western side and are here bluish and dimmer than on the eastern side.
> 3. After opposition, when the planet nears quadrature, the bluish tints are accentuated to the east, while those which turn to reddish yellow are concentrated in the west. The phenomena seem to spread to the arcs and bluish patches to the east, which become more numerous and whose tint becomes more and more pronounced as the planet approaches quadrature.
>
> Outside all hypotheses and all interpretations of the forms shown by the Martian features, the colour phenomena which I have summarized must be due less to the intrinsic quality of the materials spread over the surface of the planet than to the manner in which they reflect the light of the Sun.
>
> However, I do not want to infer that the vaporization and the condensation of atmospheric precipitates cannot play a certain role in the inequalities of colour on the planet; but I would assign to them only a secondary effect

To all these observations we can add a certain number of others, notably those of Guiot, Schmoll, Bruguière, Bressy, Léonard, Lihou, Vimont, Courtois, Leclair, Fenet, Tramblay, E. Duval, F. Loiseau, Duménil, Quénisset, Norquet, Henrionnet, Captain Noble, Denning, Antoniadi, Landerer, José Comas, Valderrama, Decroupet, Lorenzo Kropp, Stenberg, etc., but they add nothing new.[110]

---

[110] There was a conjunction of Mars and Jupiter (59′) on 13 November 1890, observed, among others, by Dutheil and Duval, at Billom. Jupiter, more brilliant and vivid, appeared gold yellow, with Mars a remarkable reddish tint.

Opposition of 1892

My aim in writing this book has been, above all, to make use of the observations and I have thought it wise to include those of the present perihelic opposition of 1892. However, we must await the full results of this opposition before trying to analyses them. Moreover, we may expect that the 1894 opposition will yield very interesting new results and I have decided not to publish another work of this kind until there are truly new developments. *Transibunt generationes, et augebitus Scientia.*

At the moment when I write the last pages of this book (August 1892) we already have a good many series of excellent observations made during the present opposition of Mars. I have one from my observatory—Juvisy—in which I am happy to have had the collaboration of observers who are as skilful as they are zealous: M.M. Guiot, Quénisset, Schmoll and Mabire. In spite of the great southerly declination of the planet (24°) and the low elevation above our horizon (17°) and the mists which have handicapped us even at meridian passages, several canals have been seen and drawn, notably the Nasmyth Pass (fine and delicate), the Indus, Ganges, Gigas, Iris, Gorgon, Titan, Eumenides, Pyriphlegethon (with a different course), Hiddekel, Gehon, Oxus, Orontes, Phoson, Lethe and Jamuna. None has been seen double; this is in agreement with what has been said above, that the geminations only occur near the vernal and autumnal equinoxes (of Mars) and not in summer or winter. Now, we have, for 1892:

Date of Opposition: 4 August
South spring and North autumnal equinox: 20 May.
South summer and North winter solstice: 13 October.

The South Pole of the planet was inclined toward the Earth. The snowy cap shrank constantly during May to August (42°–16°).

I cannot cite the observations which will be made and can only refer to the preceding comments about the micrometrical measurements which will, I believe, give an improved value for the diameter of Mars.

There is, indeed, a discrepancy between the values adopted which is inadmissible in view of the present accuracy of areographical knowledge. Here are the diameter values as given in the official astronomical journals:

Date: 1892	*Conn. Des Temps*	*Naut. Alm*	Eph. (Marth)
1 July	24.″2	24.″0	20.″17
15 July	27.2	27.0	22.75
1 August	29.4	29.3	24.66
4 August (opposition)	29.4	29.4	24.76
15 August	29.0	29.0	24.43
1 September	26.2	26.4	22.18

The *Connaissance des Temps* and the *Nautical Almanac* are virtually in accord, because both give the same values, those of the Tables of Le Verrier (11″.10 at unit distance), while Marth has adopted the value drawn from the discussion by Hartwig (9″.35). Such a discrepancy is somewhat annoying. This is why we should take advantage of the coming opposition to make new micrometrical measurements.

With my $0^{m}.24$ equatorial, with a spider's web micrometer and a power of 380×, I have made a series of measurements at the time of meridian transit, on 22 and 23 July and 4, 5 and 6 August.

These measurements have given 24″.50 for the first two dates and 24″.91 for those following.

This shows that the values adopted by the *Connaissance des Temps* and the *Nautical Almanac* are too great and give, for unit distance, a diameter of 9″.39.

To eliminate the effects of irradiation as far as possible, I took care to put the inner wires tangential to the edge of the disk.

The diameter given in Le Verrier's Tables is certainly too great.

Conclusions of the Third Period: 1877–1892

I would refer the reader to the conclusions of the first and second periods of observation of Mars. From these conclusions, we can judge the results of the progress made since 1877—results which are of the highest value. We have here a continuation of the previous series. To the 391 drawings made in the first two periods, assiduous observation of the planet has added 180 more, making a total of 571 telescopic views or areographic maps.

28. The most probable value for the diameter of the planet is that derived from Hartwig's discussion: 9″.35. For a parallax of 8″.82, the diameter = 0.530 (Earth = 1). The volume is 0.149 that of the terrestrial globe.
29. The mass is 1/3,093,500 that of the Sun, or 0.105 that of the Earth.
30. The density is 0.705 (Earth = 1) and the surface gravity is 0.376. Taken to Mars, one terrestrial kilogram would weigh only 376 g. Everything is lighter there than here.
31. The geography of Mars, or areography, has been completed by some unexpected discoveries. To the continents, seas, islands, lakes, gulfs, polar snows and irregular snows, telescopic observation has added a réseau of rectilinear tracks, going from one sea to another and which have been given the name of canals.
32. The nature of these lines has not yet been determined by observation. However, their situtation, corresponding with those of rivers arriving at known estuaries; their connection with the seas; their colours, their variations in breadth and sometimes even in course, lead us to think that they are due to a mobile element, analogous to water. Can they be true canals? Is the water the same as our? Can we assume the presence of vegetation? Future observations will undoubtedly clear up these problems.

33. In certain circumstances, near the spring and autumn equinoxes, the canals are seen as double. Perhaps the phenomenon is caused by atmospheric refraction, as happens in our atmosphere by ice-crystals which produce haloes and perihelia, and recall the double refraction of Iceland spar. However, the substance which forms the seas, lakes and canals seems to have the property of sometimes separating into two almost equal parts. Nothing analogous is known on Earth.
34. Unceasing variations are observed in the seas, lakes and canals with regard to extent, colour, and for the canals—even position.
35. The general colour of the continents is the reddish-yellow of ripe wheat. Undoubtedly it is vegetation which covers these surfaces.
36. The map of the planet has been drawn up by geometrical triangulation, as accurately as with Earth maps. 114 points have been measured micrometrically.
37. The inclination of the axis is 24°52′ and the seasons are, therefore, much the same as ours.
38. The rotation of Mars has been fixed very accurately at $24^h\ 37^m\ 22^s.65$. The length of the solar day is $24^h\ 39^m\ 35^s$.
39. The planet is accompanied by two small satellites, whose diameters do not seem to exceed 12 km for the first and 10 km for the second. The first, Phobos, goes round Mars in $7^h\ 39^m\ 14^s$, that is to say more rapidly than the planet rotates; the second, Deimos, in $30^h\ 17^m\ 54^s$, at a distance of 20,000 km (the distance of Phobos from the surface is 6,000 km). As seen from Mars, Phobos shows a small disk of 7′ and Deimos a still smaller one of 2′ 1/2. The mean apparent diameter of the Sun would be 21′.
40. Phobos and the Sun can produce light tides on the uniform surfaces of the Martian seas, and perhaps even in the canals.
41. The atmosphere of Mars is thinner and generally clearer than ours. Clouds are rare. However, beyond the polar snows—very extensive in winter, greatly reduced in summer—ice apparently exists in certain continental or insular regions. Of all the planets, Mars has the climate which most nearly resembles that of the Earth.
42. The world of Mars seems to be habitable in the same degree as with the Earth. It is cosmogonically more ancient, and its inhabitants may well be more advanced than ourselves.

Part II
Results Obtained from the General Study of the Planet

The Orbit of Mars

Distance from the Sun.—Length of Revolution.—Eccentricity.—Synodic Period.—Recurrence of Oppositions.—Variations of Distance.—How Mars is seen from the Earth.

The preceding discussions, analyses and comparisons have given all the facts which we actually have with regard to Mars. It will now be easy to complete this general study by briefly describing each of the main subjects into which our knowledge of our neighbour world is divided

First, it is important for us to have an exact and precise idea of everything concerning its revolution round the Sun, and the relationships between this orbit and that of the Earth. Here are the astronomical elements of the orbit of Mars, and all the essential data relating to the planet. There is no need to stress that the first known of these elements was the period of revolution. Over a 1,000 years ago, the position of Mars in the Zodiac was observed with an accuracy corresponding to each epoch. The oldest observation of Mars which has come down to us is to be found in Ptolemy's *Almagest*, Book X, Chapter IX. It was dated the 52nd year after the death of Alexander, or the 476th year of the era of Nabonassar, on the morning of 21 Athur, when Mars was near the star Beta Scorpii. This date corresponds to 17 January 272 before our era. But the Red Planet had been observed many centuries earlier, and the cuneiform inscription found in the ruins of Nineveh shows that 2,500 years before our era the third day of the week had already been named in its honour.

Another notable observation was the passage of Mars in front of the disc of Jupiter on 9 January 1591, witnessed by Kepler.

The reddish colour of Mars has been noted since the earliest times. It explains why the planet was identified with the God of War, and received the symbol of a shield flanked by an arrow (♂). It does not seem to have changed over 4,000 or 5,000 years.

C. Flammarion, *Camille Flammarion's The Planet Mars*, Astrophysics and Space Science Library 409, DOI 10.1007/978-3-319-09641-4_4

Astronomical Elements of the Planet Mars

Mean distance from Sun: 1.5,236,913 = 227,031,000 km.
Eccentricity: 0.0932611.
Secular variation of eccentricity: 0.000090176.
Perihelic distance: 1.3826 = 206,007,000 km.
Aphelic distance: 1.6658 = 248,207,000 km.

Period of revolution (sidereal year): 686^{d} 23^{h} 30^{m} 41^{s} = 686.979 days = 1.8808 years.
Tropical year: 686.929 days.
Synodic period: 779.94 days.

Inclination to Earth's orbit: 1°51′2″.
Longitude of ascending node: 48°23′53″.
Longitude of perihelion: 333°49′.
Secular variation of perihelion: +1,582″.43.

Longitude of south summer solstice: 356°48′.
Distance from perihelion to solstice: 36 days. (Days in original)
Longitude of north summer solstice: 176°48′.
Longitude of aphelion: 153°49′.
Longitude of south autumnal equinox and north spring equinox: 86°48′.
Longitude of south spring equinox and north autumnal equinox: 266°48′.
Distance from south summer (or north winter) solstice to south autumnal (or north spring) equinox: 160 days. (Days in original)
Distance from south autumnal (or north spring) equinox to south winter (or north summer) solstice: 199 days. (Days in original)
Distance from south winter (or north summer) solstice to south spring (or north autumnal) equinox: 182 days. (Days in original)
Distance from south spring (or north autumnal) equinox to south summer (or north winter) solstice: 146 days. (Days in original)

Obliquity of ecliptic: 24°52′.
Sidereal rotation: 24^{h} 37^{m} $22^{s}.65$.
Length of solar day: 24^{h} 39^{m} $25^{s}.0$.
Polar flattening: about 1/220
Diameter: 9″.35 = 0.530 (Earth = 1).
Surface: 0.281 (Earth = 1).
Volume: 0.149 (Earth = 1).
Mass: 0.105 (Earth = 1).
Density: 0.705 (Earth = 1): 3.91 (water = 1).
Surface gravity at equator: 0.376. $g = 3^{m}.69$.
Mean diameter of Sun: 21′2″ = 0.656 of that from Earth.
Light and heat received from Sun: 0.43 (Earth = 1).
Value of 1° areocentric on the surface of Mars: 60 km.

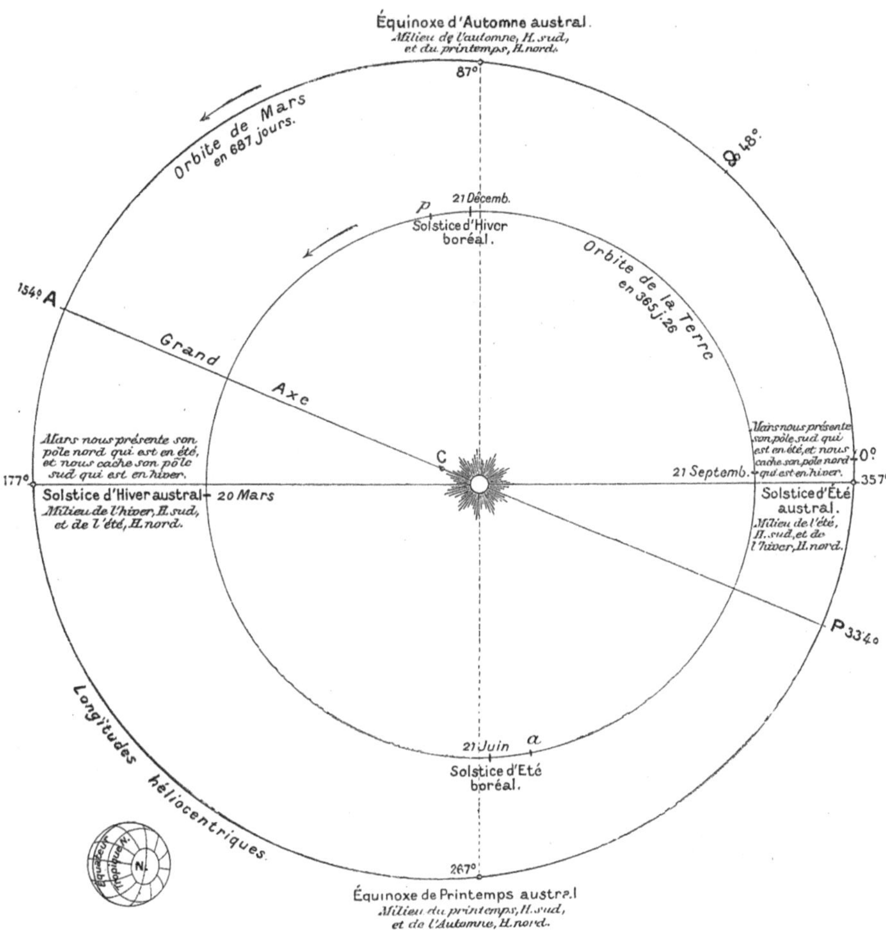

Fig. 1 The orbit of Mars around the Sun and its relations with the terrestrial orbit

The orbit of Mars is very elliptical. The distance from the Sun varies between 206 and 248 million km, so that the range is 42 million km, which is equivalent to 1/5 of the mean distance.

In Fig. 1 I have represented all the situations relating to the orbit compared with that of the Earth. The ellipticity of the orbit is shown to the correct proportions. The major axis of the orbit of Mars is horizontal, and the sunlight comes from one of the foci at the point C, marking the centre of the ellipse followed by Mars around the Sun; the distance from C to the centre of the Sun therefore represents the eccentricity. The perihelion lies at heliocentric longitude 333°489′, which the Earth passes on 27 August. The summer solstice of the southern hemisphere is not far from here, and lies at longitude 356°48′. The line of apsides therefore passes through this longitude and that which is diametrically opposite to it.

The perihelion of the Earth is at p, aphelion at a.

It can be seen that Mars presents its south pole fully to the Sun not far from the moment of perihelion, the southern summer solstice following perihelion by only 36 days. We know that on the Earth, the southern summer solstice occurs on 21 December and perihelion on 1 January; with us, the solstice precedes perihelion by 10 days. In spite of this difference between the positions of Mars and the Earth in their respective orbits, the situation of Mars relative to the Sun is analogous to ours; it is the southern hemisphere which is turned sunward at the time of perihelion—that is to say, southern summer corresponds with the minimum distance of Mars from the Sun, as is also true in the case of the Earth. Therefore, the southern hemisphere receives more heat, at the time of solstice, than the northern. The solstice of the northern hemisphere occurs at aphelion. From the point of view of the variations of the polar snows, those at the south pole, receiving more heat at the time of the solstice than do the north polar snows at northern solstice, should melt more rapidly, all other circumstances being equal.

Mars is tilted with respect to its orbit in the position shown at the bottom of the figure, and spins so as to keep its axis of rotation pointed in a constant direction. The north pole is tilted toward the Sun—and also toward the Earth—at the southern winter solstice; this happens during aphelic oppositions. At perihelic oppositions, it is the south pole which is tilted toward us.

What is the period regulating the encounters between Mars and the Earth in the same radius vector leading from the Sun, a meeting analogous to the encounters of the two hands of a watch?

The mean interval between two successive oppositions of Mars can be calculated from the following formula. The Earth moves more quickly than Mars. Its average daily advance in heliocentric longitude is:

$$\left(360°/365.26\right)\nu\left(360°/686.98\right).$$

Consequently, the encounter of the two planets on the same straight line leading from the Sun can be expressed as

$$1:\left(1/365.26-1/686.98\right)\text{days}=\left(686.98\times365.26\right)/321.72\,\text{days}=779.94\,\text{days}.$$

This is the mean synodic period. But in reality, of course, this period is irregular, because of the enormous variations in orbital speed caused by the eccentricity of the Martian orbit. This is shown by comparing the intervals between oppositions during the past 20 years:

Opposition	Interval	Opposition	Interval
1871 20 March	769 days	1881 27 December	766 days
1873 27 April	784 days	1884 1 February	765 days
1875 20 June	809 days	1886 6 March	767 days
1877 5 September	798 days	1888 11 April	777 days
1879 12 November	775 days	1890 27 May	799 days
1881 27 December		1892 4 August	

Here are the dates of the observed oppositions of Mars. The most advantageous so far as distance is concerned are those of August and September. We have seen that the perihelic longitude of Mars is that which is passed by the Earth on 27 August. The best oppositions have been: (1) 27 August 1719, (2) 1 September 1798, (3) 5 September 1877.

Opposition		Principal observers
1636		Fontana
1638		Fontana
1640		Zucchi
1644–1645		Bartoli and Hevelius
1651	April and May:	Riccioli
1653	July:	Riccioli
1655	August:	Riccioli
1657	September:	Riccioli
1659	November:	Huygens
1662	August:	Huygens
1666	18 Mar.:	Cassini, Hooke, Serra
1672	August to October:	Huygens, Flamsteed
1683	April and May:	Huygens
1694	February:	Huygens
1704	October:	Maraldi
1717	June:	Maraldi
1719	27 Aug., only 2.5 d. after perihelion:	Maraldi, Bianchini
1764	May:	Messier
1766	July:	Messier
1777	April:	William Herschel
1779	May to June:	William Herschel
1781	September:	William Herschel
1783	1 October:	William Herschel
1785	26 November:	Schröter
1788	7 January:	Schröter
1792	16 March:	Schröter
1794	23 April:	Schröter
1796	15 June:	Schröter
1798	1 September:	Schröter
1800	9 November:	Schröter
1802	25 December:	Schröter
1805	28 January:	Flaugergues
1807	4 March:	Fritsch
1809	8 April:	Flaugergues
1811	June:	Arago
1813	31 July:	Arago, Flaugergues
1821–1822		Kunowsky, Gruithuisen
1824	25 March:	Harding

(continued)

Opposition		Principal observers
1830	19 September:	Beer and Mädler
1832	20 November:	Beer and Mädler
1835	2 January:	Beer and Mädler
1837	6 February:	Beer and Mädler, Galle, Bessel
1840	12 March:	Beer and Mädler, Galle, Bessel
1841	17 April:	Beer and Mädler
1843	5 June:	Julius Schmidt
1845	17 August:	Mitchell, Arago, Main
1847	30 October:	W.Grant, J.Schmidt
1854	March:	Jacob
1856	April:	Warren de la Rue, Brodie
1858	15 May:	Secchi
1860	17 July:	Liais
1862	5 October:	Secchi, Lockyer, Phillips, Huggins
1864	30 November:	Kaiser, Dawes, Franzenau, Vogel
1867	10 January:	Terby, Williams, Huggins
1869	13 February:	Secchi
1871	20 March:	Gledhill, Burton, Terby
1873	27 April:	Green, Trouvelot, Flammarion
1875	20 June:	Terby, Holden
1877	5 September:	Schiaparelli, Green, Hall, Lohse, Cruls
1879	12 November:	Schiaparelli, Terby, Niesten, Burton
1881	26 December:	Schiaparelli, Terby, Bœddicker
1884	31 January:	Green, Trouvelot, Knobel, Denning
1886	6 March:	Schiaparelli, Denning, Perrotin, Lohse
1888	11 April:	Schiaparelli, Terby, Perrotin, Niesten, Holden, Flammarion
1890	27 May:	The same; plus Wislicenus, Pickering, Keeler, Stanley Williams

Perihelic Oppositions

1672–1689–1704–1719–1734–1751–1766–1783–1798–1813–1830–1845–1860–1862–1877–1892.

The best observed were those of 1719, 1783, 1798, 1830 and 1877.

Perihelic oppositions occur at intervals of 15–16 years, as we have seen. The mean distance between the orbit of Mars and that of the Earth is 0.5237×149 million km, or 78 million km. At perihelic oppositions the distance between the two bodies may be reduced to 56 million km.

Instead of using this period as the cycle, it is more precise to take 32 years as a double cycle. In effect, the synodic period of Mars is 779.94 days. Fifteen times this number gives 779.94×15=11,699 days. In 32 terrestrial years, there are 365×32+8=11,688 days. Therefore, the difference is only 11 days.

Now, 25 revolutions of Mars are equivalent to 47 revolutions of the Earth; the period of 47 years may therefore be substituted for the preceding ones.

Here are the minimum distances for the oppositions of the last 15 year cycle, from 1877 to 1892:

1877	Minimum distance, 2 September:	0.37666	=6,422,000 km
	Opposition, 5 September:	Diameter	=24″.8
	Meridian transit at midnight, 6 September		
1879	Minimum distance, 4 November:	0.48243	=71,882,000 km
	Opposition, 12 November:	Diameter	=19″.1
	Meridian transit at midnight, 9 November		
1881	Minimum distance,21 December:	0.60282	=89,820,00 km
	Opposition, 26 December:	Diameter	=15″.5
	Meridian transit at midnight, 27 December		
1884	Minimum distance, 30 January:	0.66909	=99,694,000 km
	Opposition, 31 January:	Diameter	=13″.9
	Meridian transit at midnight, 4 February		
1886	Minimum distance, 8 March:	0.66989	=99,813,000 km
	Opposition, 6 March:	Diameter	=14″.0
	Meridian transit at midnight, 9 March		
1888	Minimum distance, 17 April:	0.6050	=90,145,000 km
	Opposition, 11 April:	Diameter	=15″.4
	Meridian transit at midnight, 11 April		
1890	Minimum distance, 5 June:	0.48495	=72,255,000 km
	Opposition, 27 May:	Diameter	=19″.1
	Meridian transit at midnight, 26 May		
1892	Minimum distance, 6 August:	0.37736	=56,226,000 km
	Opposition, 4 August:	Diameter	=24″.8
	Meridian transit at midnight, 6 August		

Thus while the distance between Mars and the Earth is reduced to 56,000,000 km at perihelic oppositions, it does not fall below 99,000,000 km at aphelic oppositions. Oppositions occur every 2 years, while perihelic oppositions occur at intervals of 15–17 years. In the interval between successive oppositions, Mars recedes to a considerable distance, reaching maximum distance at times when the planet passes beyond the Sun as seen from the Earth; but it is then unobservable. Even if we take observing periods as extending from 3 months before opposition to 3 months after, we must agree that the distance range over this period is very considerable. Take, for instance, the year 1892:

	Dist. of Earth\Sun	km	Diameter
4 May	0.84172	125,416,000	11″.1
4 June	0.61455	91,568,000	15″.2

(continued)

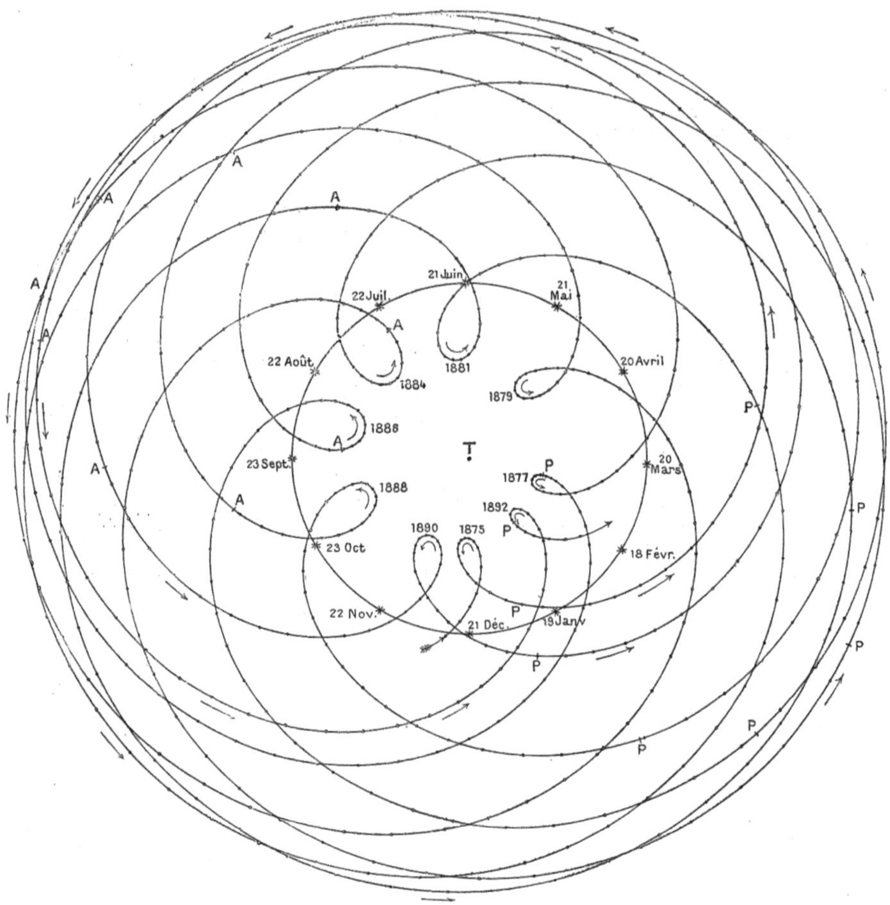

Fig. 2 Apparent movement of Mars relative to the Earth

	Dist. of Earth\Sun	km	Diameter
4 July	0,45073	67,159,000	20″.8
4 August	0.37766	56,271,000	24″.8
4 September	0.43177	64,334,000	21″.7
4 October	0.57412	85,544,000	16″.3
4 November	0.77493	115,646,000	12″.1

Mars can recede to 400,000,000 km from the Earth, with an apparent diameter shrinking to 3″.

The unceasing variation of the distance between Mars and the Earth is represented in the diagram by the apparent movement of Mars relative to the Earth from 1875 to 1892 (Fig. 2). The Earth is assumed to be central and stationery The distance between Mars and the Earth is indicated by dots for every 10 days. Mars is at perihelion at the point marked P, and at aphelion at the point marked A. The circular

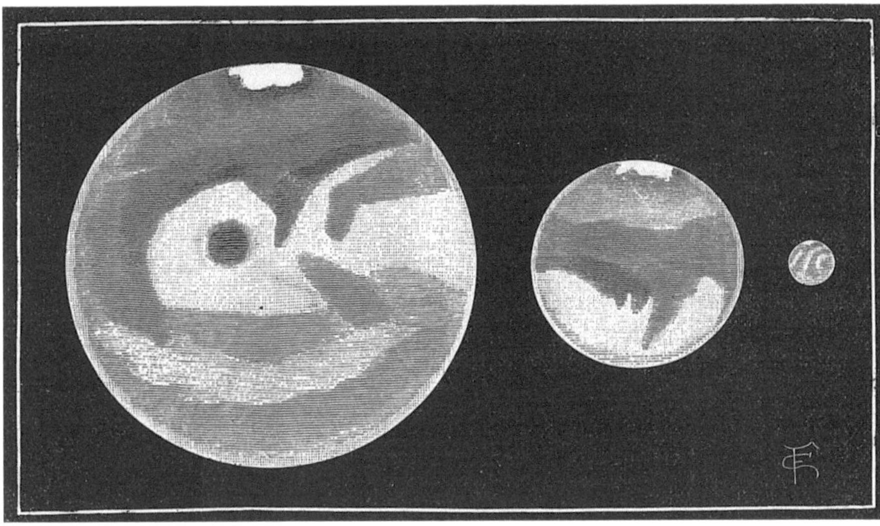

Fig. 3 Apparent dimensions of Mars at its most extreme and mean distances (2^{mm} to 1″)

curve which has been drawn around the Earth at a certain distance indicates the apparent path of the Sun, and its position in the sky for the dates indicated. From this figure it is also clear that Mars is not equally distant from the Earth at successive oppositions.

The apparent dimensions of the Martian disc as seen from Earth vary in the proportion indicated by Fig. 3, drawn to a scale of 2^{mm} to 1″. The larger disc to the left applies to perihelic oppositions. When Mars is of the size shown by the right-hand disc it cannot be observed, as it is then behind the Sun.

The angles which Mars makes with the Earth in its various positions round the Sun produce phases which reach maximum near quadrature; but these phases are not so pronounced as were indicated on the first drawings of Mars made by seventeenth-century observers. Even the greatest extent of the non-illuminated portion never exceeds 1/7 of the diameter.

I can cite a few examples:

Phases of Mars

1888.—Opposition 11 April. Quadrature 22 July							
	Deficit	Diam.	Ratio		Deficit	Diam.	Ratio
5 July	1.22	9″.58	0.127	23 July	1.16	8″.57	0.135
9 July	1.21	9″.33	0.129	25 July	1.15	8″.47	0.136
13 July	1.20	9″.10	0.132	27 July	1.14	8″.38	0.136
15 July	1.18	8″.99	0.133	29 July	1.13	8″.29	0.135

(continued)

18 July	1.18	8″.83	0.134	31 July	1.11	8″.20	0.135
21 July	1.17	8″.67	0.135	2 Aug.	1.10	8″.11	0.135
1890.—Opposition 27 May. Quadrature 21 September							
15 Sept.	1.56	10″.15	0.153	23 Sept.	1.49	9″.66	0.154
17 Sept.	1.55	10″.02	0.154	25 Sept.	1.47	9″.54	0.154
19 Sept.	1.53	9″.90	0.155	27 Sept.	1.46	9″.43	0.154
21 Sept.	1.51	9″.78	0.155	29 Sept.	1.44	9″.32	0.154
1892.—Opposition 4 August. Quadrature 9 December							
1 Nov.	1.65	12″.41	0.133	29 Nov.	1.34	9.70	0.137
5 Nov.	1.61	11″.09	0.135	1 Dec.	1.31	9″.54	0.137
9 Nov.	1.57	11″.53	0.136	3 Dec.	1.29	9″.39	0.137
13 Nov.	1.53	11″.13	0.137	5 Dec.	1.26	9″.24	0.136
17 Nov.	1.48	10″.74	0.137	7 Dec.	1.24	9″.10	0.136
21 Nov.	1.43	10″.38	0.138	9 Dec.	1.21	8″.95	0.136
25 Nov.	1.38	10″.03	0.137	11 Dec.	1.19	8″.81	0.135

The amount of the phase deficiency, which can never exceed 1″.76, is not sufficient for appreciation of the phase; one must take the diameter into account. Near maximum the proportion is 0.155, equivalent to 1/7 of the diameter.

Figure 4 represents the maximum phase at mean quadrature, such as on 22 July 1888.

It is greatest when Mars is near perihelion and least when it is near aphelion. The planet is here at its mean distance near an autumnal equinox, so that the phase more or less follows the meridian line.

It is very important to take into account the various ways in which the planet is presented to the Earth at times of opposition, according to the distance and the visual radius. For each consecutive opposition, it seems necessary to construct three projections for the 15 year cycle (1877–1892) relating to the beginning, middle and end of each period of observation. Each of these projections allows us, at a glance, to judge the latitude which the Martian globe presents to observers on Earth.

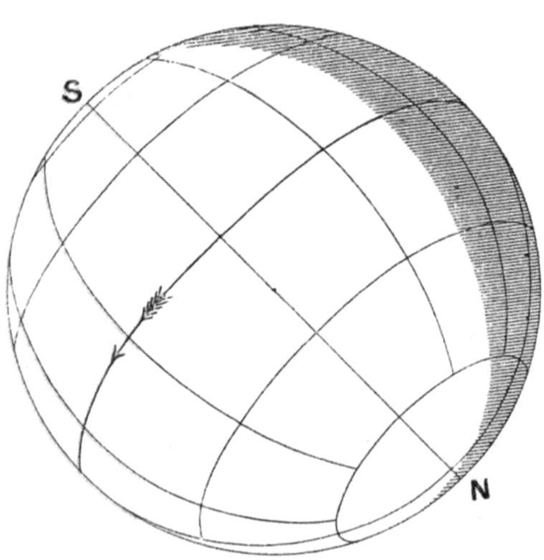

Fig. 4 Phase of Mars at its mean quadrature

For a period of observation, I have taken the period from 3 months before opposition date to 3 months afterwards. In the figures given below, the central disc represents the Martian globe with its true inclination relative to the Earth for the time of opposition; the disc to the left represents the position and relative size of the planet at the indicated date before opposition, and the disc to the right showed the aspect 3 months after opposition.

In all the sketches, the scale is 2^{mm} to 1 second of arc.

Here are the numbers corresponding to each projection:

		Lat.of centre°	Diam.″	Phase def'cy.″	Angle Earth\Sun	Dec.°	Height above Paris horizon°
	5 June	−24.1	12.5	1.7	43	−14	27
1877	{5 September (opp.)	−22.5	24.8	0.0	4	−12	29
	5 December	−28.0	10.8	1.3	41	−02	39
	12 August	−15.2	11.1	1.7	46	+13	54
1879	{12 November (opp.)	−14.5	19.1	0.0	0	+18	59
	12 February	−12.7	8.1	0.9	38	+22	63
	22 October	+6.7	10.7	1.2	43	+24	65
1881–1882	26 December (opp.)	+1.5	15.5	0.0	2	+27	68
	26 March	+4.3	7.4	1.2	42	+26	67
	31 October 1883	+16.3	7.6	0.9	39	+20	61
1884	{31 January (opp.)	+14.8	13.9	0.0	3	+21	62
	30 April	+17.6	7.4	0.8	37	+19	60
1885	24 December	+23.5	8.3	0.8	35	+07	48
1886	{6 March (opp.)	+21.9	14.0	0.0	2	+09	50
	6 June	+25.3	7.8	0.9	39	+06	47
	31 January	+20.2	8.6	0.8	35	−07	34
1888	{11 April	+21.1	15.4	0.0	2	−06	35
	11 July	+23.4	9.2	1.2	42	−10	31
	27 February	+9.8	8.4	0.9	37	−19	22
1890	{27 May (opp.)	+9.5	19.1	0.0	1	−23	18
	27 August	+7.3	11.5	1.7	45	−25	16
	4 May	−13.0	11.1	1.4	41	−22	19
1892	{4 August (opp.)	−12.7	24.8	0.0	5	−24	17
	4 November	−21.2	12.1	1.6	43	−14	27

Before opposition the phase zone is to the left-hand side of the disc (for an inverted image); after opposition it is to the right, and at opposition it is, naturally, nil (at least if we disregard differences of hundredths of a second, due to latitude differences—quantities which are negligible).

Here are projections showing how Mars is seen from the Earth according to distance and inclination (Figs. 5 and 6).

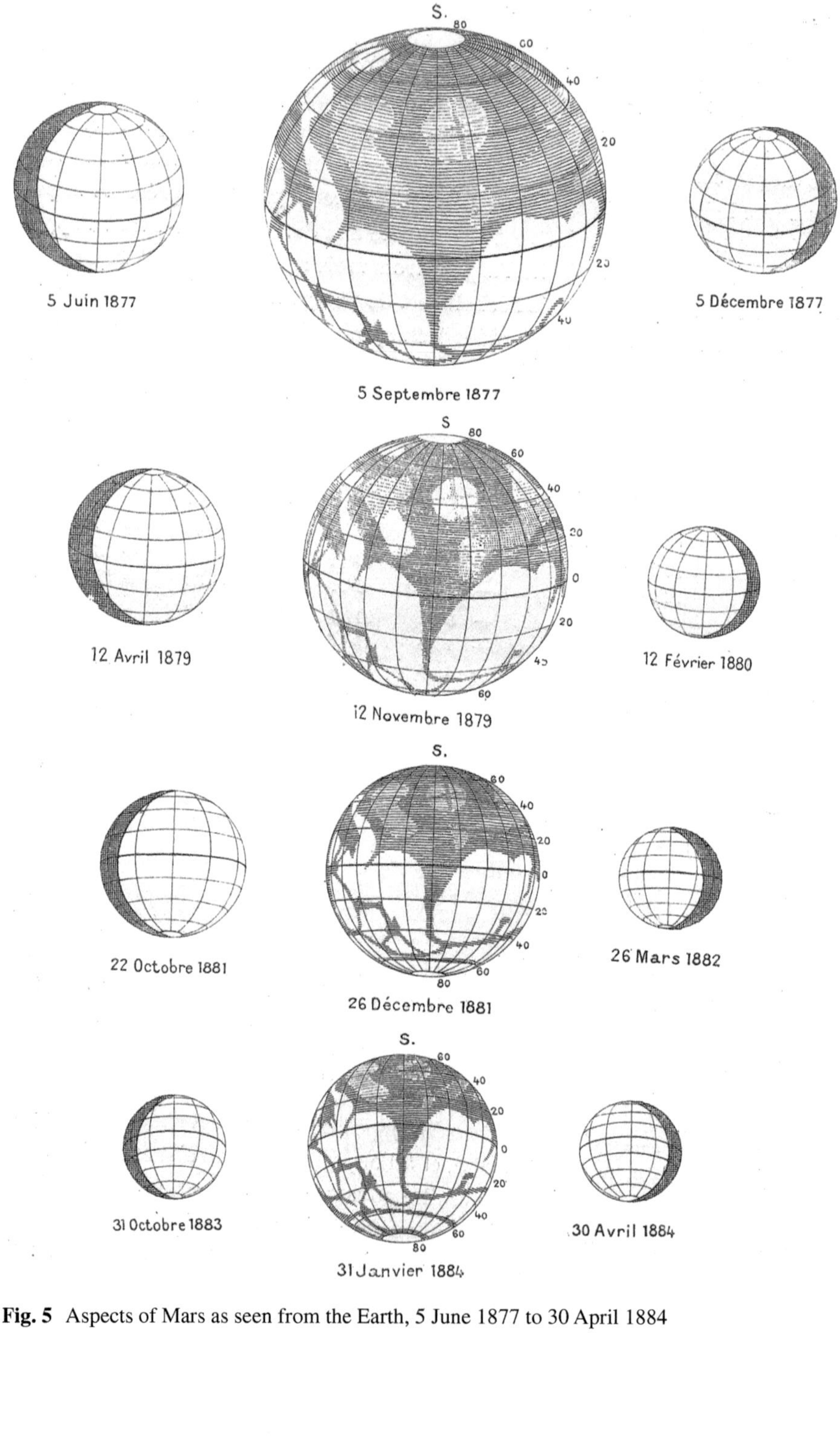

Fig. 5 Aspects of Mars as seen from the Earth, 5 June 1877 to 30 April 1884

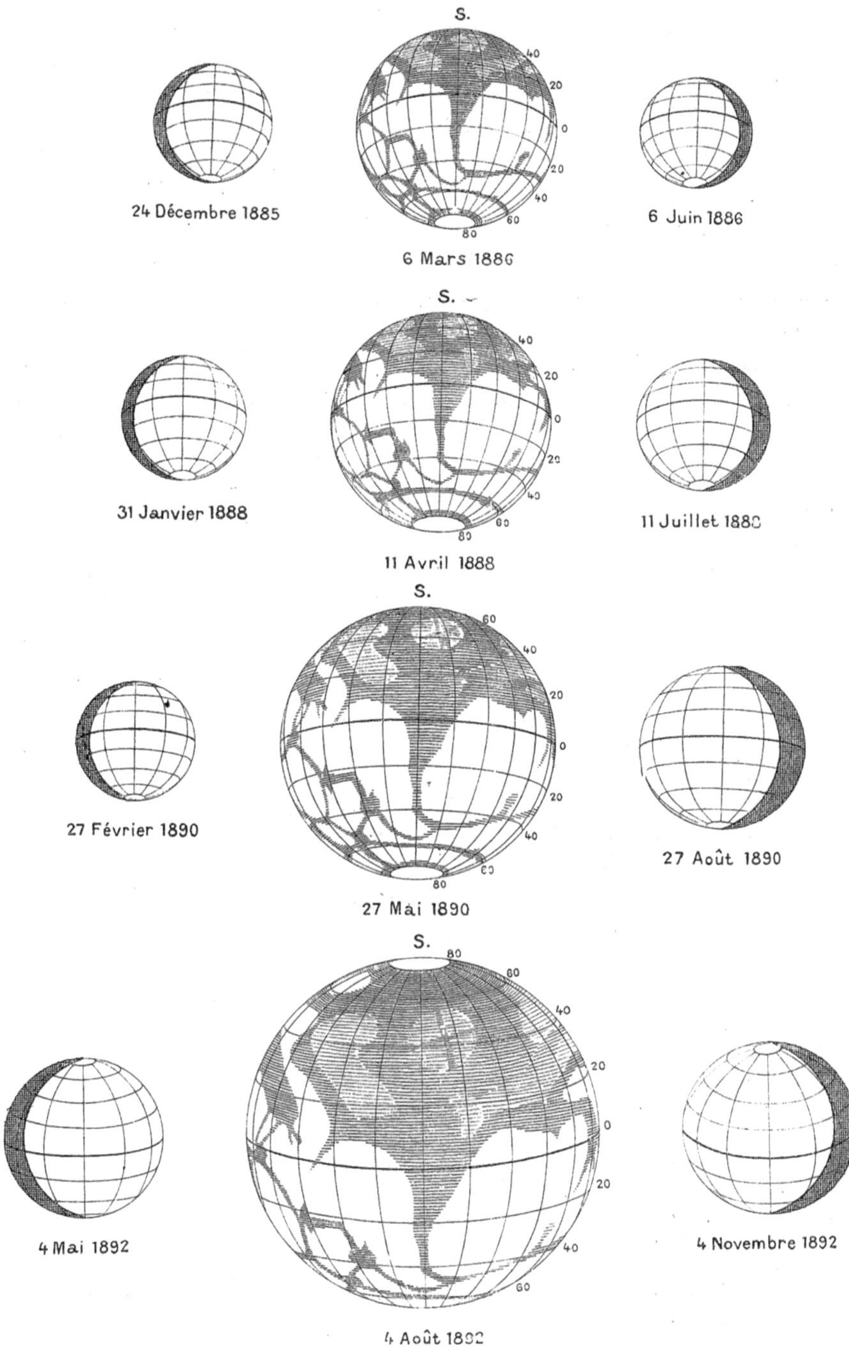

Fig. 6 Aspects of Mars as seen from the Earth, 24 December 1885 to 4 November 1892

Dimensions of the Planet

Mass-Density-Gravity

The real dimensions of Mars are calculated from the apparent size, and differ markedly depending on the observations and the adopted value of the solar parallax. According to the *Connaissance des Temps* and the *Nautical Almanac*, the angular diameter of Mars is 11″.10 at the distance of the Earth from the Sun, that of the Earth being 17″.2. This is the diameter adopted in Le Verrier's Tables.

This diameter leads to 29″.4 for the perihelic oppositions of 1877 and 1892. We will see, however, that this is too great.

Here are the measurements and determinations of the diameter of Mars, assuming unit distance 1.

The measurements made before those of William Herschel are so rough compared with more modern work that there is no point in giving them here. We will begin, therefore, with Herschel. We will give the direct measurements (M) and the determinations (D) drawn from meridian observations.

Dates	Observers	Equatorial Dia.″	Flattening
1784	William Herschel	9.13	1/16.3
1798	Köhler	9.10	1/80.8
1798	Schröter	9.84	1/81
1824	J.J. de Littrow	8.87	–
1837	Bessel	9.93	Insensible
1845	Schmidt	9.44	–
1847	Arago	9.57	1/29
1852	Johnson	8.99	Disk elongated
1854	Peirce	10.11	–
1854	Main	9.84	1/62
1856	Winnecke	9.21	Insensible
1856	Schmidt	9.73	–

(continued)

C. Flammarion, *Camille Flammarion's The Planet Mars*, Astrophysics and Space Science Library 409, DOI 10.1007/978-3-319-09641-4_5

Dates	Observers	Equatorial Dia.″	Flattening
1860	Main	9.38	1/46
1861	Le Verrier	11.10	–
1862	Main	9.38	1/37
1864	Main	9.18	–
1864	Kaiser	9.52	1/118
1864	Winnecke	9.83	–
1864	Dawes	–	Insensible
1871	Main	9.44	–
1873	Engelmann	9.25	1/71
1873	Main	9.40	–
1877	Pritchett	9.19	1/36
1877	Hartwig	9.36	1/78
1877	Hartwig. Gen. discussion	9.35	–
1879	Hartwig	9.41	1/96
1879	Hartwig. Gen. discussion	9.35	–
1879	Pritchett	9.49	1/48
1879	Young	–	1/219
1881	Downing (Greenwich 1851–1865)	9.70	–
1881	Stone (Greenwich 1851–1865)	10.73	–
1881	Pritchett	9.48	–
1892	Flammarion	9.39	–

The differences are very marked. The most probable diameter, resulting from the measurements of Bessel, Main and Hartwig (general discussion) is 9″.35. This is the value which I have adopted. The diameter given in Le Verrier's Tables (11″.10) is manifestly too large.

For a parallax of 8″.82, the diameter 9″.35, is to that of the Earth (17″.64) in the ratio of 530 to 1,000.

The true diameter cannot differ much from this, 0.530, and is a little over half that of the Earth. It is about 1/3 greater than that of Mercury (actually, 0.37), and about double that of the Moon (0.27), as shown in Fig. 1.

This diameter, 0.530, corresponds to 6,753 km.

The surfaces of the two spheres are therefore in the ratio of 28:100, and the volumes 149:1,000.

Consequently, the globe of Mars has an area of approximately 143,000,000 square km., and its volume is 161,000 million cubic km.

The polar flattening is very difficult to measure. It is probably around 1/220. The mass of the planet has been determined with absolute precision since the discovery of the two satellites. The principal values obtained by calculation since then have already been cited. The mass is:

Relative to the Sun	1/3,093,500
Relative to the Earth	0.105

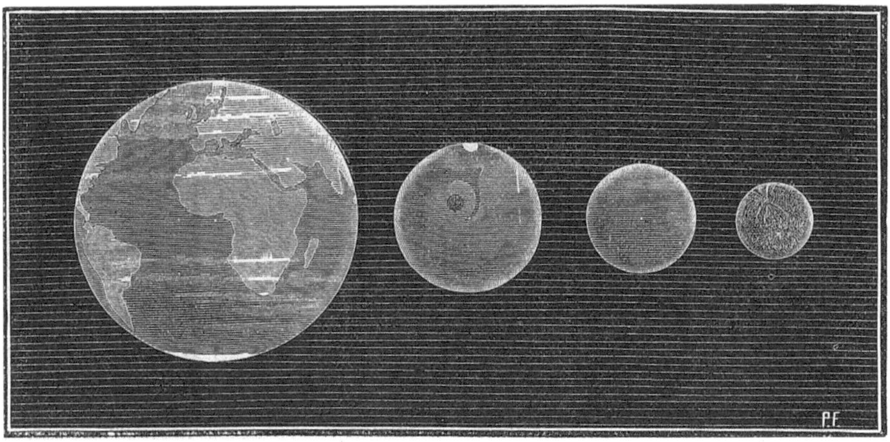

Fig. 1 Comparative sizes of the Earth, Mars, Mercury and the Moon

which means that Mars weighs about 1/10 as much as our own world.

It follows that the mean density of the planet, obtained by dividing the mass by the volume, is 105/149, or 0.705.

Taking water as unity, the density is 3.91.

We also find that the surface gravity on Mars is 0.376. By planetary standards, this gravity is feeble. Only the surface gravity of the Moon is inferior to it.

A falling body—which will, drop at $4^{m}.90$ for the first second on Eart--ill drop only $1^{m}.84$ on Mars in the same period. The velocity after one second, or gravitational acceleration g, which is $9^{m}.81$ on Earth, is only $3^{m}.69$ on Mars.

Rotation: Length of Day and Night

Let us here list all the observations made concerning the rotation of Mars; details having been already given in Part I.

Author and date	Observation period	Period
1659 Huygens	1659 November 28 to December 1	24^h
1666 Cassini	1666 March 3 to 28	$24^h\ 40^m$
1666 Salvator Serra	1666	$12^h\ 20^m$
1704 Maraldi	1704 October 14 to 17	$24^h\ 38^m$
1719 Maraldi	1719 August 5 to October 17	$24^h\ 30^m$
1779 W. Herschel	1777 April 8, 1779 to June 19	$24^h\ 39^m\ 21^s.67$
1792 Schröter	1792 March 19 to April 20	$24^h\ 39^m\ 50^s.2$
1805 Huth	1805	$24^h\ 43^m$
1822 Kunowsky	1821 December 1822 to March 1822	$24^h\ 36^m$ 40
1837 Beer and Mädler	1830 September 10 to October 20	$24^h\ 37^m\ 9^s.9$
1837 Beer and Mädler	1830 October 20 to 1832 November 17	$24^h\ 37^m\ 23^s.7$
1837 Beer and Mädler	1830 October 20 to 1835 March 12	$24^h\ 37^m\ 20^s.6$
1837 Beer and Mädler	1830 October 20 to 1837 March 11	$24^h\ 37^m\ 23^s.7$
1845 Mitchel	1830 September 11 to 1845 August 30	$24^h\ 37^m\ 20^s.6$
1858 Secchi	1856 April 25 to 1858 July 24	$24^h\ 37^m\ 35^s$
1864 Joynson	1862 to 1864	$24^h\ 37^m\ 37^s$
1864 R. Wolf	1862 to 1864	$24^h\ 37^m\ 22^s.9$
1864 Linsser	1830 September 14 to 1862 September 21	$24^h\ 37^m\ 21^s.9$
1864 Kaiser	1672 August 13 to 1862 November 1	$24^h\ 37^m\ 22^s.62$
1867 Proctor	1666 March 13 to 1862 November 26	$24^h\ 37^m\ 22^s.745$
1868 Proctor	1666 March 13 to 1867 February 23	$24^h\ 37^m\ 22^s.735$
1869 Proctor	1666 March 13 to 1869 February 4	$24^h\ 37^m\ 22^s.736$
1873 Proctor	1672 August 13 to 1862 November 1	$24^h\ 37^m\ 22^s.715$
1873 Julius Schmidt	1672 August 13 to 1856 April 21	
1878 Cruls	1877 August 16 to October 3	$24^h\ 37^m\ 34^s$

(continued)

C. Flammarion, *Camille Flammarion's The Planet Mars*, Astrophysics and Space Science Library 409, DOI 10.1007/978-3-319-09641-4_6

Author and date	Observation period	Period
1883 Marth	1704 October to 1879 November	$24^h\ 37^m\ 22^s.626$
1885 Bakhuyzen	1659 to 1879	$24^h\ 37^m\ 22^s.66$
1886 Wislicenus	1659 to 1881	$24^h\ 37^m\ 22^s.655$

For the best value of the period, I adopt the arithmetical mean of the most precise determinations:

Proctor	$24^h\ 37^m\ 22^s.715$
Kaiser	$24^h\ 37^m\ 22^s.62$
Schmidt	$24^h\ 37^m\ 22^s.603$
Marth	$24^h\ 37^m\ 22^s.626$
Bakhuyzen	$24^h\ 37^m\ 22^s.66$
Wislicenus	$24^h\ 37^m\ 22^s.655$
Mean	$24^h\ 37^m\ 22^s.6465$

or, on rounding:

$24^h\ 37^m$ 22s.65 = 88,642^s.65

The period can hereby be regarded as established with the utmost precision,[1] i.e. *to almost 1/100 of a second.*

The sidereal rotation of the Earth is $23^h\ 56^m\ 4^s.091$, or 86,164.091 seconds. That of Mars is longer by 2,478.56 seconds, or $41^m\ 18^s.56$, and is longer than our own 24 hours solar day by about 37 minutes. If we suppose that a Martian day is divided into 24 hours, as with us, each hour will last for 1 and 1 1/2 minutes longer than ours.

The sidereal year of Mars is $686^d\ 23^h\ 30^m\ 41^s$, or 59,355,041 seconds. If we divide this number by the first, we find that each Martian year is made of about 668 Marian solar days: 668.6.

It follows that the length of the Martian solar day is longer than that of the sidereal day in the proportion of 1/669, or $132^s.4$, or $2^m\ 12^s.4$. This length is therefore:

$24^h\ 39^m\ 35^s.0 = 88{,}755^s.0$.

Earlier, we have seen that the sidereal rotation of Mars is greater than our civil day by about 37 minutes. In observing the planet, the passage of a feature across the central meridian is therefore retarded by this amount, each day of the mean solar day of a terrestrial observer. The details are conveniently presented in the following table.

Rotational movement of Mars, hour by hour and minute by minute.

[1] It is curious to note that the drawings by which the rotation of Mars has been fixed are also the most rudimentary: those of Hooke, Huygens, Cassini, Maraldi, Herschel. We can therefore find out the rotation period of a planet, and even fix it to the utmost precision, without being able to distinguish the exact configurations, and by means of drawings that are decidedly vague. One point capable of being clearly identified is sufficient.

Hours	Movement°	Minutes	Movement°	Minutes	Movement°
1	14.62	1	0.24	31	7.55
2	29.24	2	0.49	32	7.80
3	43.86	3	0.73	33	8.04
4	58.48	4	0.91	34	8.29
5	73.10	5	1.22	35	8.53
6	87.72	6	1.46	36	8.77
7	102.34	7	1.71	37	9.02
8	116.96	8	1.95	38	9.26
9	131.58	9	2.19	39	9.50
10	146.21	10	2.44	40	9.75
11	160.83	11	2.68	41	9.99
12	175.45	12	2.92	42	10.23
13	190.07	13	3.17	43	10.48
14	204.69	14	3.41	44	10.72
15	219.31	15	3.66	45	10.97
16	233.93	16	3.90	46	11.21
17	248.55	17	4.14	47	11.45
18	263.17	18	4.39	48	11.70
19	277.79	19	4.63	49	11.94
20	292.41	20	4.87	50	12.18
21	307.03	21	5.12	51	12.43
22	321.65	22	5.36	52	12.67
23	336.27	23	5.60	53	12.91
24	350.89	24	5.85	54	13.16
		25	6.09	55	13.50
		26	6.34	56	13.65
		27	6.58	57	13.89
		28	6.82	58	14.13
		29	7.07	59	14.30
		30	7.31	60	14.62

This retardation is actually a little longer before and after opposition; and it can rise to as much as forty minutes for an interval of 3 months.

The diurnal difference leads to the same face of the planet being presented to the observer after 38 days.

Thirty eight terrestrial days of mean solar time gives 3,283,000 seconds.

Thirty seven Martian rotations gives 3,279,778 seconds.

The difference is only 3,422 seconds or 57 minutes—less than an hour, by which the Martian globe is retarded from the viewpoint of an Earth observer.

The rate of diurnal rotation is $350^{s}.89217$. From 1 day to the next, the retrogression in longitude amounts to about 10° for the same hour of observation.

There is always a correction to make to link this difference in sidereal rotation to the apparent or synodical rotation of Mars arising from the displacement of the Earth relative to Mars, a correction which is sometimes additive and sometimes negative, according to the position. One astronomer who has devoted much time to these useful ephemerides, M. Marth, has calculated them for each opposition, and I have made it a duty to publish them in *L'Astronomie*.

In the Martian calendar, we have, over three years, a long year or 669 days and two short ones of 668 days each, or one leap year in three-less simple than ours, because three times 668.6 does not give exactly the same number as $(2 \times 668) + 669$. As with us, the calendar would have to be reformed from time to time to give a perfect result.

On Mars, day and night follow much the same cycle as on Earth. At the equator they are of equal length—$12^h\ 19^m\ 47^s$ throughout the whole year. It is the same all over Mars at the time of the equinoxes. The length of the day increases with latitude, in each hemisphere, up to the pole at the corresponding solstices, and at each pole it attains half a Martian year, or 334 days, at the time of its summer solstice.

The 88,755 seconds making up the civil day on Mars are in the ratio of 1.26:1, to the Earth's civil day of 86,400 seconds. The terrestrial day is equivalent to 0.97 Martian days.

Geography of Mars, or Areography

All the observations brought together in this work show that the globe of Mars is characterized by a diversity of sombre patches and bright patches fixed to the surface. Over two centuries we have had concordant results. Mars is the only planet in the Solar System whose geography can be studied. Venus, Jupiter and Saturn are constantly enveloped in clouds, and on the other planets we can see nothing with real certainty.

If we consider the regions of the planet in general, we can divide them into two classes. The first class includes the bright countries, showing a colour which is ordinarily dark yellow or orange, but which can vary momentarily, and according to location, sometimes through all shades of yellow to pure white; sometimes through all tints between orange red and dark red which may be compared with a well-baked brick, or, better perhaps, well-worn leather. The second class is made up of the dark regions which constitute patches in the proper sense of the word, and whose fundamental colour appears a sort of iron-grey tinted with green, showing all gradations from black to ashy grey. In general, the regions in the second class seem to be darker than those of the first; but it sometimes happens that in the colour changes which affect certain parts of the planet, patches in Class I may become dark red and those in Class II may brighten; we cannot then decide which are the brighter and which are the darker. In fact, we are dealing with differences in colour rather than with differences in luminous intensity. Nevertheless, the distinction between the two classes is almost permanent (though there are a few exceptions).

C. Flammarion, *Camille Flammarion's The Planet Mars*, Astrophysics and Space Science Library 409, DOI 10.1007/978-3-319-09641-4_7

The secular permanence of the patches on Mars should not be equated absolutely and rigidly with that of the patches on the Moon. Assiduous observation has shown that various regions on the surface of Mars show changes of tint within certain limits, and that the rays of the Sun are reflected with a different intensity according to the time. The outlines of the sombre patches appear to be subject to displacements which, it is true, are very small compared with the size of the planet and the sizes of the patches themselves—but are at least quite definite. Moreover, the clarity of their contours is sometimes sharp, sometimes less precise. Many of the fine details are more easily visible at certain epochs than at others, even when we take into account differences in observing conditions; the details seem to be subject to relatively well-marked changes in aspect, though they are never marked enough to make the identity of the feature doubtful. Finally, Mars has an atmosphere, and this produces a variety of phenomena which we may regard as meteorological by analogy with the Earth—though it is true that on Mars the phenomena are very different.

All these changes make the study of Mars much more interesting than would be the case if the surface were unchanging. As Schiaparelli wrote[1]:

> This planet is not a desert of arid rocks. It lives; the development of its life is revealed by a whole system of very complicated transformations, of which some cover areas extensive enough to be visible to the inhabitants of the Earth. We have to explore a world containing new things, eminently well suited to arouse the curiosity of its explorers, and to provide an abundance of work for telescopes over many years. In effect, these phenomena differ so much, and are so diverse in their details, that we can recognize them as regular only after a series of rigorous and complete studies; this is the sole way to draw precise conclusions and to give plausible explanations about the modifications, and about the physical constitution of Mars.

It cannot be disguised that such studies are beset with many difficulties if they are to be exact and complete. Among the variations which occur on the planet's surface, some take place slowly (as, for example, the periodical waxing and shrinking of the brilliant polar snows), and show phases which are relatively easy to follow. But there are also changes of another kind, some accomplished in a few days, others almost sudden, and their effect is shown from 1 day to another; such is the enigmatical duplication of the canals. Then there are variations whose period clearly depends upon the planet's annual revolution. To understand the mechanism of these changes, it is necessary to make a series of uninterrupted observations for at least a period as long as a Martian year. This condition is imposed not only by the necessity of exploring the north and south polar patches at times when the inclination of the axis is most favourable from the observational viewpoint, but also, equally, because some parts of the phenomena in question are dependent upon the Martian seasons.

---

[1] *L'Astronomie*, January 1889, p. 20.

Actually, such complete control is not possible for an isolated observer; it would be impossible even for several observers, if they lived in a restricted area on the Earth's surface—Europe, for example. On the rare days of good seeing, only 2 or 3 hours can be properly utilized, during dusk or dawn. It follows that on any given day, the observer seldom has the chance to study more than a quarter of the planet really well; and as the rotation period of Mars differs from that of the Earth, the displacement of the regions available for observation takes place slowly from 1 day to another, so that the same region of the planet can be observed from 8 to 10 evenings consecutively. But to return to the same aspect of the patches at the same terrestrial hour takes the longer period of about 38 days. Consequently, a region which has been observable for 8–10 days consecutively (assuming that the Earth's atmosphere is clear) will then be inaccessible for an entire month, and at the end of this time a close inspection will sometimes show considerable changes—though it is impossible to tell exactly when they took place. If, moreover, the weather has been bad during the 8–10 days when the region has been accessible (as often happens), 2 months will pass before it can again be examined; it is even possible for an entire opposition to pass by without a particular part of the planet being favorably seen. To overcome these difficulties, the only solution is to distribute a number of observers around the Earth's surface, so that during all apparitions of Mars there will be at least one observer who will see the planet high enough above the horizon to obtain a good image.

This is not all. Effective observations of Mars can be made only when the planet is sufficiently close to the Earth. To observe the most difficult details (which are also the most interesting), the diameter must be at least 10–12 in (254^{mm}–305^{mm}). This condition is fulfilled only for 3 or months around opposition; and these circumstances occur only at intervals of about 26 months. Therefore, each opposition can provide us with information about the state of the planet for only a tiny fraction of its revolution period. Happily, this arc of the orbit is not always the same, because successive opposition points are displaced by about 48° heliocentric longitude along the Martian orbit.

It will be seen, therefore, that to follow the planet through all possible inclinations of its axis, and through all possible seasons, there must be a cycle of seven to eight consecutive oppositions—a cycle whose mean length is 16 years. If the Martian phenomena are perfectly periodical, and dependent upon the revolution round the Sun, we could hope to write a complete history of them by means of observations carried out over one of these cycles. But the periodicity appears to be only approximate, as with terrestrial meteorology.

The obstacles described above are of a purely astronomical character. Those due to bad weather and the unsteadiness of the Earth's atmosphere are much more serious. From his experiences at Milan, Schiaparelli has stated that he can hope for good seeing only 8–10 evenings; sometimes, even complete months pass by without a single satisfactory observation. Evenings with perfect images are even rarer—that is to say, those times when the maximum power of a telescope can be employed.[2]

However this may be, it may be hoped that by studying the climatic conditions in relation to the sharpness of telescopic images, we will eventually be able to reduce these obstacles to a minimum. Finally, experience has shown that the difficulty of coordinating the results obtained by different observers using different instruments is itself a severe handicap; this difficulty will be removed only when photography becomes sufficiently advanced to record details as fine as those directly observable with the help of a good telescope.

The dark and bright patches—permanent, and never changing from the one to the other—should be different in nature. On the Martian surface there are liquids

---

[2] Requirements for the Observation of Mars.—While the weather on Mars is usually fair, and its clear atmosphere presents few obstacles to the observation of the surface, the differences in tone are often slight, and the outlines of the features so vague and uncertain—even after hours of exceptionally good seeing—that satisfactory results can only be obtained with very rigorous methods of observation.

The first point, naturally, is to have a good objective—or a good mirror, for a reflector. The size of the instrument is, relatively, of secondary importance. Excellent images can be obtained with small refractors of 108^{mm}, 95^{mm} and even 75^{mm} aperture, while large reflectors of a metre and more in aperture can sometimes give only mediocre views, making the features almost impossible to identify. Therefore, if an instrument is to be useful, it must be really good.

The second point is to have the instrument at the same temperature as the surroundings. If the telescope is used in the open air, all is well. But with a telescope shut in a dome, the slit, windows and doors must be opened several hours before the start of observing. The waves of warm air which pass in front of the objective, and which increase in proportion to the power of the eyepieces used, constitutes a major obstacle to the clarity of the images.

Thirdly, it is important not to forget that when the first two conditions are fulfilled, the desired precision cannot be attained under our normal atmospheric conditions, because even when the air appears perfectly clear it is crossed by layers of heterogeneous density, which produce currents and disturb the view. Therefore we must wait, sometimes for several hours, for a fleeting moment of absolute calm and perfect seeing.

The atmosphere is most transparent during the hours before a thunderstorm.

Daylight hours, dawn and dusk, always seem to me to be better than the hours of night.

Also, to obtain the best views without being influenced by any preconceived ideas, it is important not to know the view of Mars which will be presented when observing begins; the greater one's ignorance, the more one will see. It is not long before the features will be recognized. If observations are made on several consecutive days, one cannot help knowing what will be on view, no matter how much one tries to forget it. The best course is to think of nothing, and concentrate solely upon what can be seen. In this way, sketches or drawings will have their greatest possible value. If you calculate the longitude of the central meridian in advance, and have a globe or map of Mars in front of your eyes, you will see what you expect to see, and will lose the pleasure of making your own discoveries. However, it is by no means a bad idea to work out identifications immediately after you have completed an observation.

The eye will become accustomed to instrumental conditions, and experience will bring improved results. One's first observations are not generally satisfactory. From day to day, you will see more. To make the best drawings of Mars, you must make the eye used to all aspects of the planet, over many days. Generally, one's first drawings will be without value, and will resemble the primitive drawings made by the old observers.

and solids because we can observe snows, mists and water vapour, and if the surface were entirely liquid there would be no permanent patches. But are the watery areas represented by the dark or by the bright patches?

Everything leads us to believe that the liquid areas are dark. First, water—and liquids in general—absorb more light than continental surfaces, at least if the lands are covered with vegetation. Also, each year we see the melting of the Martian snows, and this melting produces a dark border around the polar cap. In the third place, the breaking-up of the coasts into gulfs and capes agrees best with the interpretation of the dark patches being seas. In the fourth place, the observer variations—the broadening and the curtailments—agree better with the first interpretation than with the second. In the fifth place, the changes in tone so frequently observed in the dark patches, from inky black to bright grey, and the movements of these tones, also indicate a liquid element rather than solid ground. Thus, for example, in the years when the Terby Sea appears as black as ink—so dark that there have been proposals to use it as the zero for Martian longitudes, rather than the Meridian Bay. Well! In this year, 1892, over the 3 months when I have observed it, it has been vague, greyish and indefinite.

We will therefore assume that the bright regions are continents and that the dark regions are seas.

On this interpretation, the Martian geographical distribution is very different from ours.

On Earth, three-quarters of the surface are covered by water, and only one- quarter of the globe is habitable by man.

On Mars, the distribution is less unequal, since the two elements each occupy about half the total extent of the globe. There is only a slightly greater area of land than of sea—77,000,000 square km as against 66,000,000 square km of water. Discounting the polar circles, the habitable lands of Mars represent a surface area six times greater than that of Europe.

But neither chemically nor physically do these waters seem to be the same as ours. On Mars, the phenomena show no analogy with those of the Earth. We must speak of observed variations in the appearance of Mars. This is an essential point which must be incorporated into the facts given below, and which will be described in the following chapters.

What is the cause of the reddish colour of the continents?

We have seen that it is not due to the atmosphere.

It is the colour of the visible ground—that is to say, of the surface.

We can therefore assume that the surfaces of the continents are sterile and sandy, without any vegetal covering, where the general colour is reddish.

To fall in with the first viewpoint, Mars must be considered as an eternally arid desert, admitting that the atmosphere, the water and the sunlight play a role diametrically opposite to that on our own globe, and, briefly, that the combinations of these elements remain absolutely unproductive, while those which make up our immensely varied vegetable and animal life fill up the waters and develop in all places with such prodigious abundance.

To us, it seems impossible to condemn a world to a fate of this kind, above all, a world in which all the elements of life can be seen joined together—as on Mars.

The colouration of the continents is, very plausibly, due to a vegetal covering formed on the surface.

This colouration is not as red as is generally believed. It has already been compared with fields of wheat seen from the vantage-point of a balloon. It varies from one land to another, and even over the same regions. Thus in the present year, 1892, the Beer Continent, to the right of the Hourglass Sea, does not appear as red as the Herschel Continent, to the left of the Hourglass Sea; and Lockyer Land, which on my map is dark, seems this year to be as bright as the continents.

Why, we may ask, is not the Martian vegetation green?

Why should it be?—is the reply. From this point of view, there is no reason to regard the Earth as typical in the universe.

Moreover, the terrestrial vegetation can itself be reddish, and has been so for the majority of the continents; the first terrestrial plants were lycopods, whose colour is a Martian reddish-yellow. The green substance which gives our vegetation its colour—chlorophyll—is made up of two elements; one green, the other yellow. These two elements can be separated by chemical processes. It is therefore perfectly scientific to admit that under conditions different from those on Earth, the yellow chlorophyll can exist alone, or be dominant. On Earth, the ratio is 1 to 100. The contrary could be true on Mars.

The most probable cosmological theory regards Mars as being in a later stage of evolution than the Earth, and more advanced in its career.

Helmholtz has calculated that the Sun's condensation from the primordial nebula produced a temperature of 28,000,000 °C. Applying the same principles of the Earth and Mars, Schiaparelli finds that the heat of condensation should be 8,988° for the Earth and only 1,795° for Mars. Therefore, Mars should long since have become cold at its centre.

We know moreover that the internal heat of the Earth's globe has no influence upon the surface temperature, nor upon the phenomena of vegetable and animal life.

The antiquity of Mars explains very naturally the greater scarcity of waters upon the surface, and the probable levelness of the continents.

We have seen that the seas are of Mediterranean shallowness. The intermediate regions are sometimes dry and sometimes flooded, or perhaps mist-covered. As islands, note the snowy Hall Island, lying in the De la Rue Ocean, at longitude 47° and latitude 22° (refer to my map, Fig. 31, Chap. 1), which is sometimes visible when covered with snow or clouds, and sometimes invisible; Schiaparelli calls it Proteus. Secondly, note Dawes island, also called Jacob Land and Argyre, lying above Hall Island. Nix Atlantica and Nix Olympica seem to be peaks sometimes covered with snow, lying in wide lands, the former at longitude 267°, latitude 17°, the latter at longitude 128°, latitude 21°N.

We have already seen that chains of mountains are rare. Everything seems to be level. However, there are some peaks, as shown by the whitish patches observed on the terminator. The right-hand coasts of the Hourglass Sea, up to Herschel Strait, seem to be cliffs rather than beaches; because for one thing the extensions of the sea always occur at the left-hand coast, and moreover the frequently seen white borders indicate snow, white ice or cloud.

The northern hemisphere of Mars is higher than the southern; the seas seem to be mainly in the south. The cause may be due to the effect of the Sun's attraction on the Martian and terrestrial hemispheres which are closest to the Sun during the half-period of revolution of the line of apsides at the critical epoch of the solidification of the crust.

The Atmosphere of Mars—Martian Climatology and Meteorology: The Conditions of Life on Mars

The existence of the atmosphere of Mars has been rendered absolutely certain by the totality of the observations. The evidence is of varied nature and strength.

First, we note that the polar caps, whose extent varies with the seasons and in proportion to the amount of solar heat received, are conspicuous. It is not important whether these snows are chemically the same as ours, or are different. The very fact of their existence proved that on Mars there are movements of vapours which condense as white deposits and then vanish under the influence of heat, to re-form again during the cold season. Vapours are sometimes invisible, as in our own atmosphere; water vapour is sometimes visible in the form of light cloud, sometimes condensed as variable snows, and incontestably as liquid in the seas. The Martian polar snows are by themselves sufficient to prove that the planet is surrounded by an atmospheric envelope.

True clouds—immense, opaque agglomerations such as those which extend in the Earth's atmosphere—are not often seen on Mars. But we can observe mists, rather light in appearance, sometimes semi-transparent, which frequently veil vast countries, particularly in winter. These mists—these hindrances to visibility—are the second evidence of the existence of an atmosphere.

The third evidence is offered by the luminous gradation of the disk, from the centre toward the edge. The interior circumference of the Martian disk is whitish, and the sombre patches disappear when they reach this pale ring. As the sunlit hemisphere is exactly turned toward us at the time of opposition, it is natural to attribute this circular whiteness to an increase in the thickness of the atmosphere relative to our line of sight, and to the sunlight reflected from the thicker atmospheric layer. We could, it is true, suppose that during the night the globe of Mars is covered with a layer of white frost, which melts when the Sun rises high, but as this whitish ring extends over the aqueous regions as well as the continents, and as the patches fade and disappear upon reaching the border, this second hypothesis is not tenable. In passing, we may note that the contrast between the white border and the general yellow-ochre tone of the continents is better seen during the night than during the day

C. Flammarion, *Camille Flammarion's The Planet Mars*, Astrophysics and Space Science Library 409, DOI 10.1007/978-3-319-09641-4_8

(this is an observation which I made very recently—on 1 July 1892—when the ring was most evident from 3 o'clock in the morning until sunrise). The atmosphere of Mars is certainly the cause of this effect. It could be due to light mists, to refraction by layers of atmosphere lying beyond the tangent of our visual line of sight around the edge of the globe—refraction which lifts up the images and slightly increases the size of the globe by a whitish ring.

This luminous degradation, of varying intensity and size, is due to the transparency of the Martian atmosphere; it can, it is true, be explained by reflection from mountains, as in the case of the Full Moon, where the limb is as bright as the centre. This was the explanation adopted by Zöllner[1] who assumed the surface of Mars to be studded with mountains, as with the Moon—mountains whose slopes were inclined to the horizon at an angle of 76°. But this seems rather forced, and as the presence of the Martian atmosphere can provide a better explanation, Zöllner's theory may be rejected.

Observation establishes that the patches on Mars—even the darkest of them—become, in general, invisible when they reach 53° from the centre of the disk; sometimes they may be distinguished up to 60° or even 65°, but this is extremely rare. The very white features can be followed further out; the polar snows are visible right up to the edge of the disk, their brilliancy piercing the atmospheric veil. These white regions seem to be even more luminous when near the edge of the disk than when near the centre, as, for example, the islands of Thyle, Argyre and Hellas (Schiaparelli, 1877). On more than one occasion these regions have been mistaken for polar snows. However, this year Hellas, or Lockyer Land, which is more favourably presented than usual, is on the contrary particularly bright.

Spectral analysis has now come upon the scene to confirm the existence of the Martian atmosphere, and has moreover revealed absorption lines due to water vapour. The researches of Huggins, Vogel and Maunder are in agreement about this; notably, for the absorption line at wavelength 628 [nanometers], and for line 656. There are a dozen groups which may be satisfactorily identified. We must, then, acknowledge that water vapour exists there, and is undoubtedly the same chemically as water on Earth. This does not mean that other substances in the planet's atmosphere do not also exist.

The existence of the Martian atmosphere is therefore demonstrated conclusively. This atmosphere differs from ours in several respects.

First, it is not charged with clouds such as those that we breathe. We do not see immense veils, as on Earth, lasting for days, weeks or months. In general, the Martian atmosphere is transparent, and it very seldom prevents us from seeing the surface features. When we can see nothing on Mars, the trouble nearly always lies in our own atmosphere.

Seen from afar, the Earth must be more difficult to observe than Mars; its thick atmosphere envelops it in a veil which is seldom transparent.

What has been said above applied particularly to the Marian atmosphere in summer. At this time, the sky is almost permanently clear over all latitudes. It is the same, in

[1] *Photomerichsche Untersuchungen*, p. 127; Leipzig, 1865.

general, at the equator. But the atmosphere is often veiled above the hemisphere which is experiencing its winter. Thus, for example, at the present moment (July/August 1892) it is summer in the southern hemisphere of Mars; the summer solstice occurs on 13 October. In the northern hemisphere, on the other hand, it is autumn, with winter approaching, and—well! one can see appreciably less detail in this hemisphere than in the other.

It seems that cold produces veils in the Martian atmosphere, while heat clears them away.

The eastern limb of the disk generally appears whiter than the western. This is the morning or sunrise meridian; perhaps we have here light morning mists which later dissipate.

Seen from Earth, what effects will be produced by clouds floating above a planet?

When clouds are projected against the dark patches of Mars, they show up in guise of diffuse lines of variable form. They can become as brilliant as the brightest regions on the planet. In other cases, they can become less bright, but always brighter than the background against which they are projected.

"These tones probably indicate not the particular colour of the clouds," wrote Schiaparelli, "but we see them as dark against brighter backgrounds, and the result is not greater atmospheric transparency, but a veiling of the details. This is what I have observed, notably in Mare Erythraeum and the land of Noachis."

The atmospheric mists brighten progressively over the equatorial lands from the southern summer solstice to the following autumnal equinox. The atmosphere appears brighter above the interior seas during the months immediately following the southern solstice.

In the zone between south latitude 10° and 30°, which receives the full force of the Sun's rays at the time of the solstice, nothing can be observed analogous to the rainy zones and the equatorial calms which, in our seas, follow the Sun's movement in declination. It may even be said that we see no clouds at all. Martian meteorology does not in the least resemble ours. We can hardly ever observe tempests or cyclonic air-movements.

At the solstices, one hemisphere seems to be subject to evaporation and the other to condensation. At intermediate times, a zone of evaporation seems to be limited to the south and to the north by two zones, or rather caps, of condensation.

The declination of the Sun and the distribution of the lands and seas influences the breadth of these zones following the seasons. The seas have a sky which is generally relatively clear; on the other hand, atmospheric humidity tends to condensation over the islands and lands.

Here are the observations: atmosphere generally clear, few clouds; mists during winter, perhaps in the morning and probably at night; and often, without doubt, white frost.

And here is the theory to be derived from the observations:

Since gravity at the surface of Mars is much weaker than that on Earth (0.376), all bodies there weigh proportionately less; the atmosphere being no exception. If each square metre of the surface of Mars supported the same atmosphere as ours, the pressure of the atmospheres would be reduced in the same proportion; that is to say, on Mars a barometer, instead of reading 760^{mm} at sea- level, would register only 285^{mm}. This is the pressure we experience in a balloon at an altitude of $8{,}000^{m}$, which is the height of our highest mountains. On the summit of Mont Blanc, the pressure is 424^{mm}.

It is certain that the atmosphere of Mars in not analogous to our own, and that the water there does not exist under the same conditions; but the surface of the planet can rise above freezing-point, even bearing in mind the greater distance from the Sun, and we are not observing any icy globe. On the contrary, we see limited snow-areas on Mars, and these limits vary with the temperature. If we consider a hemisphere of Mars during its summer, it no longer has snow at its pole, as with Earth; the snow-area is much less. Such is what we see from one time to another in various regions; the temperate regions also have their snow melted.[2]

According to the observations upon which these calculations are based, we should therefore believe that the atmosphere of Mars is less dense than ours; that the clouds are less common, that the air-currents are of lower intensity, that winds are never very strong, and that storms are absent. The conditions of density and pressure are very different from those on Earth. The zero point 0°, at which water solidifies, is not the same as it is here. The atmosphere is not the same either chemically or physically. The mean temperature should be higher than on the Earth. The observed effects correspond to a degree of surrounding warmth which is greater than might be thought.

If we represent the Earth-Sun distance by 1,000, that of Mars will be 1,524. The solar light and heat received varies in inverse ratio to the squares of these numbers. Mars therefore receives only 43/100ths—that is to say, less than half—of the solar heat received by the Earth. Because of Mars' great orbital eccentricity, this proportion is a little greater at perihelion and a little less at aphelion, in the ratio of 5 (aphelion) to 6 (mean) and 7 (perihelion).

Obviously, then the atmosphere of Mars behaves differently from ours, since the meteorological effects which we observe on Earth are due chiefly to the temperature.

Water vapour can by itself lead us to an interesting result. We know that a molecule of water vapour is 16,000 times more effective than a molecule of air in shutting in the heat received from the Sun. As he percentage of water vapour in the Martian atmosphere is greater than in ours, a higher temperature results.

---

[2] Very recently (*Knowledge*, June 1892), in reply to a letter from W. H. Pickering about the polar ices of Mars, A. C. Ranyard wrote: "We are, it seems, forced to admit either that the polar caps of Mars are not made up of snow, or that the mean temperature of the equatorial and temperate regions of Mars is above zero, i.e. warmer than would be indicated by the distance from the Sun. This fact can be explained if we admit that the Martian atmosphere is denser than ours. But perhaps the caps are not made up of snow. White crystals of carbon dioxide could evaporate at a temperature much lower than the greatest cold that we experience on Earth. I believe, however, that it is simpler to admit that the density of the Martian atmosphere corresponds to that of the planet, and that it is in consequence less than ours; but there is abundant water vapour, and perhaps other vapours which contribute to the storing-up of the warmth received from the Sun."

Yet water is not the only substance to possess this property. Vapours of sulfuric, formic and acetic acid, ethyl iodide, chloroform and carbon bisulphide can all act in the same way. Each substance with analogous properties in the atmosphere of Mars could suffice to explain this remarkable climatology.

To sum up: things happen on Mars as though the mean temperature were almost the same as it is here, whether because thermometry really differs there, or because Nature has there adapted itself to a lower temperature. On Mars, a lower temperature would be the normal average, and able to produce greater effects than it would for us.

The climate and the conditions for life on Mars do not, then, seem to diverge so much from those of the Earth, as do those that prevail in the various parts of our own globe.

"But when we consider how easily Man can adapt himself to tropical or Arctic climates," writes an English astronomer,[3] "and when we compare the temperatures in which the Greenlanders and the Esquimaux live with those in the countries of the Papuans or the negroes of Central Africa, we are led to think that only a slight change in the human organism would suffice for adaptation to conditions on Mars:

The same author adds that the inhabitants of Mars should be larger and stronger than ourselves, because the weakness of the gravity would make a man 5^m in height as agile and light as Earthmen of less than 2^m. Such large bodies could also maintain a higher temperature than ours. They would also need larger eyes because of the reduced light.

All conjectures as to the form of the inhabitants of Mars are evidently premature. Interesting as they may be, they have no place in a technical work such as this.

[3] E. LEDGER, *The Sun, its planets and their satellites*.

The Seasons of Mars

The revolution period of Mars round the Sun, or its sidereal year, is, as we have seen earlier, $686^d\ 23^h\ 30^m\ 40^s$, or 686.979, or 687 days to within an accuracy of 2/100. The orbit is markedly elliptical, with an eccentricity of 0.093. The major axis, or line of apsides, from perihelion to aphelion, lies in the direction 334° (perihelion) to 154° (aphelion). The summer solstice of the southern hemisphere lies at heliocentric longitude 356°48′, and the line leading from this solstice to the opposite solstice lies in the direction 365°′48′ to 176°48′ (see Fig. 1, Chap. 4).

The inclination of the Martian equator to the plane of the orbit is 24°52′. The seasons are therefore essentially of the same type as ours, since the inclination of the Earth's equator is 23°27′. The difference is only 1°25′. If our planet had the same inclination as that of Mars, our seasons would be scarcely changed, though they would be a little more pronounced.

If, in the orbit of Mars, we draw a line perpendicular to the line of the solstices, passing by the Sun, we will find the positions of the equinoxes, or times when the Sun lights up a full hemisphere of Mars reckoning from the poles. These two points are respectively at heliocentric longitudes 86°′48′ and 266°48′.

This line of the equinoxes divides the orbit of Mars into two unequal parts, because of the orbital eccentricity. As the southern summer solstice is in the neighourhood of perihelion (at an angular distance of 23° or 36 days), the section to the right (in Fig. 1, Chap. 4) subtends a shorter arc than the section to the left; moreover it is covered more rapidly, according to the Law of Areas. It follows that there are strongly-marked inequalities in the lengths of the seasons.

(1) To go from the equinox of southern spring to the southern autumnal equinox, the planet takes less time than it does to go from autumnal equinox to spring equinox. This section of the orbit is divided into 306 days, while the other section requires 381 days. The difference therefore amounts to 75 days.
(2) The section most rapidly covered is that from southern spring equinox to the following summer solstice, because this section includes the rapid motion past perihelion. The length of this section is only 145 days.

C. Flammarion, *Camille Flammarion's The Planet Mars*, Astrophysics and Space Science Library 409, DOI 10.1007/978-3-319-09641-4_9

(3) The season which comes next in length is that from southern summer solstice to autumnal equinox; it amounts to 160 days.
(4) The two other seasons are, respectively, those from southern autumnal equinox to the following winter solstice, opposite to the first and therefore the longest—199 days, while the second lasts for 181 days. Taking fractions of days into account, we have:

Lengths of seasons on Mars, in terrestrial days	
1. South spring equinox to the following summer solstice	145.6
2. South summer solstice to the following autumnal equinox	160.1
3. South autumnal equinox to the following solstice	199.6
4. South winter solstice to the following equinox	<u>181.7</u>
	687.0
Warm season in the southern hemisphere	305.7 days
Cold season in the southern hemisphere	381.3 days

Here are the dates of the solstices and equinoxes for the last Martian cycle, from 1877 to 1892.

Calender of Mars

Dates of seasons, solstices and equinoxes of the planet Mars

S. Hemisphere	N. Hemisphere	Dates	Longitude noon	Interval days	Opposition
Summer solstice	Winter solstice	27 Sept. 1877	357° 1′	160	5 Sept. 1877
Autumn equinox	Spring equinox	6 Mar. 1878	86° 58′	199	
Winter solstice	Summer solstice	21 Sept. 1878	176° 46′	182	
Spring equinox	Autumn equinox	22 Mar. 1879	267° 0′	145	
Summer solstice	Winter solstice	14 Aug. 1879	356° 25′	161	12 Nov. 1879
Autumn equinox	Spring equinox	22 Jan. 1880	87° 1′	199	
Winter solstice	Summer solstice	8 Aug. 1880	176° 48′	182	
Spring equinox	Autumn equinox	6 Feb. 1881	267° 3′	146	
Summer solstice	Winter solstice	2 July 1881	357° 5′	160	
Autumn equinox	Spring equinox	9 Dec. 1881	87° 3′	199	26 Dec. 1881
Winter solstice	Summer solstice	26 June 1882	176° 49′	182	
Spring equinox	Autumn equinox	25 Dec. 1882	267° 5′	145	
Summer solstice	Winter solstice	19 May 1883	356° 31′	160	
Autumn equinox	Spring equinox	26 Oct. 1883	86° 35′	200	
Winter solstice	Summer solstice	13 May 1884	176° 52′	181	31 Jan. 1884
Spring equinox	Autumn equinox	10 Nov. 1884	266° 32′	146	
Summer solstice	Winter solstice	5 Apr. 1885	356° 33′	160	
Autumn equinox	Spring equinox	12 Sept. 1885	86° 38′	200	

(continued)

(continued)

Dates of seasons, solstices and equinoxes of the planet Mars					
S. Hemisphere	N. Hemisphere	Dates	Longitude noon	Interval days	Opposition
Winter solstice	Summer solstice	31 Mar. 1886	176° 54′	181	6 Mar. 1886
Spring equinox	Autumn equinox	28 Sept. 1886	266° 34′	146	
Summer solstice	Winter solstice	21 Feb. 1887	356° 34′	160	
Autumn equinox	Spring equinox	31 July 1887	86° 40′	200	
Winter solstice	Summer solstice	16 Feb. 1888	176° 57′	181	11 Apr. 1888
Spring equinox	Autumn equinox	15 Aug. 1888	266° 37′	146	
Summer solstice	Winter solstice	8 Jan. 1889	356° 36′	160	
Autumn equinox	Spring equinox	17 June 1889	86° 41′	199	
Winter solstice	Summer solstice	2 Jan. 1890	176° 32′	182	27 May 1890
Spring equinox	Autumn equinox	3 July 1890	266° 39′	146	
Summer solstice	Winter solstice	26 Nov. 1890	356° 40′	160	
Autumn equinox	Spring equinox	5 May 1891	8 6° 44′	199	
Winter solstice	Summer solstice	20 Nov. 1891	176° 34′	182	4 Aug. 1892
Spring equinox	Autumn equinox	20 May 1892	266° 41′	146	
Summer solstice	Winter solstice	13 Oct. 1892	356° 41′		

In this table, apart from the dates of the equinoxes and solstices, I have calculated the intervals, in days, which correspond to the times when the planet passes the heliocentric longitudes for each position at noon; and I had added these with respect to the opposition dates. It is seen that the 1877 opposition was reached 22 days before the summer solstice of the southern hemisphere; that of 1879, 89 days later; that of 1881, 17 days after the autumnal equinox; that of 1874, 103 days before the winter solstice; that of 1886, 15 days before; that of 1888, 55 days after; that of 1890, 37 days before the autumnal equinox; and that of 1892, 76 days after. The series of the oppositions thus travels over the whole cycle of the seasons, and at each opposition the observer is presented with a different aspect.

Since the inclination of the globe of Mars is 24°52′, and the heliocentric longitude of the planet at its solstices is 356°48′ in the southern hemisphere and 176°48′ in the northern; it follows that the inhabitants of Mars, at the summer solstice in the southern hemisphere, see the Sun at the zenith at latitude 24°52′ south, longitude 176°48′, while at the northern summer solstice the zenith position is latitude 24°52′N., longitude 356°48′. The first of these points lies in the constellation Leo, and the second in Aquarius. Instead of the Tropics of Cancer and Capricorn, the analogous circles of the Martian celestial sphere are the Tropics of Leo and Aquarius. It is superfluous to discuss the real names used by our neighbours in the heavens.

The lengths which I give below for the Martian seasons are expressed in terrestrial days. It is no less interesting to calculate them in Martian days. Now, we have seen that the Martian sidereal date is $24^h\ 37^m\ 22^s.6$, and the civil day $24^h\ 39^m\ 35^s$. There are thus 668.6 civil days in the Martian year.

668.6 Martian days in the tropical year are distributed as follows

Lengths of the seasons, in Martian days		
1. From the south spring equinox to the following summer solstice	142	298
2. From the south summer solstice to the following of autumnal equinox	156	
3. From the south autumnal equinox to the following solstice	194	370
4. From the south winter solstice to the following equinox	176	

The axial inclination of Mars is very little different from that of the Earth, so the seasons are of essentially the same type, though almost twice as long. The temperatures, however, depend upon the constitution of the atmosphere. If, for example, Mars were devoid of atmosphere, the ground would not be warmed at all, and would remain frozen even in midday, as are the summits of Mont Blanc, the Andes or the Cordilleras—more so, in fact, because of Mars' greater distance from the Sun.

But Mars is enveloped in an atmosphere containing considerable amounts of water vapour, and in which other vapours or gases may also exist. From this, it follows—as is confirmed by observation—that the temperatures (and thus the conditions there) are comparable with ours. Does the very considerable orbital eccentricity play a role in the intensities of the seasonal variations in each hemisphere?

First, for comparison, let us examine the seasons of the Earth.

Seasons and Climates in the Two Terrestrial Hemispheres

The Earth's orbit is not a circle, but is a definite ellipse. The eccentricity—that is to say, the distance of the centre of the elliptical orbit from the focus, relative to the semi-axis major—is 0.01677. Adopting 149,000,000 km. (parallax = 8″.82) for the semi-axis major, or mean distance, we see that this value is 2,498,730 km. The Earth is therefore 4,997,460 km nearer to the Sun at the perihelion than at aphelion. We have for these distances:

Perihelion distance	0.98323	=146,501.270 km
Mean distance	1.00000	=149,000,000 km
Aphelion distance	1.01677	=151,498,730 km

Since the eccentricity is about 1/60 of the mean distance, the Sun is about 1/30 times nearer to us at perihelion that at aphelion. As the solar light and heat is radiated in all directions on the surface of a sphere which continually increases in size, the intensity of this light and heat falls off because of the increase in size of the surfaces on this sphere—that is to say, because of the square of the distance. Consequently, the hemisphere exposed to the Sun receives, at perihelion, 2/30 or 1/15 more heat than it does at aphelion.[1] The ratio is almost 106.5 to 100.

The Earth passes through perihelion on 1 January and through aphelion on 1 July. The winter in our hemisphere occurs when the Earth is at its closest to the Sun, and the summer when it is furthest way. This means that our winters are less cold than

[1] $(61/59)^2$ = almost $(62/60)^2 = (31/30)^2 = 96/90$ or 16/15.

they would be if they occurred with the Earth on the other side of the ellipse, while the summers are less warm. It is the contrary in the southern hemisphere. It seems, therefore that south pole ought to have more winter snow than the north pole, but less in summer.

We should not immediately conclude that the winters in the southern hemisphere are colder than ours in the northern hemisphere, because the absorption of the solar heat receive depends largely upon the nature of the surface upon which it falls. If the surface is an ocean, for instance, the waters of our seas can never rise to a temperature of more than 29°, even in the tropics (Humboldt, *Mélanges*, p.441). If the surface is forest, the temperature can go up to 36° or 40°; but the surface is dry sand, it can attain 60° or even 70°. The temperature distribution is therefore regulated by the conducting powers of the waters, by the wind régime and by the atmospheric humidity. Thus we ought never to rely upon an exclusively geometrical estimate. The southern hemisphere of the Earth is essentially aquatic.

Thus, at the winter solstice of our hemisphere, which actually occurs near perihelion and is only 11 days away in position, the terrestrial globe receives, during the daytime, the maximum amount of heat that it can ever receive from the Sun. At the December solstice, it is the south pole which is turned sunward, and it is the southern hemisphere which has the longest days. Geometrically speaking, the southern summers ought therefore to be hotter than ours. At the June solstice, it is our hemisphere which is exposed to the Sun, but the Earth is then at its greatest distance from the Sun; thus we ought to have cooler summers, while the southern hemisphere ought to have colder winters.

But here we must take note of another consideration.

While the Earth is closer to the Sun at perihelion than at aphelion, and so receives more heat and if we consider a complete half of the orbit—for instance, the half contained between the September equinox and the March 8 equinox—we find that the quantity of solar heat received is the same for the two halves: the Earth receives no more heat between 21 September and 20 March that it does between 20 March and 21 December.

Undoubtedly there would be a difference if the seasons were equal in length. But the movement of our planet along its orbit is not uniform. The angular speed of the Earth in its orbit varies in inverse ratio to the square of the distance, *with precisely the same relation holding for both light and heat*. It follows that the same quantities of heat are received from the Sun for the same arcs traversed when it is the section of the ellipse which is being considered.

The lengths of the seasons are unequal because of the variation in orbital speed of the Earth. Its movement is most rapid at perihelion, slowest at aphelion. If it receives more heat in the first position than in the second, it passes the position quickly and stays under the influence of the solar rays for a shorter period. Here are the actual lengths of the seasons, taking our hemisphere as a starting-point:

(Northern Hemisphere)		
Spring	$92^d\ 21^h$	$186^d\ 11^h$
Summer	$93^d\ 14^h$	
Autumn	$89^d\ 19^h$	$178^d\ 19^h$
Winter	$89^d\ 0^h$	

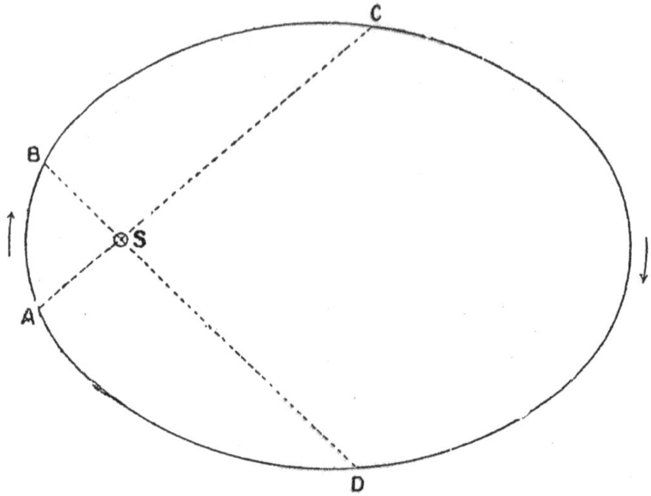

Fig. 1 Equality of heat received for equal arcs

The opposite applies in the southern hemisphere.

The Sun therefore stays 8 days longer in the northern hemisphere than in the southern. The Earth takes 8 days longer in going from the March equinox to the September equinox than from the September equinox to that of March. If we examine the globe as a whole, we find that the compensation is exact; the Earth, despite its changes in distance from perihelion to aphelion, stays longer in one section of the orbit than in the other, so that in each section it receives exactly the same quantity of heat.

Thus, the quantity of heat which the Earth receives from the Sun in each part of the year is proportional to the angle described during the same interval of time by the radius vector from the Sun. This relationship may be linked with Kepler's Second Law, according to which the arcs swept over by the radius vector are in proportion to the times taken to cover them—replacing times by angles. Thus, if in going from A can B (Fig. 1), in the section of the orbit closest to the Sun, the Earth receive no more heat in covering this 90° section than a does in traveling from C to D, also in angle of 90°, the principle is clear. However, the time taken is greater for the latter arc than for the former.

The quantity of heat received by a planet, at each point in its orbit, varies exactly as the heliocentric longitude, so that the same quantities of heat are received from the Sun over equal angles, no matter what position in orbit is considered. For instance, let us cut the Earth's orbit by a diameter passing through the Sun; we will then have 180° of longitude of each part of that line, and the same quantity of solar heat will be received along each segment. Then, passing from one equinox to another, say from A to B over aphelion (Fig. 2) or from B to A over perihelion, the planet as a whole will receive the same quantity of heat, whether it be over the perihelic or the aphelic sector.

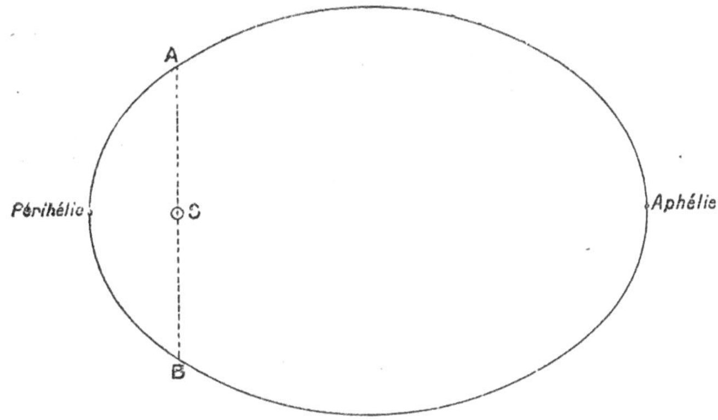

Fig. 2 In passing one equinox or the other, be it from A to B by the perihelion or B to A by the aphelion, the Earth receives from the Sun the same quantity of heat

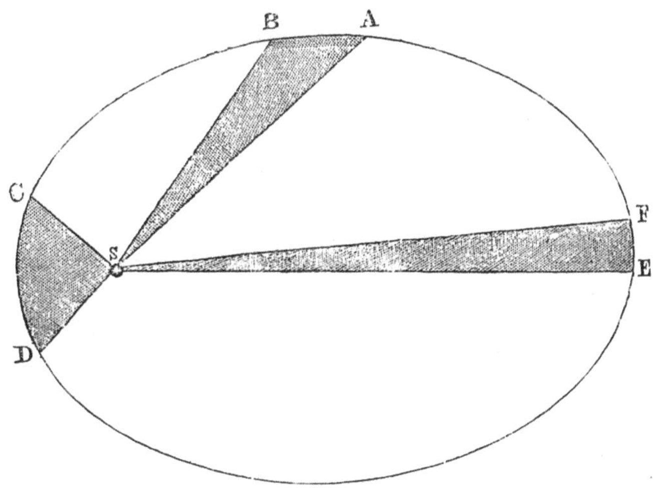

Fig. 3 The quantity of heat received is proportional to the arcs

It therefore follows that *over a full year*, each hemisphere will receive the same quantity of solar heat, whatever may be the positions of the equinoxes in the orbit and whatever may be the eccentricity.

If we next consider the arcs covered over equal times (Fig. 3), for example in a month at perihelion (CD) and a month at aphelion (EF), the quantity of heat received will be proportional to the arc; in the first case the arc is 90°, and in the second it is 60°; the proportion will be 15; that is to say, in the ellipse described, the planet will receive, in 1 month, 15 times more heat at perihelion than at aphelion.

The equality referred to above, by which principle each hemisphere receives the same amount of heat over the course of a year, does not affect the contrasts due

Fig. 4 The eccentricity of the orbit of Halley's Comet

to the differing distances at perihelion and at aphelion. The Earth receives more heat at perihelion than at aphelion, inversely as the square of the distances.

The hemisphere turned toward the Sun during the perihelic sector is favoured in the same proportion. We have seen that this proportion is 1/15.

This seems to be much too great. We can imagine ellipses elongated in such a way that the proportion would become 1/10, 1/4, 1/2, 2, 5, 10, 20 and more. Halley's Comet, for example, has an orbital eccentricity of 0.9673 (Fig. 4); the semi-axis major being 18.00, the eccentricity is 17.41. The aphelion distance is 35.4112, the perihelion distance is 0.5889; the comet is therefore 60 times closer to the Sun at perihelion than at aphelion, and so receives 3,600 times more heat. Let us imagine that we have a planet whose axis lies along its elliptic and parallel to the line of the major axis; snows would never melt in the aphelic hemisphere passing into shadow and perihelion. Of course, no planet has an orbital eccentricity of this order. However, that of the minor planet Æthra is 0.38; several others exceed 0.30; that of Mercury is 0.2056, and that of Mars is 0.09326.

Let us examine *each hemisphere separately* or the summer session (from one equinox to anther), and for the winter season.

The axial inclination of the globe introduces an element of differentiation between the winter and the summer. (Winter is here to mean the 6 months between the autumnal equinox and the spring equinox, and summer as the other section of the orbit.) The eccentricity and the position of the line of equinoxes cause little change, but it is not the same with respect to the inclination. It is evident that if the Earth had its axis perpendicular to the orbit, as with Jupiter, there would be no seasons at all.

The inclination of the Earth's equator to the elliptic is 23°27′, and varies over the course of centuries between 21°58′36″ and 24°35′58″.

Though it is true that the heat received by the Earth *as a whole* from the spring equinox to the autumnal equinox is equal to that received from the autumnal equinox to the spring equinox, it would be wrong to conclude that the heat received by each hemisphere during the winter is equal to that which it receives during the summer.

In fact, the ratio is very unequal. If we take 100 to represent the total quantity of solar heat received by each hemisphere over the course of a year, 63/100 will be received in summer and 27/100 in winter.[2]

[2]See SIR ROBERT BALL, *The Cause of an Ice Age*, 1892.—Wiener, *Ueber die Stärke der Bestrahlung, Zeitschrift der OEsterireichischein Gesellschaft für Meteorologie*, 1879.

The same ratio can be expressed in another way. If we take unity to be the mean heat received from the Sun each day for one hemisphere of the Earth, the annual quantity will be represented by 365. Of these, 229 units apply to summer and 136 to winter. The result is a marked contrast between the climates of the two hemispheres.

If we consider the northern hemisphere, we find that the 229 units of summer are spread over 186 days, which gives a daily mean of 1.21; the 136 heat-units of winter are spread over 179 days, giving a daily mean of 0.75.

Therefore, we have for the northern hemisphere:

Mean heat received daily in summer (186 days)	=1.24
Mean heat received daily in winter (179 days)	=0.75

The opposite applies in the southern hemisphere.

The northern hemisphere therefore has a temperate climate. The southern hemisphere, on the other hand, has longer winters and shorter summers, undergoing much rougher conditions.

After this introduction, we can now consider the effect of the orbital eccentricity upon the Martian seasons.

Seasons and Climates in the Two Hemispheres of Mars

The orbital eccentricity is 0.09326, the semi-axis major is 1.52369; this is therefore represented in the Martian orbit by the number 0.14209945.

For the distance of Mars from the Sun, we have:

		In km. for parallax = 8″.82 = 149,000,000 km		
Perihelic	1.3815929	205,857,000	21.173	
Mean	1.5236913	227,030,000		42,346
Aphelic	1.6657907	248,203,000	21.173	

The difference between perihelion and aphelion is, as can be seen, 42,346,000 km. That is to say one fifth of the mean distance, 5.36. Therefore, the eccentricity (half the total distance between the two foci) is 1/10.7, instead of 15/60 as for the Earth; the difference between perihelion and aphelion is 1/5.36, the difference between the heat received at these two extremes is the square of this latter number, or almost one-half: 1/2.3. While on Earth the Sun's heat is 1/15 more at perihelion than at aphelion, on Mars the ration is more like 1/2.

The axial tilt of Mars is much the same as ours; it is the southern hemisphere which is turned toward the Sun at perihelion. Therefore, the southern hemisphere should have, in much more marked degree, hotter summers and colder winters than those of the northern hemisphere.

Because of the greater eccentricity, the differences in the lengths of the seasons are much more marked on Mars than on the Earth. We have seen that the Earth takes $186^{d}\ 11^{h}$ to go from the March equinox to the September equinox, and 8 days less, or $178^{d}\ 19^{h}$, to go from the September equinox to that of March. On Mars, the difference is much greater.

The warm season of the northern hemisphere includes 381 terrestrial days or 370 Martian days, while the cold season lasts for 306 terrestrial days or 298 Martian days. The difference in Martian days is 74 (out of 668) in favour of this section of the orbit, while for the Earth it is only 8 (out of 36). The difference between Earth and Mars, with respect to orbital eccentricity, is therefore very great. On Mars, the two halves of the year, separated by the equinoxes, are in the ratio of 19 to 15.

Applying to Mars the formula given for the terrestrial seasons, we find results which are analogous to those of the Earth with respect to the distribution of heat received by each hemisphere.

If 100 represents the total quantity of heat received by Mars over the course of one revolution round the Sun, this quantity can be divided into two very unequal parts. Each hemisphere receives, during the summer, 63/100 of the total heat, and only 37/100 during the winter. This is the same ratio as for Earth. There is only a negligible difference. Actually, the axial inclination is only a little greater.

The Martian year contains 687 days. If we represent the heat received by each hemisphere as 687 units over the whole year, each hemisphere will receive 436 units in summer and 251 in winter.

Next, note the great inequality of the lengths of summer and winter in each hemisphere. For the southern hemisphere, summer lasts for 306 days and winter for 381. The mean daily heat received by this hemisphere is therefore 436 units spread over 306 days, which gives, for each day, 1.42.

In winter, the southern hemisphere receives 251 units spread over 381 days, which gives, for each day, 0.66.

The difference is considerable. If the meteorologic conditions of Mars were the same as those on Earth, this hemisphere would be experiencing an Ice Age. It is probable that the snow there is much less deep; it melts completely after the summer solstice.

Since the seasons in the two hemispheres are symmetrically opposite, the northern hemisphere of Mars has a summer of 381 days and a winter of 306 days. The daily heat received by this hemisphere during its summer is therefore 436 units spread over 381 days, which gives 1.14 for each day. In winter, this hemisphere receives 251 units spread over 306 days, giving a daily value of 0.82.

The climates of the northern hemisphere of Mars are therefore much milder than those of the south, where conditions are much rougher. The situation is basically the same as for the Earth, but is vastly more accentuated, because, at the greatest eccentricity of the Earth's orbit, the maximum contrast is 1.38 to 0.68.

Considering the difference between perihelic and aphelic distance, the southern hemisphere receives, at its solstice, about one and one-half times as much heat as the northern hemisphere. It seems that in summer the southern snows ought to shrink more markedly than those of the north if the geographical conditions are the same. But, as with Earth, there is more sea in the southern hemisphere than in the northern.

What can observation tell us about this?

This is what we must next examine.

Observations of the Polar Snows of Mars

As early as the year 1781, William Herschel had already noted the variations of the polar snows, corresponding to the planet's seasons. In March, June and July of that year the south polar cap showed up as six times greater in diameter than during September and October. In the former months, it extended down to latitude 60 °, and measured the arc of a great circle equal to 60 °. (This states degrees in the original)

The years 1781 and 1783, when Herschel made these observations, were years of perihelic oppositions, when the south pole of Mars was tilted toward us. The north pole could be observed only imperfectly.

It was the same in 1798, during observations made by Schröter, who also noted that the south polar snows diminished considerably in extent between July and the end of October. He drew it as eight times as large in July as at the end of October. The values given by Herschel and Schröter are not very different:

Southern Snow 1781 (Herschel)

	Areocentric arc:
June	60°
October	10°

Southern Snow 1798 (Schröter)

	Areocentric arc:
July	50°
November	6°

But we must look to the opposition of 1830 before we have any really accurate measurements. This was again a perihelic opposition, and again it was the south pole which was under observation.

Southern Snow, 1830 (Beer and Mädler)

Date of observation	Latitude assuming The pole°′ to be central	Radius of cap°′	Diameter°′
31 August	83 37	6 23	12 46
10 September	84.15	5 45	11 30
15 September	86.25	3.35	7 10
2 October	86.50	3.10	6 20
5 October	87.7	2.53	5 46
20 October	85.59	4.1	8 2

In 1830, the southern solstice fell on Mars on 18 September.

As we have seen, these observations were made near the time of the solstice and about a month afterwards. They indicate only the minimum of the south polar cap. The observers assumed that it must be considerably larger at times well away from the summer solstice.

The first comparison which we can make between the northern and the southern snows is furnished by the observations of Beer and Mädler in 1837; these were made at aphelic opposition, when the planet's north pole was presented. These epochs are much less favourable for observations than the first, because the planet is much further away from the Earth; but Beer and Mädler partly made up for this by using a more powerful instrument than on the first occasion, producing twice as high a magnification and collecting six times as much light.

In all these observations, without exception, wrote Mädler, from 12 January to 22 March, the white cap of the north pole has been visible with a degree of clarity which we never recall having seen for the south polar cap; at the same time it was considerably larger than the south cap of 1830, and it was so brilliant that one could almost believe that in this site the planet was covered by another planet. A dark patch surrounded it.

We could recognize a trace of the southern snow. Because of the inclination of this pole, then invisible from Earth, this snow would be visible only if it came down to latitude 55°. Therefore, in February and March 1837 the southern snow was much more extended than in September and October 1830. We have the following proportions:

S. snow 1830. Summer solstice. Diameter	6° ±
» 1837. Winter solstice. Diameter	70° ± ».

But the observers must have mistaken the snowy southern islands for the pole As for the northern snow, Beer and Mädler found:

1837	12 Jan.	0.27 of the diameter of Mars
	or	31°.4 of the globe of Mars
	or	74°.3 limiting latitude

Thus in 1837, the northern snow, at the time of its summer, was much more extensive than the southern snow in 1830 at a corresponding time during its summer but it was much smaller than the southern snow of 1837 during its winter (assuming that the observers had actually seen the snow at the south pole, which is not likely). They concluded that:

The snow of the south pole varies in extent to a much greater degree than that of the north pole.

But these observations are not sufficient to decide the matter definitively. We can only affirm, provisionally:

	Herschel	Schröter	Mädler
South pole variation	10°–60°	6°–50°	6°–70°

The maxima are very uncertain. The observations resumed in 1839, and this time concerned the northern snow:

Northern Snow 1839

Date	Latitude°′	Radius°′	Diameter°′
26 February	78.33	11.27	22.54
1 April	80.48	9.12	18.24
16 April	82.20	7.40	15.20
1 May	81	9.0	18.0

While the southern snows shrank to a diameter of 6°, the northern snows still had 15° as a maximum; that is to say, a surface area five times greater.

Beer and Mädler estimated that the minimum of each cap occurred 1/18 of a year after the relevant summer solstice, corresponding to 12 July and 12 January on our calendar.

During the opposition of 1862—with respect to the presentation of the planet, this was analogous to those of 1783, 1798 and 1830—Lord Rosse, J. Norman Lockyer and William Lassell observed the polar snows. Their observations yielded the following results:

Southern snow 1862

Summer solstice: 9 September			
22 July	48 days before solstice:	36°	(Lord Rosse)

Series of Lassell					
13 September	4 days after solstice	20.0	21 October	43 days after solstice	8.6
20 September	11 days after solstice	14.5	23 October	45 days after solstice	7.6
22 September	13 days after solstice	13.0	25 October	47 days after solstice	8.0
24 September	15 days after solstice	9.0	27 October	49 days after solstice	8.2
25 September	16 days after solstice	11.1	4 November	57 days after solstice	8.0
27 September	18 days after solstice	9.3	5 November	58 days after solstice	9.3
29 September	20 days after solstice	10.4	15 November	68 days after solstice	7.1
11 October	32 days after solstice	7.6	17 November	70 days after solstice	5.5
13 October	34 days after solstice	10.6	18 November	71 days after solstice	8.2
15 October	36 days after solstice	9.4	22 November	75 days after solstice	9.1
17 October	38 days after solstice	9.1	8 December	91 days after solstice	7.5
18 October	39 days after solstice	9.3	11 December	94 days after solstice	9.5
Series of Lockyer					
17 September a.	8 days after solstice	18.0	3 October a.	24 days after solstice	9.5
17 September b.	8 days after solstice	13.8	3 October b.	24 days after solstice	8.8
23 September a.	14 days after solstice	10.0	3 October c.	24 days after solstice	10.0
23 September b.	14 days after solstice	12.9	9 October	30 days after solstice	11.0
23 September c.	14 days after solstice	12.9	11 October	32 days after solstice	8.4
25 September a.	16 days after solstice	11.3	15 October a.	36 days after solstice	7.3
25 September d.	16 days after solstice	10.0	15 October b.	36 days after solstice	6.8
25 September b.	16 days after solstice	9.3	18 October	39 days after solstice	7.5

The irregularities of these determinations prove the difficulty of making a precise estimate of the extent of the polar cap. They also show that the cap does not mark the exact pole; rather, it is eccentric to the pole, and turns around it, so being presented to the observer at different angles. For instance, in 1862 the limits in diameter of the south polar snow were:

From 36°±48 days before the solstice, to 6°±70 days after the solstice.

In 1873, I observed the northern snows myself. They were then very white and well defined (June 1873). They extended to latitude 80°, and sometimes seemed to project from the disk, because of irradiation. As for the southern snows, this is what I wrote:

> The southern region is visibly marked by a white streak near the limb. Can it be that the snow has come down to latitude 40° Latitude south? It is more likely that clouds are responsible. (*Comptes Rendus*, p.278.)

The opposition of 1877 was analogous to that of 1862: perihelic, and exhibiting the south polar cap. Schiaparelli, at Milan, made the following observations:

Southern Snow, 1877.

Summer solstice: 26 September		
Date	Days	Deg.°
23 August	34 days before solstice	28.6
28 August	29 days before solstice	23.9
3 September	23 days before solstice	26.0
10 September	16 days before solstice	23.9
11 September	15 days before solstice	20.2
12 September	14 days before solstice	17.4
13 September	13 days before solstice	16.9
14 September	12 days before solstice	17.4
15 September	11 days before solstice	14.1
15 September	11 days before solstice	16.1
16 September	10 days before solstice	16.1
18 September	8 days before solstice	19.1
20 September	6 days before solstice	18.5
22 September	4 days before solstice	14.7
24 September	2 days before solstice	13.8
25 September	1 day before solstice	11.5
26 September	Solstice	11.5
30 September	4 days after solstice	12.5
1 October	5 days after solstice	13.7
2 October	6 days after solstice	11.8
4 October	8 days after solstice	12.7
10 October	14 days after solstice	10.4
12 October	16 days after solstice	9.5
13 October	17 days after solstice	9.3
14 October	18 days after solstice	7.6
27 October	31 days after solstice	7.0
4 November	39 days after solstice	7.0

It can be seen that from this series that the diameter limits of the south polar snow in 1877 were:

From 29°, 34 days before solstice, to 7°, 39 days after solstice.

When the patch was not circular, Schiaparelli gave the diameter of the snow as being that of a circle of equal area.

At the end of October and the beginning of November, the white point had become so small that Schiaparelli expected to see it vanish from one day to the next. But this did not happen. During November, December and part of January, it continued to glitter; even in the middle of December—growing to nearly 15°. But the existence of this polar cap had become more and more difficult to confirm, because of the increasing obliquity of vision, the advance of the shadow, and the clouds which began to appear and could easily be confused with the polar cap as the autumn drew on. The minimum of the polar snow seems to have been reached at the end of November—that is to say, about 2 months after the solstice.

This minimum of 7° is decidedly smaller than anything which occurs on Earth. Here the polar ices never melt; they never shrink down to less than 84° latitude around the 70th degree of western longitude, or latitude 75°, near the 110th degree of eastern longitude.

In 1879, the skillful Milan astronomer repeated the observations of the south pole, which, strange to say, is much better known than our own poles. His results:

Southern Snow, 1879.

Summer solstice: 14 August		
Date	Days	Deg.°
12 October	59 days after solstice	7.6
17 October	64 days after solstice	11.5
18 October	65 days after solstice	9.5
21 October	68 days after solstice	8.0
22 October	69 days after solstice	6.7
23 October	70 days after solstice	9.5
28 October	75 days after solstice	3.8
8 November	86 days after solstice	4.0
10 November	88 days after solstice	4.6
11 November	89 days after solstice	11.0
17 November	95 days after solstice	6.1
18 November	96 days after solstice	11.5
27 November	105 days after solstice	9.5
28 November	106 days after solstice	5.7
28 November	106 days after solstice	4.4
29 November	107 days after solstice	4.3
21 December	120 days after solstice	5.5
26 December	134 days after solstice	12.0
2 January (1880)	141 days after solstice	14.3

These observations were not begun until 59 days after the solstice, and were made under poor conditions, because of the obliquity of the pole (which was very great,

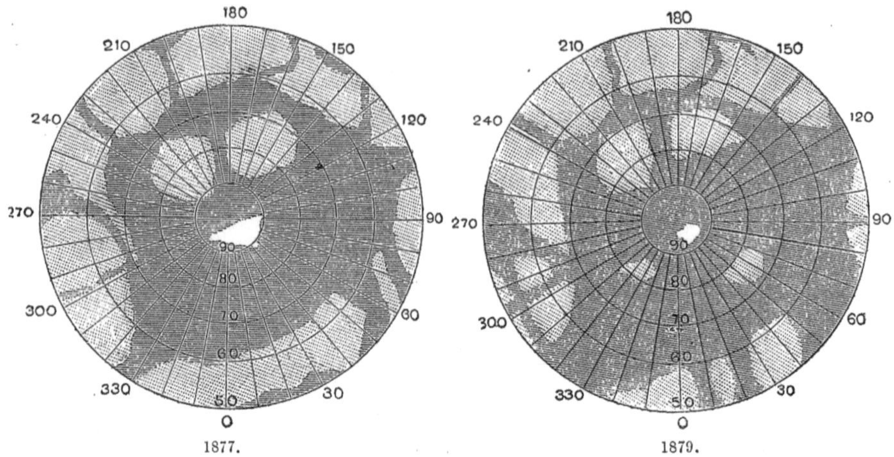

Fig. 5 The South pole of Mars at the epoch of the minimum of polar snow

particularly on 28 October). Nevertheless, Schiaparelli believed that he could say that the minimum took place in the second half of November three and one-half months after the southern solstice. There were great fluctuations in the estimated dimensions.

The minimum size measured was 4°. On several occasions the white patch seemed to project beyond the disk, by irradiation, which doubled the apparent size. The real diameter was therefore reduced to 2°, or 120 km.

From the observations made by Bessel in 1830, Kaiser, Lockyer and Linsser in 1862, Hall and Schiaparelli in 1877, and now this new series of 1879, the position of the southern polar snow, when at its minimum size, is:

Distance from geographical pole	=5.4°
Longitude	=30°

The double figure (Fig. 5) shows the south polar snow at its minima of 1877 and 1879.

In 1881, the south polar snow lay on the circle of the terminator, during summer, until 11 September, when it became tilted into the invisible hemisphere. It was therefore impossible to observe it during the period from October 1881 to April 1882. Sometimes white patches were seen; these were not the polar snows, but the southern islands. In June the south polar snow measured 25°; in July, 10°. The solstice fell on 25 June. Seven months after the solstice, and January and February, the south polar snow could not have been greater than 10° in diameter, because I was unable to see it. Therefore, it had not increased further.

From 1881 to 1888 the north pole could be studied in its turn, while the south pole remained almost hidden.

During the observations from 30 September 1879 to 24 March 1880, Schiaparelli had noted five very curious branches, apparently coming from the invisible north pole, and arranged in a ring at polar distance between 30 and 40°.

During the following opposition (1881–1882) it could be confirmed that these white patches did emanate from the pole, because the pole itself could be seen—exactly at the boundary of the visible hemisphere. From 26 October 1881 to 25 January 1882 no permanent polar snow could be confirmed at the site of the pole; but scattered snows could be seen, coming from it and forming eight different branches, of which some occupied positions observed in 1879. These branches shortened and became whiter during January, and were gradually concentrated toward the pole. On 26 January (I had been unable to observe on the preceding days, because of bad weather) the pole was marked by an immense snow-cap, almost round, measuring about 45° in diameter—proving that the branches referred to above had come together in a single mass. This condensation of the snowy branches in a single polar mass has no terrestrial analogy. This phase of concentrations occurred on 25 January. Subsequently Schiaparelli found, for the diameter of this north polar snow:

Northern Snow, 1882.

Summer solstice: 25 June		
Date	Days	Deg.°
26 January	150 days before solstice	45
28 January	148 days before solstice	45
4 February	141 days before solstice	40
17 February	128 days before solstice	30
27 February	118 days before solstice	20
10 March	107 days before solstice	30
4 April	82 days before solstice	20
10 April	77 days before solstice	27

It was impossible to continue the observations up to the solstice.

The precision was not high, because of the obliquity and because the boundaries were not always sharp. The polar cap is not only the result of the suppression of the snowy branches noted above, but rather their contraction toward a central nucleus, and to their growth in breadth at the same time as their diminution in length, so filling in the gaps—which however sometimes remain visible at certain small points. As the spring equinox of the northern hemisphere fell on 8 December 1881, and the northern summer solstice on 25 June 1882, it may be seen that the greatest dispersion of the polar whiteness, in the form of branches thrown outward toward the equator, took place several months after the winter solstice; but the greatest intensity of the snow, with regard to surface brilliancy, occurred only a month or more after the spring equinox, or only 5 months before the summer solstice. The maximum therefore corresponds to a regular cap of about 45° diameter, concentric to the north pole, which then slowly diminished with the approach of summer.

The Milan astronomer tried to represent graphically (Fig. 6) these characteristics of the north polar snow, deduced from his observations of November and December 1881. The eight branches resemble streaks sketched here around the polar region.

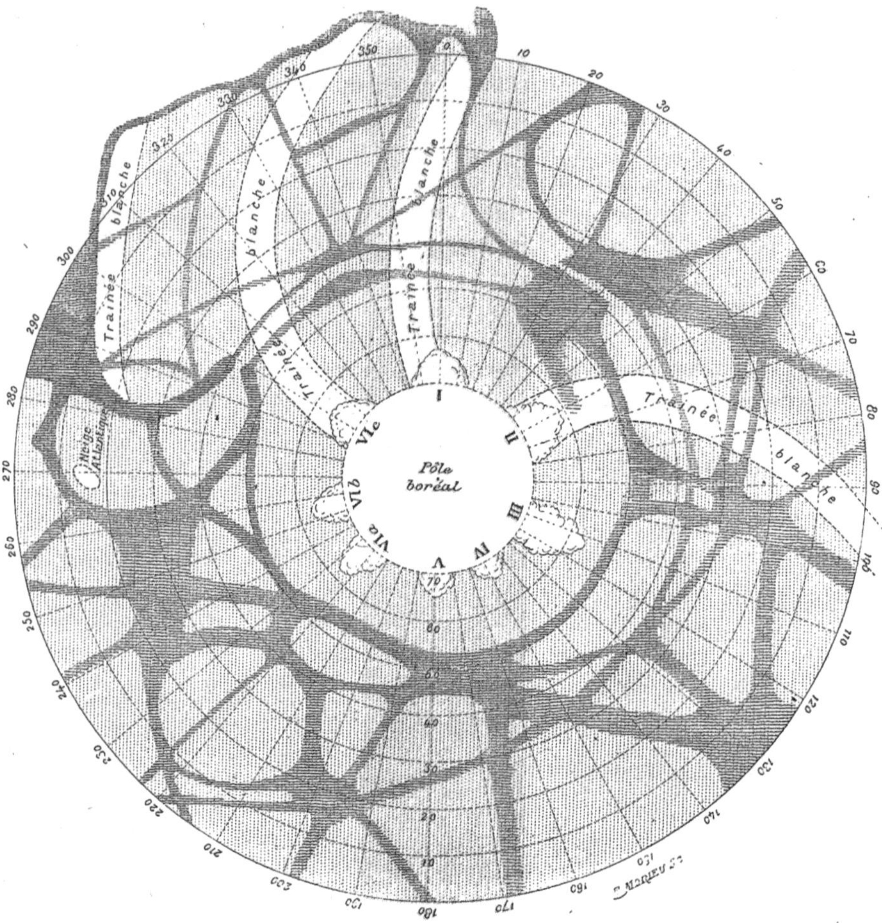

Fig. 6 The north pole and the northern hemisphere of Mars in 1879 and 1881. Streaks of snow

Moreover, three whitish streaks, well-formed and of uniform breadth, came, in their turn, reaching the equator at longitudes 5°, 95°, and 350°. It may be noted that another, less certain streak crosses the Hourglass Sea and extends along its coasts. We have a series of spirals, coming from the north pole and running obliquely toward the south-west. We must also assume that this direction is related to the rotational movement of the planet, and the theory of prevailing winds. These streaks of snow produced during the spring indicate regular currents; but they are different from those of our own air inasmuch as they are less dependent upon chance circumstances. When these ribbons of snow cross canals, they cease to be visible on these canals, as if the snow falling on the water melts there, while on dry land it persists. We may assume that these whitish streaks, which remain visible for weeks or months, cannot be clouds, but are snow—though no doubt of slight depth.

We have now come to the opposition of 1884, when measurements by Schiaparelli yielded the following results:

Northern Snow, 1884.

Northern solstice: 13 May		
Date	Days	Deg.°
20 January	114 days before solstice	36
15 February	89 days before solstice	31
23 March	51 days before solstice	23
2 May	11 days before solstice	15

The series made in 1886 and 1888 gave analogous results:

Northern Snow, 6

Northern solstice: 31 March			
Date		Days	Deg.°
1886	18 January	62 days before solstice	25
"	26 February	33 days before solstice	10
"	14 March	17 days before solstice	6
"	28 March	3 days before solstice	6
"	21 May	51 days after solstice	5
"	1 June	62 days after solstice	9
1888	7 May	81 days after solstice	12
"	2 June	107 days after solstice	11
"	July	Scarcely visible	

The patch diminished in brilliance rapidly in July 1888, because of the enormous obliquity of solar illumination, soon to be followed by entry into the polar night. The north pole entered the shadow on 15 August, the day of the equinox.

From all these observations, it seems that the north polar snow is centred on the pole, while, on the other hand, the south polar snow lies to one side.

At the opposition of 1890 both the poles were presented to the Earth. In June, July, August, and September both the northern and southern snows could be seen at the same time, but just at the edge of the disk; the northern patch was the better-marked, and also more tilted toward us. On 30 July I estimated it as about 20°. The Earth passed through the plane of the Martian equator on 23 September. At the end of this period, I found that the southern cap was the better seen.

Southern Snow, 1890

Summer solstice: 26 November		
Date	Days	Deg.°
24 September	63 days before solstice	30
2 October	55 days before solstice	25
9 October	48 days before solstice	25
22 October	35 days before solstice	20
13 December	17 days after solstice	10
20 February	86 days after solstice	8

Let us now summarize these observations.

Variations of the south polar snow			
	Maximum°	Minimum°	
1781	60	10	W. Herschel
1798	50	50	Schröter
1830	50	6	Beer and Mädler
1837	70	6	Beer and Mädler
1862	36	6	Rosse and Lassell
1877	29	7	Schiaparelli
1879	29	4	Schiaparelli
1880	30	8	Flammarion
Variations of the north polar snow			
1837–1839	31	15	Beer and Mädler
1882	45	15	Schiaparelli
1884	36	15	Schiaparelli
1886	36	5	Schiaparelli

If we eliminate the 1837 results, which depend upon a white patch not identified with the pole (then invisible), the variations at the two poles seem to be very much alike, contrary to the opinion of Beer and Mädler. Indeed, in 1882, 150 days before the solstice, Schiaparelli measured the north cap as 45°. This value does not coincide with the absolute maximum, which can attain 50° or even 60°. Moreover, in 1886 Schiaparelli found that the northern cap had shrunk to 5°. These limits are equal to those of the south polar snows.

It is very possible that the years on Mars are not all alike, though undoubtedly each hemisphere has its warm and cold spells. Moreover, considerable transformations in appearance can sometimes be seen from one day to the next, attributed to enormous falls of snow. Recall the photograph taken by Pickering on 9 April and that of the following day, in which the second view shows the south polar cap as much vaster than the first.

There are thus meteorological variations on Mars which occur at different rates, as with us. The weather changes also on Earth, it to a lesser degree.

As we have seen, a summary of the observations of the polar snows does not lead us to the conclusion at the orbital eccentricity leads to greater changes in the southern snow was than in the northern. In any case, the difference is slight.

Certainly it is warmer at the south pole at its summer solstice than at the north pole at its corresponding solstice, because the former occurs at perihelion and the latter at aphelion. The south polar snow ought to be more completely melted. This is indeed so at the geographical pole; but there always remains a residuum, 120 km in diameter, at 324 km from the pole; the longitude is 28°, in the middle of the sea. Probably there is a vast, high island there; otherwise the ice would melt completely. On the geographical north pole on the other hand, the ice remains at the centre point; but there is no large sea there.

The southern and northern hemispheres of Mars are thus very different from the point of view of the distribution of land and sea. As with Earth, the land and water are subject to the usual laws relating to surface relief, so that the northern highlands are well above the mean level. This difference has a definite climatological effect. It makes the seasons and climates more temperate in the south, more extreme in the north—that is to say, it counteracts the effects of the orbital eccentricity, and may even neutralize them completely. We must also assume an indirect influence upon the distribution of temperature regions, and upon sea and air currents.

To sum up: the seasons on Mars are strictly comparable with those of the Earth, despite the greater distance from the Sun. At each pole the snows melt in summer, more completely than ours. The winters seem to be more severe. Evaporation and condensation take place more rapidly than with us. The meteorological regime there, seems very temperate. Apart from winter mists, the atmosphere remains constantly clear.

Conclusions

This analysis, comparing the seasons on Earth and on Mars, leads us to the following conclusions:

The seasons on Mars are of about the same intensity as ours, but are almost twice as long.

The warm season lasts for 381 days in the northern hemisphere, and the cold season in the southern hemisphere lasts for an equal time.

The cold season lasts for 306 days in the northern hemisphere; the warm season in the southern hemisphere is of the same length.

Each hemisphere receives 63/100 of its annual heat during its summer, and 37/100 during its winter.

The seasons in the southern hemisphere are more marked than those in the northern hemisphere.

The polar snows on Mars vary according to the seasons. They attain their maximum from 3 to 6 months after the relevant winter solstice, and are reduced to minimum from 3 to 6 months after their summer solstice. As on Earth, not all years are alike.

In the two hemispheres, the polar snow in winter can attain a diameter of 45–50°, and in summer can be reduced to 4° or 5°. We cannot finally decide upon the effects of the orbital eccentricity on each hemisphere. More complete observations are very desirable.

Beyond the polar ices, snowfalls are observed in the temperate regions, and even up to the equator. In the northern hemisphere, streaks are seen to spiral outward from the pole, indicating atmospheric currents influenced by the rotational movement of the planet.

The north polar cap appears to be centred on the pole. The southern hemisphere's cap is 5.4°, or 340 km, away from the geographical pole, at longitude 30°; therefore at minimum, the south pole is entirely uncovered. The polar sea is free.

The climatology of Mars shows strong analogies with that of the Earth; conditions seem to be rather better. The greater distance of Mars from the Sun, and the thinness of its atmosphere, are more than compensated for by physical conditions which are more favourable than ours.

The theory of the secular variation of terrestrial climates due to the eccentricity of the orbit, proposed by Adhemar and supported by James Croll on other grounds, is not confirmed from an examination of Mars. Mars has an orbital eccentricity five and one-half times greater than ours at present, and the Earth's orbit never becomes as eccentric as that of Mars. Therefore, Mars is very well suited as a control study. It also has its southern summer at perihelion, and its southern winter at aphelion, so that in the southern hemisphere the summers are shorter and hotter than those in the north, while the winters are longer and colder.

The theory referred to above assumes that the south terrestrial pole is becoming colder year by year, because there are 8 days less sunshine each year. For Mars, this rate would rise to 74 days. We might think, then, that the summer would be too short to melt the ice formed at the south pole during winter. Yet this is not so, as we have seen. The south polar cap melts completely after its summer; the same is true of the north cap. There remains a residuum 120 km in diameter, placed eccentrically to the south pole, and undoubtedly lying upon an island.

As with Earth, the southern summer solstice of Mars occurs near perihelion. For Earth, the semi-revolution of the line of apsides is achieved in 10,500 years; the southern summer solstice—and therefore the northern winter solstice—will have occurred at perihelion in the year 1248 A.D. On Mars, the semi-revolution of the line of apsides takes 9,866 Martian years. Of this period, at present (1892) 4,235 Martian years have elapsed since the last equality of the seasons; 5,631 years remain before the seasons are again equal. Actually, the summer solstice in the southern hemisphere of Mars occurs 36 days after perihelion, at heliocentric longitude 357°; the longitude of perihelion is 334°.

The winter cold at the south pole of Mars should be much greater than that at the Earth's pole. The polar night is almost twice as long as ours; it lasts for 338 days, instead of 182, and the atmosphere is undoubtedly much less dense. Yet after some months following the summer solstice, the snow has melted.

For the south pole, this melting of the ices can be attributed to somewhat tepid sea-water, and to currents analogous to our Gulf Stream; but this explanation cannot apply to the north pole, because there is no vast sea there. We can assume that there is less water or less water-vapour on Mars than on the Earth, less cloud, a lesser quantity of snow, and a much lower thickness of ices. Perhaps, also, the length of summer—twice as long as on Earth—amply suffices to melt all the ices. There are limitations to the formation of snow; and the Sun remains below the horizon, at each pole, a half a year.

In short, the climatological analogy with the Earth, is supported by all the observations and studies of Mars, which throw special light onto the general knowledge of our own globe.

Changes Actually Observed on the Surface of Mars

As a world, Mars greatly resembles the Earth, and this is why we are so interested in it—perhaps to a greater degree than with any other planet. On the other hand, the differences are great enough to avoid monotony, and in this respect their study is highly significant. The question of the changes which take place unceasingly on the surface poses some of the most curious problems of contemporary astronomy. This is been recognized ever since the year 1876.

The observations prove that considerable variations really do take place in the geographical aspect of the planet. However, we should not admit the reality of the observations until we have made sure that they are not due to differences in visibility, observation or interpretation.

Every observer of Mars knows that it is difficult to be certain about its surface details, and how hard it is to interpret a drawing with absolute precision. Each observer sees in his own fashion, and draws in his own fashion. Quite apart from differences due to the power of definition, no two people's eyes are the same. One observer may see small details particularly well; another may note a cloudy feature which escapes his colleague. Of the 572 telescopic drawings or maps which we have passed in review, not one gives the complete representation which could be seen by an observer studying the Martian surface from a balloon.

Critical analysis of all these drawings convinces one of their inadequacy. The distance is too great, our atmosphere is too dense, and our instruments are not perfect enough.

Can we then say that these representations of Mars are valueless? No; the best of them give us a general idea of the planet. But we must not take any of them literally.

Undoubtedly there are hours of perfect calm and complete transparency, which give excellent telescopic images, and even recently I have experienced spells of this kind (the nights of 15 and 16 July 1892, 31 July, 1 August, 5 and 6 August, and 12 and 13 August); but one is far from seeing everything, or seeing this curious topography exactly.

C. Flammarion, *Camille Flammarion's The Planet Mars*, Astrophysics and Space Science Library 409, DOI 10.1007/978-3-319-09641-4_10

Imagine a tapestry in which the design is formed by a medley of human figures, some laughing and some crying, alternating with the figures of animals, plants and flowers covering the whole of the space; the whole accompanied by criss-cross alignments in various tones, making up an ensemble of large geometrical patterns.

From afar, one could distinguish only these large geometrical patterns; circles, squares, oblongs, stars, rectangles, triangles, polygons etc.

From closer up, we could also note that the background is covered with straight lines, crossing each other at various angles.

Closer still, we could distinguish the plants, animals and men.

At last, we would recognize the figures of humans, animals and plants; faces which laughed, faces which were crying. The general aspect of the scene would then be lost to the observers eye.

This is the story we find in the observations of Mars.

It is therefore important that we should first be convinced of the astonishing discrepancies which exist between drawings made by the best observers with the best instruments. We have had them before our eyes throughout the course of this book. Let us now recall a few examples which demonstrate the point particularly well.

Differences Due to Observers

Here (Figs. 1 and 2) are two drawings made on the same evening and almost at the same time, with the help of excellent and strictly comparable instruments, by two competent, careful and skilled observers (and, moreover, these are two of their best drawings, made under the best conditions). They represent almost the same hemisphere of the planet, on 18 October 1862 at 8^h 13^m for the small drawing, made in Italy by Secchi, and 8^h 0^m for the large one, made in England by Lockyer. The first time is reckoned from the Rome meridian, the second from that of Greenwich.

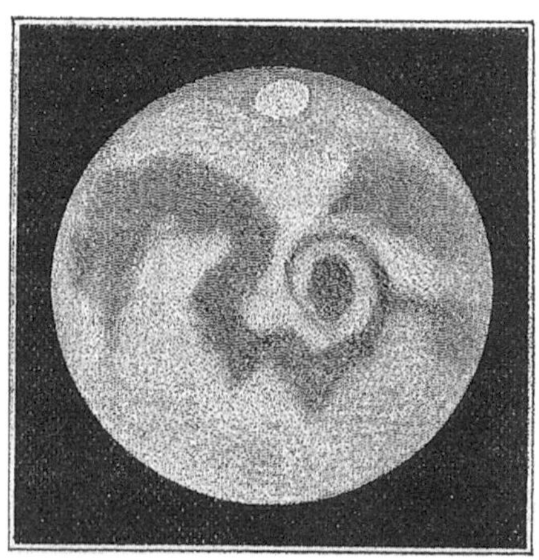

Fig. 1 Drawing of mars made on 18 October 1862, at 7^h 33^m (hour of Paris), by Secchi

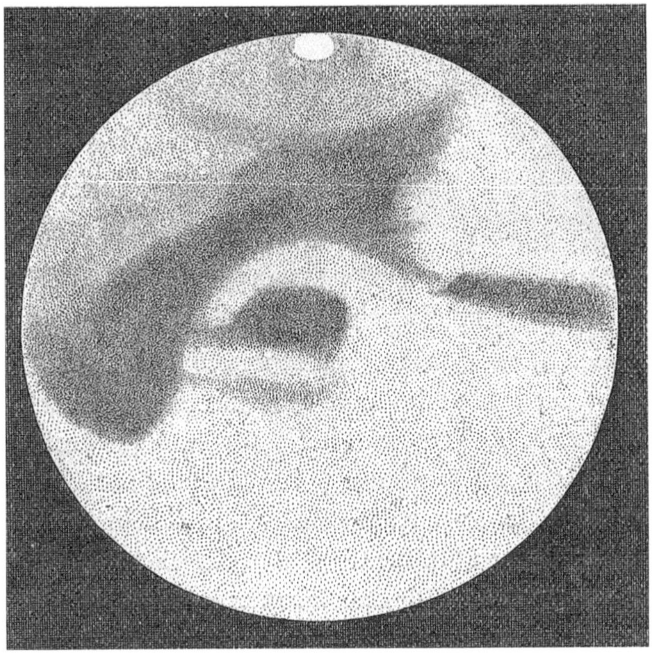

Fig. 2 Drawing of Mars on the same night, at $8^h\ 9^m$, by Lockyer

Consequently, by Paris time, the first was made at $7^h\ 33^m$ and the second and $8^h\ 09^m$. The time difference is 36 minutes; the round spot surrounded by whiteness, which we see near the centre of the disk, has advanced 36 minutes further to the left in the large drawing relative to the small one.

These two drawings are strongly characterized by this circular patch. The Roman observer saw it elongated north–south, while the English observer drew it as elongated east–west. Moreover, Secchi drew a shady curve recalling the shape of a cyclone—a turbulent atmosphere; and indeed he went so far as to say, *La crederei una gran burrasca in Marte*. Lockyer shows simply a calm sea, which he called the Baltic. The impression given by the two observers is totally different. Compare the details on the two drawings, and you will see what is meant.

Also, compare the two drawings made on the same day, and at nearly the same time: 4 June 1888, one by Schiaparelli at Milan (Fig. 3) and the other by Perrotin at Nice (Fig. 4), both with the aid of excellent instruments. Without doubt the two drawings generally confirm each other. We have on both (1) the Hourglass Sea (a little further to the left in the second drawing than in the first, because of the rotation of Mars), (2) its prolongation, called Protonilus; (3) a gulf leading to Astaboras; (4) the small line named Astaboras; (5) the double canals of the Phison and the Euphrates, showing up as straight lines from Protonilus to join the upper sea; (6) the Hiddekel, running from Ismenius Lacus to the Forked Bay; (7) the Gehon, coming from a point well away from the lake to join it at the same bay; (8) the lower

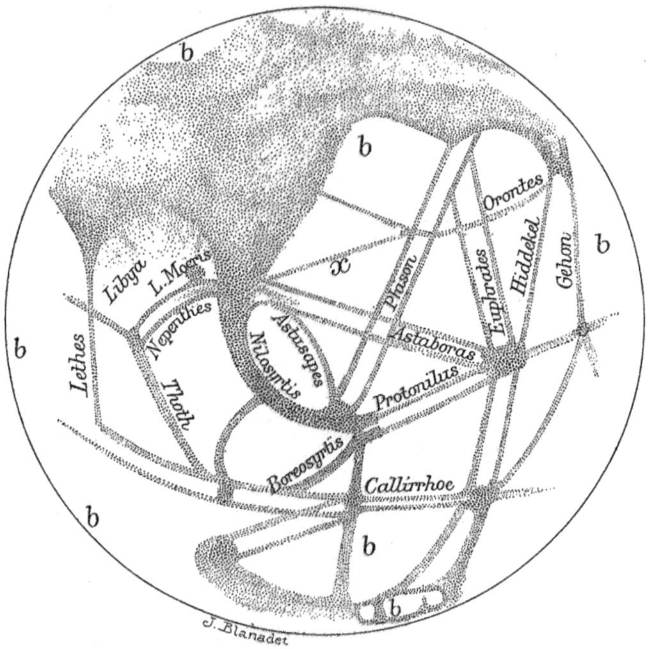

Fig. 3 Drawing of Mars made on 4 June 1888, by M. Schiaparelli, at Milan

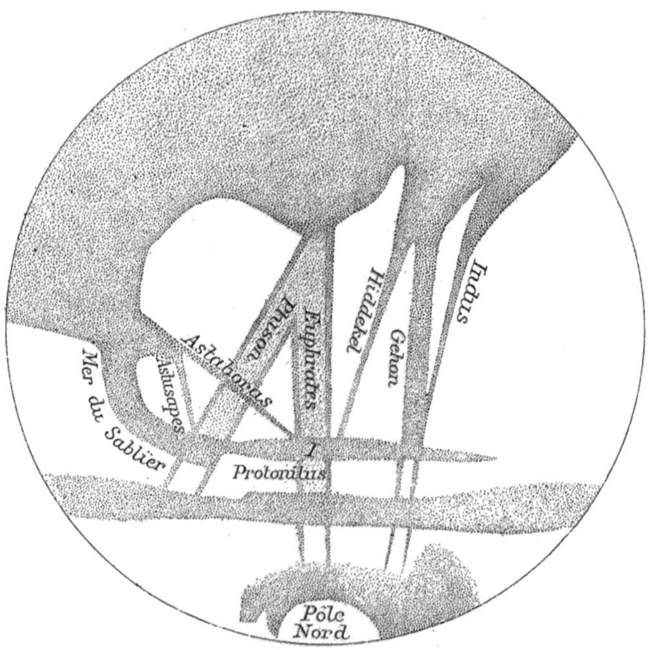

Fig. 4 Drawing of Mars made on the same night, by M. Perrotin, at Nice (about an hour after the preceding one)

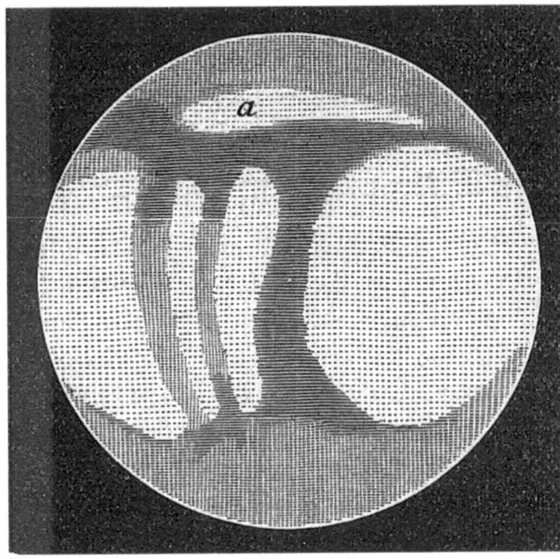

Fig. 5 Drawing of Mars made by M. Holden, at the Lick Observatory, with the Grand Equatorial, on 27 July 1888, at $8^h\ 0^m$ (*a*, whitish island)

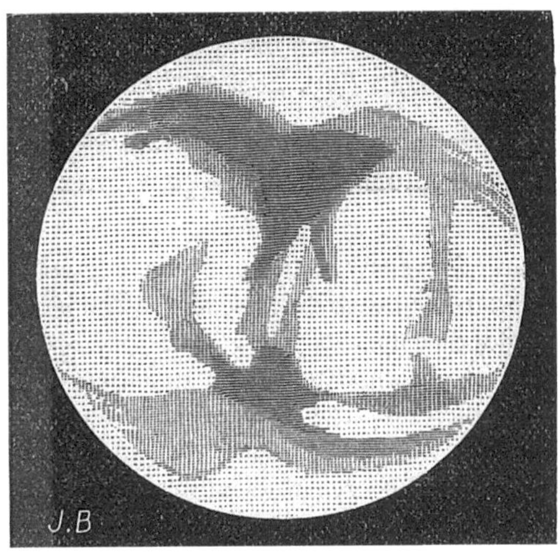

Fig. 6 Drawing of Mars made by M. Keeler, at the same Observatory with the same instrument, the same hour, about a quarter hour later

polar cap; (9) the Euphrates, going almost as far as the polar cap; (10) a sombre opening in this cap, to the left. The agreement is surely unquestionable, proving that these extraordinary straight lines really exist.

But what differences there are in the rendering! The Gehon is incomparably broader in Perrotin's drawing than in Schiaparelli's (it is true that it is nearer the centre of the disk); the bay of the Hourglass Sea, near Astaboras, is more marked and more important; it is the same with the Prontonilus, made up of a single shading in the Nice drawing and of a double canal in that made at Milan. We might believe these changes to be real if the drawings have been made on two different dates. On the contrary, they show that in astronomical observations, as an actual life, when we are dealing with shadings, each of us sees them somewhat in his own manner.

Now consider, if you will, a third example of the same order. Figures 5 and 6 show two drawings made with the same instrument—and this instrument is the most powerful in the world: the great $0^m.91$ equatorial at the Lick Observatory—by two different observers, on the same day and at an interval of only a quarter of an hour. Both were made on 27 July 1888. The first was made by Holden, the second by Keeler. In each of these two sketches, the Hourglass Sea crosses the middle of the disk, from above to below. But what differences in aspect! It is not to be believed! Thus, at the same time, and at the same instant, an observer sees the aspect as in Fig. 5 and another sees it as in Fig. 6. If, from this, we conclude that real changes have occurred on the planet, we will be making the greatest mistake.

The drawings made on the preceding and following days show that Holden really saw—to the left of the Hourglass Sea—the two grey vertical streaks which are drawn here, while Keeler did not see them; he saw something else.

We have already noted several other similar cases, such as in Fig. 102, Chap. 3.

These examples, to which it would be superfluous to add, prove without doubt that *each observer sees in his own way and draws in his own fashion*; and from the discrepancies in the drawings, we ought not to conclude that real changes are taking place in the surface of the planet[1].

And to these personal differences between observers, we must add those which are definitely due to instruments.

Instrumental Differences

Instruments play a role that is certainly not unimportant.

First, even with their modern instruments, whose powers of definition are much greater than those of our predecessors, we can almost never see the details we want to represent with true clarity. It outlines are not precise, and the images are somewhat vague. We try to draw what we see as faithfully as possible, but the shades, the tones, the outlines, the details cannot be rendered identically. As we must, however, define them on our drawings, we have an inevitable cause of greater or lesser discrepancies

This trouble is evidently reduced to a minimum when the observer knows in advance what he ought to find on the planet—when he knows, from the ephemerides, what is the longitude of the central meridian, and what aspects should be presented. This is found notably in the observations made by Terby of Louvain during the opposition of 1888. Terby had received Schiaparelli's drawings, and had at once set out to verify and confirm the work of the Milan astronomer. His instrument is an excellent 8-in. ($0^m.20$) equatorial by Grubb of Dublin, while that of the Milan Observatory (since 1886) has been an 18-in. ($0^m.46$) of perfect quality, from the

[1] *The view of each observer plays a major role*. For example, I can easily read the minutes on my watch dial, though I am somewhat myopic, but cannot easily distinguish objects which are far away. Other eyes are quite different. The best double-star observer that we have, S. W. Burnham, has discovered many pairs separated by less than a second of arc, but has never been able to see the nebulosity in the Pleiades, etc.

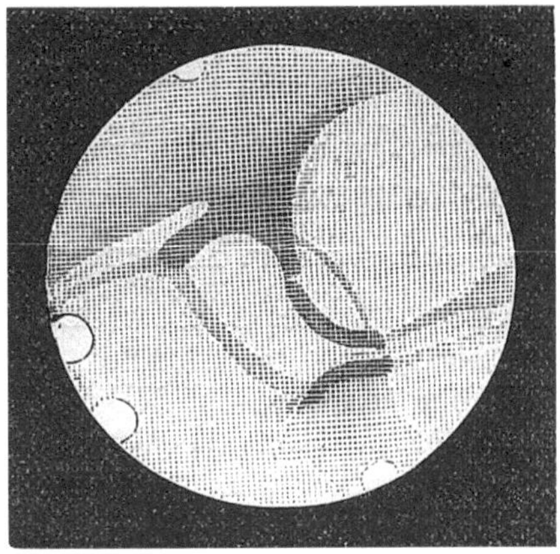

Fig. 7 Drawing of Mars, made by M. Terby, with the aid of an objective of 0^{m}.20 (compare with *fig*. 266, obtained by M. Schiaparelli with the aid of an objective of 0^{m}.46)

workshops of Merz in Munich. Well—let us compare the drawings published above by Schiaparelli (Fig. 3) and Terby (Fig. 7), the results of several evenings' work, and in which they have recorded all that they could see. Taking into account all that has been said earlier about the differences in eyes and methods, we must attribute a major part of the differences to instrumentation. There are far fewer details in the second drawing than in the first.

The dissimilarity will be still greater if the observer does not know in advance which face of the planet will be on view, and if he simply draws what he can see without prior knowledge. Personally, this is a principle from which I have never departed, for fear of self-deception.

The size of the objective, but above all its quality in definition, are two important considerations. The optical power of the equipment used plays a major role; the power of the eyepiece plays another—often, a lower magnification will give clearer and more complete images than a higher one. The state of our atmosphere may be adjudged another cause of discrepancies, and is linked with the power used.

However the size of the instrument is not as important as is sometimes believed. The most powerful refractors in the world, those at Mount Hamilton in California, Nice and Pulkowa, have not given images comparable with those obtained by Schiaparelli at Milan with his 0^{m}.46 objective, or even the 0^{m}.21 which I used until 1886. Undoubtedly the atmospheric waves and temperature differences counteract the advantages of magnification, and decrease the clarity of the images. Then eye and the method of observing are of prime importance.

Terrestrial Atmospheric Conditions

The state of our atmosphere plays a considerable part with regard to the clarity of the images. Sometimes it seems perfect, and yet the images are unsteady and blurred, because the layers of air lying above the observer are at different temperatures, and slide one against the other in the manner of atmospheric rivers. The temperature of the instrument itself is equally important, and becomes more serious with larger objectives. The instrument and even the observer's eye must be at the same temperature as the air, and conditions are best when the temperature is constant up to a great height.

Moonlight and light mist are not generally harmful. The images are sharper when seen through light clouds, which pass in front of the object and cut down the brilliance. I have noted that for Mars, in particular, drawings are often easiest to make during the daytime, even in full sunlight, an hour after sunset or before sunrise, and during the clearness of the dusk which proceeds the true night.

The diversity of these conditions, according to the positions of observatories and their altitudes above sea level, is therefore a cause of variety in the observations.

Various Presentations of the Martian Globe

Next we come to a reason for differences in appearance which is due to the planet itself.

The globe of Mars is not always present in the same fashion. Without taking into account the distance from the Earth, which is always varying because of the movements of both planets, and without taking into account the variation in apparent size which this produces, not only for the disk, but also for all the geographical configurations, let us consider the axial tilt, 24°52′. Mars does not often present its axis of rotation perpendicularly to our line of sight, south pole at the top and north pole at the bottom, with the equator lying horizontally along the middle. Sometimes the upper pole is inclined toward us, and the lower pole is averted; in this case, the equator runs well below the middle horizontal line which devised the disk into two equal parts. Sometimes, on the contrary, it is the lower pole which is tilted toward us and the upper pole which is hidden; the equator is then above the centre. The aspect of all the configurations varies considerably because of this, to such an extent that some of them become absolutely unrecognizable.

This is shown by the three globes given previously (Figs. 17, 20, and 28), Chap. 1, and also by the various projections presented earlier in the present book.

If we consider a characteristic feature, for instance the Hourglass Sea, we find that under southern presentation it appears as an enormous patch in the form of a V, whose point almost touches the lower edge of the disk. In the opposite case, it appears as a delicate canal occupying the centre of the disk and becoming broader toward the top like a tunnel, or even—because of the adjacent lower sea—like an hourglass. The aspects are very dissimilar, and one finds it hard to believe that he is dealing with the same face of the same planet.

Before comparing several drawings against each other, it is therefore important to bear in mind the inclination of the planet at the time of observation, and note the effects of perspective upon the shapes of the features. As far as possible, estimates should be made by assuming the features to be at the centre of the disk, seen face-on; disregarding the variations due to the foreshortening of the sphere in regions near the limb.

Some of the variations in the numerous drawings of Mars that we have examined are due to this cause, to which we must add the movement of rotation, which as of the same order as the effect described above.

The areas of the features, their positions on the disk, their respective distances, and above all the thin lines such as canals, are apparently broadened near the centre of the disk, and diminish near the edges, because of the effects of perspective.

Here, next, is a fifth cause of the differences.

Atmospheric Variations of Mars

This fifth cause is not so important as the four preceding ones, because the atmosphere of Mars is generally clear. However, the effect is far from negligible, because there are variable clouds and mists which sometimes completely mask vast areas under their white veil and completely alter the normal geographical configuration.

No observer doubts that the atmosphere of Mars gives rise to precipitation analogous to our snow. This sky is much less cloud-covered than the Earth's, even in winter, as is easily shown by a comparison of the observations. However, great cloud-sheets do occur, and among the drawings there some which show more than half the disk hidden under a whitish veil. Sometimes the clouds are scattered, and disappear quite quickly. As an example, I can cite the apparent squall noted by Schiaparelli in which a fixed feature on the planet-a well-known circular lake—had been seen earlier. But there are also the observations of clouds made by Lockyer on 3 October 1862, from 10^{h} 30^{m} to 11^{h} 23^{m}. Refer back to Fig. 31, Chap. 2; the region which extends from *x* to *y* shows up as white. In the drawing made at 11.23 (Fig. 32, Chap. 2), on the other hand, *v* is seen as a sort of grey gulf, showing that the cloud which had covered it had disappeared. This is the Main Gulf (or Lacus Mœris). Lockyer concluded that this observation demonstrated the presence and variation of clouds on Mars.

The same impression is given from the appearance of a drawing by Phillips, made at Oxford on 15 October 1862, and which shows the whole coastline masked by a border of white clouds (Fig. 41, Chap. 2).

These clouds on Mars have been the subject of a special study by Trouvelot, who has several times—thanks to his great perseverance—had the good fortune to see clouds form gradually before his eyes, in a period of from 1 to 2 hours, over areas where there had been no previous sign of them—particularly along coasts (see, notably, Fig. 50, Chap. 3).

On 14 October 1877, Schiaparelli wrote that a storm had occurred from 4 to 10 October, almost completely covering the Mare Erythræum and Noachis with clouds.

From 21 to 25 May 1886, Perrotin had the impression that clouds or mists extended the Hourglass Sea.

I have had the same impression during several observations, but more rarely than would have been the case in an atmosphere such as our own.

The atmosphere of Mars is not only clearer, but also calmer and more pacific than ours. With regard to the snowy-white streaks which were observed in November and December 1891, in the form of spiral bands coming from the north pole, visible on the continents but not over the seas, Schiaparelli regarded as being due to air-currents, regular and less spread around by chance circumstances than those in our atmosphere (*Memoria terzo*, p.88).

To obtain the best use of the Martian features, it is therefore necessary not only for the Earth's atmosphere to be clear, but also for the same conditions to apply all over the atmosphere of Mars in the Earth-turned hemisphere at the time of observation. Atmospheric absorption over Mars can be very much in evidence, because it effaces, or at least enfeebles, all the patches near the edge of the disk; but when the atmosphere is clear—as is most often the case—the geographical configurations can be clearly distinguished. Mists, fogs or light clouds are sufficient to modify the aspect of the disk, masking small or large regions and producing a bright white patch (that is to say, a cloud seen from above, illuminated by the Sun) instead of the dark tone of the seas or the reddish hue of the continents.

Can clouds sometimes appear dark when seen from above? In general, no. However, it is not impossible. Black vapor (smoke), raised above highlands, would—it seems to me—appear darker. It may be that such black vapours exist on Mars. The reflecting power depends on the state of the surfaces of these mists.

However this may be, we have here a certain cause of the apparent variations in aspect of the Martian features.

Here, then, is a list of causes which we must first eliminate if we want to find out whether we can place any real credence on these changes observed on the Martian globe. Differences of eye; of method; skill in observation; skill in drawing; instruments; atmospheric conditions; differences in the way in which the Martian globe is presented to us according to inclination and rotation; variations due to the planet's own atmosphere;—differences which we can classify in the following order, according to importance:

1. The eye of the observer.
2. The method of observation.
3. Interpretation of the drawing.
4. Instrumental differences.
5. Terrestrial atmospheric conditions.
6. Variations in the inclination of Mars.
7. The Martian atmosphere.

Are these causes sufficient to explain all the observations variations? No.

Real changes take place on the planet's surface—changes which are in no way analogous to those which take place on the surface of the Earth. The *extent* and *tone* of the dark patches unquestionably vary.

I am not here referring to the periodical changes in the polar snows, according to the seasons; these changes are known, have been measured over a long period, and can be explained. I am referring to variations of extent of the sombre patches which are regarded as seas, lakes or watercourses.

Undoubtedly we need irrefutable proof before accepting these variations. This proof has been given throughout the present book, and I will now summarize it.

We are dealing with phenomena absolutely unlike those of our own world, as we have concluded after deep thought—and because there is no other explanation.

I have discussed and accounted for the apparent variations on Mars. Now we come to the real changes.

One of the most persevering and assiduous observers of Mars, Schröter of Lilienthal, whose observations extended from 1785 to 1803, concluded from his studies that there was nothing stable on the surface of our neighbour world, and that the patches we observe are atmospheric in nature. One of the earliest observers of Mars, Maraldi, expressed the same opinion in 1710, with regard to the instability of the patches of Mars. Up to a point, their observations and their drawings justified this conclusion. Schröter made 230 drawings of the planet; 65 of them have been reproduced in the present book. It is easy to understand why they led their author to the idea that the Martian features are atmospheric in nature, similar to those of Jupiter. (Schröter recognized, by the way, that the clouds seen from above could appear darker than the ground or the waters.)

Let us begin this comparative study with the most characteristic sea on Mars, the Hourglass Sea, of which we have drawings going back to the year 1659. The region extends to the left of the Hourglass Sea, below the Flammarion Sea, which has been given the name of Libya, is particularly remarkable from the point of view of its variations in appearance, and it is henceforth impossible to doubt that we are seeing sudden flooding and drying. The breath of the Hourglass Sea varies unquestionably, and it often overflows to the left. Here is certain and secular evidence.

Changes Observed in the Hourglass Sea

In Huygens' first drawings of Mars, made in 1659, the Hourglass Sea appeared so broad that it occupied a large part of the disk (Fig. 9, Chap. 1).

It was the same in 1672, in a sketch made by the same astronomer (Fig. 19, Chap. 1).

One has the same impression in examining a sketch made by Maraldi in 1719 (Fig. 24D, Chap. 1). These three epochs (1659, 1672, 1719) were epochs when Mars was well presented; it was near perihelion, and climatic conditions and axial inclination were favourable, and the conditions for observation excellent with the inclination as given in Fig. 20, Chap. 1. The Hourglass Sea was then, unquestionably, very broad, even allowing for the uncertainties of the observations and sketches within wide limits. On the other hand, a drawing by William Herschel in 1777 (see *Fig.* 17 in our *fig.* 30) shows the Hourglass Sea as very narrow, with a constriction at *ef.* In this year Mars was near aphelion, and with an inclination like that shown in *fig.* 28.

The Hourglass Sea is again broad in a drawing made by Schröter on 18 November 1785 (see *Fig.* 2 in our Fig. 32, Chap. 1) (at this epoch the feature was just to the right, but showed the same inclination, as in the projection given in Fig. 17, Chap. 1). The same astronomer on 9 September 1798 (Fig. 42, Chap. 1), showed the Hourglass

Sea as very broad—the feature was again just to right, but a little more inclined toward the upper pole, compared to the projection; yet it was very narrow on a drawing made on 20 November 1798, by the same observer (Fig. 46, Chap. 1), and on a sketch made on 24 October 1800. 1798 was a year of perihelic opposition, but since opposition took place on 1 September, and the observation was not made until 30 November; the planet was already very far away.

To the right of the Hourglass Sea, on the drawing made on 24 October 1800 (see *Fig*. 161 in our Fig. 49, Chap. 1) there is a black disk corresponding to the Meridian Bay; it is about 90° from the Hourglass Sea. This point was very dark in 1800, as Beer and Mädler later saw it in 1830.

An observation made by Schröter on 2 November of the same year shows the Hourglass Sea as very broad. The axis of the planet was perpendicular to the line of sight.

This great variety in the drawings, which has been often commented on in the present book, at once gives the impression of considerable variations in the breadth of the Hourglass Sea, and Schröter himself concluded that we were dealing with atmospheric phenomena—clouds or mists—rather than stable geographical configurations. However, it would indeed be strange to recover the same forms after an interval of years, as for example from 1719, Maraldi's drawing (see Fig. 24D, Chap. 1) to 1798 (*Fig*. 56 in our Fig. 41, Chap. 1), Schröter's.

The observations become more reliable as we approach our own era, and we find examples of changes which are much more certain.

During the famous opposition of 1830, when Beer and Mädler inaugurated real areography, we find one interesting drawing relating to the Hourglass Sea; this is no. 6 in the author's plate (reproduced as our Fig. 66, Chap. 1). It was made on 19 September at $10^h\ 6^m$. The Hourglass Sea was very broad.

It was shown as very narrow on an excellent drawing made by Warren de la Rue on 20 April 1856 (*fig*. 76).

It was again found to be perceptibly broader in the drawings made by Secchi in 1858 (Figs. 15 and 16, Chap. 2).

It was equally broad in the drawings by Lockyer and Kaiser, in 1862 (Fig. 8). On the other hand, Dawes showed it as very narrow in 1864.

Another factor interposes itself here. Let us consider for a moment Dawes' drawing of 26 November 1864 (Fig. 9). We remark, for the first time, the feature, which rises up like a leaf from the stock of a flower, on the left bank of the Hourglass Sea, and at the same time we can see that the bank is indefinite and misty. Well! This is often the case in this region. See, for example, the observation made by Lockyer on 3 October 1862, at $11^h\ 51^m$, which gives the same impression, as we have already noted in our discussion of Fig. 8.

Let us also consider the drawing made by Schiaparelli on 28 October 1879, which is so remarkably similar to that of Dawes. The region concerned is equally veiled and cloudy (Fig. 10).

In these two drawings, which correspond to the same aspect of the planet (1864–1879), the Hourglass Sea is very narrow, and the region bounding it to the left is hazy or marshy (this is Libya). The Hourglass Sea was even narrower as shown in the 1877 drawings.

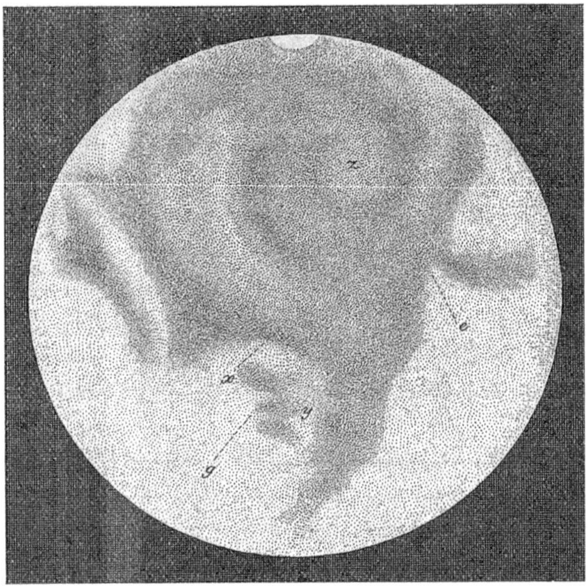

Fig. 8 Drawing of mars, by M. Lockyer, on 3 October 1862. At *x* and *y*: Libya

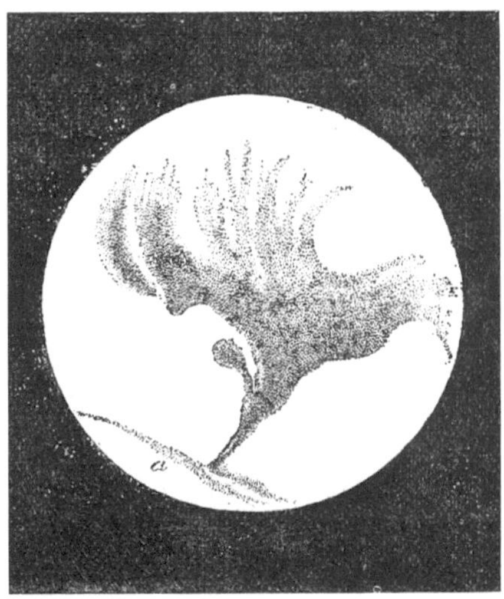

Fig. 9 Drawing of Mars, by Dawes, on 26 November 1864

To the left of the Hourglass Sea we must note the Main Gulf, also called the Lacus Mœris. This little lake was shown strongly on Dawes' drawings of 1864. In 1877, Schiaparelli represented it as a curved prolongation which he called the Nepenthes. In 1888, the Nepenthes was presented in a very different form, as a curved double canal, lying above the Lacus Mœris, which for its part was reduced to insignificant dimensions and was near the sea. At the same time (2, 4 and 6 June) a slight tint, showing flooding, covered the south of Libya as far as the Flammarion Sea, as was also the case in 1882. According to Perrotin, the flooding was at its most extensive in May. This variability in the tint of Libya has been known for a long time; what happens there

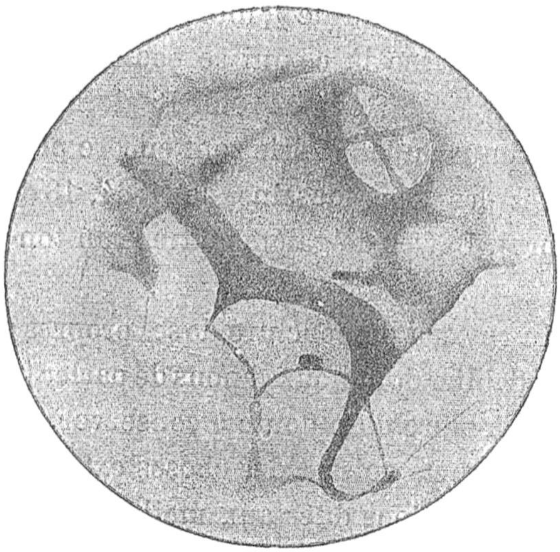

Fig. 10 Drawing of Mars, by M. Schiaparelli, on 28 October 1879

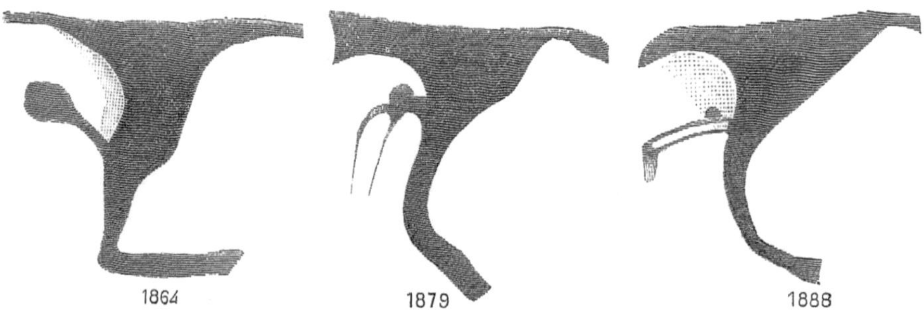

Fig. 11 Changes on Mars. The Lacus Mœris in 1864 (Dawes), in 1879 and 1888 (Schiaparelli)

also happens in Deucalionis Regio, and with Pyrrhæ Regio, Protei Regio, Tithonius Lacus, etc. But the changes followed in Mœris Lacus are perhaps even more worthy of attention. Compare the three drawings of 1864, 1879 and 1888 (Fig. 11). In 1864 (26 November), Dawes too traced a long grey tone along this region.

From his own observations, Schiaparelli confirmed these changes in the Hourglass Sea (see Fig. 80, Chap. 3).

Thus, no shadow of doubt remains. Extraordinary things are happening on our neighbour planet, even at the present moment.

Here, then, we have a series of observations made by the best astronomers; some of the discrepancies can be attributed to the cases listed above, vagueness and uncertainty of vision, difficulties of making representations by drawing, etc.; but these great differences in breadth evidently indicate real differences in the aspect of the planet, because they exceed the limits of differences of appreciation. Moreover, the series of observations made year by year by Schiaparelli since 1877 absolutely confirms this variability in extent. These variations are not apparent; they must be regarded as real.

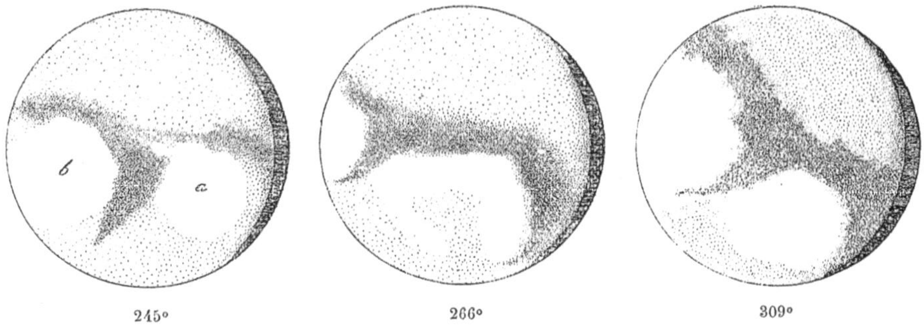

Fig. 12 Three drawings of Mars made by Schröter on 8 December 1800, demonstrating an unknown sea to the left of the Hourglass Sea (between *a* and *b*)

Here now is extra evidence of variation, not less obvious. This is given by the excellent observations made by Schröter.

Changes Observed in Gruithuisen Bay (Syrtis Minor)

First consider Fig. 12; three drawings made by Schröter on 8 December 1800, at 5^h 19^m, 6^h 45^m and 9^h 43^m. These three drawings confirm others of the same period, notably those of the preceding 1, 2, and 4 November. According to calculations by Van de Sande Bakhuyzen, the longitude of the central meridian in the first drawing was a 245°. Note, on it, a grey streak crossing the planet from east to west, and a triangular patch going down in a point toward the north; this patch bears a strong resemblance to the Hourglass Sea, but it cannot in fact be the Hourglass Sea, because the position is wrong. In this position, on my chart, we find a bay of the Flammarion Sea, the Gruithuisen Bay, going down from year 255° as far as the equator. To explain this appearance was to assume a lengthening and broadening of the bay, all which Schiaparelli has called Syrtis Minor.

The following sketch, made 1^h 26^m later, confirms the first drawing, and is also in agreement with those of 1, 2, 3 and 4 December; it shows the feature advancing toward the left edge of the disk, by virtue of the planet's rotation, and the Hourglass Sea coming into view on the right. The drawing made later, at 9^h 43^m, shows the Hourglass Sea in the middle of the disk; the central longitude is 309°.

This unknown sea, lying near 240° or 250°, is indicated—with its varied degrees of size—in a large number of drawings made by Schröter, notably in his drawings of 12 September 1798; 16 September; 18 October; 25 October; 1 November 1800; 2 November; 24 December 1802. Figure 13 shows his drawing of 1 November, where *b* is the Hourglass Sea.

One of the drawings which most resembles Fig. 13, is that made by Schiaparelli on 28 October 1879. It is given here (Fig. 10). To reproduce the unknown sea drawn by

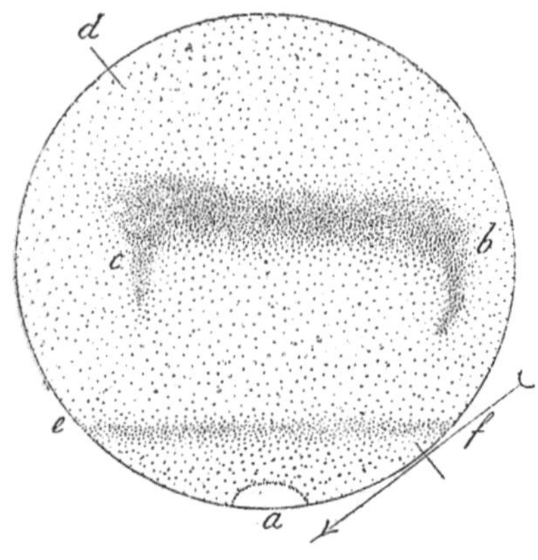

Fig. 13 A similar drawing, 1 November 1800

Schröter, it is only necessary to lengthen the triangular base, which lies left of the centre. At the same time, note, on the first drawing of 8 December 1800 (Fig. 12) the white region above *a*, which makes an indentation in the sea and which is presumably due to clouds. This adds to what has been said above about atmospheric variations.

It cannot be doubted that Schröter really did observe the sea, the size of which he gives in his drawings. This region of Mars seems to be particularly subject to changes of appearance of varying degrees of importance. Without giving extra drawings—for fear of wearying my readers—I will simply state that a comparison of the maps by Green and Burton confirms another part of this conclusion: that this region is formed of a medley of vague, indecisive features.

From a comparison of the old and modern drawings, it is possible to conclude with certainty that this region is subject to considerable variations in aspect. These variations could be of slight importance in themselves, but actually they are very extensive. Suppose, for example, that under certain meteorological conditions, the region could be covered with black smoke?—this would suffice to explain the changes in appearance. But as all the patches have the aspect and the tone of seas, it is not impossible that they are true floods. Be this as it may, the variability of this region (Schiaparelli's Syrtis Minor and Lethe) is certain.

Three other drawings by Schröter, on 1798, 19 September at 7^h 31^m, 20 September at 7^h 27^m and on the same evening at 9^h 48^m, show, at longitude 190°, a feature where today we see nothing—a point of sea coming down from the Maraldi Sea. This is reproduced here (Fig. 14). See also the drawing made by Secchi on 1 December 1864.

Changes in Herschel II Strait and Meridian Bay

Following up what has been said above, one of the surest ways of resolving the question of these problematical changes is to compare observations made at analogous oppositions, during which the globe of Mars is presented to us with the same

Fig. 14 Drawings of Mars by Schröter, in 1798, showing the point of a sea in a region where one is no longer seen

inclination. In taking this precaution, we eliminate one cause of discrepancies, due to variations in the inclination of the planet.

The oppositions of 1798 and 1800, 1815, 1830, 1845, 1862 and 1877 are of this kind. On each occasion the planet passed through opposition when near the perihelion of its orbit, that is to say under the best conditions for observation, and with the same inclination relative to the Earth. Thus, for example, in 1830 the southern solstice fell on 18 September and opposition on the following day; in 1862 the southern solstice fell on 9 September and opposition on 5 October; in 1877 the solstice fell on 26 September, and opposition had occurred on 5 September. In these different years, therefore, observations were made under conditions which were very similar, and the planet was presented to observers in the same position.

By taking into account the differences in the instruments used, as well as differences in the visual acuity and appreciation of the observers, we should therefore be able to draw up analogous representations of the planet. So let us now give a more detailed comparison between excellent observations made in 1830, 1862 and 1877.

In 1830, Beer and Mädler made 17 series of observations and 35 drawings between 10 September and 20 October. The feature which they noted as the most characteristic on the planet was a small black round spot, attached to a large grey patch by a strongly curved and winding arc. This was the sharpest point on the disk. It had already been noted on Schröter's drawing on 24 October 1800 (*fig.* 161 in our Fig. 50, Chap. 1). It was also shown on two drawings by Kunowsky, in 1821 and 1822 (Fig. 65, Chap. 1).

This small, round black patch was specially observed and drawn by Beer and Mädler, and chosen by them as the usual point in measuring the rotation period. Six drawings, on 10 and 14 September and 14, 19 and 20 October, represent it with perfect clarity, and well separated, with the winding arc of which it formed the extremity (Fig. 15).

The instrument used for these observations was a 3¾-in. (108^{mm}) Fraunhofer equatorial.

In 1862, Lockyer undertook the same series of observations with the aid of a 6¼-in. (158^{mm}) Cooke equatorial, and again saw the same aspect, with a little extra detail,

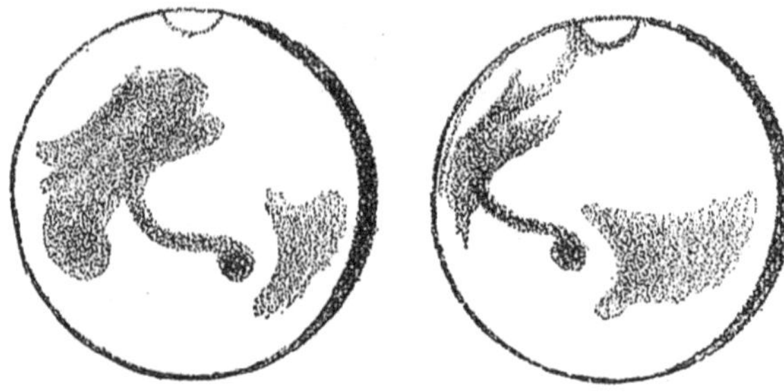

Fig. 15 The undulating ribbon drawn in 1830 by Beer and Mädler

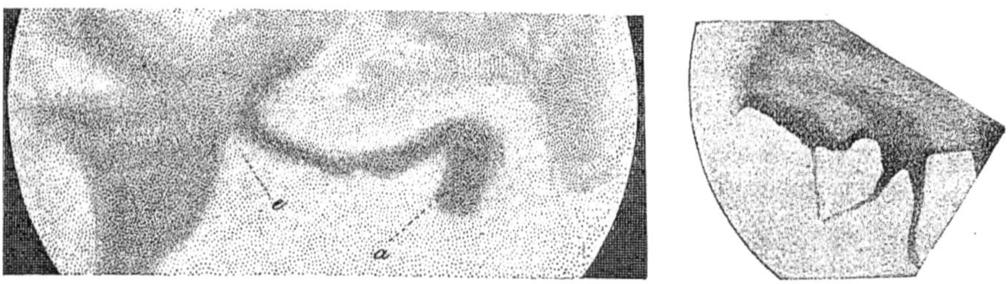

Fig. 16 (Left) The same ribbon observed in 1862 (25 September), by Lockyer. (Right) The sam, observed in 1879 (28 Nov.), by M. Schiaparelli

corresponding to the greater aperture of his telescope. But we can see one notable difference between his representations and those referred to above. Instead of a round black disk attached to a winding arc, Lockyer drew a rectangular patch continuing from a broadish ribbon (Fig. 16). The drawings by Kaiser, made during the same opposition, agree well with Lockyer's, confirming the broadening of the arc shown by Beer and Mädler, and also the rectangular form. Compare, for instance, with the drawing of 31 October (Fig. 2.45, Chap. 2).

If we now continue the comparison of the same feature, and come to the 1877 drawings, we find a still more striking difference. With a 218^{mm} Merz equatorial, more powerful than the two telescopes used in the observations described above, Schiaparelli observed details which had been previously unknown. But if we simply consider the appearance of the feature concerning us here, we return neither to a round disk attached to a thin ribbon, as in 1830, or to the appearance in 1862; we see instead a vast sombre streak, which is not detached from the upper patch, and with which is associated an intermediary grey tint. Never, or almost never, could this configuration be taken for a black circular disk, better suited than any other, because of its isolation and its clarity, to serve as the reference point for measurements of rotation. Moreover, in the observations of 1830 and 1862, this point was connected with a vast triangular sea—the Hourglass Sea—but in 1877 this sea was reduced to almost nothing by a tongue of land; which penetrated and almost cut off its upper part. The whole of this country, which Schiaparelli has called Deucalionis Regio, Mare Erythræum and Iapygia, is therefore certainly variable, and so is the normal patch shown by Beer and Mädler, which today we call the Meridian Bay. With the

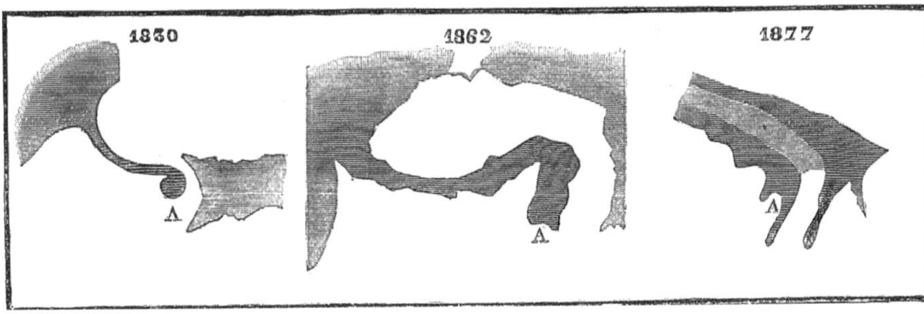

Fig. 17 Variations observed on Mars (the Meridian Bay)

preceding drawings, compare for instance that of 20 October 1877 (Fig. 25, Chap. 3). This, already cited in evidence in my *Astronomie Populaire* in 1879, from which is reproduced Fig. 17, clearly demonstrates the variation.

While in 1830 the Meridian Bay was so clearly detached, and appeared black against a bright background, in 1877 it was confused with the adjacent marches; and if one wanted to choose a characteristic black, sharp, circular point as a zero for the planet's rotation, as in 1830, it would not have been this one. One would have chosen the circular lake lying 90° to the east, which was *detached*, in 1877, *as a black disk, absolutely sharp*.

The observations of the Meridian Bay in 1879 gave yet another result which was sensibly different; there was a narrowing which caused a singular modification in appearance, recalling the aspect as it had been in 1862.

Thus the region of the Meridian Bay also is subject to obvious variations. This has been known for a long time.

On six drawings in made by Mädler in 1830, the curvilinear ribbon of which the circular disk forms the extremity flows out from the left to a sea, almost at a right angle, and seems to continue in a darker tint, following the undulations in the coast (see the two drawings in Fig. 15, cited above). This sea was named Dawes Ocean in the charts by Proctor and Green. Schiaparelli called it the Mare Erythræum. The 1862 drawings are in complete agreement with this for more, but there is nothing like it in the drawings of 1877 and 1879. Therefore what Schiaparelli has called the Mare Erythræum, above the ribbon which forms its border, seems to be nothing but a plain which was doubtless covered with water (or something else) in 1877 and 1879, particularly from 330° as far as 5°, but which in 1830 looked exactly like a continent—as it also did in 1862, and again, almost identically, in 1879.

This region, occupying about 35° in length from east to west and 20° from north to south (14° to 34°), that is to say, about 2,000 km in length by 1,000–1,200 km in breadth, seems to be alternately uncovered and submerged. The observed variations cannot be attributed to clouds, in view of the sharpness of the 1830 drawings between 10 September and 20 October.

Briefly, let us examine this region again.

Here (Fig. 18) is a disk of Mars drawn in 1890, on of which several double canals are shown. The upper horizontal one, Herschel II Strait, has never before been

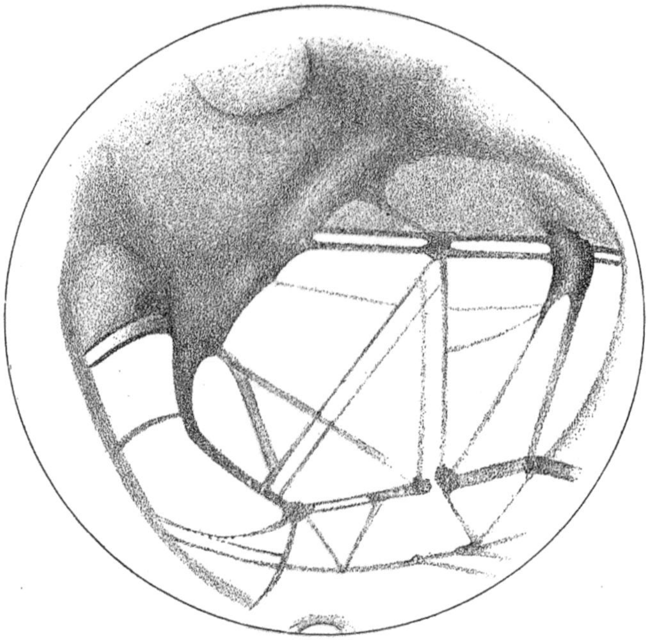

Fig. 18 The Herschel II. Strait in 1890

Fig. 19 The same region in 1888

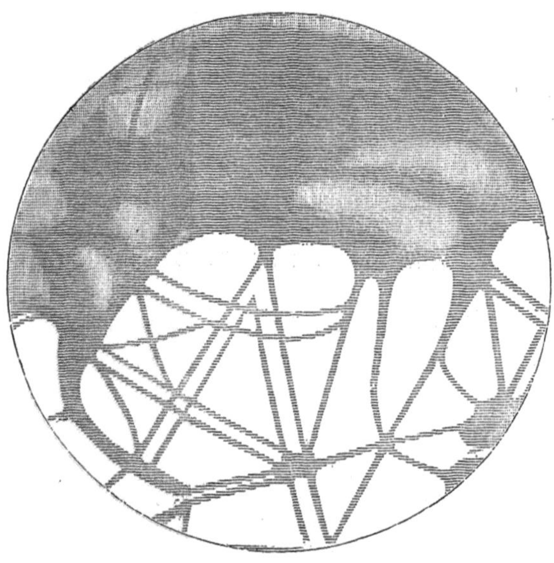

considered as a double canal. For comparison, let us take the drawing made in 1888 by Schiaparelli (Fig. 19).

The topographical aspect is completely transformed. Instead of being sinuous, the coastline is straight and double, divided by a white longitudinal trail. Also double—as usual—is the Meridian Bay; a small lake below is also double.

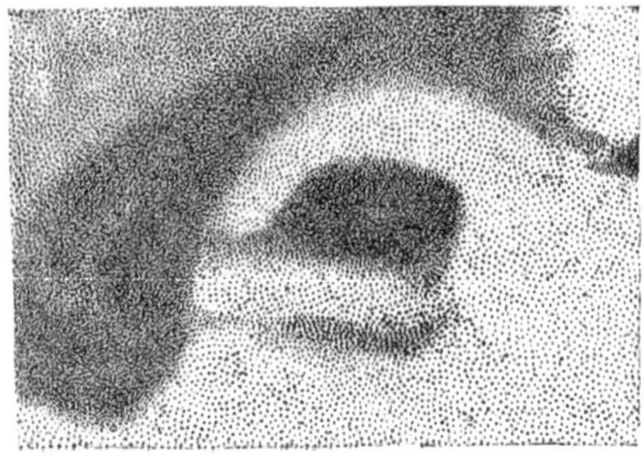

Fig. 20 The Terby Sea or Lake of the Sun in 1862 (Lockyer)

This region, like the preceding one, has been notable ever since 1879 for its observed changes, and I have already mentioned an analogous doubling seen briefly in 1877.

This tendency toward doubling is the phenomenon which we are particularly anxious to explain.

If these double canals are the two sides of a band of water, as could be inferred by comparing the aspect with that of a strait—brighter in the middle sector than along the borders—we still have to explain how the transformation comes about. Even if we assume that a sand bank has been raised up, the water could still run from one part to another without necessarily producing straight borders.

Changes Observed Around the Terby Sea

In the 1862 drawings, one feature noted (among others) was an oval patch, elongated east–west, attached to the neighbouring sea to the left by a narrow thread which was always visible. This patch, recalling the form of an eye, was shown excellently in Lockyer's drawings of 17 September and 18 October 1862; also on those by Kaiser on 24 October and 23 November, and those by Dawes in the same year. There can be no doubt about the existence of the strait connecting the circular lake to the neighbouring sea. The drawings of 18 October (Lockyer) and 23 November (Kaiser) are quite sufficient to make its existence certain (Fig. 20).

Now compare these drawings with the 1877 results—for example, the drawings by Schiaparelli and Green (Fig. 21) from September to December. The strait has completely disappeared, though the Eye and its surroundings remain easily visible and are not veiled by clouds. On my drawings of 1877, the lake is no longer attached to the sea. Here we have an unquestionable variation. Note that the Italian observer made every effort to recover the feature, but was unable to see the slightest trace of it. However, the streak was recorded as early as 1830, on the map by Beer and Mädler. Thus the change is absolutely proved. Below, I reproduce Green's drawing, which is in agreement with all the others made in the same year. The strait became visible once more in 1879, but was incomparably thinner than it had been shown in 1862 (Fig. 21).

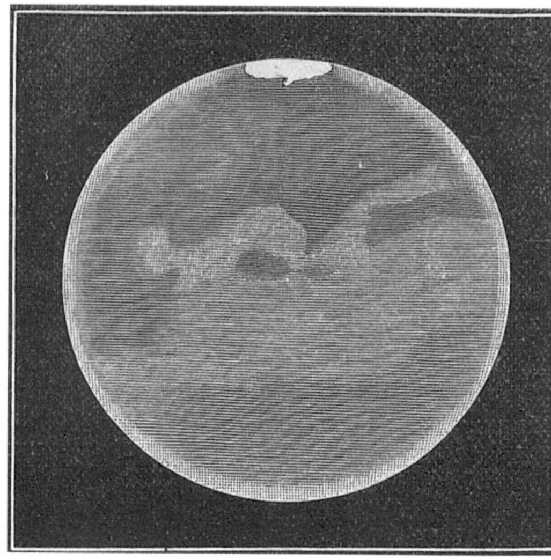

Fig. 21 The same sea in 1877 (Green)

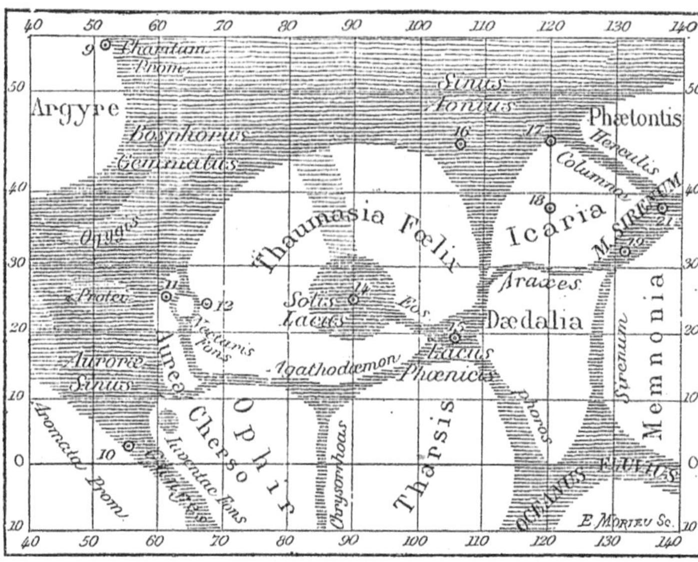

Fig. 22 The Lake of the Sun in 1877

This circular lake is 17° in length by 14° in breadth, or 1,020 km by 840 km; that is to say, its surface area is a little greater than that of France.

This study leads us to the certain conclusion that real changes take place constantly on the surface of their neighbour world.

According to Schiaparelli in 1877, the lake is circular (Fig. 22); a tributary joins it to the right to the little Phoenix Lake, and a second broader but paler tributary joins it above to the southern sea. Schiaparelli examined this region with very special

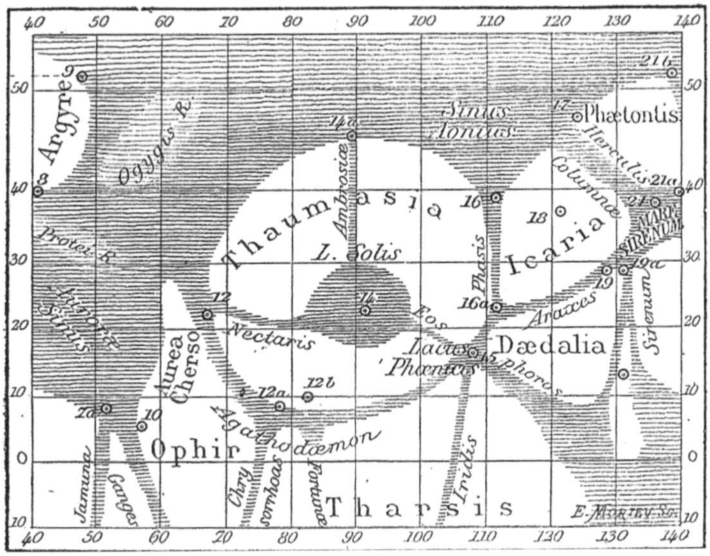

Fig. 23 The Lake of the Sun in 1879

care, because the appearance was sensibly different from that shown in the drawings made by Dawes, Lockyer and Kaiser in 1862 and 1864; the lake had then been oval, elongated in the east–west sense. In 1877, on the contrary, it was perfectly circular, with a slightly undulating border, and sometimes it seemed even to be rather elongated in the vertical sense. Moreover, in 1862 and 1863 a broad tributary was seen flowing from the lake leftward into the neighbouring ocean. Instead of this, the Milan observer saw the area suddenly clear, and in 1877 discovered the little circle which he named the Fountain of Nectar.

Mars again approached the Earth in 1879, and Schiaparelli observed the region once more. Obvious changes were found. The tributary referred to above, which had been invisible in 1877, was now perceptible again, although very thin, and was given the name of the Nectar canal; the Aurea Chersonesus was broadened, and the Chrysorrhoas had changed its position; instead of coming down vertically along the 86° meridian, it began at 78° and ended at 77°. The lake was slightly elongated toward the Nectar canal, which gave it the form of a pear, whose tail extended from 15° to 20°. The upper tributary was incomparably less broad than in 1877, and was given the name of Ambrosia. The Phoenix Lake was very reduced in size. The Juventæ Fons was sought in vain (Fig. 23).

With new studies in 1881 came new transformations. The lake was decidedly elongated in the east–west sense, concentric with the outline of Thaumasia. The Phoenix Lake had become the centre of numerous tributaries. The Agathodæmon gave rise to a lake which had already been indicated in 1877, but had developed greatly, and was named Tithonius Lacus. This view correspondence to those of 1862 and 1864. The Juventæ Fons, which had disappeared in 1879, was again visible.

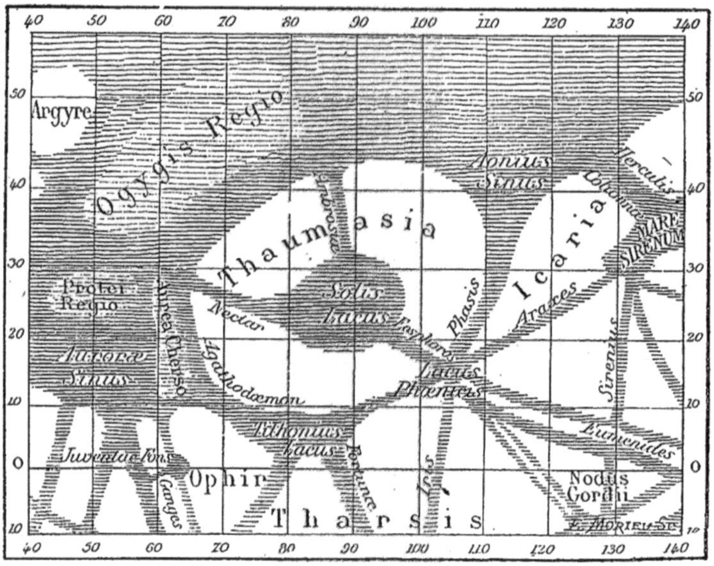

Fig. 24 The same region in 1881

Fig. 25 The same region in 1890

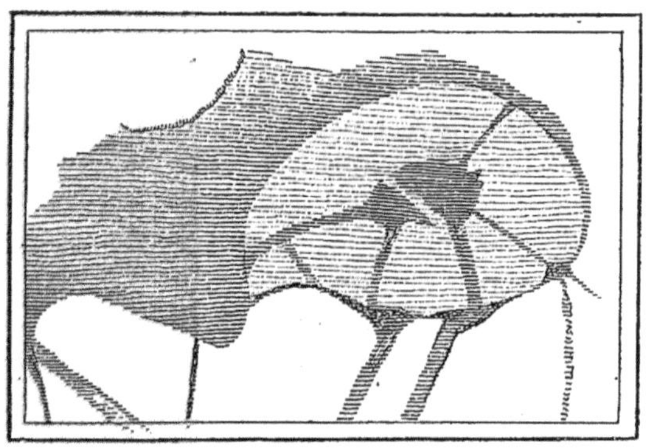

"*Che il Lago del Sole cambi di forma e di grandezza*," wrote the eminent observer, "*e cosa indubitable.*" Its colour was very dark, and was darker when rotation carried it to the edge of the disk than when it passed over the central meridian. This, as in other cases, was undoubtedly because the surrounding regions became whiter.

The Araxes was clearly shown, going right from the Mare Sirenum to Lacus Phoenicis, and was no longer winding, as it had been an 1877 (Fig. 24). Thus we have here a lake (or something which resembles a lake); it was oval in 1862 and 1881, and round in 1877. All its surroundings showed equally notable changes.

These three drawings are enough to establish, without question, the state of the planet during the observation period. Well, let us come now to 1890 (Fig. 25).

The lake is split in two; the little Tithonius Lacus below is also divided in two; the large tributary from the lake, which has been compared above with the tail of the pear, comes from the north-east instead of from the south-east (in all these drawings the north is at the bottom); the Ambrosia is inclined to the right of the meridian instead of to the left; the Chrysorrhoas canal isn't double as far as the Lunæ Lacus; beyond, as far as the Mare Acidalium, that doubling persists. Two new tributaries, previously unknown, come in from the Solis Lacus.

Here, then, is the state of the problem. There is no way of disputing that real, incontestable and considerable changes, take place on the surface of Mars.

The question is not lacking in interest. Beyond the fact that it is in itself fascinating to find out all we can about what is happening on Mars, is no less interesting to realize that there is a great general resemblance in constitution between Mars and the Earth; the atmosphere, the waters, its snows, its continents, its climates, its seasons—and yet Mars differs from the Earth in a most bizarre manner—with respect to its geographical configurations. It differs in its double canals, and more so in the sudden doubling of its lakes, some of which are as large as France.

How can we explain these variations?

The simplest hypothesis is that the surface of Mars is flat and sandy, and that the lakes and canals have no banks; in other words they are shallow, with only a slight depth of water, so that under suitable atmospheric conditions floods or even marshes can easily retreat, widen, overflow and even change in position. Since the atmosphere is thin, evaporation and condensation are easy. We may therefore assume that there are floods of various sizes and various durations. For example, the separation of the Solis Lacus in 1890 could be due to a diminution or a displacement of the water in the lake, the line of separation being regarded as a sandbank which was uncovered by the falling water-level.

This explanation accounts for part of the phenomena observed. But it does not explain one particular aspect: the doubling.

Note first that it does not seem that there will be less water when the tributaries from the lake are more numerous, and when the tributary to the left forms part of the arms of the sea.

Displacement of waters due to the tides? This would be periodic, and would last for only a few hours; it would not be characteristic of entire seasons.

Should we prefer to assume that the sandbank is raised above the water-level, and that in general the displacements of the waters are due to upheavals in the ground?

It is equally difficult to accept this interpretation, first because such instability of the ground would seem to be most extraordinary, and secondly because such disturbances would, in general, have to be rectilinear; thirdly because after several years the aspect reverts to its original state. In any case, such displacements of water could not explain the crucial, even characteristic tendency towards doubling.

We must recognize that it is therefore extremely difficult, if not impossible, to explain these transformations by any natural forces known to us. But it should be added that we do not know all the forces of nature, and there are some, very close to us, about which we are more or less ignorant. People who live in the tropics, and who come to Paris in winter for the first time and who have never seen snow or leafless trees, are amazed at our climate. It is something new to them to handle solidified water of this sparkling brilliancy, and they question whether those black skeletons can be trees which, several months later, will be covered with luxuriant foliage. Consider an inhabitant of Venus, who has never seen snow. Arriving on Earth, he would hardly understand the nature of the white patches which cover our poles. We Earth-dwellers can understand the snows of Mars; but we cannot explain these variations of coastlines, these displacements of water, these rectilinear canals and their doubling. We on Earth have nothing analogous.

Changes in the Canals

Let us again consider the small maps below (Figs. 26, 27, 28 and 29).

In 1877 the Hourglass Sea was very narrow, and no canal had been seen as double. Among others, there was one canal which had been named the Phison. In 1879 the sea was larger, the Nile seemed to have changed its course, and two canals were seen instead of only one. In 1882, another change had affected the course of the Nile, and it was now double over part of its length; the canals of 1879 were also

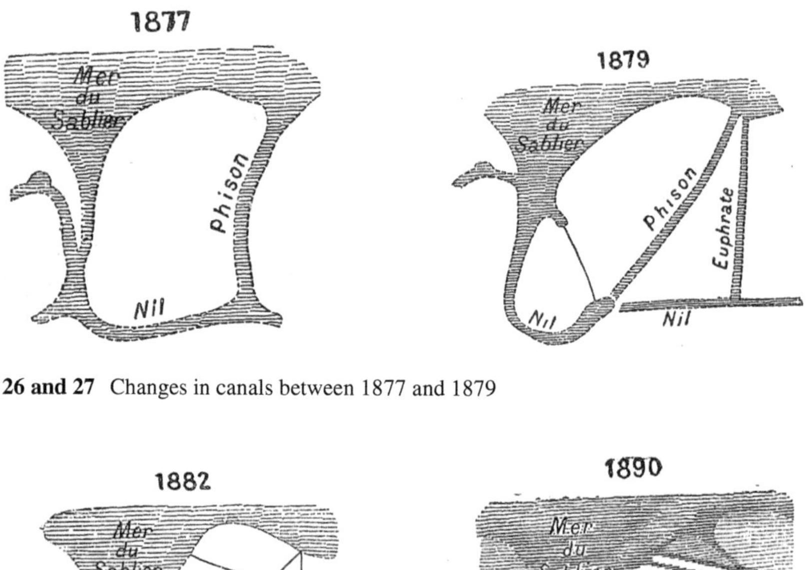

Fig. 26 and 27 Changes in canals between 1877 and 1879

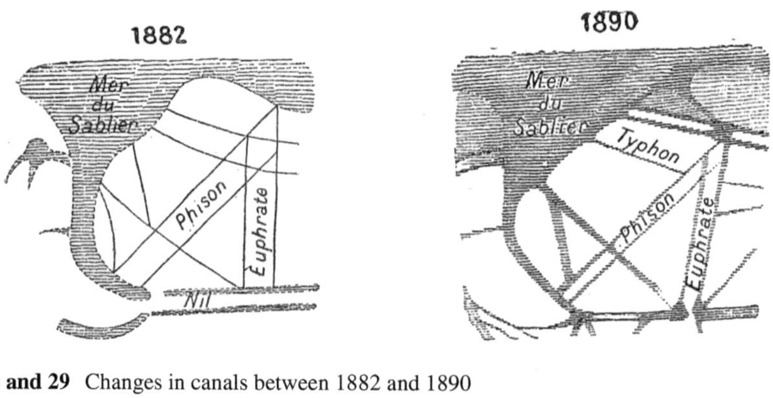

Fig. 28 and 29 Changes in canals between 1882 and 1890

shown as double, and five others were discovered. In 1888 the Euphrates, the Phison and the Nile (now called the Protonilus) showed double, as in 1882, but there was further doubling—of the Astaboras, which came down obliquely from the Hourglass Sea to a small lake (Ismenius Lacus), and also of another neighbouring canal. Still further changes were evident in 1890 (Fig. 29); the Euphrates and the Phison were shown as double, as was a part only of the Protonilus, but now the Astaboras was not; the 1888 canal had disappeared, and the upper strait was divided into two along its length.

The same conclusion may be drawn from an examination of the course of the water which comes into the Meridian Bay (Hiddekel, Gehon, Orontes, Edom) and from the Hydaspes and the Indus, as represented by the drawings made by Secchi in 1858, Kaiser in 1864 and Schiaparelli since 1877. I will not give further examples of drawings, which are already too numerous to engage the reader's full attention; but I must at least comment that the Hiddekel, broad and obvious in 1877, was completely invisible in 1879 and was replaced by a river of different form (the Orontes). It reappeared in 1882, the Orontes debouching further from the sea; it showed up in another form in 1888, with the Orontes flowing into the Meridian Bay at the same mouth. Note also that the Hydaspes was shown as very broad in Secchi's drawings of 1858 (3 and 5 June). The Indus varies in the same way. We have a particularly remarkable fact; that as a result of these variations, the Hourglass Sea seems to reproduce a form which has been repeated at least three times in different years. It lies at 95° to the mouth of the Indus, and 40° to the left of the prolongation of Syrtis Minor; it was seen also on the early drawings by Schröter (Fig. 30).

It is very hard to deny that these canals, which vary in such a way, represent some mobile liquid element. All, without exception, debouch into a sea, a lake or another canal, and in consequence water is a reasonable assumption. At the points of intersection between canals we often see a patch which gives the impression of a

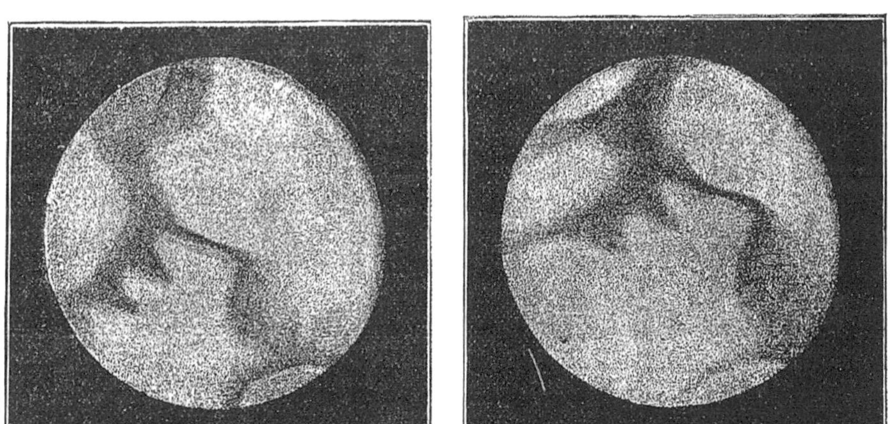

Fig. 30 The Hydaspes, observed by Secchi in 1858

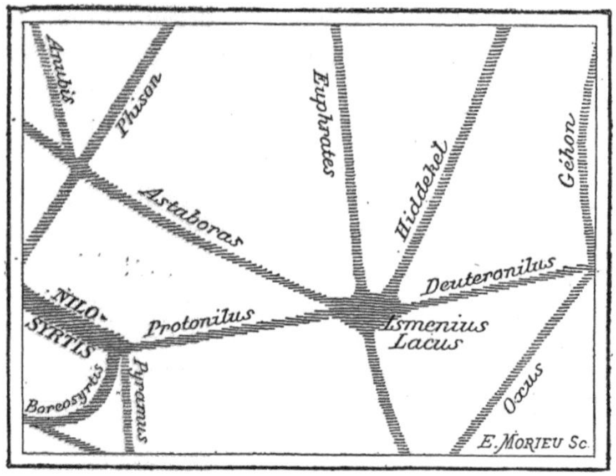

Fig. 31 Lake formed by the intersection of several canals

lake, as shown in Fig. 31. The aspects of these patches change in a manner analogous to the canals, doubling with them and in the same sense. Moreover, during the winter, long streaks of snow are seen crossing these lines; these snows are melted, as if the snow were falling upon water.

Can the cause be geometrical crevasses naturally produced on the globe of Mars? Perhaps; but by themselves, water-filled crevasses could not explain the variations observed.

Then, too, the canals are sometimes completely invisible even under the best conditions of observation. This disappearance occurs near the planet's southern solstice.

Their breadths are very different. Thus the Nilosyrtis sometimes has a breadth of 5°, or 300 km while others measure only 1° or 60 km.

Some are of tremendous length; more than a quarter of a meridian, over 5,400 km.

All of them change in breadth.

All, or nearly all, can double.

The reader is at least conversant with all the facts relating to these strange formations, as given in the dissertation by Schiaparelli published above, and it would be superfluous to repeat them.

These are the facts, which are truly extraordinary and which it is difficult to fit into any interpretation.

To sum up: it is difficult to accept changes in the level of the ground—seesawing motions—or that the surface of the globe of Mars is mobile, rising and falling; that the seas easily and frequently take the place of lands, and vice versa. Such conclusions are difficult to accept, first because we cannot understand why the surface of a planet should be so unstable and also because we find a recurrence in the geographical configurations observed over two centuries. We cannot imagine a planet swelling and shrinking, even partially, in the manner of a gaseous sphere covered by a thin skin. Undoubtedly our own Earth is far from being stable; from century to century the seas have taken the place of the lands and vice versa, and every day, without exception, the Earth's surface stirs at one point or another. But instability reaching the proportions indicated by the phenomena on Mars, and occurring from 1 day to the next, would be comparable to the sea sweeping in to Paris and then retreating back to Cherbourg again, and this is very difficult to accept. Keep in mind, moreover, that the general Martian geography remains stable.

Let us examine the curious problem of the canals on Mars a little more closely.

The Canals, the Rivers, the Geometrical Réseau of the Continents and the Circulation of the Waters

We have reached the most delicate part of this book, and I am well aware of the difficulties.

Should we accept this vast geometrical réseau which extends over all the continents? If so, can we find an explanation for it?

One can easily understand the doubts which have been cast upon Schiaparelli's assertions. First, these straight lines lead from one sea to another, cutting each other, and appearing so unnatural that it was difficult to accept them, without full confirmation. Such confirmation was long delayed; and when it came, there were intimations that the idea of lines had been put into the observers' minds by the maps in front of them so that the observed features were due to auto-suggestion. The canals were discovered in 1877 and, more conclusively, in 1879, and were seen doubled in 1882. It is true that some of them could be found on older drawings. The Nectar was shown in 1830 on the map by Beer and Mädler, the Hydaspes on the drawings by Secchi in 1858; canals are also found on the drawings by Dawes in 1864, Burton and Dreyer in 1879, Niesten in 1881, Knobel in 1884; but it was only in 1886 and 1888 that Schiaparelli's curious réseau was confirmed, at least for the most part, by Perrotin and Thollon at Nice and by Terby at Louvain; then in 1890 by Stanley Williams in England, Pickering in the United States, etc. Very many observers have looked for them in vain, even with the most powerful instruments; personally, I have seen only the broadest of them (Nilosyrtis, Ganges, Indus) and then only in 1892, with the $0^m.24$ equatorial at the Juvisy Observatory. With the same instrument, a particularly keen-signed observer, Léon Guiot, has seen many more of them. It is only fair to add that during these last oppositions the planet has remained low over the Paris horizon, and has not cleared the thick layers of the lower atmosphere.

To me, it seems very difficult not to accept the full accuracy of Schiaparelli's observations. The Milan astronomer is an excellent observer; moreover, his results have now been verified by a number of different observers. I therefore believe that we should regard this strange réseau of lines as real, at least in its essential outlines. Certain details remain doubtful.

C. Flammarion, *Camille Flammarion's The Planet Mars*, Astrophysics and Space Science Library 409, DOI 10.1007/978-3-319-09641-4_11

Can we find an explanation?

There is nothing analogous on the Earth. And we have, unfortunately, only terrestrial ideas from which to reason.

Assuredly, hypotheses are not lacking; but we want to discover the true cause.

That which we have examined above, presented by M. Fizeau of the Académie des Sciences, and which regards the canals as open crevasses and immense ice-fields, does not seem to be admissible, for the simple reason that we can observe the ices of Mars, and know that they are limited to a definite distance from the poles. The seas and the continents are completely different in appearance, the first greyish, and of varying degrees of darkness, the second yellow-ochre, more or less red, and contrasting sharply with the white ices. Do we want to imagine that the ices on the continents are reddish, while those at the poles remain white? But if the planet were in a glacial epoch, why should not the seas, surrounded by frozen continents, have solidified also? For the most part these seas are small, and presumably shallow. If the continents were made up of deep glaciers and crevasses, everything on Mars would be frozen. Mars, therefore, is not glacial.

It has also been suggested (see *L'Astronomie*, 1888, p. 384, an article by E. Penard) that these enigmatical lines could represent fissures; geological cracks caused by the cooling of the planet. The valley of the Rhine, between the Vosges and the Black Forest, is the result of this kind of action. The course of the Rhine appears similarly rectilinear when seen from afar. This hypothesis is undoubtedly more plausible and more acceptable than the preceding one, but against it we have the regularity of these immense straight lines, and their equally regular intersections. We have to admit that the Martian globe is, according to this theory, completely cracked over the whole of its continental surfaces. This is not impossible. Water could easily penetrate all the cracks. But it must be said that the appearances of the lines do not favour this natural hypothesis. On the globe, could Nature trace such straight lines, cutting each other in such a fashion?

Schiaparelli has asked himself whether, on Mars, we could have a geological situation recalling the geometrical forms proposed in terrestrial orography by Élie de Beaumont, but he has come to no firm decisions.

M. Daubrée (see Société Astronomique de France, meeting of 7 May 1890, and *L'Astronomie*, 1890, p. 213) has attempted to model the Martian canal réseau by means of deforming a rubber ball, wrinkling it and breaking its spherical skin by contraction. Nothing that he has been able to obtain resembles the canal réseau, though he has obtained imitations of terrestrial mountain chains, continents and seas. But he has been able to imitate the Martian réseau by an opposite process: by distending a spherical crust and observing the resulting cracks. A plastic cast, made from a putty and paraffin mould, was applied to a rubber balloon which was gradually distended by the introduction of water under pressure. Rectilinear breaks were produced, often parallel, two by two, cutting at various angles to each other and recalling what is seen on Mars.

We can, again, suppose—as does M. Armelin of the S.A.F.—that the continents of Mars are sandy beaches, and that rain water streams along the surface producing watercourses which can change in position from one season to another. But the length of these lines, their geometrical precision, and above all their intersections, would seem to present insuperable objections to this theory.

Truly, the more we look at these maps and drawings, the less we feel that they can be the chance work of Nature.

In trying to give a rational explanation of the canal mystery, one question above all stands out at once: do these lines give the impression of being watercourses?

To discuss this, let us return to the question asked before the discovery of the canals by Schiaparelli in 1877. Before that time, I wrote that certain gulfs of the Martian seas, certain bays lengthened into points in the interiors of the lands, give the idea of mouths of great rivers.

It may be appropriate to come back to this idea.

For instance, look again at the drawing made by Dawes on 20 November 1864 (see Fig. 54, Chap. 2). Note, in the region marked *a*, the Forked Bay discovered by Dawes himself—the point adopted for the Martian prime meridian, and for which I have proposed the name of the Meridian Bay.

The appearance of this gulf gave this eminent observer the impression of two very broad river-mouths. He himself sought out the rivers themselves, but could discover no trace of them.

Now, among Schiaparelli's canals we can note two, the Hiddekel and the Gehon, which were precisely in the position of issuing from these mouths (see Fig. 1).

Why should not these dark lines be the rivers indicated by these mouths? Why should they not spread out on to the tableland, forming a shallow sea?

Let us continue on this theme.

On the drawing by Dawes we note, to the right of the forked Meridian Bay, another mouth, still more greatly lengthened.

This is the mouth of the Indus, with which comes the Oxus.

Here we have no need to imagine canals or mysteries. We have, very naturally, a river—very broad, wider than the Mississippi, the St. Lawrence or the Amazon—but this is no objection, particularly on a flat terrain.

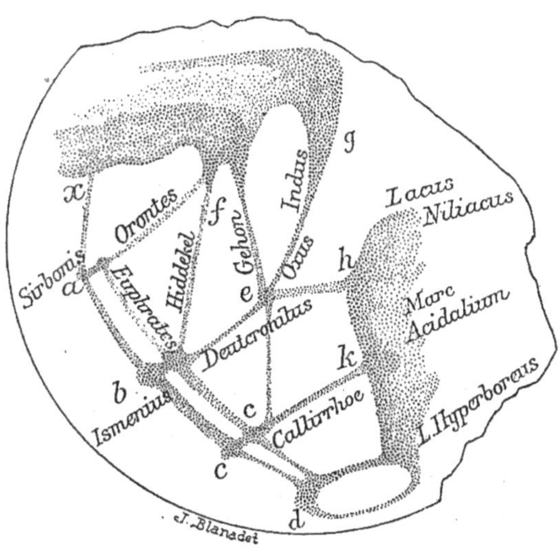

Fig. 1 Rivers of Mars.—The Gehon, the Hiddekel and the Orontes leading to the Meridian Bay. The Oxus and the Indus

Thus a careful discussion of Dawes's observations, and a comparison of them with Schiaparelli's, leads me to put forward the following explanation:

The Hiddekel, the Gehon, the Indus and the Oxus are, quite simply, rivers.

Equally, I suggest that the same explanation applies to the dark lines which issue from gulfs or mouths, such as the Lethe, which flows from the Flammarion Sea at Gruithuisen Bay (Green's map); the Titan, which flows from Herschel II. Strait at Schmidt Bay; the Ganges or the Channel, which flows from Christie Bay; the Chrysorrhoas, which flows from the variable Tithonius Lacus; the Araxes, the Sirenius, and in general all those which debouch from gulfs and mouths.

All these rivers run from north to south, and, by successive transformations, arrive at the north polar snows, where there are continents higher than the equatorial seas into which they flow; these equatorial seas, moreover, extend to the south pole, where there is no great land-mass.

These stretches of water are subject to *considerable variations*. Sometimes the Indus is much wider than the Hydaspes, sometimes narrower. For instance, Secchi, in his 1858 observations, constantly observed and drew (1) the Meridian Bay, a simple gulf lengthened to a point; (2) the mouth of the Indus and (3) the Hydaspes, without ever showing the Indus itself. We may recall what was said at the end of the last chapter. Comparisons made with other regions lead to the same conclusion.

At this point there is one important characteristic to be noted.

The existence of numerous rectilinear watercourses crossing the Martian continents strongly indicates that the continents are vast plains.

The frequent widespread inundations along the banks provide further confirmation.

We must therefore assume that the continents are almost completely flat.

But the term river is not applicable to a very broad watercourse, that is to say a stretch of water lacking in depth.

Thus it may be that the streaks are made up of rivers, or at least primitive rivers, which give rise to branches and streams which are then lengthened until they pass from one sea to another, giving the same result as that suggested by Daubrée. The cause of the lengthening would presumably be areological.

Mists of a special nature could extend along these watercourses. Refraction phenomena could produce doubling under certain circumstances. These mists could play a major role in explaining the observed phenomena.

On the Earth, water exists in five different states: *solid ice*, *flakes of snow*, *the liquid state*, *the cloudy state* and *the invisible state of transparent vapour, spread in the atmosphere*. It may be that on Mars, water exists in six states, and that the cloudy state is divided into two others.

On the Earth, the clouds and the mists differ sensibly from each other. Mist is stationary; clouds move. Wind passes across mist without sensibly displacing it; wind carries cloud along. In my books, *L'Atmosphère* and *Voyages en ballon*, I have given proof of this distinction.

The sixth state which I postulate has some resemblance to our mists, and represents a kind of exaggeration of them; I assume it to be a state of vapour which is so dense that it is almost liquid. Assume a sheet of thick mist, viscous, evele, dark, which continues the seas along the rivers, even sometimes totally or partially replacing the lakes. This is a transformation if water which we can accept. Under certain atmospheric conditions, water could pass from the liquid to the viscous state, then to the cloudy state, and then to invisible vapour.

This could happen if the water were not completely liquid, and was condensed by its weight into stable basins, as if its molecules were separated—so forming a viscous fluid, heavier than the air, and subject to forces other than gravity. Imagine, for a moment, that these molecules have a tendency toward aggregation, but nevertheless obey other forces in the usual way—for instance electricity and planetary magnetism, as well as unknown forces (because we do not pretend that we know all the forces of Nature). These waters, perhaps liquid at the centres of the seas but fluid, in the state of viscous vapour or gas, near the borders and at higher levels, such as in rivers or canals—these waters could extend or dry up according to the atmospheric conditions of heat, electricity, etc. The streaks of vapour would be essentially variable in appearance, thickness and density. If, under certain conditions, these molecules become electrified, they could repel each other (as we see in the phenomena of positive and negative electricity), and produce the doubling observed. The canals, lakes and aqueous regions can change in position. They are kinds of dense mists, which meekly obey the forces acting upon them; the proximity of the seas, the dampness of the ground, they hygrometric state of the air, temperature, electricity, etc.

True, we must now assume a very calm atmosphere. But does not this seem to be the state of affairs on Mars?

Such an hypothesis would explain the variations of extent and tone, the doubling, the disappearances and reappearances according to the seasons, and all the innumerable changes in aspect which are admittedly hard to account for by water in the same state as ours. Neither, need these Martian waters be the same chemically and physically as ours.

That they are similar in some respects is indicated by the appearance of the snows, white like ours, and which melt or evaporate under the action of the Sun—just as we discover on Earth. This analogy is made more probably still by the absorption lines in the spectrum of Mars, which correspond to lines of water vapour. But it is equally probable that these waters differ from ours.

Who knows if on Mars there is a different combination of elements, instead of our own hydrogen, oxygen and sodium chloride?

For one thing, the density is not the same as it is here. A cubic metre of terrestrial water weighs 1,000 kg; a cubic metre of Martian water would have a density of 0.711, and would weigh only 711 kg, assuming that the Martian water would be less dense than terrestrial water in the same ratio as the density of Mars is less than that of the Earth. The specific gravity would be 3.91 instead of 5.50. Yet the difference in gravity is greater still, and 1000 terrestrial kg taken to the surface of Mars would weight only 376 g.

The conditions are therefore completely different from those we know on Earth. It is the same with the atmosphere, whose pressure plays so important a role in the transformation of water. If the Earth's atmosphere disappeared, the seas would immediately evaporate to give birth to a new watery atmosphere, until the pressure became high enough to allow water to exist in the liquid state. Continuing to remove the atmosphere would end by making the seas dry up completely.

If Mars had the same atmosphere as the Earth, this atmosphere would nevertheless be less dense than ours, in the proportion of 376–1,000. Instead of reading 760^{mm} at sea level, a barometer would read only 286^{mm}. This is the barometric pressure at the summits of our highest mountains, at an altitude of nearly 8000^{m}. Such an atmospheric layer is very rarefied, even at sea level, and it seems that under such conditions the Martian atmosphere cannot be very extended.

Yet it could contain substances—gases and vapours—which do not exist in our own air.

We are, however, faced with one great difficulty, the geometrical and rectilinear pattern of the réseau does not appear natural. The more we look at these drawings, the less it seems that we can attribute them to blind chance. However, let us remember that we are far from knowing all the forces of Nature.

On the other hand, are we justified in rejecting the hypothesis of intelligent action on the part of possible inhabitants of our neighbour world?

As we have seen above, the actual conditions on Mars are such that it would be wrong to deny that it could be inhabited by human species whose intelligence and methods of action could be far superior to ours.

Neither can we deny that they could have straightened the original rivers and built up a system of canals with the idea of producing a planet-wide circulation system. In view of our absolute ignorance in this respect, it would be wrong to reject the possibility; such an attitude would be unscientific.

The hypothesis of an intelligent origin for the canal system appeals to our spirit. We are forced to take into consideration all the evidence we have. True, there are abundant objections. Is it likely that the inhabitants of a planet could construct works as gigantic as this? Canals 100 km in breadth? Can we think so? And to what end?

Well!—in the hypothesis of a human origin for these markings, we can find the explanation in the state of Mars itself. First, we have seen that the materials there are much less heavy than they are here. Secondly, cosmological theory indicates that Mars is older than the Earth. It is therefore natural to conclude that it was inhabited sooner than the Earth, and that its humanity, whatever it may be, is more advanced than ours. While the crossing of the Alps, the Isthmus of Suez, the Isthmus of Panama and the Channel Tunnel between France and England appear as colossal scientific enterprises, indicative of our time, they will be child's play to humanity of the future. When we sing the praises of our nineteenth-century progress, such as railways, the telegraph, applications of electricity, photography, or the telephone, we ask ourselves what would be our reactions if we could see the material and social progress of, say, the twentieth or twenty-first centuries. Not even the least optimistic of prophets will doubt that serial navigation will become the ordinary mode of travel, so that the so-called frontiers between different peoples will be effaced forever; the inflamed hydra of war, and the unqualified folly of permanent armies, ruinous and opprobrious in an intellectual society, will be wiped out before the rise of humanity believing in light and liberty! Is it not logical to admit that since it is more ancient than ours, the Martian humanity should also be more perfect, and that the fruitful unity of its peoples will result in great developments?

We do not know the nature of these long, evele traces across the continents, or even whether they are uniform in breadth, so that we certainly cannot prove that they are canals full of water. A thousand conjectures could be made. There could be drainage works for the water which is becoming rare on Mars; we might prefer to picture a series of collective cultures on a globe which as attained a period of peace; we can recall that in an interesting article in the Times, Proctor suggests that the inhabitants of Mars could be engaged in vast engineering works, so that the lines spread in all directions, keeping a constant and significant distance from each other. At a meeting of the Royal Astronomical Society of London, Green, the skillful observer of Mars, noted this interpretation, adding that he did not wish to introduce any hint of brevity into a scientific matter of such importance; the geographical appearance of Mars merited the greatest attention, and it would be of the highest interest to verify the existence of the canals. Maunder, at Greenwich, commented that it was very strange that the canals could change in position, and that they were sometimes visible and sometimes not; they were not canals in the true meaning of the term, but were more probably the boundaries between districts of different degrees of darkness. Be this as it may, Nature could have been copied. The great inundations, frequent and always threatening, wearing away the level continents, could have produced the impression of artificiality. It seems that at present it is no more possible for us to give an explanation of the Martian réseau than it would be for an inhabitant of Venus to explain the réseau of our railways simply by natural forces.

The globe of Mars has been almost smoothed over the centuries. Water exists only in small quantities, so that the mountains are broken up very slowly by the rains; rivers carrying evele to the sea-bottom do so gradually, while at the same time the amount of water diminishes, penetrating the interior of the globe and being fixed in the rocks in the form of hydrates. The whole globe ends up by being slowly eveled. There is nothing surprising in the idea that on Mars, the efforts of civilization have been above all directed to spreading out the life-giving waters on to the old continents.

These rectilinear streaks, connecting the Martian seas with each other, appear to be intentional. Is it water which flows there? Yes, without doubt—in principle; but there must be another form of water associated with it, which I have described above: elongated mists above the streaks, which make them appear broader to our eyes, and make them subject to considerable apparent changes.

Perhaps, also, the phenomena of vegetation are added to this circulation of the waters.

The doublings come into a different category. It is difficult to believe that new canals can really be formed from one day to the next, similar and parallel to the first; it is preferable to think that they are due to the mists referred to above, and involve a double refraction in the Martian atmosphere. With these temperature conditions (the Sun's heat passes easily through the Martian atmosphere to warm the soil), evaporation must be very intense, and a large quantity of rapidly-cooling water vapour must be held constantly above the water; this could give rise to special phenomena of refraction (see too *L'Astronomie*, 1889, p. 461; an article by M. Meisel). To me it seems rational to involve the effects of refraction, above all because all trace of a canal may suddenly disappear to give way to new lines nearby.

M. de Boë has attributed these doublings to secondary images formed in the eye of the observer, as does indeed happen in looking at a straight line drawn in ink on a white box and placed to either side of the point of precise vision. But can we admit that the observers of the canals do not know when the image is in focus?

Whatever may be the true explanations—those given here are certainly premature, and are preliminary hypotheses only—the considerable variations observed in this aquatic réseau provide evidence that the planet is the site of energetic vitality. To us, these variations movements take place in silence, because of our great distance; but while we calmly observe the continents and the seas, slowly carried round in front of our eyes by the planet's rotation on its axis, we can ask ourselves upon which of these shores it would be most agreeable to live. Can there not, be there at this very moment, storms, violent tempests, volcanoes, social tumults—all the battles in the struggle for life? Similarly, the astronomers of Venus, armed with optical instruments analogous to ours, look at the Earth and see a calm smoothness under a tranquil sky; yet they certainly cannot doubt that in these continents gilded by the Sun, and these azure seas which are cut up into delicate gulfs, there is ambition, greed and barbarity; often added in the form of willful storms in the fatal inclemency of an imperfect planet.

We can however hope that since the world of Mars is older than our own, its inhabitants may be wiser and more advanced than we are. Undoubtedly it is the spirit of peace which has animated this neighbouring world.

The explanation of rivers modified by the inhabitants of Mars, and of a rational system of the distribution of the waters over continents which have been leveled over the centuries, does not, to me, seem absurd. Mists can easily form above the canals, and double atmospheric refraction, recalling that of Iceland spar, could explain the doubling. After all, we are discussing a new world, and we should deny nothing.

Let us nevertheless admit that there are some natural phenomena which are unknown to us, and which may be simpler than we expect. Our knowledge is inadequate. It would be a great error to claim that Science gives us the last word, and that we are in a position of knowing everything; it is no less puerile to claim that we are familiar with all the forces of Nature. On the contrary, the Known is a tiny island in the midst of the ocean of the Unknown. Moreover, our senses are very limited; our power of perception is still lacking; our science remains, and will always remain, fatally incomplete.

Assuredly, these bizarre phenomena carry us to another world, very different from that upon which we live, although offering some sympathetic analogies. From the viewpoint of the atmosphere, seasons, climates and meteorological conditions, Mars appears to be just as habitable as the Earth—or even better—and could well be actually inhabited by a human race very much superior to ourselves, since in all probability it is older and more advanced. Is the industry of these unknown beings shown by the rectilinear network of canals, which double at certain seasons? What causes the sudden and enigmatical changes which we observe from here? Undoubtedly we will need many years yet to discover exactly what goes on in our neighbour world in the sky.

This is why we must ask the question so precisely and discuss it methodically. All our work is carried out with this in view.

Summary of the Conditions for Life on the Surface of the Planet Mars

As a general conclusion from the study of the planet, I cannot add to the pages at the ends of the three parts of this book. It would be superfluous to repeat them here, and I ask the reader to read these articles again. To them I can add, in conclusion, the following résumé:

As was commented by William Herschel, Mars seems to be the planet in the Solar System which most nearly resembles the Earth. We can repeat today what the great observer wrote over a century ago, on 1 December 1783, about the inhabitants of Mars: Its inhabitants probably enjoy a situation in many respects similar to ours.

1. With regard to density, it is true that Venus more closely resembles the Earth, but as yet we have no certain knowledge of its rotation period, its seasons or its geography.
2. Observations of the surface of Mars establish that the glove has a mean temperature, climate and seasons very little different from those of the Earth, and, moreover, equally varied. The temperature conditions there lead to effects analogous to those of terrestrial meteorology.
3. The duration of day and night is $24^h\ 39^m\ 35^s$.
4. The years are almost twice as long as ours, and last for 687 Earth days or 668 Martian days. It follows that the seasons also are almost twice as long as ours. But the axial inclination is almost the same, and the relative intensity of the seasons too, is nearly the same.
5. The Martian atmosphere is optically clearer than ours. It is more rarefied, and, without doubt higher. Clouds and rain are rare, and violent tempests have never been observed.
6. There is rather more land there than sea. The seas are finely cut up into elongated Mediterraneans. The shores are joined beaches, in general exposed to the floods, or to banks of mists bordering the waters. It seems that the seas are shallow.
7. The diameter of Mars is roughly half that of the Earth, and measures 6,753 km. The waters occupy 66,000,000 square km, and the lands 77,000,000 square km. The habitable area appears to be five or six times that of Europe.

C. Flammarion, *Camille Flammarion's The Planet Mars*, Astrophysics and Space Science Library 409, DOI 10.1007/978-3-319-09641-4_12

8. From Mars, the Sun has a diameter of 21′, instead of 31′, as from Earth.
9. Two tiny moons circulate rapidly in the sky
10. The globe is older than the Earth, and seems to have been almost completely levelled. The northern hemisphere is, however, higher than the southern. We have not found great chains of mountains, but only a few elevated plateaux.
11. The canals appear to be surface fissures produced by geological forces, or perhaps even to the straightening of old rivers by the inhabitants, with the aim of distributing the waters over the surfaces of the continents.
12. It is possible that Mars is actually inhabited by a human species analogous to our own. More light-weight, no doubt; and older, and much more advanced. However, major differences exist between the two worlds. We do not yet have enough information to speculate about the possible forms of the human, animal, vegetable and other types of life on Mars. But the habitation of Mars by a race superior to ours seems to me to be very probable.

In life we experience some charming hours; exceptionally agreeable sensations, for instance, and good humour which touches the sky and envelops us. Well!—in the midst of these hours of enchantment, there are very few which bestow on the spirit a more complete satisfaction, an emotion nobler and more elevated, than the observation of Mars, on a clear summer evening. It is to be regretted that so very few people experience it. See before you a world—another world—with continents, seas, coasts, gulfs, capes, islands, river estuaries, snows glittering with whiteness, golden lands, dark waters—all of this placed before you at the eye-end of your telescope, giving night and day to different regions, passing from spring to summer, summer to autumn, presenting a miniature of the Earth as seen from space… here is scope for a contemplation which will bring us face to face with the greatest of all mysteries, the mystery of universal and eternal Life. Here we face sublime Truth, and come to terms with the Creation itself. Seen in such guise, the Earth becomes a province of the universe—we sense unknown brothers in other father lands of the Infinite!

To this sentiment we may perhaps add that of the beauty and grandeur of the conquests of modern Astronomy. The novelty of it has always excited particular attraction. Yet this is the first time since the origin of mankind that we can discover a new world in the heavens, similar enough to our Earth to awaken our sympathy; this is the first time that a book such as the present one could have been planned and written. Many years will undoubtedly pass by before we can gain much information about our other neighbour, the planet Venus, as we have already gained about Mars.

But what does science hold for the future? Will Martian and terrestrial humanity ever be in touch with each other? Generations pass. Progress continues on the march of its ascent…

My own youth and adolescence in a scientific and literary career have been devoted chiefly to the defence of the concept of the Plurality of Worlds, and to showing that astronomy does not stop at celestial mechanics; it should also concern itself with the *conditions of life*, past or future, in the immense Universe. I am happy to have lived long enough to help with the birth and development of physical astronomy, to contemplate the first world to be explored in the heavens, and to have had the privilege of writing this history. Perhaps my readers have shared my own satisfaction. Yet the preceding pages represent only a coarse and humble prelude to the discoveries reserved for our posterity.

Appendix: The Opposition of 1892

As the last pages of this book go to press, some interesting observations have already been made at the Juvisy Observatory, during the current opposition. I am happy to be able to add here some of these drawings, which need no description (Figs. A.1, A.2, A.3, A.4, A.5, A.6, A.7, A.8, A.9, A.10, A.11, A.12, A.13, A.14, A.15, and A.16).

C. Flammarion, *Camille Flammarion's The Planet Mars*, Astrophysics and Space Science Library 409, DOI 10.1007/978-3-319-09641-4

Fig. A.1 Drawings of Mars made in 1892 at the Observatory of Juvisy; by M. Schmoll

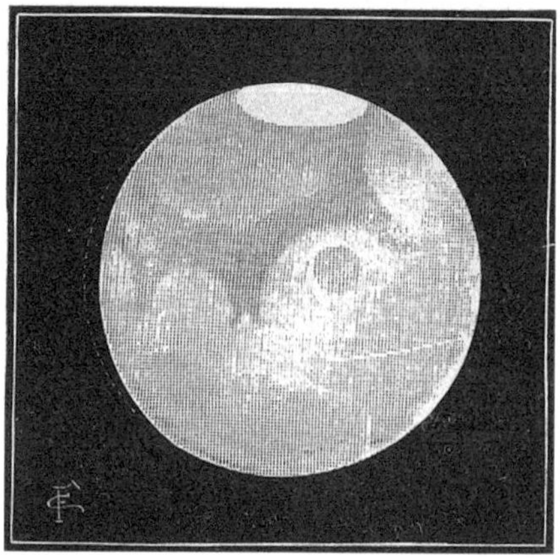

Fig. A.2 Drawings of Mars made in 1892 at the Observatory of Juvisy; by M. Mabiere

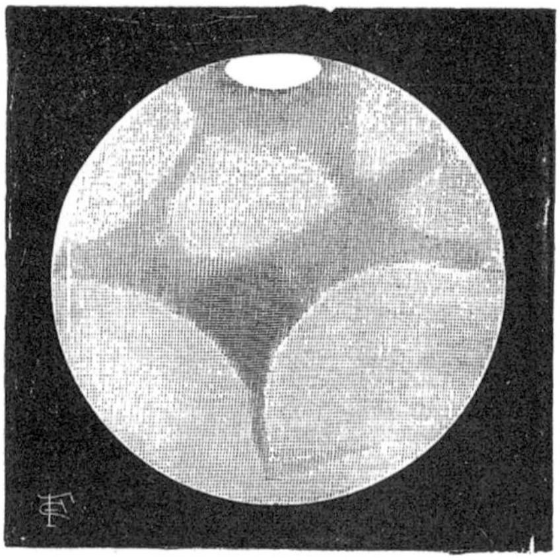

Fig. A.3 Drawings of Mars made in 1892 at the Observatory of Juvisy; by M. Flammarion

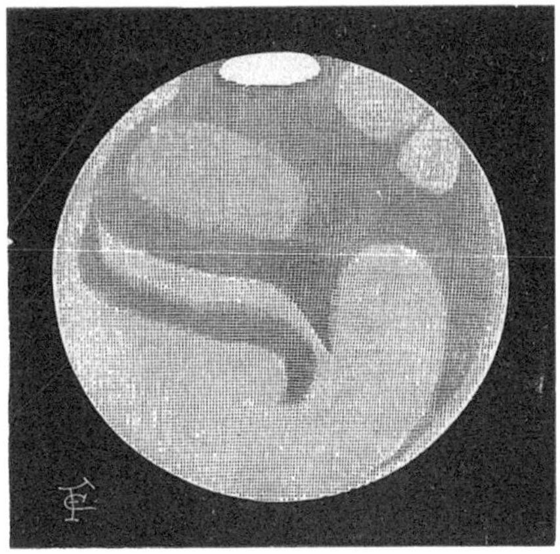

Fig. A.4 Drawings of Mars made in 1892 at the Observatory of Juvisy; by Léon Guiot

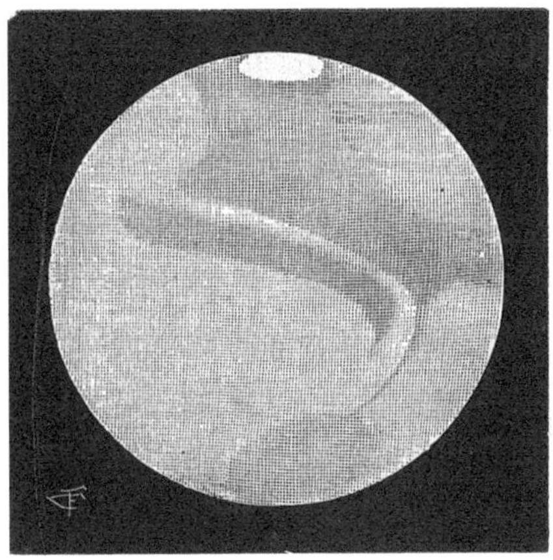

Fig. A.5 Drawings of Mars made in 1892 at the Observatory of Juvisy; by Léon Guiot

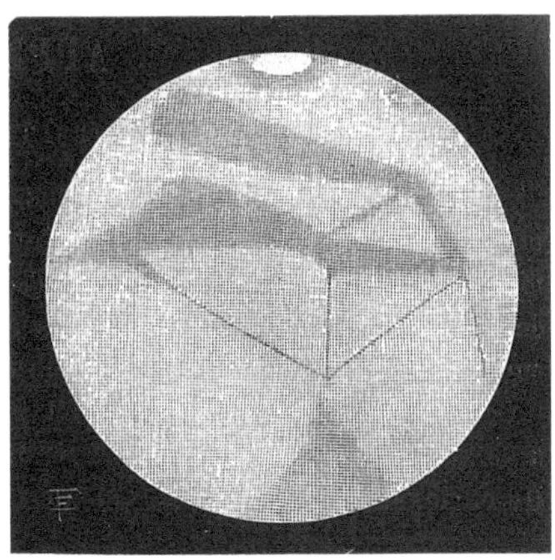

Fig. A.6 Drawings of Mars made in 1892 at the Observatory of Juvisy; by Léon Guiot

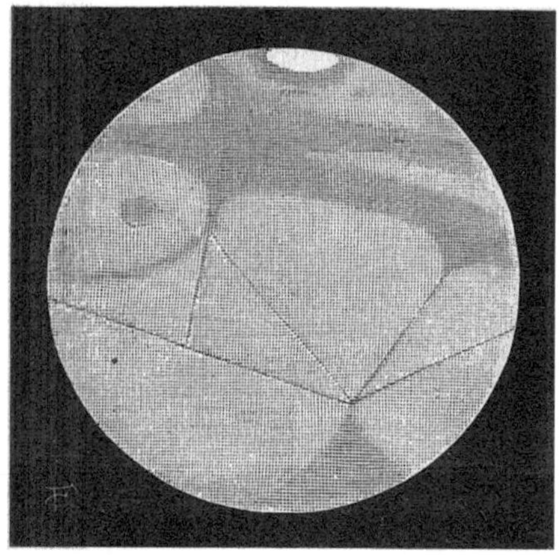

Fig. A.7 Drawings of Mars made in 1892 at the Observatory of Juvisy; by Léon Guiot

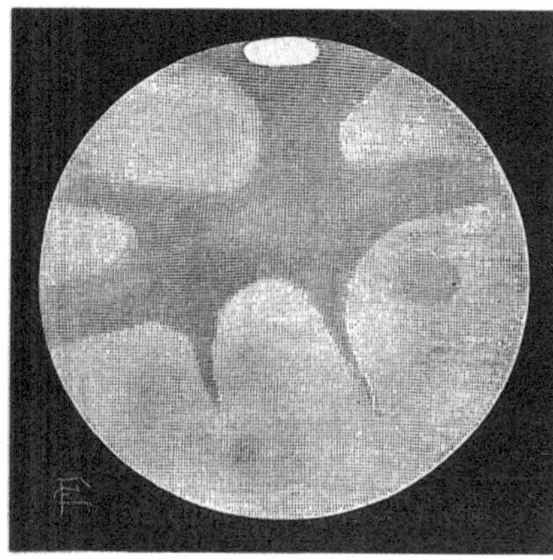

Fig. A.8 Drawings of Mars made in 1892 at the Observatory of Juvisy; by M. Flammarion

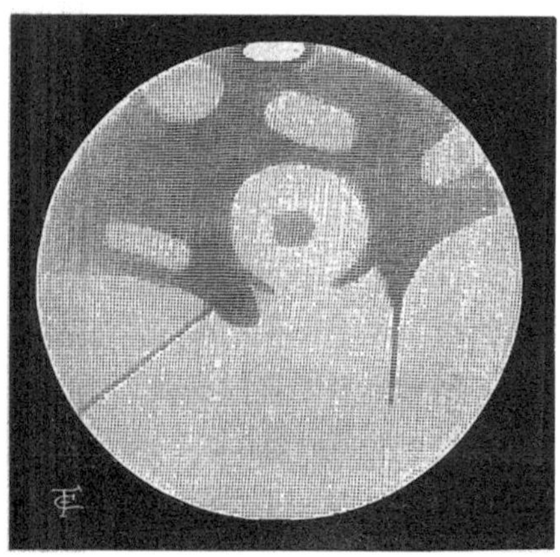

Fig. A.9 Drawings of Mars made in 1892 at the Observatory of Juvisy; by Léon Guiot

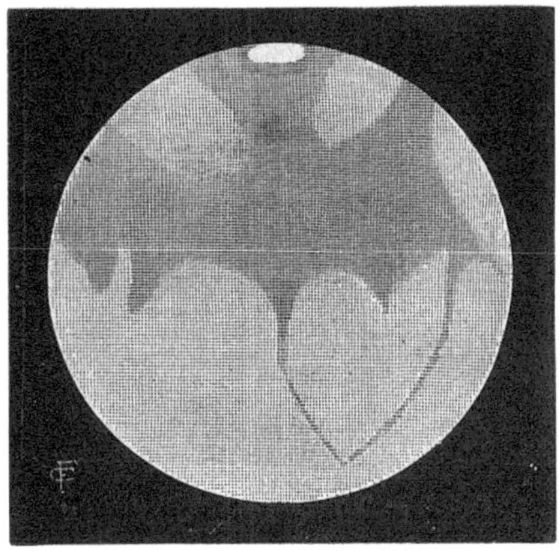

Fig. A.10 Drawings of Mars made in 1892 at the Observatory of Juvisy; by Léon Guiot

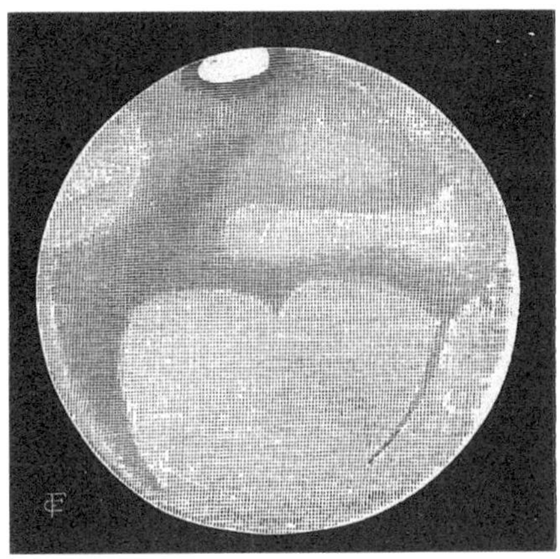

Fig. A.11 Drawings of Mars made in 1892 at the Observatory of Juvisy; by Quénisset

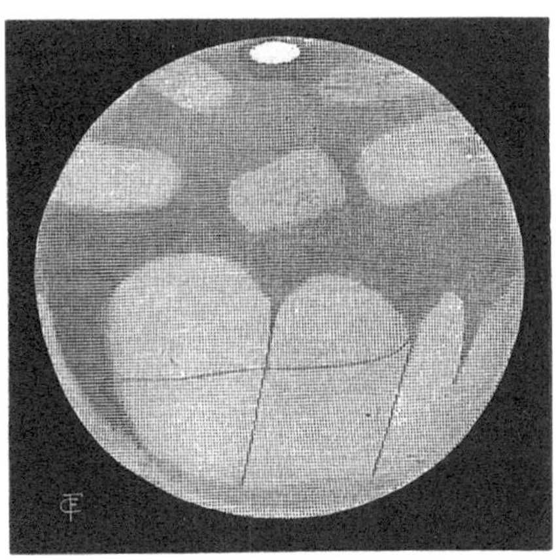

Fig. A.12 Drawings of Mars made in 1892 at the Observatory of Juvisy; by Léon Guiot

Fig. A.13 Drawings of Mars made in 1892 at the Observatory of Juvisy; by Quénisset

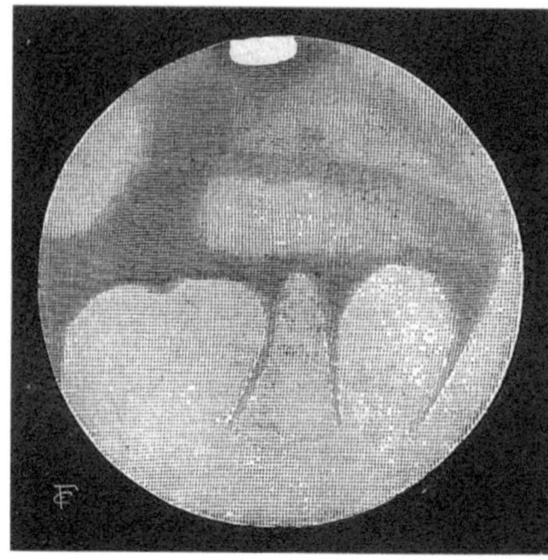

Fig. A.14 Drawings of Mars made in 1892 at the Observatory of Juvisy; by Léon Guiot

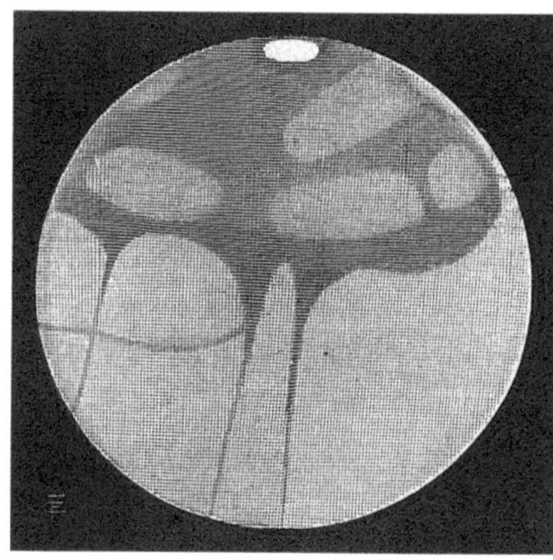

Fig. A.15 Drawings of Mars made in 1892 at the Observatory of Juvisy; by Léon Guiot

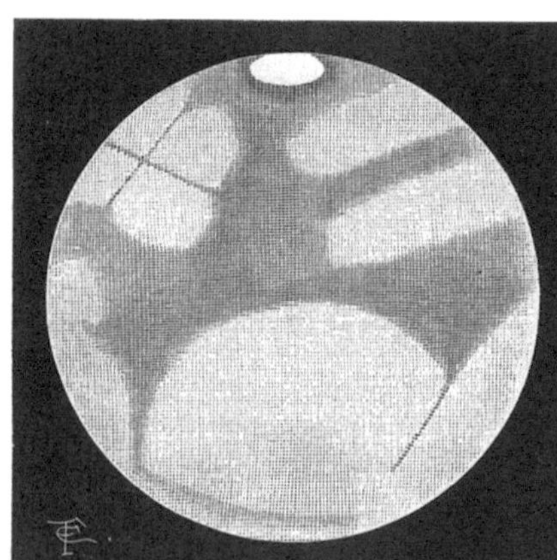

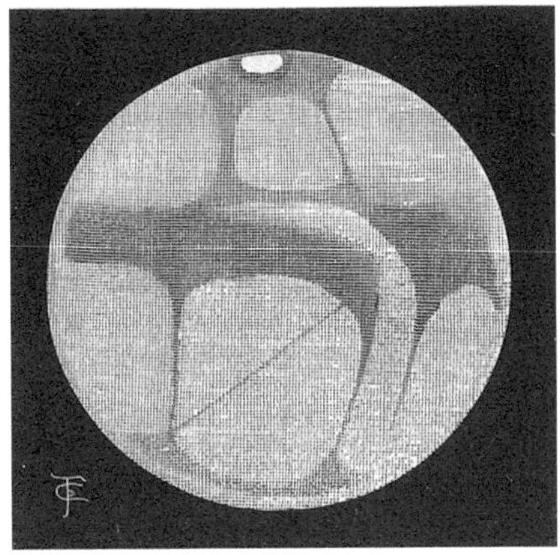

Fig. A.16 Drawings of Mars made in 1892 at the Observatory of Juvisy

Author Index

C. Flammarion, *Camille Flammarion's The Planet Mars*, Astrophysics and Space Science Library 409, DOI 10.1007/978-3-319-09641-4

Subject Index

C. Flammarion, *Camille Flammarion's The Planet Mars*, Astrophysics and Space Science Library 409, DOI 10.1007/978-3-319-09641-4

MIX
Papier aus verantwortungsvollen Quellen
Paper from responsible sources
FSC® C105338

If you have any concerns about our products,
you can contact us on
ProductSafety@springernature.com

In case Publisher is established outside the EU,
the EU authorized representative is:
Springer Nature Customer Service Center GmbH
Europaplatz 3, 69115 Heidelberg, Germany

Printed by Libri Plureos GmbH
in Hamburg, Germany